William Julius Mickle

General Paralysis of the Insane

Second Edition

William Julius Mickle

General Paralysis of the Insane
Second Edition

ISBN/EAN: 9783337140472

Printed in Europe, USA, Canada, Australia, Japan

Cover: Foto ©berggeist007 / pixelio.de

More available books at **www.hansebooks.com**

OF THE INSANE.

BY

WM. JULIUS MICKLE, M.D., M.R.C.P. Lond.,

*Honorary Member and Member of Several Learned Societies;
Medical Superintendent, Grove Hall Asylum, London.*

SECOND EDITION.
ENLARGED. REWRITTEN.

To *BAYLE, WILLIS, HASLAM, CALMEIL.*

LONDON:
H. K. LEWIS, 136, Gower Street, W.C.
1886.

CONTENTS.

CHAPTER		PAGE
I.	Synonyms. Definition. Discovery. Division into Simple and Complicated. Division into Stages	1
II.	Symptoms in the several Stages	5
III.	Particular Semeiography.	
	A. Psychical symptoms	29
IV.	B. Symptoms partly psychical, partly somatic	71
V.	C. Somatic symptoms.	
	(a) Motor	82
VI.	(b) Eye-Symptoms: some motor, some sensory, (and including psychical elements), also trophic, ophthalmoscopic, &c	106
VII.	Somatic symptoms, continued.	
	(c) Sensory	120
VIII.	(d) Vaso-motor	133
IX.	(e) Trophic: including Hæmatic	138
X.	(B. resumed): Seizures (special).—Meningeal Hæmorrhage, and Hæmatoma	159
XI.	Temperature. Circulation and Pulse. Respiration. Lung-disease. Digestion. Sweat. Urine. Saliva	174
XII.	Course of General Paralysis. Remissions. Arrest. Duration. Terminations.	199
XIII.	Diagnosis	217
XIV.	Causes	245
XV.	Morbid Anatomy	278
XVI.	Morbid Anatomy, continued.—Microscopical	298
XVII.	Pathology	324
XVIII.	Pathological Physiology	343
XIX.	Prognosis.—Recoveries	386
XX.	Therapeutics and Hygiène	390
XXI.	Varieties of General Paralysis	404
XXII.	Cases	414

GENERAL PARALYSIS OF THE INSANE.

CHAPTER I.

Synonyms—Definition—Discovery—Division into Simple and Complicated—Division into Stages.

Synonyms.—General paralysis of the Insane. Progressive general paralysis. General paralysis. General paresis. Paralysis of Insane. Paralytic dementia, or Dementia paralytica. Paretic dementia.
Allgemeine progressive paralyse der Irren. Progressive paralyse der Irren. Paralyse der Irren. Paralyse.—Paralytischer Blödsinn. Allg. prog. gehirnlähmung.
"Arachnite chronique;" or, "Arachnitis ou phrénésie chronique," Bayle, 1822, the principal motor symptoms of which he epitomized as "paralysie générale et incomplète;" "Méningite chronique," Bayle, 1825 and 1826 (same summary of motor symptoms*); or "Aliénation ambitieuse avec paralysie incomplète."†—"Paralysie générale incomplète," Delaye, 1824.—"Paralysie générale des aliénés," Calmeil, 1826."—"Folie paralytique," Parchappe, 1838-1841, and Jules Falret, 1853.—"Périencéphalo-méningite chronique diffuse," or, "encéphalite chronique diffuse," or, "encéphalite diffuse," Calmeil, 1841.—"Paralysie progressive" (practically = motor side only), Requin, 1846.—"Paralysie générale chronique," Rodriguez. "Paralysie générale progressive," Lunier, 1849; Brierre de Boismont, 1850.—"Periencéphalite chronique diffuse," Calmeil, 1859.—Polyparésie. Démence paralytique. — "Anoia paralytica," Fischer. — "Insania paralysans," Kjellberg.
Definition.—General paralysis ‡ is a disease of the nervous system, especially of the brain, marked clinically by

---

* A. L. J. Bayle, "Traité des Maladies du Cerveau et de ses Membranes," 1826, p. 538, 542.
† *Ibid.*, p. 568.
‡ The following abbreviations are frequently used in this book:—*g.p.* for *general paralysis; G.P.* for *general paralytic* (patient).

(I.) some general affections of motility, viz.: ataxy, and, finally, paresis, usually following a definite order and course of development, and especially obvious in the apparatus of speech and of locomotion; also, but in less degree, by (II.) sensory disorder or defect; and marked also by (III.) mental symptoms, which constitute, or invariably tend to, dementia, but often consist in part of exaltation of feeling, or even expansive delirium. Finally, it is evidenced by (IV.) certain organic changes in the encephalon and its tunics, often in the spinal cord and membranes also, and sometimes in some sympathetic ganglia as well.

With this definition, it necessarily follows that I adhere, as did Bayle, Duchek, Falret, and others, to the " doctrine of unity " in general paralysis. Yet, on the other hand, many, especially among the earlier writers, have deemed it to be merely a complication, or, some, even a termination, of insanity; thus viewing general paralysis in accordance with the terms of the older "doctrine of duality." Of these mention may be made of Pinel, Esquirol, Dubuisson, Georget, Delaye, Calmeil (1826), Foville *père*, Griesinger, and Billod.

History of the Discovery of General Paralysis.—The limits of space prevent more than a passing reference here to the history of the discovery of general paralysis. Suffice it to say, the light of the first knowledge of the disease gleams in the pages of Willis (1672); that Haslam (1798-1809) caught flashes and sparkles of the truth in his discerning appreciation of some of the chief features of the affection; that Ph. Pinel, Esquirol, and Georget did the same with relation to some other, but more equivocal, features thereof; and that the discovery was completed by Bayle (1822, 1825, 1826), who gave the first full account of the affection, and who was closely followed by Calmeil (1826) in this investigation. Perfect, in 1787, and Guislain in his work of 1826, had not a knowledge of the affection. The chief merit here belongs to Willis, Haslam, Bayle, and Calmeil, especially the first three.

Mode of Commencement.—More often gradual in its onset, general paralysis is stated, and particularly by the older writers, to begin frequently by sudden symptoms of cerebral congestion, or by an acute maniacal attack. This I have occasionally seen. The more frequent modes of commencement are detailed below, when treating of the earlier symptoms.

Course and stages of general paralysis.—A division of g.p. has been made into *simple* and *complicated*.

In that clinical phase of the disease sometimes designated *simple*, the semeiotic phenomena are mainly of two orders, the mental and the motor, the latter being represented by inco-ordination, and by paresis tending to terminate in helplessness, to which, also, disorder and impairment, or even obliteration, of special senses and general sensibility usually add themselves.

To the above, one or more of the phenomena pertaining to a special group, and for the most part characterized by their suddenness and severity, are superadded in what has been sometimes termed the *complicated* form of g.p. In that group of phenomena are comprised the various epileptiform seizures, ranging from those which resemble the epileptic *grand mal*, to the quasi-syncopal which are analagous to the *petit mal* of epilepsy; in it also are comprised the apoplectiform, and what I have termed the simple paralytic seizures—all of which so often break in rudely upon the course of g.p. They occur in the most varied degrees, and the epileptiform and paralytic vary extremely in extent and situation in different cases. Acute muscular atrophy; hysteriform and tetaniform seizures, have been included here also. Most varied have been the attempts to assign the earlier-named of these to this or that lesion. In fact Calmeil, treating of the complications as consisting of comatose, apoplectiform, or convulsive attacks, hemiplegia, or contractions, associates them with a great variety of morbid lesions due to recrudescences of the inflammatory process in its already-existing seats; or to the invasion of new places by it : a patho-anatomical thus always coexisting with a clinical complication. And Voisin very well confines the word "complication" here to phenomena whose point of departure is the cerebro-spinal axis. Several other details as to the *course* of g.p. will be given in Chapter XII.

Stages.—The division of g.p. into stages is of limited value and only in a very restricted sense can any rule be laid down. The variety in the sub-divisions into stages indicates the incomplete adequacy of any one of them. A division of the course of g.p. into stages being mainly founded upon the mental phenomena, there are many cases to which it cannot be applied, so diverse are the order of evolution, the duration, course, and association of the mental symptoms. Sometimes three, sometimes four or more stages are described. Thus, at the commencement of the disease we may speak of a sort of incubative stage, or period of mental alteration; then of a period of decided, and often obtrusive, mental alienation, or of distinct dementia; next, of a period in which usually occurs a progressive decline of the mental and physical powers, chequered by more sudden changes and modifications in the domain of both. If another, and fourth, period be described, it corresponds to the latter portion of the period last-named, in a protracted case. A division into four stages, therefore, is:— First, a stage of mental alteration:—Second, a stage of dementia only, or, in most, of decided mental alienation, often with active mental symptoms:—Third, a stage of chronic mental disorder or failure, during part of which there is usually a remission, of at least any active mental symptoms of the preceding stage, and:—Fourth, the stage of fatuity, impaired sensibility, and complete helplessness; the stage of entire prostration of the mental and motor, of the sensory and nutritive powers.

Or, a division into three stages may be made, either by omitting the last one, of those just mentioned, as a separate stadium and joining it to the third; or, by omitting the first as a distinct stage, and either joining it to the second, or excluding it as being in reality prodromal in nature.

Nevertheless, stages equally exact, though far less striking than these, can in reality be constructed out of the phases through which the motor signs pass. Even at a very early date (1829) had the elder Foville described a broad distinction between two periods of the chief motor troubles; in the first of which, he said, there is vigour but some tension of movement, and in the second some relaxation or resolution of the same.

Such are the stages; but the majority of the cases diverge more or less considerably from this type; any one stage may be absent, and in some no distinct stages whatever can be traced; a gradually increasing ataxy and paresis, with progressive dementia—or with early exaltation, and later on with dementia—and sensory failure, being the chief outlines.

An example of typical general paralysis may be described as running its full course in four stadia, of which one precedes the decided mental alienation; but, inasmuch as three stages, only, are more commonly described,—the first of which commences at the first moment of decided mental alienation, —with the view of obviating confusion, four *periods* in g.p. will be described in this work; the first of which is sometimes termed prodromic, corresponds in the psychical sphere to a mere mental alteration, yet issues from the earlier stages of evolution of that same morbid process which eventually operates a lethal effect upon the somatic and psychical powers;—and the last three of which periods correspond to three *stages* of the disease when it has become established or confirmed.

As examples of division into stages made by others may be mentioned Bayle's (1822-5-6), into stages of ambitious monomania, mania, and dementia:—Dr. E. Salomon's, into stages of, (*a*) chronic lepto-meningitis; (*b*) chronic diffuse periencephalitis; (*c*) degeneration of cerebral cortex; (*d*) true atrophy of same; these corresponding to successive phases of mental change and failure:—and Dr. H. Obersteiner's, into a stage of hyperæmia,—of initial symptoms; next, a stage of exudation,—of motor phenomena, maniacal excitement, paralytic attacks, mental weakness; lastly, a stage of connective tissue overgrowth,—of lasting paralytic attacks, and complete mental decay. Dr. Aug. Voisin, in the "simple" form, makes five periods: the first prodromic; the second intermedial; and the other three being stages of the disease in its confirmed state;—the intermedial period, when present, consisting mainly of some form of insanity lasting not longer than two years, with a possible remission (at maximum) of two years more; beginning with the onset of mental trouble—of distinct *délire* or dementia—and ending with the supervenient appearance of some particular somatic signs: (or, if not so ending within the time specified, being held as pertaining to simple insanity and not to general paralysis). The division made use of by Dr. E. Mendel * is convenient. In typical g.p.:—1, a prodromic stage:—2, the depressed or first stage of the disease:—3, the maniacal stage:—and, 4, the stage of dementia. The predominance of one of these stages with absence of one or two of the others characterizes the non-typical and clinically different forms.

CHAPTER II.

The Symptoms of General Paralysis in the Several Stages.

Later on, the symptoms of g.p. will be described separately and fully. Here I will only briefly mention the

* "Die progressive Paralyse der Irren," 1880, p. 1.

chief symptoms in the several stages. First the sensory and motor, then the mental, symptoms will be described in each period; and although the thermometrical, the circulatory, urinary, and other coincident phenomena are semeiotic, it will be more convenient to speak of them separately and afterwards.

Of the symptoms occasioned by the morbid processes in general paralysis some are in the first rank, are characteristic, though in a free sense only, and are found in almost all cases; others in the second rank and accessory; others in a third rank and relatively infrequent and accidental; while there are still other and more remotely dependent consequences.

I. *First period, the stage preceding frank mental alienation, and sometimes termed prodromic. Stadium prodromorum.* —"*Medico-legal stage*" of Legrand du Saulle.

In dealing with the precursory or premonitory or prodromic signs and symptoms of this affection, it is obvious that under this head have usually been included two groups of facts which have not quite the same relation to the disease in its established mode. In the first place, there are some conditions that have been set here, and which by their long duration, semi-independence, and existence apart from obvious striking disturbances, can scarcely be said to be part of the affection (g.p.) itself. Secondly, there are signs and symptoms which arise from the first disturbing effect on the organism of the first phases of those circulatory and nutritive disorders of the cerebro-spinal system, which, should the patient survive, will end in the production of the functional disorders and of the organic lesions of g.p. in its fully-developed form.

These latter might fairly be considered as a part, a phase, a period, or stage, of the malady; and so, in truth, they are. But it happens with comparative rarity that they are accurately observed, scientifically recorded, or even understood as threatening insanity or dementia. The mental change is of a degree short of decided mental alienation in a practical or medico-legal sense; and as a rule the patients do not, at this time, come under medical, or at any rate under expert, observation. Moreover, some of the symptoms at this time are apt to be mistaken as "causes" of the subsequently invading frank disease. In many cases one must be content with such items of history as inobservant friends of the patients may be able to give—fragmentary, biased, or irrelevant as these may be—and, after passing them through the furnace of scientific analysis, take them for what they are worth. Thus, some points are apt to become unduly emphasized; some, equally important, but less striking, to be ignored, or to pass unobserved. The customary tendency, however, is to defective rather than to prejudiced observation.

For these reasons, I shall, in this edition, treat of all the signs, symptoms, and conditions preceding the established affection under the head of the prodromic stage, or stage of the precursory or premonitory signs and symptoms.

The duration of the prodromic stage is extremely variable. But the more accurately the histories are given by observant friends the longer, usually, the prodromic stage appears to be; and many months, and, in many cases, several years, of some perceptible change, have preceded the point of time at which the mental derangement or dementia is understood as such. Morel spoke of patients whose whole previous lifetime seemed to be a premonitory stage of g.p. Undoubtedly this is true occasionally, but only so far as concerns a facile, changeable, irregular and spasmodically excessive and energetic activity of the nervous system; and this a nervous system which is extremely obnoxious to the results of injurious impressions. No doubt, some of the cases supposed to be examples of an extremely long prodromic stage were cases of monomania on which g.p. eventually supervened.

Although the mental prodromata are usually the earliest, yet in them there is so great a diversity that they are often of little value with regard to a diagnosis of impending g.p., of however much value they may be in forecasting a mental disorder or dementia. Therefore, the precedent somatic symptoms, so valuable in diagnosis, will be first treated of; and of these the sensory are usually the earliest and most marked.

A. Physical Symptoms.

Sensory symptoms. Impaired hearing, or sudden or transitory deafness, may occur; or amaurosis, appearing and disappearing for several years before the incidence of g.p., or showing itself anew two or three months prior to the first symptoms of the latter, either to cease again, or to remain persistent (Lélut, Lasègue, Parchappe, Calmeil).

Severe headache is occasionally observed, and by some the cephalalgia is spoken of as being deeply-seated, vague, or like that of a contusion. In confusion, and urged by distress, the patient may knock his head for relief against the wall; but in many of these cases he is really not in a prodromic stage, but in the established disease of the quiet demented form. The feeling of being "stunned" is frequently spoken of. Dull morning headache, suddenly disappearing, and followed by a subjective feeling of lightness and exaltation, is noted by some.

So, also, may neuralgiæ affect the head; or the spine, or the limbs, and affect different parts in succession;—especially when symptoms of the spinal order are comparatively prominent, or precede the cerebral symptoms, may the neck, back, or limbs be painful. Such patients are usually spoken of, self-styled, and treated, as suffering from chronic rheumatism. These severe pains and cramps of the limbs are often owing to spinal sclerosis or chronic spinal meningitis, and have usually been present in the marked cases I have seen of posterior spinal sclerosis in g.p. Visceral neuralgiæ are sometimes spoken of.

De Crozant * asserted that a general, almost complete, cutaneous anæsthesia precedes the disorders of motility, that it is temporary only, and coincides with the acute period or that of the invasion of the malady, thus belonging to the disease as it is becoming established, rather than to prodromic symptoms. The anæsthesia he attributed to a sort of compression of the brain, brought about by the commencing morbid alteration and turgescence in g.p. But de Crozant's attempt to establish insensibility of the skin as a criterion of g.p. was rejected at an early date, as by Guislain,† although a few examples of it at the onset were reported.‡

Other abnormal sensations occasionally experienced are those of pricking or formication of skin, or as of electric shock in the head, or of modifications of the muscular sense, such as of undue corporeal buoyancy or heaviness.

Sparks before the eyes, colour-blindness, and disorders of colour-vision, are stated to occur in some cases at this time, so also is the "globus" sensation about the throat, or a

* Société de Médecine de Paris, Fév. 26, 1846. "Revue Médicale," Oct., 1846. "Annales Médico-Psychologiques," T. ix., 1846, p. 433.

† "Leçons Orales sur les Phrénopathies," par J. Guislain. Gand, 1852, T. i., p. 339.

‡ As by Brierre de Boismont, with M. Brochin. "Ann. Méd.-Psych.," 1859, p. 326.

sensation of pressure or fulness about the hypochondria or stomach, or abnormal, subjectively-produced, sensations of heat or coldness of one or more limbs.

Vaso-motor and other symptoms. Sometimes the prodromic symptoms consist of seizures of cerebral congestion,[*] or of mere palpitation, flushing and heat of the face and head, with aural tinnitus; or of cracking sounds, perceived as if in the ears or head. A heated state of the head, flushing of the face either alone or alternating with coldness of the same; or palpitation, with injected eye and lively look, may occur from time to time. So also may vertigo, or a momentary sensation of being stunned, or ringing or blowing sounds in the ears. Sensation of being "stunned" or "absences" especially occur after excitement or overwork, and among other symptoms here Verga[†] mentioned epistaxis. Vertigo, whether slight or severe, and occurring independently of gastric disorder, is not at all rare at this preliminary stage.

The gastro-intestinal functions may be disordered (bulimia, anorexia, constipation); so may the menstrual. Symptoms of the asthmatic order may arise.

Motor signs. In the prodromic stage are sometimes[‡] ocular palsies (strabismus, ptosis, diplopia), or facial palsy.

Mr. Phillips[§] asserted that extreme contraction of the pupils, and Mr. J. T. Austin[§] that a fixed and unsymmetrical condition of the pupils, is frequently prodromic; whereas Prof. Griesinger[||] observed that the pupils are sometimes irregular for years before the onset of the disease, and Dr. Mendel[¶] states that pin-head pupils are not rare in the prodromic or early stage, particularly in the ascending form. Dr. L. Lunier[**] observed trembling of the limbs at this period.

In its earlier course the usual ataxy and paresis of g.p. ordinarily escape observation or fail to be recorded, and, indeed, they may be absent. In the first period, corresponding in the psychical sphere to that of mere mental alteration, the motor signs, if present, are for the most part

---

[*] In 60 per cent. "Annales d'Hygiène," &c., T. xiv., 1860, p. 428.
[†] Abstract, "Journal of Mental Science," April, 1873, p. 158.
[‡] "Annales Méd.-Psych.," 1859, p. 295. "Annales d'Hygiène," 1860, p. 430.
[§] "A Practical Account of General Paralysis," by Thos. J. Austin. London, 1859, p. 65.
[||] "Mental Pathology and Therapeutics."
[¶] *Op. cit.*, p. 150.
[**] "Ann. Méd.-Psychol.," Jan., 1849.

simply ataxic; movements of the lips, tongue, and sometimes of muscles about the face and eyelids, being accompanied and interrupted by fibrillary tremor or twitch, and voluntary movements effected in these several parts being sometimes preceded and followed by less ample involuntary movements in the same and in the opposite direction, producing a faint quasi-spasmodic action, and this inco-ordination becomes more marked in the subsequent stages, or, if not now present, appears later on.

Occasionally, even in a state of voluntary rest, the involuntary spasmodic twitches of the lips, face, eyelids, or occipito-frontalis may occur. Rarely, spasmodic movements about the mouth, or even teeth-grinding, may be evinced.

The tongue may be protruded in a sudden, jerking, and momentary manner, but even when this is not the case the motor disorder may sometimes be recognised in fine irregular contractions of the protruded organ; or its movements may be dull and heavy.

In some very few cases a slight inco-ordination of gait, the tabic form of gait in the bud, is to be seen, but as a rule the coarser movements of the limbs, such as the locomotor, are free at this period, and many of the patients can run and dance apparently as well as ever. This is irrespective of cases with prior tabes dorsalis. Yet the finer movements, such as those concerned in writing, may suffer; and occasionally, besides omissions, etc., indicative of the mental state, a shaky and erratic penmanship bears witness to the motor involvement.

The pupils are now in some cases contracted, irregular, sluggish, or unequal.

A general relaxation of the patient's energy is sometimes described, but this pertains, rather, to the groups in which dementia, hypochondriasis, or melancholia predominate at the first, although even then the ataxic condition, if present, is revealed by indications similar to those above-named. There may be a momentary weakness or sinking in one or more limbs.

Epileptiform or apoplectiform seizures are not infrequent forerunners of established g.p. Sometimes they precede it by years, even by so many as seven years; but, if prodromic, commonly occur shortly before, or sometimes form the invading stroke of, the confirmed disease;—mark, that is to say, the transition from the prodromic to the first stage of the established disease. In several cases I have found them form the first thing that arrests the attention of friends, and others have noticed the same. In one case, afterwards under my care, an apoplectiform attack, with temporary loss of speech, preceded by twenty-eight months the time at which a well-known competent observer, watching the case, was first able to diagnose it as g.p.— Dr. W. B. Goldsmith * has recently emphasized the occurrence of convulsion or loss of consciousness as early symptoms in g.p.

B. MENTAL SYMPTOMS,

In prodromic. stage of g.p., first period, period of mental alteration. Some embarrassment of speech, a slow and difficult search for words, omission or substitution of words, is occasionally the first symptom noticed. Thus far we have spoken mainly of somatic prodromous conditions; but in addition, or alone, some mental modification is often observed, and ordinarily consists in an alteration of the affective or moral powers, as shown either in demeanour or action ; or, on the other hand, there may be some exaggeration of a naturally choleric and emotional disposition. Some peculiarities or eccentricities of conduct or habits are now often noticed by those associated with the patients.

As in the case of the physical signs, so also in that of the mental symptoms, it is difficult to seize upon and limn any types that can be pronounced to be characteristic of the several stages of the disease. One need only briefly indicate, therefore, the phases of mental disorder most frequently observed in each period ; premising that the greatest variety, as well as the greatest changeability, is constantly to be met with in each, and that the mental symptoms in a given case may vary almost from day to day, while yet, on the whole, a progressive march is observed, unless a remission of prolonged duration occurs.

In g.p. of the simplest form there is solely, or chiefly, an increasing dementia. Yet in agreement with Guislain † it may be said that usually there are two orders of phenomena here, the permanent and the transitory ; the former consisting of the gradual enfeeblement of conception, memory, and all the mental faculties; the latter, of various forms of mental derangement. Truly, he admitted the existence of cases without obvious mental disturbance, but even in these drew attention to the childish manners and the expression of astonishment.

* "Archives of Medicine," Aug., 1883. † *Op. cit.*, T. i, p. 327.

Appetites and desires. Loss of the genital faculties and desire has been noticed as a prodromic symptom in some cases.* On the other hand, sexual excess sometimes coincides with the development of the first germs of the disease, and when deemed to be a cause is often only a prodromic symptom in reality. In one case, at this period, a patient, who soon afterwards came under my care for a short time with symptoms only justifying a provisional diagnosis of impending g.p., was the subject of excessive sexual desire and capacity for one week only, during which term he engaged, it is said, in about fifty acts of marital coitus, with full seminal ejaculation on each occasion. Not only may there be this extreme lasciviousness, as formerly noticed by Billod and Guislain: in some rare cases there is the grossest kind of self-defilement with fæces.

Excess and coarseness in eating and drinking may now astonish the patient's friends.

Affective and moral faculties; æsthetic feelings. The language of the patients is sometimes changed in character, is coarse and foul. Sometimes there is a species of coarse jesting, or unwonted familiarities or liberties with servant maids; and on several occasions wives have asserted to me that this latter was the very first thing unusual or surprising, noticed by them, in their husbands, who afterwards became G.Ps. Possibly, this may not have been the first deviation, but, touching their *amour propre*, was the first they noticed.

Some lie and dissimulate (as it were) in the most imperturbable manner.

Sometimes from forgetfulness and mental weakness and disorder, the patient openly and coolly commits theft. This also occurs in the next period of g.p.

Thus, as well as the intellectual life are the affective and moral usually touched; nay, their disorders may be so obvious as to appear primary, the moral powers suffering more notably in one case, the emotional in another. Of all the prodromic symptoms, perversion of the moral sense is the most important. This perversion is closely linked with, and often largely based upon, mental failure. Brierre de Boismont † long ago drew special attention to perversions of

* Baillarger. "Gazette des Hôpitaux," Juillet 16, 1846, No. 83, p. 329.
† Société de Médecine de Paris, Fév. 20, 1846. " Revue Médicale," Avr. 1, 1846, T. i, p. 617; and Dec., 1846, T. iii, p. 605. " Gazette Médicale," Mai 22, 1847, p. 391. Acad. des Sciences; Séance du 24 Sept., 1860. " Annales d'Hygiène publique et de Médecine Légale " T. xiv, 1860, p. 405. " Annales Médico-Psychologiques," 1861, p. 88.

the moral and affective faculties in the prodromic stage of g.p. Even several years (six or seven) beforehand there may be unwonted acts of indelicacy, impropriety, or debauchery, and with these there may be a placid apathy and an utter indolence. Great irritability of temper may occur two or three years before the outbreak, or threat of suicide,* or failure of the usual determination of character, as well as the failure of memory, or of the clearness or precision of judgment already referred to.

This disease, therefore, is often to be feared when sudden apparent moral falls or delinquencies—of which theft † is one of the most frequent—occur in those hitherto without reproach. In the history of many a case do we find that some moral or other mental change in the patient, some perversion of the affective sentiments, has been noticed long before the acknowledged onset of the disease—a view in which Guislain ‡ also participated.

Intellectual failure. If the purely intellectual powers are involved there is mainly a species of inattention. Repeatedly may the patient evince the same particular forgetfulness or blunder; or the conduct of duty or of business may become irregular, fitful, and ill-judged. Sometimes the patient forgets what he was about to say, or makes mistakes in changing money, or in the simplest calculations. Or, again, with forgetfulness and inattention to duties, there may be progressive mental confusion and stupidity, with or without drowsiness and heaviness of head. Memory fails, especially for recent events and recent mental acquisitions, and the power of attention progressively lessens. The intellectual debility may be the most marked feature; and this impairment of memory and of business capacity—probably in part from mental confusion—when occurring at this period, is sometimes the first of a linked series of conditions and occurrences. For, from this intellectual and mnemonic impairment and confusion may originate falsity in money accounts;—from this latter fine or imprisonment;—from, or with, this, mental depression; and this soon followed by an acute maniacal attack, forming, or after a remission passing into, the onset of the first stage of the established disease (second period). In some cases, impairment of memory and the various results to which it gives rise, are the principal or

---

* " Annales d'Hygiène," *loc. cit.*
† For theft, etc., by general paralytics, see Chapter III.
‡ " Leçons Orales sur les Phrénopathies," T. i, p. 325.

only points actually noticed by the friends of the patient before the disease is established.

There may be incapacity to carry on the usual avocations, or to fix the attention. Temporary absence of mind, forgetfulness of duties, meals and appointments, and of the consequences of froward acts, are often predominant features.

Mental confusion. Sometimes the chief symptoms noticed in the first period are the stupid, dazed, but yet excitable, states of the patient. There seems to be sometimes a condition of general mental confusion, and when better the patients often speak of themselves as having been "confused in the head," "light-headed," or "giddy." Vivid, and usually disagreeable, dreams may occur.

Mental depression. Or the patient may be depressed, dull, worried about trifles, yet angry if opposed: or full of vague fears and hypochondriacal fancies; and, indeed, langour, ennui, and melancholy are frequent* initial phenomena, the patients often having the sad consciousness of their state. As I mentioned years ago, depression when precursory, or of the first stage, is melancholic rather than of the hypochondriacal form so often observed later on. Lowness of spirits, mental depression, dulness, sadness, and morbid fears, therefore, are sometimes found, and sometimes increase to a form of melancholia or hypochondria, with which the first stage of the confirmed disease appears, as the depressed form of g.p.; so that the early symptoms seem as if part of the established disease, and the one stage blends imperceptibly with the other, an exemplification of the fact that most of the phenomena of this so-called prodromic stage are essentially parts in the symptom-assemblage of g.p.

In the history of cases any mention of depression at the onset, or as a precursory condition, is wanting in the majority of cases: this is because the depression, anxiety, forebodings, are usually not extreme; pass as lowness of spirits, as a melancholy, but not as insanity, not as melancholia or hypochondria; and are thought lightly of, and soon forgotten. Yet in some cases they become so marked and developed into insanity, recognised as such, that the patients find their way into consulting rooms or asylums, are entered under the head of melancholia or hypochondria; if they recover, are discharged as recovered melancholiacs or hypochondriacs; eventually to turn up again there, or elsewhere,

---

* Dr. J. A. G. Doutrebente, Thèse, 1870 "Recherches sur la Par. Gen. Progressive."

manifestly as G.Ps.; or, should their affection have been thus early diagnosed, we hear of recovery of the hypochondriacal form of g.p.: or, not recovering, they linger on until the form of mental disorder changes, or until the physical signs of g.p. become evident, if they were not so before.

In the great majority of cases Dr. J. G. Kiernan,* also, found the " early mental state " to be one of depression; and this of course is peculiarly true if symptoms of comparatively little intensity are included. Dr. E. C. Spitzka,† in the initial period, lays stress upon the existence of the physical indications of the act of weeping without co-existing emotional depression. I find a rapid alternation of emotion, and the weeping like April shower.

Expansive mental symptoms. Here are elevation, "abandon," recklessness. There may be an expansive, busy, speculative frame of mind; a restless, fitful, yet energetic application to business, though perhaps with some loss of the usual foresight and acumen; a rashness of action, and brusque manner, and a forced and noisy laugh, with unnatural loudness of conversation. To these warnings may be added that of the danger-signal, insomnia. But a tendency to sleepiness, and, in some cases, especially somnolence after meals, is now quite frequent, whether it exists alone or whether it is associated with defective night-sleep and restlessness. In some cases the falling asleep occurs in public places.

The expansive state in this stage, and especially in the next stage, often leads to absurd and extravagant projects, by which the patient, if possessed of any means, impoverishes himself or family, if they are carried into effect; and he often orders large amounts of merchandise entirely useless to him.

Sometimes a naturally lively and choleric disposition is morbidly exaggerated in this period, but usually it is an *alteration* in the mental condition, in the demeanour and action, that is observed. Gaiety and self-assurance are frequent; and in one group replace an habitual, or a prodromic, dulness and mistrustfulness. Together with an expansive state of feeling there are sleeplessness, restlessness, or a speculative turn displaying itself in ill-considered or absurd projects for the bringing down of golden showers. The restlessness and ill-directed over-activity may take the form of a superabundant, unwonted, philanthropy and generosity,

---

* " Alienist and Neurologist," Jan., 1885, p. 65.
† "Æsculapian," Vol. i, cited by Kiernan.

either rudely planned and busily proclaimed, or rashly carried into actual execution. An expansive egotism, an inflated view of the patients' own position, powers, and aptitudes, is often evinced, and blends this period with the next. In a soldier this may show itself in his unusual and unseasonable familiarities in speaking to his superior officers, or in offering in excited and loud tones of voice to lay wagers with them; and, besides excitation, in this, as well as in the inattention to orders, is revealed the strain of hebetude. The egotism may, however, be of a more disagreeable nature.

Sometimes there is a relaxation of all energy; on the contrary, there is sometimes an extreme activity, a restless wandering to and fro, and by night as well as by day, a petulant issuing of contradictory orders to subordinates or to family, a loquacity with self-contradiction, and furious passion at the slightest opposition or roused by the most trivial matters. Although the irritability of temper, so often observed, usually falls short of this extreme degree, it may yet lead to acts of violence. Thus, I have known cases in which wives now had to rush from home, for their lives, to escape the sudden irritable violence of their husbands. More frequently the irritability leads to less dangerous, but destructive, acts.

With the wandering habits, or inclination to travel, far too often are there linked a squandering of means, a buying and selling unnecessarily, imprudently, and at great loss, even while the patient is extolling the acuity and cleverness of his dealings. Or he dispenses undiscriminating largesses.

Thus conditions often termed prodromous may really come to constitute part of the first stage of the developed disease (second period), the transition from the one to the other being imperceptible. Much of the present description, therefore, applies to the next stage as well.

This last remark applies also to that which Dr. Em. Regis has stated in dilating upon the functional exaggeration at the beginning of g.p., and supposed to be occasioned by irritation of the brain as an ultimate result of inflammatory action. The excitation of function may exhibit itself in one or a few only of the functions; or may extend to all, and is usually, for the time, persistent, intense, impulsive. The excessive activity may be intellectual, emotional, motor, sexual; may exhibit itself in extreme alcoholic excess; or may be vegetative, the pulse, respiration, and temperature, hunger, and thirst being all in excess of the normal standard.

Summary. Much that is done, and especially much that is left undone, indicates a grave enfeeblement of mind; or ideas are acted upon, without reasoning, as they arise in consciousness, or there is action of a semi-conscious kind, such as is exemplified in an irrational and objectless exposure of the person. But, in another group, the failure to distinguish between the things desired in fancy and actual possessions, as well as an expansion of the *ego*, or an impulse, lead the patients to appropriate as their own the property of others, as already described. There is, also, no fixity, no tenacity, in the purposes which engage their mind; the old schemes are abandoned and derelict, and the new may command but a flagging and flickering interest. Neglectful of personal appearance, of regularity in meals, and of cleanliness in eating them, they pass hither and thither in slovenly and incongruous garb.

In a word, some or others of the following phenomena are frequently observed :—The patients are restless, agitated, irritable in temper, or obstinate, speculative, volatile, mobile in resolution ; they buy, sell, or give away, without any adequate object, policy, or necessity, and usually at a loss; even lose articles of their clothing, seek interviews with personages higher in the State, or pester them with correspondence; for their extraordinary acts, or thefts, offer still more extraordinary, silly, and varying excuses ; and occasionally exhibit salacity, or an open and unconcerned self-exposure.

But besides, or instead of, the more or less expansive symptoms just described there may be other, and differently associated, changes in the moral, emotional, and intellectual faculties ; in the conduct, demeanour, energy, and outward life generally.

Thus, in one group, differing from those already described, the prevailing characters are mental confusion, failure of perception, silly childishness, stupidity, forgetfulness, heaviness of head, drowsiness, and incapacity for the usual avocations. If there have been prodromic sensations as of being "stunned" or vertiginous, they may continue. The forgetfulness and general mental failure, and possibly a sensory failure, may occasionally display themselves in " foul " or uncleanly habits. Moreover, outbursts of some excitement, of insubordination, or even of destructiveness, may chequer the progress of the now incipient, slow-creeping dementia.

In still another group the patients are sad, morose, depressed, confused, forgetful, inattentive, neglectful of their duties or privileges, hypochondriac or melancholy in feeling, yet, perhaps, irascible if roused; troubled and distressed about contingencies either trifling or imaginary; although they are apt to be apathetic about matters of real importance. If we include here the milder symptoms, these, with moral failure, are the most frequent precursors. Often overlooked or forgotten, lowness of spirits is perhaps brought to recollection by careful inquiry.

The nervousness and restlessness are observed in several of the groups. Sullenness and taciturnity, when observed, are usually associated with other symptoms.

Occasionally, the principal phenomena now are suspiciousness, avoidance of society, taciturnity, and a miserly disposition.

Invasion; or the period of invasion. The initiatory phenomena already described often pass by gradual increase or modification into the first stage of the disease in its established condition; the one stage blending almost imperceptibly with the other.

In other cases there are distinct invasion-symptoms; although it is unnecessary to constitute, of these, a separate stage.

The acute invasion-symptoms may assume various forms.

Thus, the disease may be ushered in by an attack of cerebral congestion, varying in intensity from swimming and aching of the head, mental confusion, ringing in the ears, and pricking about the body, to coma, with, perhaps, paralysis and convulsions. This may be immediately followed by the fully-developed first stage of the confirmed disease; or the patient may drag on for a time without fully-pronounced mental alienation or dementia.

Or an acute maniacal attack may abruptly end the initial stage, and form the beginning of the established disease; a tumultuous transition from one period to the next.

Or, without any history of motor forerunners, there may be protracted mental alienation without distinct physical signs of g.p.: this I have seen last for weeks, months, one-and-a-half-year, upwards of two years; the motor signs of g.p. eventually manifesting themselves.

These are the most striking forms of invasion; but

dementia, or any clinical kind of mental derangement, may mark the onset; and it is only necessary to make formal mention of the fact at present, inasmuch as the clinical aspects of these cases at the time of invasion will be fully described under the head of the next stage.

In Bayle's view, cerebral congestion, either of a sudden or of a gradual and persistent nature, was always the first step; and this was successively followed by the three stages;— ambitious monomania; mania; dementia.

Drs. Jules Falret[*] and L. V. Marcé[†] described four clinical varieties, at the onset of g.p.; two in which the physical symptoms predominate, namely the *paralytic*, and *congestive;* and two in which the mental, namely the *expansive*, and *melancholic*.

II. *Second period. First stage of Established general paralysis.*

A. PHYSICAL SYMPTOMS.

(a). *Motor.* The motor signs may now be well marked; on the contrary, they may be absent during the maniacal excitement so frequent at this time, for, as so long well known,[‡] maniacal excitement may mask the motor signs, although in the later stages it only serves to heighten the manifest impairment of speech. The affection of speech (partly psychical, however), and lingual and labial ataxy are now usually decided, even if slight. Naturally the consonants form the chief difficulty; especially in pronouncing linguals or labials, or in uttering the syllables of a long word, does the patient fail; a stoppage, a faint stuttering, is observed, as of one somewhat in liquor; with an effort is the word uttered; and at the same time may often be seen tremulous twitches of the upper lip or of the facial muscles, as in one about to weep. There is a stumble, as it were, in articulating some words, an occasional lingering pause, a quivering or thickness in speech, now and then with indistinctness, or elision of one or more syllables, and ofttimes with an utterance slow and circumspect: or with these conditions any voluntary movement of lips and tongue may be several times involuntarily repeated. Hence, when the movements involved are those of speech, there results an occasional embarrass-

---

[*] " Recherches sur la Folie Paralytique," &c., Thèse, 1853.
[†] " Traité pratique des Maladies Mentales," 1862.
[‡] A. L. J. Bayle, *op. cit.*, pp. 497 and 503. Baillarger, " Annales Méd.-Psychologiques," 1847.

ment of it, very faint as a rule at this period, only to be detected by close examination, or even altogether masked, so to speak, by the state of motor restlessness and erethism, with loquacity, into which the patient has sometimes fallen. Chiefly is this embarrassment evidenced by an occasional pause or hesitation, followed, maybe, by an emphasized or somewhat explosive utterance of the next syllable; or there is the elision of a syllable, or the act of slurring over it, without complete omission; or its incomplete stuttering repetition.

There may be tremor of the tongue, which may be jerked in and out in a convulsive manner when the patient is desired to show it. Even now, so early, there may be grinding of the teeth, champing and masticatory movements about the jaws and cheeks.

Inequality of the pupils, or irregularity of their shape, constant or intermittent; myosis; iridal sluggishness to light; are frequent now: the retina may be hyperæmic; the conjunctiva in some cases injected; and, later, perhaps, an irregular dilatation of pupils may occur.

The occipito-frontalis is often gathered together or twitching; the eyebrows are often raised, the lower lines of facial expression partly obliterated; and the features in some degree florid, or showing dilated venules; or, on the other hand, the skin may become coarse, and of muddy, or greasy, or parchment-like appearance.

The expression of emotions without a corresponding emotional state, mentioned by Spitzka* as sometimes a very early symptom, is essentially connected with the incongruity of facial expression and emotional state which I described years ago (1st edition), and which will be noticed under the next stage.

The ataxy may show itself in an impairment of the manual dexterity. Later, the movements of the hands become somewhat lessened in adroitness and exactitude, and there may even now be some collapse of the general figure. Especially when there is some maniacal agitation may the movements of the extremities still remain strong and free, but on the other hand they usually exhibit a species of slight inco-ordination, and sometimes paresis.

The handwriting is sometimes shaky, irregular, hieroglyphical; letters, syllables, or words are jumbled or omitted; or the patient fails to complete his epistle.

* "Insanity, its Classification, Diagnosis and Treatment," 1883, p. 209.

Usually the gait is still free, but in other cases it is somewhat awkward, the steps are long and slightly irregular, sharp turning round may be attended with awkwardness and momentary uncertainty, or even with swaying of the frame. Trembling movements of the limbs, varying in degree and frequency, also attest the ataxia. The gait varies with the relative proportions of ataxy and paresis present.

If the spinal symptoms precede, or commence with, the cerebral the gait may be decidedly tabic or ataxiform.

Occasionally there is a general subsultus, a quasi-shivering, unconnected with extraneous febrile complication, or any impression of external cold.

(β). *Vaso-motor disorders and Cephalic Hyperæmia.* Attacks of redness of the face and heat of the head may occur in all degrees, from merely temporary flushing and hyperæmia, to marked congestion of the cephalic region and brain. These may be due to altered vascularity of, and other incipient changes in, the cervical sympathetic, but more often to cerebral causes, and they are more severe in the later stages.

(γ). *Sensory Symptoms.*—Local colour blindness has been observed by some, including Dr. Batty Tuke, in g.p. A gradual diminution of visual power is occasionally observed even thus early, or it may be more sudden and transient, and perhaps owing to circulatory changes, œdema, &c.; rarely is there visual hyperæsthesia. Hallucinations and illusions of sight are not infrequent at this stage. Auditory hallucinations, too, are sometimes revealed, and may be vivid and almost unceasing. Auditory hyperæsthesia is occasionally observed. The muscular sense, the organic, and the tactile often undergo a strange perversion, the patients having muscular fatigue, weakness, lassitude, or most extraordinary sensations as to the position, expansion, shrinking, or flight of their body or limbs.

It is in this stage, especially, that Dr. Aug. Voisin* has indicated the importance of loss or diminution of the sense of smell, which he looked upon as an almost constant and persistent symptom, and affecting either both sides or one side alone. Although the sense of smell fails in some cases, my experience differs somewhat from his, and for details see Chapter VII. Hallucinations of smell are comparatively infrequent.

* " L'Union Médicale," Août 4, 1868, p. 180 and " Traité de la Paralysie Générale des Aliénés," 1879, p. 39.

Cutaneous hyperæsthesia, and hyperæsthesia of cranial nerves other than those already mentioned, may be observed now, but chiefly at an earlier moment, yet anæsthesia, either real, or only apparent and due to vivid mental preoccupation, is not rare. But any prodromic neuralgia scarcely ever continues into this stage.

(δ). *Less usual phenomena.* Othæmatoma may now make its appearance, and so, too, occasionally, in their various degrees may the apoplectiform, or epileptiform, or other seizures, yet to be described.

B. MENTAL SYMPTOMS

Of this first stage of the established disease. Second period of the four. In this stage the mental symptoms may be mainly those of—(1) dementia; or of (2) expansive delirium; or of (3) acute maniacal excitement; or of (4) a hypochondria; or of (5) a melancholia; or, in a few cases, of (6) "stupor," or (7) of circular insanity.

Yet a species of mental disharmony, confusion, and dementia are of the essence of the mental affection throughout g.p., and are analagous, at first, to the ataxy, and then to the paresis, in the motor sphere.

All the following forms will be described in detail in one of the chapters on special symptomatology.

1. *Symptoms of Dementia predominant.* Here are to be found all grades of mental failure and loss from disease. In some examples there is merely a gradually increasing impairment and eventual loss of attention, memory, ratiocination, volition. In some, there are added, restlessness, untidiness, meddlesomeness, a tendency to self-stripping, destructiveness, and "foul" habits; in a few are rare paroxysms of excitement, or some morbid perception or idea of danger; some terror and attempt to escape from imaginary ill. In some, are paroxysmally the phenomena of *silent excitement.*

2. *The form with expansive or ambitious delirium predominating.* Even here, the expansive symptoms usually alternate with querulousness, irritability, hypochondria, melancholy, mental confusion.

Yet in this form the emotional exaltation of g.p. reaches its highest pitch. Joy, gaiety, hilarity, fill the patient, and with felicity his cup runneth over. The countenance is beaming, the eye bright, the figure expanded, the movements are lively, free, and gladsome. Nay, further, the imagination, gradually freeing itself from the control of reason, may conjure up visionary schemes for the enriching of the multitude, for the ennobling of mankind, and for the regeneration of the race; all to be effected by the force and supreme energy of the patient himself.

With this are usually manifold, varying, absurd, and self-contradictory delusions. In some cases the delusions are of wealth; and a patient without means asserts in one breath that he possesses a thousand, in the next breath a hundred thousand, pounds, or has numerous horses, carriages, houses, servants, jewels, and suits of clothing.

Or he asserts that he is judge, bishop, general, or king; or of colossal stature, and capable of herculean feats of strength;—or of grand mental endowment, artist, poet, author; or that he is several or all of these. Or that she has surpassing beauty, lovely children, gorgeous dress and jewellery.

Other symptoms may crop up, and at times hypochondriac ones may make appearance.

Some patients show merely a quiet, pleased self-satisfaction, and gentle exaggeration of the pleasant parts in their past history, or of their future expectations, seen in roseate hue. A childish self-complacency, and silly irrelevant laughter on trivial provocation, may be found in the same cases; and, indeed, may occur with or without marked expansive symptoms.

3. *Maniacal symptoms prominent.* In these cases, it is merely that, with more or less of the expansive symptoms just described, there are maniacal excitement, restlessness, and, often, destructiveness, and complete loss of feelings or indications of delicacy and propriety, or of any regard for social amenities, the habits, indeed, being degraded, the language foul, the person untidy and unclean. No sharp line, therefore, separates this from the cases in which the expansive symptoms predominate, and the two conditions might be described together. Occasionally the maniacal excitement is excessive and protracted.

4. *Hypochondriac symptoms prominent.* Here there is an oppressive condition of hypochondriacal feeling and idea. The delusive beliefs usually are to the effect that the patient's hollow viscera are closed or obstructed, so that he cannot swallow, or cannot defecate; or that his eyes or nose, his throat or gullet, his head or heart, his liver or bowels, are diseased, or rotten, or wasted away, or "gone"; or

that his whole frame has dwindled to child-like or infant size.

5. *Melancholic symptoms prominent.* These may be of any shade of mental depression, from mere sadness to deep distress—with or without suicidal bent—or to stuporose melancholia. Harmonizing with the dejected delusions, are usually hallucinations of a distressing and painful character. Delusions of a terrifying nature often exist; the patients, stricken with anxiety and terror, declaring themselves to be persecuted, insulted, hunted, or about to be shot or killed, and begging that their lives be spared. With others it is more a querulous anxious worried restless disturbed state; the patient, in distress, trying to get away and reach the persons whose hallucinatory voices he fancies he hears calling to him.

Though in many cases hypochondriac symptoms may occur at the prodromic, or in the early, stages, or mingle then with the melancholic, yet, as I* pointed out some years ago, melancholia, when it occurs as a temporary phase, is associated, rather, with the commencement of the disease, hypochondria with the middle part of its course, though each may be found in any part.

6. *Stuporose symptoms.* Stuporose symptoms may now occur, and at the time, or at first, the patient is usually supposed to offer an example of melancholia with stupor.

7. *Symptoms like circular insanity.* This form is rare, like the stuporose symptoms will be described later on in detail, and need only be formally mentioned here.

III. *Second stage of established general paralysis. Third period of the four.*

A. PHYSICAL SYMPTOMS.

(a). *Motor signs. Speech.* In this stage the articulation becomes more imperfect and shaky than before. The words are jumbled together, or there are slowness and stuttering, and an increase of the former hesitation to a momentary arrest of speech, usually ending in an explosive utterance with, perhaps, the elision of syllables. Or, with a mumbling and drawling utterance the voice is thick, hoarse, coarse, or muffled; and the accompanying twitches of the lips, face, and tongue very marked.

Tongue. As a rule the tongue is protruded only in a jerky, partial and momentary manner. The reflex activity and

* " British and Foreign Medico-Chirurgical Review," Apr., 1877, p. 457.

sensibility of the soft palate, pharynx and larynx are generally lessened; deglutition is impaired; the patient may be choked by food.

Pupils. The pupils have now become sluggish, irregular in shape; and usually *unequal* as first described by Dr. Baillarger.* This inequality may be due to contraction of one pupil, or to dilatation of one pupil, the other being still of about its normal size; but often one is contracted and the other dilated; they vary from time to time as to irregularity and relative size.

Face. The features have now a flabby, and often a greasy appearance, and the lower lines of facial expression are partially effaced, but the forehead is corrugated, and the facial balance distorted and awry. The physiognomy, the expression of the emotions, are much altered owing to this relaxation of the lower part of the face, and to relative overaction, or even some momentary twitching spasmodic action, of the muscles about the eyebrows and forehead, which tends to produce an unwonted expression of apparently unfelt astonishment or regret. Thus, at the same moment one part of the face may seem to express one emotion, another part a different emotion.

Trunk, limbs. The body is often bent awkwardly forward, or to one side; this may occur with some hemiparesis, and with or without a degree of rigidity. The gait is slow, unsafe and swerving; or, later, even zigzag with the feet kept wider apart than usual; or, if hurried, is more unsteady still. The steps are short, groping, or even dragging; or the gait may be of ataxiform character, or occasionally has partly a "spastic" impress. The patient stumbles too easily over an obstacle; may sway and fall in attempting to turn quickly, or when starting off in rapid walk under impulse or excitement.

Manual dexterity fails, as shown in defects of the act of buttoning the clothing, in the writing, or in any manual operation connected with the former occupation of the patient.

The superficial reflexes of the limbs are more often lessened; the so-called tendon reflexes may be normal, increased, or diminished.

Silent Excitement may occur, with its motor elements. See Chapter III. By this time the ataxic symptoms, so prominent in the early periods, have often become overshadowed

* " Gazette des Hôpitaux," Mai 14, 1850. 3°. série, T. ii, No. 57.

by the paretic, the muscular force is diminished, but, on occasion, considerable muscular strength can be put forth. Even those who argue that throughout "general paralysis" there is no true paralysis, admit the lessening of the muscular power therein,* in the great majority of cases; a condition sufficiently explained by the lesions of both the cellular and tubular elements of the cerebro-spinal system. At least, a paretic or paralytic element, whether general or local, now exists, or occurs, as an almost certain, or certain, clinical feature. At last the patient is more paralysed and helpless, or is bedridden. This occurs sooner if the spinal have been early and highly marked in proportion to the other symptoms.

Various motor conditions. Seizures. Epileptiform, apoplectiform, and paralytic seizures are now frequently seen. Violent muscular tremor or subsultus may affect part or the whole of the frame;—or there may be a symptomatic paralysis agitans—a true tremor cöactus; or, again, choreiform movements about the head and upper limbs; or athetosis.

If at last he lies in bed, the head and neck are often bent forward for hours together; or almost constantly is his head kept raised away from the pillow, the patient gazing stupidly here and there, or from time to time looking fixedly, but unintelligently, before him, the forehead at the same time being corrugated and the eyes widely open. In a few cases the lids may droop owing to temporary ptosis, supervening or not on local spasm. Often the legs are now more or less contracted and rigidly flexed, the forearms and hands flexed and lying across the chest, though the limbs perhaps are at times straightened by the patient, the speech is extremely limited, articulation indistinct, expression vacant, the tongue protruded with great difficulty, or not at all; the features often have a swollen relaxed puffy appearance, or grow thin earthy and coarser; all the natural discharges are passed involuntarily under the patient, while in many the teeth are frequently, or almost incessantly, and noisily ground together, not only by day but also making night hideous with the sound. Or strange inarticulate guttural noises are made.

(β). *Sensory symptoms.* Atrophy of the optic discs, following, or not, upon hyperæmia with exudation, exists in some cases; in some, white atrophic pallor is evident, and im-

* Dr. J. Christian, "Annales Méd.-Psych.," Jan., 1879, p. 32.

paired sight or blindness coexists. Or vision is confused; or a degree of "psychical blindness" may be present.

Even in this stage, however, many a patient starts back with a preternatural facility and blinks when a darting movement is made towards his eyes, or he starts unduly at a sudden touch.

The other special senses may fail. Smell and taste often seem to become diminished, or lost, and cutaneous sensibility, also.

Cutaneous hyperæsthesia, neuralgic pains, and sensorial hallucinations are now comparatively rare.

The palatine, faucial, and laryngeal sensibility is apparently lessened.

As a rule, there is general and progressive loss of sensibility, affecting the surface of the limbs and other parts, and pinches, pricks, injury, heat, or cold, are often but little heeded.

(γ). *Eating.* The habits and manners of the patients, as in eating, become degraded.

(δ). *Cachexia.* Cachexia exists; and the blood is altered in the manner hereafter to be described.

Bed sores are apt to form; and the skin to assume a dull, earthy, parchment-like, or else a greasy look. Diarrhœa, boils, carbuncles, vesicular blebs, or zoster crops, may now present themselves; or mucous hæmorrhages take place, or *othæmatoma.*

B. *Mental symptoms of the Second stage of the confirmed disease. Dementia.* In this second stage the failure of mind is very obvious, and the patients can no longer produce new ideas and new delusions. They must live upon the now disconnected or fragmentary morbid ideas of the past. Shreds of their former delusions are now repeated almost mechanically, and, towards the close, with little or no involvement of the emotions; the loss of memory and general mental failure are extreme; exaltation is now far more rare than it was; some self-satisfaction may be observed, but more often are the patients dull stupid and unemotional; or heavy sullen morose and irascible; or querulous worried and cast down. Yet, indeed, there is often, rather, a merely mechanical expression of these feelings; the inattention, apathy, and indifference of mental failure, for the most part, excluding all else. Falling into a chronic state of dementia, these patients may survive for prolonged spaces of time.

The mental state often passes from one phase to another; to-day repeating mechanically some fragmentary exalted delusions; to-morrow the patients are fatuous, or morose and irascible, and these delusions forgotten.

All sense of propriety or of shame is commonly lost. The habits of the patients are usually filthy; at times their dress is untidy. At last, helplessly and hopelessly demented, their rapidly failing powers of utterance are perhaps devoted to incoherent reviling, ribaldry, and obscenity; especially if, as often with soldiers, the language during the former life has been much of that type.

At irregular points of the downward journey may occur congestion and heat of the head, with increased mental dulness, heaviness, and stupidity, and often insomnia; restlessness, excitement, self-smearing, furious teeth-grinding, increased ataxic disorder, and often paralytic helplessness. The mental state is often worse after the decided apoplectiform or epileptiform seizures.

IV. *Fourth period. Third stage of the established disease.*

A. PHYSICAL SYMPTOMS.

The ataxic symptoms, and the paralytic, reach their greatest measure. Any movement is attended with the utmost trembling and shakiness; or, if the paralytic element predominates, the movements are feeble, ineffectual, and of small range. Seated upon a chair, the patient leans or tumbles forwards, or to one side; placed upon his feet, he stands still in an awkward attitude, or he stumbles awkwardly, or falls with the attempt at locomotion. Therefore, he is usually bedridden. And now contractions of the limbs, particularly flexed contractions, oftentimes make their appearance, and induce an unwonted deformity. These are sometimes transitory, and relax under the patient's muscular efforts, and sometimes are persistent. Now it is, especially, that bedsores, that blebs, boils, or herpetic eruptions about the extremities may appear; and now that the skin, although clean, often becomes dull dirty or greasy in appearance.

Finally, the wretched patient is completely prostrate, bedridden, helpless, inert, of "wet" and "dirty" habits, and his cutaneous and general sensibility, as well as his power of motor co-ordination, are almost abolished. With special senses blunted or lost, almost or quite unable to swallow with any safety, he lies wasting; afflicted, sometimes, with diarrhœa, pulmonary lesions, or contractions of the limbs; and still, perhaps, grinding the teeth.

If the general exhaustion from the primary disease does not carry him off, then some of its secondary, or indirect, effects, such as the pulmonary, intestinal, cystic, or renal, affections, will yield him a happy deliverance by death.

B. MENTAL SYMPTOMS.

The third stage of confirmed general paralysis (fourth period) is simply a profound degree of the preceding stage. The mental faculties, and coherent speech-power are practically almost abolished, and the merely vegetative existence soon comes to an end. But not before we can verify the failure or loss of sensibility, both general and special, the obliteration of all moral feeling, and the retention, at the most, of merely a thin, spectral, semblance of some former phases of idea or emotion.

CHAPTER III.

PARTICULAR SEMEIOGRAPHY.

The symptoms already mentioned in the course of g.p., and others, will now be described in detail and separately analyzed. In the preceding sketch of g.p. I placed the somatic signs first, in order to fix attention on their diagnostic importance. Here the order will be reversed, and the psychical symptoms precede.

The symptoms vary much in their *relative* importance and frequency and intensity. There is no single symptom which is pathognomonic. The only two always present in an established case are dementia and ataxy (or a mixture of paresis and ataxy). The most important and frequent symptoms will be described first. The description of less frequent, and comparatively accessory, symptoms and clinical occurrences in g.p., will be deferred to future chapters.

MENTAL SYMPTOMS.

1. *Dementia in general paralysis; and the form in which dementia predominates, clinically, throughout.* Failure of the faculty of ideation and of the logical power; also of the faculty of the reproduction of images—of ideas. Amnesia.

Cases without expansive delirium, excitement, or depression, may with convenience be grouped separately, and this without any prejudice to the view that clinically the mental basis of g.p. is essentially a weakness, a pseudo-dementia or a dementia.

In a few cases of g.p. a dementia begins, progresses, and practically includes or conditionates the entire range of mental symptoms throughout the whole course. It is a question whether these are not examples of g.p. in its most pure and simple form. Although embraced under the present head, these are not more referred to here than are the cases in which dementia, although the predominant symptom, is associated with other mental phenomena such, indeed, as may occur in other forms of dementia.

Mention has already been made of the incipient dementia in the prodromic stage; of the marked dementia in the later stages. With this dementia, whatever its degree, there is associated more or less failure of the moral qualities; more or less impairment or disorder of the instincts, sentiments, emotions. In the earliest stages, the dementia comes on in an insidious manner; but may advance by sudden leaps after apoplectiform seizures.

At their onset, the symptoms of dementia are those described when treating of the prodromic stage. It is unnecessary to reproduce them here. Suffice it to bear in mind; that impairment of memory is a leading feature, that the ordinary duties of life receive imperfect attention; that blunders and forgetfulness impair the patient's business capacity; that mental confusion accompanies the stupidity, and so does, often, headache or heaviness of head, and flushing of the face and scalp; as well as drowsiness; that vertigo, independent of gastric disorder, is a frequent accompaniment; that with the intellectual failure, is failure, also, of the moral and æsthetic feelings, and open breaches of the law, of morality, of public or social propriety; that weakness of the emotions, or a sombre tinge of the same, usually coexist; that fits of abstraction, sudden cessation, without objective cause, of the work or conversation in which the patient is engaged, surprise his friends; as do, also, the unwonted irritability and angry explosions about the lightest trifles; while heavy dull headache is apt to alternate with vertigo, or with the feeling of being "light-headed." It is rare, however, that the symptoms are all, or always, of so easily tractable or mild a type as these; for there may be insomnia, restlessness, propensity to wander aimlessly and listlessly, to collect rubbish, to be fussy fidgety meddlesome;

occasionally, terror and excitement, or furious outbursts of passion; and congestive attacks, or simple paralytic attacks, the seizures of either kind being manifested by hemiparesis. These patients in early dementia may become the dupes of knaves; be induced to sign documents or "back" notes to their own detriment, or to engage in some felonious enterprize; become the "plucked pigeons" of sharpers; perhaps wed courtesans.

As I have stated elsewhere,* in this form of g.p. there are:—
" General intellectual weakness and loss of memory, increasing to fatuity if the patient survives long enough, and evinced by blunting of perception, of apprehension, and of reasoning power, by confusion of thought and of verbal expression, and a slowness and dulness of mental operation, the patient, also, paroxysmally becoming more 'lost.' Incoherence of speech. Long periods, or paroxysms, in which the patient is restless, meddlesome, prone to self-stripping, destructive to clothing and bedding, or apt to wander about listlessly. The habits soon become degraded. . . . In a few there are delusions of impending danger or of injury, together with terror, in the earlier stages; and at this time the patients may attempt to escape from the vaguely apprehended evil; in a few, also, insomnia or even hallucination, is noticed. . . .

The emotional display and emotional facility are less marked in this than in any other of the symptomatological forms of g.p. Still, at an early period, there may be found a slightly expansive state of feeling, a shadowy gaiety of self-consciousness, which, however, soon disappears; or, on the other hand, an irascibility or petulance on the slightest occasion."

The hypochondriac or melancholic tinge, if present, is only slight and transitory.

Early abolition of speech, with persistence of inarticulate cries, is not infrequent in this form. Occasionally the paretic signs are steadily or rapidly progressive; but many of the cases are long-lasting.

At first the patients may be conscious of their state, or vaguely so, but do not press for cure. Later on, they cease to be able to apprehend their intellectual condition and their state of impending wreck. Relaxation of energy; failure of decision of character, of promptitude and determination, are obvious.

* "British and Foreign Medico-Chirurgical Review," Apr., 1877, p. 449.

As the dementia advances not only can nothing new be assimilated by the patient's mind, but his previous mental treasures have nearly dissolved away; there is merely the mechanical repetition of a few fragmentary phrases, the relics of former mentation. The feelings of propriety and shame are lost; and, in many, now and then are paroxysms of heat of head, mental dulness and heaviness, restlessness, furious teeth-grinding, self-smearing with fæces, destructiveness, insomnia; and the peculiar paretic grasp and pull and shove.

At last he may be reduced to a few inarticulate cries; to sniffing, lip-smacking, or sucking or clucking sounds;— emotions faded out, moral feelings obliterated, the intellect a void, to say nothing here of the emaciation, contracted limbs, and helplessness.

As a rule, if one makes sufficient allowance for exceptional cases and deviations, and for various complications, the progress of the dementia can be fairly followed through the speech, in tracing, successively, the unwonted slowness of speech, pauses, search for words, drawling, substitution of a common for a less familiar word; omissions of essential words or syllables; hemming; irrelevant repetition of words or sentences; cessation of speech, or a fragmentary utterance owing to forgetfulness of the idea whose expression was begun; echo-speech; gradual or partial substitution of inarticulate cries for articulate speech; or total loss of the latter.

As there will be occasion to mention again, the psychical weakness affords to the deliria of g.p. something of their special impress, and it is needless to insist upon the point here except to mention the contradictoriness, thereby usually induced, between the expressed belief and the conduct and life of the patient, and the frequent incongruity of his morbid notions. In these deliria, the patient, as it were, may forget, even at the moment of their utterance, to act up to or in harmony with his assertions as to enucleated eyes, putrefied frame, wealth, physical prowess, titles, dignities. His conduct disproves his words, and he cannot perceive that it does so.

Here an important matter is to be noted. During the remissions, it is some psychical failure, some impairment of memory, attention, energy, determination, and of capacity for sustained or complex mental effort, that remains not wholly cleared up, and forms the uncrossed barrier to the *restitutio ad integrum*. Here a few simple arithmetical questions will often puzzle the patient, and the replies betray the mental state.

Amnesia. The most important phase of mental failure is that affecting the power of representation, in consciousness, of past impressions. The impairment of memory in g.p. in some cases undergoes remission; but when this latter is very marked the impairment of memory previously existing has usually been factitious rather than real, and due more to a confusion of thought than to actual amnesia. Nevertheless, the amnesia may undergo a considerable and real remission independently of any fallacious appearance of the kind just mentioned. In too many, and most, cases, however, the impairment of memory is, roughly speaking, progressive, to a complete obliteration of the stored images of the past.

In the earlier stages, the failure of memory partly accounts for many of the strangest of the symptoms. Such as the quiet, imperturbable, apparent lie or dissimulation of the G.P.; as when, in the presence of many others, he hurls a stone through a window, and, very soon after the crashing of the glass is heard, denies the act in presence of the witnesses thereto. Here, as a rule, is not a conscious falsehood; the act, performed quasi-automatically, has not been, except dimly, an object of consciousness, has not been registered, and cannot be reproduced (represented). In some cases, therefore, it is owing to mental failure the patient seems to utter lies in the coolest and most imperturbable, unabashed manner. For it is not that he is *splendidè mendax*, but that he is *inops mentis*.

The memory for recent events suffers most and earliest; the patient who can relate some older experiences with sufficient accuracy, has, perhaps, forgotten that he was visited yesterday by some member of his family, and he mistakes the identity of his doctor and attendants. At last, even if not incapacitated therefor by paralysis, he is so forgetful as to need to be fed, dressed, and every species of attention.

2. *Defect and Disorder of Moral and Æsthetic sense; of Religious feeling.* The defects and perversions of the moral feelings, combined with an incipient or a more advanced mental failure, lead to acts that are frequent at any stage of g.p. up to the time that the whole mental life is lost in the night of absolute fatuity; acts that if of trifling nature would be deemed childish, ridiculous, contemptible, in the

same;—if more serious would be held as immoral, illegal, or even felonious. It is their occurrence in the prodromic stage, or else in the earlier stages of the established disease, that gives them their medico-legal importance. And many an unfortunate G.P., at this early stage, has been agitated and disturbed in mind by a legal trial and conviction and imprisonment, whose illegal acts were the direct outcome of his mental defect and disorder from organic brain-disease, and the course of whose lethal malady has thus been hastened, and the least ray of hope of recovery thus shut out.

In speaking of the *prodromi*, special stress was laid on the impairments and perversions of the moral feelings, of the character and disposition; and it was mentioned how these might be the first changes noticed in the patient, and might even precede the pronounced g.p. by years.

A coarseness and brutality of language is common in g.p. In some cases it is the very first thing noticed by the patient's friends; for his conversation is apt to become rude, brusque, revealing an overbearing, undiscerning, selfish and conceited tone of mind. When angered, these patients will often pour forth a flood of insulting and reviling language, intermingled with oaths, strong expletives, or obscene phrases. Fortunately, it often is not thus; and the grotesquely absurd, but generously worded, proposals of the expansive G.P. only create a smile.

Offences against sexual morality, public decency, or social propriety. Some patients make open love to their maid-servants, or to the nearest female, and these unwonted familiarities in presence of wife, children, or other witnesses, I have mentioned as sometimes an early symptom. Some rush into debauchery of all kinds, and precipitate the course of their malady by libations at the altars of Venus and Bacchus.

The perversion of sexual feelings, or the removal of these from normal control, leads the G.P. into many scandals; charges of indecent acts or assaults, of adultery, attempted rape or pæderasty. Nor are the charges always ungrounded. G.Ps. sometimes do openly urinate, or expose their genitals, in public places, or make off-hand and indecent proposals to persons of the other, or of their own, sex; or use obscene expressions in the most open, and, to the public, apparently shameless manner. Sometimes they attempt to take indecent liberties with females, or to commit rape upon them. Steady domesticated men, with wife and children, sometimes, in the earlier stages, are misled into illicit sexual relations, or bigamy: the unmarried are sometimes duped into unfit and disastrous marriages; it may be, with worthless schemers, or even with a casually-met prostitute, as in a case by Dr. Legrand du Saulle.*

The self-exposure above-mentioned is sometimes obviously due to confusion and dementia, as in the case of a married woman who sat down by the roadside and began to undress, mentioned by Dr. W. H. O. Sankey; † or where the patient onanises publicly, as the case of Chorinsky reported by Hagen. Dr. J. G. Kiernan ‡ mentions a G.P. and "hitherto respectable physician who suddenly indecently exposed his person. Fined for this, he immediately exposed it on leaving the court-room and the fine was doubled but he again repeated the offence." Dr. Legrand du Saulle § has insisted upon the frequency of this self-exposure in G.Ps.—Dr. Mendel ‖ also observed cases in which G.Ps. onanised in public, or attempted pæderasty; and Dr. Dagonet mentioned examples of open exposure of penis, and solicitation of females. Westphal ¶ mentioned a case; here a larval epileptoid basis preceded g.p.; the patient indecently fingered little girls. Dr. Edgar Sheppard ** described the case of a G.P. clergyman, who put up at a hotel in London, and immediately tried to bring in three prostitutes. One of my patients walked into a hotel, where he was unknown, ordered a bed, and openly invited the manageress to share it then and there.

Self-defilement, self-daubing or smearing of clothes and walls with fæces, and various filthy practices, are usual at the last; but may occur at any time, even at the onset.

Failure of the social sentiments, of knowledge or recollection of the usages of society, lead the patient to coarse, ungraceful manners in eating; to slovenliness, and personal uncleanness.

Some make silly, objectless, assaults. Some become

* "Gazette des Hôpitaux," No. 104, Sept. 11, 1883, p. 826.
† "Lectures on Mental Disease," Second Edition, 1884, p. 256.
‡ "Alienist and Neurologist," Oct. 1884, p. 672.
§ Loc. cit., p. 825.
‖ Op. cit., p. 123.
¶ "Archiv für Psychiatrie," vii Bd., p. 622.
** "Lectures on Madness."

destructive; I have known almost the first thing that drew attention to a military G.P., to be his apparently wanton, objectless, destruction of his kit. They do not attempt to conceal acts of this kind, or to escape the punishment and inconveniences that swiftly follow such offences, but on the contrary so do them as almost to ensure instant detection.

The moral perversions or defects spoken of in the preceding sections, and in those yet to follow, are often prominent in the prodromic stage, and play a part in the phenomena of the next stage; whilst in the last stages moral feeling and culture are obliterated with the advance of dementia, and as a necessary consequence of the disintegration of mind which occurs.

Theft, frauds, defalcations, "false entries," &c., by general paralytics. If the patient steals, the objects taken are often useless and valueless to him. Quietly, and in the most natural way, does he carry off articles from under the very eyes of the owners, or even seek the assistance of strangers or of the police for that purpose; if stopped, he coolly confesses his act, or says the stolen articles have been given or lent to him, or belong to him, or he contradicts himself; the meanwhile often displaying an utter absence of any feeling of shame. Baillarger mentioned three cases, all females. One stole a variety of articles from shops, and, making no use of them, left the tickets still attached. She, as well as the other two, fell into the hands of the police, and thence into prison, before reaching an asylum. In some cases, however, as in one related by Dr. Maudsley, the theft is carefully planned, and a disguise assumed. On two occasions. this patient robbed his fellow-passengers in a railway train at night, with some coolness and skill.

One of my patients, in full day, had walked off with flowers in pots from the window of a London mansion, and when taxed with the act had said the flowers were a gift to him from the owner. Another, wandering hither and thither, brought home at night a collection of linen from despoiled clothes-lines. Several of my patients, non-commissioned officers, while in the prodromic stage appropriated, or failed to account for, moneys passing through their hands in virtue of their office; whereupon the necessary court-martial and degradation to the ranks precipitated the course of that inconspicuous or latent intellectual and moral defect and disorder—due to organic disease of the brain—of which the primary defalcation had been an early or the first manifestation.* (See also "housebreaking.")

Some of the thefts by G.Ps. seem to be due to expansive delusions as to possessions and rights; others to be impulsive; others due to such a degree of mental failure and confusion that the patient is scarcely conscious of what he is doing; and often it is nothing more in reality than an absent-minded, and practically automatic, involuntary, act. In this last case it is merely a first degree of what is often seen later in G.Ps., in the asylum, viz., the tendency to collect and pilfer trifling, useless, and valueless objects, as buttons, bits of paper, pebbles. Mendel mentions a patient who had a special predilection for stealing ivory articles from the shops, but before going out would tell his relatives from which shops he would steal.

As the robberies, so the frauds and defalcations usually reveal extreme mental weakness, in the way they are perpetrated, and in the patient's manner and conduct when detected and charged with knavery. But the fraud may be somewhat elaborately planned.

G.Ps. often go to hotels and run bills they are unable to meet; or engage cabs, and drive about all day with no set purpose, and without the means to pay; one of my patients, in penury, who did this, after the day's drive offered the cabman a worthless rusty old saucepan as his fare.

In carrying out frauds and forgeries G.Ps. are the easy dupes of designing knaves; self-motived, these, and incendiarisms, as a rule are a result of impulse and of dementia

* For other cases of theft by general paralytics see:—J. C. Prichard, "On the Different Forms of Insanity," 1842, p. 156. Lélut (2 cases), "Annales Médico-Psychologiques," T. i. Baillarger (3 cases), ibid., T. v., 1853, p. 479. "Propension au vol chez les Malades atteints d'un premier degré de Paralysie Générale." Parot, ibid., p. 481. Billod, ibid., 1850, p. 626 (case of "N"). Hippolyte. Devouges, ibid., 1857, p. 532, et seq. Brierre de Boismont, "Ann. d'Hygiène publique," &c., 1860, p. 409. A. Sauze (4 cases), "Ann. Méd.-Psych.," 1861, p. 54. Maudsley, "Responsibility in Mental Disease," 1874, p. 75. "Lancet," Nov. 13th, 1875, p. 693. Wilkie Burman (6 cases), "Jl. Ment. Sci.," Jan. 1873, p. 536; (4 cases), ibid., July 1874, p. 246. W. Julius Mickle, "Journ. Mental Science," Apr. 1872, p. 41; also cases 1st Ed. Fabre (2 cases), "Ann. Méd.-Psych.," March 1874, pp., 198 and 207. F. Darde (6 cases), "Du Délire des Actes dans la Paralysie Générale," 1874. W. A. Hammond, "Treatise on Insanity," 1883, p. 598 (3 cases). Mendel, op. cit., p. 123. V. Magnan's case—a G.P. induced two policemen to help him to roll away barrel of wine from wine-shop door. Simon's case of fisherman, "Die Gehirnerweichung der Irren." Legrand du Saulle (3 cases), "Gaz. des Hôpitaux," Sept. 11th, 1883, p. 826. "Gaillard's Medical Journ.," Oct. 1882 (2 cases). Granger, "Buffalo Med. and Surg. Journ.," July, 1883 (cited "Alienist," &c., Oct. 1884, p. 671). W. H. O. Sankey, "Lectures on Mental Disease," 2nd Ed., 1884, p. 256.

as said by Legrand du Saulle* (but see *infra*). At last the moral sense, affections, religious and æsthetic feelings, are well-nigh, sometimes completely, obliterated.

G.Ps. so often impoverish themselves and their families, in the expansive moods, that examples of this are unnecessary.

Other illegal criminal acts by G.Ps. Here there is not merely failure or perversion of moral sense, but the intellectual weakness, the failure of memory, the morbid self-consciousness, the deranged emotions, the perverted ideas; some or all, variously combined or associated in different cases, are concerned in the unwonted illegal acts of G.Ps.

Suicide. As I stated in the 1st edition, if G.Ps. "attempt suicide or homicide there is usually no fixity in the reasons they assign; no precaution, no concealment, no persistent aim in the accomplishment of the act itself. There is the same obvious childishness in the suicidal acts or attempts, the same mental moral or affective enfeeblement, as in the expansive form." On several occasions, I have known fussy, blubbering G.Ps. rush to the w.c., and make some sort of attempt to strangle themselves with their braces, but usually not before they had openly declared their immediate intention of so doing. One with great determination rammed "bed-flock" into his mouth and throat. Several had made suicidal attempts before they came under my care. Six days before admission, one patient made a resolute attempt at night to hang himself by means of sheets; and, when rescued, made a violent attack upon the orderly who had prevented the suicide.

Refusal of food by hypochondriac G.Ps. is usually not the outcome of a distinct suicidal intention; but is, rather, that of the delusive ideas of these patients; particularly those of inability to swallow, obstruction in some part of the alimentary canal, or absence or wasting of some organ; though occasionally the refusal is owing to ill-temper; or, again, to disdain of homely fare.

Incendiarism by G.Ps. Dr. v. Krafft-Ebing reported one example, and Dr. Legrand du Saulle* a motiveless act of incendiarism, while Dr. Spitzka† mentioned a case in which the patient attempted thrice in one night to set fire to his house.

G.Ps. will often burn holes in their clothes, or cast into the

* "Gazette des Hôpitaux," Sept. 18th, 1883, p. 852.
† *Op. cit.*, p. 205.

fire articles of clothing, or of furniture, or anything ready to hand. In the expansive phases they may attempt to burn goods or houses; the latter sometimes under the belief that they are theirs, and with the avowed intention of erecting much larger and finer ones instead.

Housebreaking. One of my patients, before admission, stole a large ladder, and secured the help of a friend, acting unwittingly, to carry it a long distance. Afterwards he broke into an uninhabited house, and made a fire in the middle of one of the rooms with some building-material, and placed bricks on it. When arrested, he declared that he was frying fish, had rented the house, and was on the point of opening it, that evening, as a fish-shop.

Capital crimes. Murder. Treason, etc., are rare in G.Ps. But there is always the chance of a pliable G.P. being induced to crime by a conspirator or murder-monger. Thus, in the case of Count Chorinsky;* he assisted his paramour to poison his wife, procuring the poison, making concealment, and taking various precautions to enable the proofs of an *alibi* to be set up. Legrand du Saullet† speaks of a G.P. who armed himself and asked his way to the house of a doctor, at the same time openly expressing his intention to kill the latter.

Testimentary acts by G.Ps. Above, I have cited a case in which a G.P., of property, met in a railway train, and subsequently married, a common prostitute. Very soon he died, leaving her all he had. Others, in the melancholic phases, might, no doubt, disinherit those who have been dearest to them.

3. *Anomalies of self-consciousness.* These are connected with the weakness of thought, of logical connexion, and of reproduction of impressions; all of which have already been discussed. They are also allied to the hypochondriacal delirium. Disorder of self-consciousness in g.p. reveals itself sometimes in the feeling, by the patient, that his identity is changed; that he is no longer himself, but died long ago; or he feels that he has a double personality, or speaks of himself in the third person. Thus, one of my patients, referring to himself, used to say, "they're tearing him to pieces—he's very bad" (*i.e.*, physically); spoke of himself as "him," or "the parts here;" and described himself as having no name, and as being "made up of all sorts

* Hagen's account, Erlangen, 1872.
† *Loc. cit.*, p. 852.

of things." Another patient thought he had been changed into one of the lower animals—" a beast."

As the disease advances, self-consciousness is gradually weakened, narrowed, and at last fades out.

The dreams of G.P.'s exercise an influence upon the subject of their morbid ideas, of which they seem sometimes to be the starting point (*vide infra*). Sleeplessness, or on the other hand, somnolence, may occur at any stage.

4. *Anomalies of the general feeling ; cœnæsthesis, feeling of well-being or malaise.* For convenience this is placed here rather than with the sensory affections.

This sum and complex organization of all the impressions on the system collected, correlated, and expressed in the central nervous apparatus, gives each sane person under ordinary circumstances the consciousness of health or of disease. In the prodromic stage, the patient about to become G.P. has often the feeling of disease, and forecasts that he, as he expresses it, is "getting softening of the brain," or "is going mad." If there is a depressed stage, later, the same notions may occur during it, and his other expressions as to bodily weakness, disease, and decay are striking, though changeable. Even in the expansive delirium the same unpleasant and overcast state of the cœnæsthesis may alternate, or may mingle strangely, with the light and pleasurable feeling of vigorous health and redundant energy which then, for the most part, holds sway, and which is the antithesis of the preceding oppressed and suffering state of the "general feeling." With remissions of the disease may or may not return a perception, on the part of the patient, of the state of disease and of something of its danger to life. These briefly sum up the chief anomalies, in g.p., of this phase of sensational consciousness.

5. *Emotions.* Combined with some of the mental defects and perversions above mentioned; and perhaps alternating with opposite and genial conditions; may be a readiness to ill-temper, intense irritability, and furious outbursts, which, his disease not yet recognised as such, make the patient a tyrant-terror to all his dependents and belongings.

The emotions are often brought into play with preternatural ease; an occasion yielding moderate pleasure to the healthy, perhaps may cause irrepressible joy, even to tears, in the G.Ps.; while slight chagrins and vexations may cause him to sob like a child. The emotional facility and irritable weakness show themselves, also, in the rapid changes from one form of emotion to another, and in the shallowness of each; in a word, the feelings are facile, changeable, shallow.

A lively sentiment of pure sexual love is stated by Moreau (de Tours)* to sometimes precede the sexual excesses of G.Ps.

The description of the emotional changes in g.p. is inextricably blended with that of the next group of symptoms —the deliria, q.v.

The deliria of general paralysis. We now come to speak of symptoms flowing from *disorder* of the more purely intellectual, sensorial, and emotional faculties. Necessarily these, especially the first, are phenomena of the established disease, for so soon as mental disorder of the kind referred to exists the patient is "insane;" and for examples we look chiefly to its first stage.

It has been said (Salomon) that the distinguishing characteristic of this stage is the confusion, owing to his defect of judgment, that the patient makes between his ideas or beliefs and his desires, so that to him they are the same: or, as Billod† long ago suggested, for him the mirage is such that he believes he possesses that to which he aspires, and holds as realised his dreams of happiness. Esquirol‡ appears to have particularly noticed the contentment and indifference of the patients, the facile diversion of their attention, and their easy-going acquiescence in their enforced detention.

Yet dementia is the mental groundwork of general paralysis.

Here I shall speak, successively, of *expansive delirium, mania, melancholia, hypochondria, stupor,* and *"circular"* symptoms, as they severally occur in g.p.

In describing the several semeiological forms, viewed merely from the psychical side, it would be useless and cumbersome, in this place, to make, like some observers, several divisions of each of the above forms, as, for example, to make a quadrified division of the expansive form.§

Sub-divisions of this kind are all the more unnecessary as none of the phenomena are persistent, as they interchange rapidly and frequently, and are in reality but phases of the same morbid mental condition.

---

* "Psychologie Morbide," p. 267: cited by Legrand du Saulle.
† "Annales Méd.-Psych.," 1850.
‡ "Des Maladies Mentales," par E. Esquirol, 1838.
§ See paper by Author, "Journ. Mental Science," Apr. 1878, p. 27.

Nor, in describing them, shall I place under separate heads the delusions and the morbid emotional conditions. So often are these intimately, and even inextricably, bound together as essential co-factors of the clinical phenomena, that, from a clinical standpoint, their separation would reduce what should be a living, breathing, picture, to merely disconnected, dissected, dead, dry details.

6. *Expansive or ambitious delirium;* * *and the form of g.p. in which it predominates.* By many writers extraordinary stress has been laid upon the description of the extravagant delusions and the emotional exaltation of g.p., and so striking are these phenomena that attention has been too much withdrawn from other, and equally common, facts pertaining to the mental order of symptoms.

It is rarely that the expansive delirium comes on suddenly, or bursts into full bloom, as the initial mental disorder. Usually, it has been preceded by mental failure, showing itself in various ways to the skilled observer; and often also by a period of depression, melancholic or hypochondriacal. Moreover, it comes on somewhat gradually as a rule; the patient becomes more rash and speculative, has a larger idea of his own powers and abilities than has been customary to him, enters into all sorts of engagements in an over-sanguine spirit, becomes more restless, passionate, and excitable than he has been wont: what he says and does bears the impress of exaggeration and of inflated self-feeling. Soon the teachings of reserve and prudence are parted with, and, should he fall into evil hands, the patient is the easy dupe and prey of every unscrupulous intimate, and every swindling bubble-monger; for by this time he has come to relax rather to the stranger than to the tried and trusty friend. At this stage the delusions lead to the most extravagant projects on the part of the patient himself; who, if at large, sends telegrams containing preposterous orders, such as to lade a ship with wine, "for sale in the uninhabited parts of the earth," and thus, without extraneous aid, these patients often fritter away their means on impracticable or ill-advised business-schemes and speculations.

When the exaltation is highly developed we find the patient with an inflated mien, an expansive, smiling, benevolent expression, an impressive bearing of good-fellowship and friendship; overflowing with good-nature, thanks and

* Present in 52 of Bayle's 85 cases, and in 25 of Calmeil's 62 (early treatise) "Ann. Méd.-Psych.," 1859, p. 328.

compliments; pleased with everything and everybody; delighted with, and sounding the praises of, his surroundings, however mean and irksome they may be; loquacious, singing, restlessly moving about, dancing, or capering; eating rapidly and with avidity; covering his hat with some fantastic or tawdry ornaments; destroying his clothing and stitching it together again, to improve it, as he says; or, when the symptoms verge towards, or attain, the maniacal, passing the night in re-arranging his bedding a thousand times, or in tearing it up with the avowed benevolent intention of making a dozen, or an hundred, beds out of the one supplied to him.

With this there may be hallucinations of sight or of hearing, in harmony with the extravagant notions. These notions usually run upon the possession of enormous wealth, high titles, prestige, position, great muscular powers or sexual capacity, marriage to members of royal houses, and perhaps to many of them; the distinguishing features of these delusions being that they are neither fixed nor systematized, but on the contrary are multiple, varying, ridiculous, and self-contradictory; and betoken an abrogation of the power of judgment, while they often *culminate* as if in a crescendo movement of the expression of magnificence. The patient is not only " possessed," but inflated, with greatness. The methods of language fail him here as he rides uplifted on the mighty wave of feeling; or, to him borne on this swelling tide of exultation, the very heavens appear to open, and he holds converse with celestial beings, and has ecstatic visions of eternal fields.

The exalted delusions announced by him are the indicia of a progressive delirium, which is in contrast with the fixed monomania of pride or ambition sometimes seen in ordinary non-paralytic insanity.

To read the descriptions of some writers one would suppose that such as these were the almost constant and abiding mental symptoms of general paralysis. This, however, is rarely the case, even in the expansive form. More often symptoms like the above alternate with hypochondriacal feeling and ideas; or with a whining, moaning expression of peevishness and distress; or with childish and unnecessary fear and terrors; or with sullen irritable states of feeling; or, finally, with a condition of simple dulness, stupidity and confusion.

Frequent, also, in place of pure grandiose delirium, is that condition of quiet, pleased, smiling, self-satisfaction seen in some, but not associated with decidedly exalted delusions. These are the optimists who tell with complacency, though without marked exaggeration, of their possessions, however slender the latter may be; dwell upon the looks and personal qualities of their common-place wives or husbands and children; live amidst a few reminiscences of the past flattering to their vanity; but who, if questioned, exhibit none of the above extravagant delusions; though a soft brightness lights up for them the beauties of their now narrow mental world. Immersed in contentment and complacency, they are "well" and it goes "well" with them, say they, with hesitating tremulous utterance, breaking health, neglected work, and shattered prospects; thus proclaiming a most striking proof of that utter prostration of perception and of judgment which already announces the imminent ruin of the whole edifice of their mental life.

Here come cases where the expansive ideas are found only on one point—some personal trait—as in examples by Billod and Hammond—a transitory condition in my experience.

In the above, I have spoken of the extravagant delusions and of the expansive state of emotion as coexistent, if present. Usually co-ordinate or welded together as they are, they may be found separate, and, at least temporarily, independent of each other's presence. Thus, on the one hand, when these two symptoms are unyoked, G.Ps., for spaces of time, may exhibit exaltation of feeling unaccompanied by any exalted delusion, and in vain will one seek for the large delusive idea correspondent to the exaltation or beaming joy in which the patient is immersed. Feebleness of intellect, defect of memory, blurring of intellectual perception, looseness of ratiocination, feebleness of logic, and distraction and wandering of thought, there may be; but not in these particular cases, and at this time, any ambitious delirium.

On the other hand, we find patients who at times, and some who as a rule, give utterance to exalted delusions; but do not manifest any joy or emotional exaltation; and this, perhaps, at a comparatively early stage. Thus, a private soldier without the faintest trace of emotion expressed to me the delusions that he was 6ft. 2in. in height, had rowed stroke-oar in the inter-university boat-race, was in Lord's cricket club, was a M.A., an artist, was Frederick

William, Mark Antony, Julius Cæsar Borgia, Pope of Rome at £100,000,000 a year. His wife's titles were equally a medley, and he asserted that she was descended from the two kings, was an empress, and he himself an emperor. "Has 170,000 ships, and the same number of gymnasia, and of dogs (corrected as "8 million" dogs), and of breweries, distilleries, and tanyards; and carries on every trade in the wide world." "Has invented all sorts of machines, was at Netley with his brother the Prince of Wales, went to the planet Uranus or Hell ('it's all humbug about the fire'), went to heaven in a balloon."

Still more striking is it to hear patients expressing exalted delusions of ambition, pride, and pretension, in conjunction or blent with hypochondriac delusions, and sometimes with the expression of no joy or satisfaction, but, perhaps, with that of a pained and hypochondriacal state of feeling; though sometimes here the patients alternate from moment to moment between depression and satisfaction. Examples of this will be found further on in the description of hypochondriac symptoms. And here I may mention a sergeant who talked to me in a sober, sedate, and emotionless way, as I made the following note: "Says his head is bad, will get a new head when he goes to heaven, it was full of diamonds and has been taken off. His heart and lungs have been taken out. . . . Will be a giant 125 feet high, and made so, in sun or heaven, by 7 quarts each of brandy and whisky, and 3 of beer. Has four ducal titles . . . is first field-marshal, has millions and millions of gold watches."

Sometimes the largeness of idea is observed even in the melancholic G.Ps., who, in speaking of their afflictions woes and sins, use terms of, perhaps, grotesque exaggeration. This betrays the same fundamental psychical weakness as the expansive delirium; but it is unnecessary to erect it, as Voisin does, into a separate variety of the expansive delirium ("délire d'exagération"). Indeed, there is usually with it no exaltation, and it is merely a formal moulding of the depressed delirium.

Sexual ideas often crop up in the expansive form. The patient has many wives or husbands, or is about to marry into noble or royal families, has thousands of children, or immense sexual power and organs. Or they sing delightedly of the beauties and charms of their spouses. It is the romantic, the ideal, the purer, phase of the sexual relationship that more especially is apt to engage their thoughts.

The position in life, and the sex, modify the form of the delusions. The woman's delusions concern, rather, her beauty, her dress and jewellery, her lovely and numerous children, her many husbands or lovers, or their position, wealth and titles. The engineer or craftsman has amazing skill, and has invented apparatus of almost miraculous powers. The soldier has a hundred gold and lace and scarlet uniforms; and is, in one breath, captain, major-general, commander-in-chief.

If thwarted they may become irritable and violent, yet G.Ps. with pronounced expansive delirium can often readily be diverted by a few dexterous suggestions on the lines of humouring their self-satisfaction.

Like the other delusions, these tend to dwindle to a few remnants of the original formation; to be, perhaps, eventually reduced to one or two fragmentary statements, and finally disappear. One such patient, at last, was left for a long time with two such fragments; the one, "I fought at Alma;" the other, "I want a piece of paper to write for Kilmacune estate." Once, seizing and trying to gulp down a lump of meat, he was nearly asphyxiated and quite unconscious; as the blackness of face was disappearing under artificial respiration, and he was regaining consciousness, the first words were "piece—paper—Kilmacune estate."

The various seizures may interrupt the expansive symptoms, and be followed by those, predominantly, of some other order.

It is strange how amidst the wildest flight of expansive delirium, some act or word utterly incongruous therewith appears. The patient who says he possesses millions, decorates himself with some worthless bits of coloured paper, or begs a pipeful of tobacco. Again, some accurate statement may be maintained throughout, though jostling with the most absurd ideas on the same general subject. An ex-Colonial Governor, who died under my care, would at the very same moment speak or write of his imaginary claims to great sums of money from the Government, and of the overdue quarterly payment of his (actual) Government annuity of £139. The imaginary claim of hundreds of thousands varied much from time to time in its exact amount; the real £139 never varied to the last; nor were they ever merged in one another or confused.

7. **Mania in, and the acute maniacal form of, general paralysis.** An acute maniacal attack of some duration may occur at the transition from the prodromic to the first stage; or attacks of this kind, but of a more ephemeral duration, may be found throughout the second and third periods. Yet a more persistent condition is now referred to.

In the acute maniacal form, add to much in the foregoing sections the symptoms of extreme mental and motor agitation; with insomnia, and restlessnesss at night as well as by day, and occasionally hallucinations of sight or of hearing, or even illusions of some of the other special senses. An incessant motor activity possesses the man; he talks, shouts, sings, stamps, seizes and destroys surrounding objects, tears his clothes and bedding or soils them, smears himself and his room with his ordure, or drinks his urine; sometimes even neglects or refuses his food, or collects pebbles, rags, and other rubbish. At times he may be most irritable, destructive, or most dangerously hostile, threatening, or violent, and assault others far more powerful than himself. His temperature ranges higher than normal, his eye is quick and bright, and his countenance vividly injected.

Silent excitement. In some cases the patient is intensely restless, pushing, pulling, disordering, upsetting or destroying surrounding objects, as if unconsciously, mechanically, and without apparent aim; but resistant to interference, and to the necessary tending. He seizes objects, which he incessantly pulls to, and pushes from, himself, pulling again and again upon them if they are fixed; drawing them over him, and again thrusting them from him, if they are moveable. He grasps tightly, resists passive motion, tears or soils clothes or bedding, and, though perhaps very helpless, is difficult to manage. In an extreme degree, this condition is mainly found in the second stage; then hours may be spent in this monotonous and aimless handling, pulling, thrusting, rubbing of objects, and never-ceasing movements of the limbs and trunk, until the perspiration breaks forth on the heated face, the fumbling hands; yet the patient still works on with stupidly earnest gaze. Speech or phonation may not be entirely in abeyance, there may be a muttering in low tones.

Dr. Jules Falret (*op. cit.*), who named this state "agitation silencieuse" spoke of it as a subdivision of his "maniacal" variety. But in many cases its associations are, rather, with the depressed states of g.p. in the last two periods. Yet noisy, may be succeeded by silent, excitement, and, in its turn, the latter may again make way for the overt form of maniacal disturbance.

To resume. Short attacks of furor may occur, and recurring maniacal seizures, or one of longer, or even of protracted, duration. The maniacal symptoms often appear to grow out from the expansive ones, which, indeed, at one time or other are usually accompanied by, at least, some symptoms of a mildly maniacal type. Thus, the expansive and maniacal might be treated of together, but there are conveniences in dealing with them separately. The maniacal are the G.Ps. who give trouble by their constant tendency to strike their attendants or fellow-patients, their readiness to engage blindly in struggles, and so to act as to require separation for their own and others' safety.

In the maniacal state are observed cerebral disorders of speech. The rapid and tumultuous flight of ideas overwhelms the power of language to give expression to the thoughts; the powers of selection of objects of ratiocination, and of concentration of attention, are lessened or even abrogated. Scarcely has the expression of one idea been commenced ere another urges itself into consciousness, and forces expression, and another in its turn;—thus the speech is fragmentary, disconnected, changing from subject to subject, and this for hours together; or at the *acmé* of excitement the language becomes a gibberish, a logorrhœa; or even an absolutely incoherent, chaotic, jumble of words; or consists of these mingled with articulate, but unintelligible, or meaningless, utterances; or even with inarticulate cries.

Nevertheless, the disorders of articulation, usual in g.p., are often in abeyance during this condition of excitement. If they have been noticed previously they are now apt to be much slighter, or to disappear; if they have not been noticed, or the patient comes under observation only when already maniacal, their absence obscures the diagnosis, associated, as it more or less is, with the absence, also, of analogous symptoms. Occasionally, sexual excitement, erections, emissions, or masturbation are observed. In restless agitation, with advanced dementia and helplessness, the latest vocal utterances of a formerly uxorious G.P. were the fragmentary stereotyped phrase "pretty girl"—the relic of former endearments.

Marked maniacal excitement may exist for weeks together. I have seen it in an excessive form last for half a year. Under such circumstances, the disease seems to run a steady and malignant course of gradual but rapid exhaustion and destruction of the powers, both psychical and somatic. This brings us to the extreme pole of this phase, namely, to the maniacal, galloping form of g.p., in which is excessive and continual restlessness and violence;—raving, resisting, violent, sleepless, with dry tongue, sordes, and heated skin, the patient soon passes into exhaustion, temporary partial collapse, and early death. These, however, are comparatively rare and extreme cases.

As a rule, the maniacal form of g.p. is that one which offers, perhaps, the most frequent, protracted, and marked remissions. Sometimes these patients appear to have recovered; rarely, however, is the recovery permanent. Thus, in one case admitted acutely maniacal, apparent recovery ensued; discharged, he was reported as keeping well for 8 months. Re-admitted with maniacal g.p., recovery again ensued; again discharged, he kept moderately well, mentally, for 10 months.

Nocturnal insomnia and restlessness may exist in any of the forms and at any stage of g.p., even in advanced dementia; but extreme insomnia is, rather, an appurtenance of the maniacal condition in g.p. than of any other.

8. *Melancholia in g.p., and the melancholic form of g.p.* The melancholic symptoms are relatively more frequent at the onset than at any other time. Depression in the prodromic stage may pass into distinct melancholic symptoms at the commencement of established g.p., or the dejection may exist throughout the whole course of the malady. Writing in 1859, Calmeil declared that for about a decennium previously melancholic symptoms had occurred at the commencement of g.p. almost as frequently as the "monomania of pride;" and in 1873 Dr. Lunier stated that the symptomatology of g.p. had become a little modified since 40 years previously; and it seemed to him that the depressed chronic form, advancing with some slowness, had become more frequent nowadays.

G.Ps. may have the more ordinary melancholic delusions, depression, sadness, or even suicidal impulses;[*] some dread or terror of impending evil weighs upon them, and

[*] Castiglioni found about 15 p.c. suicidal (quoted by Dr. Harrington Tuke, "Journ. Ment. Sci.," Oct., 1860, p. 95).

the ideas run on persecution, poisoning, or spiritual perdition. These melancholic delusions have already been briefly referred to in writing of the stages. Dr. A. J. Linas,† indeed, denying Baillarger's earlier views as to the special nature of either its ambitious or hypochondriacal delirium, showed that in general paralysis depression might take not only the hypochondriacal form but also assume every shade of depressive delirium, and I may add that cases illustrative of this may be found even in the pages of the earliest writers on the subject. Yet has it not in general paralysis the comparative cohesion and fixity of ordinary melancholia. The delusions betray too great an incoherence and mutability. If so be that these patients attempt suicide or homicide there is usually no fixity in the reasons they assign; no precaution, concealment, or persistency, in the endeavour. This is true of most cases. But at the onset the suicidal attempt, though not the outcome of a reasoned and deliberate choice, may be sufficiently desperate and determined, and successful.

With the suicidal, there may be also homicidal inclinations or attempts, and that either simultaneously or alternately.

With the melancholic ideas hypochondriacal may also flourish.

Hallucinations of hearing and of sight are common in this condition. Hallucinations and illusions of smell may occur; so may those of all the five more special senses, and even in the same individual. Some patients assert that their beds are poisoned; others hear themselves accused by imaginary voices of being the most wicked in the world. The food of some tastes and smells horribly to them, stenches arise from their companions, reptiles within them speak, frightful objects menace them.

Sometimes, owing to refusal of food, forcible feeding becomes necessary. But attention to the digestive and excretory organs usually clears up that symptom.

In one G.P., under my care, together with marked mental enfeeblement, there were protracted delusions of annoyance and of injury; depressed and weeping, he declared that his comrades "were against him," and at times he had the delusion that they desired to murder him.

Another suicidal G.P. spent his days in fear alarm and suspicion; and was querulous, restless, anxious, worried, distressed, and unable to engage in any occupation. He

† " Recherches Cliniques sur les Questions les plus Controversées de la Paralysie Générale," Thèse, Paris, 1857, No. 193, p. 35.

evinced visual hallucinations, and also expressed melancholic delusions, as that "every one was against him," and that poison was constantly administered to him in his food. Feeling his life to be in danger, he besought protection against his enemies. He was full of delusions about his treatment before admission, and querulously made groundless charges against those who had been concerned in his detention.

Another had delusions as to being hanged or shot; and suicidal and homicidal tendencies. Later on, when under my care, there were depression of spirits, insomnia, incoherence, confusion of ideas, impairment of memory, hallucinations of sight and of hearing, delusions that he must be hanged or shot, and that he was "warned" daily to that effect. Subsequently, were epileptiform seizures, many attacks of paralysis, occasional wild and noisy excitement, or self-burial in bedclothes, as if in terror.

Another was depressed throughout, and often whining, moaning, and readily weeping, timid, emotionally disturbed and distressed by trifles, even by being moved, fed, cleaned, or merely touched.

Another stood sad and dejected, refused food, had delusions that he was about to die, that he had been illtreated, entirely deprived of food for several days, sent into hospital in order to be cut up by doctors; had been poisoned and could not live; that horrible food was given him in a military hospital, and his 300 comrades there stank. He said he had attempted suicide because of the terrible life others had led him. Subsequently, said he was dead, or a shadow, and had been in hell.

Melancholic delusions of the religious order, or relating to imaginary poverty, or even a melancholia agitans, may be observed.

According to one observer weeping in g.p. is usually sympathetic with intestinal ailment; another thought g.p. from alcoholic excess to be characterized by visual hallucinations, delusions of persecution and of poisoning, with homicidal and suicidal tendencies.

Several observers, and among them Dr. Wm. Wood,[*] have noticed the greater relative frequency of the melancholic form in *female* G.Ps. One observer found melancholia more frequently among those G.Ps. who were phthisical. In

[*] "Brit. and For. Med. Chir. Rev.," July, 1860, p. 198. Also Baillarger, Jules Falret, & *al.*

some cases the associations of *silent excitement* are, rather, with depression (hyp. or mel.) than mania; and to the several varieties of the melancholic condition in g.p., admitted by Voisin—who, more carefully than any preceding writer, described melancholia in g.p.—one might add an occasional melancholic condition of which a silent excitement is a prominent feature. The various forms of melancholia may be psychical symptoms of g.p., and Voisin enumerated as varieties of "délire lypemaniaque" therein: (a) the form like agitated melancholia; (b) that with stupor; (c) melancholia with religious ideas; (d) that with ideas of persecution; (e) the hypochondriacal.

Dr. Casimir Pinel, * observing the alternation of expansive and depressive deliria in some G.Ps.—a transformation which may even occur on the same day,—noticed it oftenest in the morning, and attributed it most frequently to the influence of dreams overnight, less frequently to that of nocturnal hallucinations. This applies to the hypochondriac, as well as to the melancholic, delirium. He asserted the almost constant coexistence of the hypochondriac and lypemaniac deliria. Their coexistence is frequent, it is far from invariable.

In a few cases, in the first stage of the established disease, the morbid ideas and feelings remind one of persecutory monomania modified by psychical weakness. The patients affirm that they are persecuted and injured in every possible way, that every one about them is inimical. Occasionally, their mental attitude is one of hostility and revenge, and they threaten the most cruel and vindictive punishments to their enemies, to those who thwart their morbid impracticable wishes, or detain them against their will. These ideas and feelings are fickle, changeable, and not fixed, nor the delusions systematized. I have mentioned this because some have asserted that cruelty and vindictiveness are never to be seen in the G.P., a view which clinical observation has shown me to be quite untenable as an invariable rule.

8. *Hypochondria in g.p., and the form in which hypochondria predominates.* With this form the name of Baillarger † is particularly associated. He it was who first declared hypochondriac delirium to be a characteristic symptom of

---

* "Annales Médico-Psychol.," 1859.
† "Gaz. des Hôpitaux," Feb. 3, 1857, p. 55; *ibid.*, May 9, 1857, p. 218; *ibid.*, Oct. 13, 1857, p. 477; "L'Union Médicale," 1857, p. 385; "Ann. Méd.-Psych.," T. vi, 1860, p. 509.

g.p., saying that as ambitious delirium was that special to excitation in g.p., so the hypochondriacal delirium, even if less frequent and overt, was that special to depression in the same disease. Moreover, that this hypochondriacal delirium was of a quite special stamp; and, presenting the characters he indicated, was found only in g.p. No doubt this is too exclusive a view. The hypochondriac delirium, he also stated, was of evil omen in g.p.; when it was present the tendency to gangrene was enhanced, and occurred at an earlier period than usual; if it was prolonged the patient fell into marasmus. He also deemed the hypochondriac delirium a valuable precursory sign; for when it was observed in cases classed as melancholia, the patients, in some instances, were afterwards recognised as being the subjects of g.p.

A hypochondriacal state of ideation or feeling of depression very often occurs in the first period, that which is sometimes termed prodromic, but the pronounced hypochondria of G.Ps. is apt to occupy part of the second stage of the *confirmed* disease (third period), and less often of the first stage (second period). When marked depression occurs at the commencement, and thence is continued into the first stage of the established disease, the delusions, as I long ago showed, often assume by preference the melancholic form, and in the soldiers under my care were usually of a terrifying nature.

On the other hand, as I* have stated elsewhere:—" If occurring for the first time during the middle stages, the hypochondria coexists with a shattered understanding, and the delusions are usually of the following nature:—The patients say of themselves that they are 'dead,' 'have no head,' 'have no throat and cannot swallow,' 'their intestines are gone,' 'their eyes have been extirpated,' 'their testicles are wasted,' or that they 'have been castrated,' or that 'part of their body is dead.' With this there is often a refusal of food and obstinate resistance to feeding, a wild howling when they are interfered with, an inability to walk or stand, a huddling up in bed, and general spasmodic tremor when locomotion is attempted or when they shrink from contact." This last sentence embodies the first description, I think, of the condition referred to in it.

Having thus, from my own practice, as just quoted, illustrated certain clinical phases of hypochondriacal delirium

* "British and Foreign Medico-Chirurgical Review," April 1877, p. 457.

in g.p., I may now add other cases, from the same, taking the following forms.

In one; after a period of exalted delusions with considerable maniacal agitation, hallucinations, insomnia, and restlessness, the patient for a time is usually silent and of apathetic appearance, but now and then suddenly and paroxysmally breaks forth into the reiterated shrill shout, "come here my cavalry;" or, as if in a paroxysm of terror, "oh God have mercy on my soul."—Later, the physical signs continue well-marked, and the face muddy, sallow, and relaxed, with apparently enlarged and flattened features, which are almost expressionless—except of apathy. Several months later he says, in one breath, that "the world has nearly starved him," and he "is nearly dead;" and, in the next, that he "is first class," "is the Crystal Palace," "is born of the virgin Mary," "is Adam and Eve,"—all this with a dull, unhappy look, and weeping soon coming on. "Is aged 15, entered army 10 years ago at age of 5 or 10 or 12." Still later, he sometimes denies any exalted ideas, and at other times, when more than usually quiet, dull, heavy, stupid and feeble, he says, if closely questioned, that he is "nearly dead," and to almost every question replies "I don't know."—Garnishing his words with coarse additions, he says at a still later period that "he is nearly dead;" that "his legs, his wig (he wore none), his chest and belly are gone to hell;" that "his little eyes are blind," "his little nose" contains no mucus, his "little penis is of no use to him;" and that he "can't take food and will be dead by night-time." Immediately afterwards he says he has no money, but adds that he "owns all the world and the regiments belong to him."—Subsequently, epileptiform seizures took place, and were either universal, or chiefly on either side. A few apoplectiform seizures. Depression, fear, taciturnity, mutism. Later on, occasional loud inarticulate cries, produced as if with convulsive effort. Knee-jerk absent. Gait somewhat ataxic. Distressed by slightest pinch of legs. Usually mute, obstinate, of lugubrious expression. Died in terrific convulsions. Lesions of g.p. more marked in left brain; and unusually at base of brain; chronic spinal meningitis: softening of lower cord; some atrophy left side of spinal grey.

The tendency to believe the bodily organs or limbs diminished in size, as noticed in this case, is not so rare, in my experience, as has usually been supposed.

In another case, the patient, after a period during which exalted delusions with some excitement and violence were evinced, became restless and sleepless at night, and more dull, heavy and confused in mind; the face and head became flushed and heated; the gait and various movements of the limbs, as well as of speech, displayed a great increase of ataxic irregularity and trembling. In anxious perturbation the patient strove to tear off his clothing and get under the water-tap under the delusion of imaginary fæcal self-defilement and false sensation of fæcal lumps about the nates. Saying he was not well—indeed was "very bad and very dirty" (not so)—and that there was nothing to eat, he refused food. This continued; nervous and frightened in appearance, he resisted most strenuously his being fed and all the necessary tendings of his nurses; struggled to get here or there without apparent aim or object; grasped tenaciously, and pulled or pushed at, any surrounding object as the door-handle or the clothing of a bystander. —Later on, trembling as he stood, or as he advanced with tottering, jerky, slow, and apparently feeble steps, he still would raise a shaky hand to push against one, or to seize a hand, grip it hard, and thrust away or pull towards himself its owner; in everything taking the contrary course to that desired of him. Held by one attendant, he was fed by another, the meanwhile bellowing between each mouthful. I might add many other cases to these.

Ofttimes a striking event with regard to the hypochondriacal delirium of g.p. is its appearance preceding or in sequence to the more common expansive delirium, and then, indeed, it often is intermingled, or alternates, with the latter. Some express expansive and hypochondriacal ideas almost in the same sentence, or pass with quick fickleness from one to the other; as where the above patient deeming himself at death's door, with obstructed passages and annihilated organs, and without any pecuniary resources, yet, in the same breath, declares himself to be the Crystal Palace, and claims proprietorship of the world. From the coexistence and blending of the expansive and hypochondriacal ideas, also, arise some of the strangest of the notions of G.Ps.; such extraordinary delusions, for example, as some of those just mentioned, or as where patients assert that they have grown enormously, and their stature now reaches 12 or 50 feet—this, however, is chiefly expansive— or that their eyes have been taken out and blinded in order to get at the diamonds in their heads.

Limits of space preclude the insertion here of a case with marked exalted, hypochondriacal, micromaniacal, and melancholic symptoms.

The frequent delusions that the mouth, throat, or bowels are closed may induce the patient to refuse food, and enforced alimentation may become necessary.

One of the earliest hypochondriacal symptoms in a magnificently-built Grenadier Guardsman, under my care, was that his generative organs had wasted away. Later, when in one of his expansive moods—which,. indeed, alternating with depression, were the more frequent as a rule—he prayed for testes fifteen inches long.

Very similar to several referred to in my own practice was Baillarger's illustrative case *:—At first incoherence and maniacal excitement, followed by physical signs of g.p., and, later on, some notions of grandeur together with dementia, and eventually hypochondria. This case died, six weeks after, with deep and rapid gangrenous eschars over the sacrum, feet, toes, and left shoulder-blade. There were adhesions and pallor of cerebral cortex. The dark and fluid blood in the heart, &c., was intermingled with bright spangles, like fat.

But not all the cases of this form have the same symptoms; some, corresponding in part to such as Dr. Falret described under the "depressive" variety, have a melancholic appearance, are apathetic, immobile, speak but little of their own accord, wear an expression of indifference and lack of mobility; the features, not concentrated, as in melancholiacs, towards the median line, are as if pendent and without any tension, and the face appears to be enlarged and flattened; a special facies, somewhat like that of double facial palsy, and expressing silliness and absence of ideas rather than preoccupation; while, occasionally, spasmodic twitches of the face are observed, especially during speech. Speak to them and they smile, their countenance expands, and expresses general satisfaction and great feebleness of intellect. Rather than sadness, they exhibit apathy, or even have a vague and general contentment; perhaps say they want for nothing, and often have some ideas which reflect satisfaction, or even certain ideas of grandeur, which sometimes suddenly light up their countenances. This state, as a rule, occupies only one period in the course of general paralysis, and is usually preceded or followed by excitement. Such, in abstract, was Falret's description of a condition, which he agreed with Baillarger in assigning more frequently to the female sex.

Patients of the above groups do not often volunteer their statements. Left to themselves, they are generally silent; interfered with, roused, or questioned, they may reveal the absurd hypochondriacal notions; in fact, as noticed by Dr. Legrand du Saulle,† and originally implied by another, the

* "L'Union Médicale," 1857, p. 385. "Gaz. des Hôpitaux," Oct. 13, 1857, p. 477.
† "Annales Médico-Psychologiques," 1861.

hypochondriacal delirium of G.Ps. often must be sought for to be verified. It is, however, extremely doubtful whether the proportion of cases presenting the special hypochondriacal symptoms is as great as he says (sixteen to one) in paralytic, as compared with non-paralytic, insanity, with depression;—nor, perhaps, as large as that given by Dr. Moreau, of whose fifteen examples of the special hypochondriacal delirium thirteen were G.Ps., or doubtfully G.Ps. Yet, before deciding negatively and definitively as to g.p. in any such given case, one must be able to follow it for several years. Sometimes a "recovered" hypochondriac is to be met with some months or years afterwards as a G.P. in an asylum.

In cases of the hypochondriacal form Baillarger said the prognosis was of the worst, and Falret held that the cases with the "melancholic" form of onset were often short. But, in my own experience, while the general results are the same, and chiefly in early typical cases, yet *some* of the cases presenting these marked hypochondriacal symptoms at some part of their course are amongst those of comparatively long duration; and on turning to Austin's work I find that his experience was similar, except in those cases where the "melancholic" delusions (under which he evidently classes the hypochondriacal) were of a very dreadful character.

Already has reference been made to the assertion of Baillarger that there is a gangrenous diathesis, a tendency to easy mortification of the tissues, in these cases. I have seen this in some typical examples, but not to so unusual an extent, or not, in some others where the hypochondria was temporary and intercurrent; nor have I found, as he did, the inequality of the pupils markedly greater here than in other cases of g.p.

With regard to Dr. Baillarger's view as to the *special* character of the hypochondriacal delirium in g.p., it is sufficient to say that it has given rise to discussion, some favourable, some hostile. To my mind, the matter stands very much as it does with the exalted delusions and exaltation of g.p. In the one case as in the other, these symptoms, whether hypochondriacal or expansive, have been viewed as special to, and characteristic of, g.p. This view, which had so long a reign as regards the expansive symptoms, but never an undisputed authority as to the hypochondriacal, must equally be abandoned for both. In the one case as in the other, the special kind of morbid ideas is found not only in g.p., but in other mental affections, also. As with the one so with the other, it may be entirely absent throughout the course of g.p. Like as the one so has the other almost always a special recognizable clinical expression and surrounding. Each owes the moulding of its form, and its impress, to the co-existing psychical lapse and degeneration, and each has its special clinical aspects, and its importance in diagnosis. In each, the morbid ideas are numerous, changeable, absurd, and often self-contradictory; indeed, the transformation they undergo may be rapid. In neither are they reasoned; in neither co-ordinated with the past and present into a morbid ideational system, logically reflected, in a persistent way, in the life and actions of the patient. In neither are they firm, or defended with tenacity.

It is not contended that the two groups of symptoms run on entirely parallel lines. In the nature of things, there are some clinical differences over and above the subjects of the respective deliria. For example, not only is it true that hypochondriacal paretics do not press their sufferings upon attention, like the sufferers from simple hypochondria, who are earnest in making known their imaginary disease and in seeking relief; but also the former do not so readily express their delusions as do the expansive G.Ps.; nay, even may be silent and taciturn. Voisin * mentions an interesting case in which the slight somatic signs had disappeared; hysteriform seizures had occurred; and in which he made the diagnosis solely by eliciting, with difficulty, from the patient the baseless idea that she had an aneurysm.

Like as I refrained from splitting up expansive delirium, for description, into several varieties according to the particular subjects of the morbid ideas; so here will few sub-divisions be proposed.

The most common variety is one of which mention has already been made and examples have been given; and in which the patients declare some of their tubular viscera to be closed, or some of their organs to be absent or entirely destroyed. Others say this or that part of them is rotting; others declare themselves to be excessively feeble and helpless, when such is not the case; some assert that they are suffering from this or that ordinary disease. Others deny their personal identity, or deny their own existence; they have ceased to live, they say.

* Traité, &c., *op. cit.*, p. 100.

Delusions of belittlement. Others have the delusion that they have become dwarfed, or that some part or organ of their body has become far smaller than normal. This condition, which I term delusions of belittlement; which is often called micromania, but for which a better name, than the latter, was the one originally proposed by Materne—"délire des petitesses,"—rarely exists alone; is usually of not very long duration, and easily escapes observation unless patients are closely examined. It is not accurate, as a recent writer has done, to widen the application of the term, and place it in coterminous contradistinction to the ordinary delusive grandeur. I have given one example, and the same symptom recurs in a case yet to be mentioned. Voisin states that he has met with it in g.p. only. It is not, however, pathognomonic of g.p. I have known it occur temporarily, in a nascent form, in a sane child, who had suffered from malarial disease (ague), and gastro-intestinal disorder.

The foregoing include, therefore, morbid ideas and sensations as to:—

Obstruction of hollow viscera.

Absence, or destruction of some organs or parts of the body.

Rottenness of this or of that organ.

Imaginary debility and helplessness, or ordinary disease.

Denial by patient of his identity, or of his existence.

Delusions of belittlement.

Necromīmēsĭs. To these, I would add a condition hitherto undescribed, or, at least, of which I have never met with any description; and which I propose to term *necromimesis*. Its associations, however, are equally (possibly rather) with melancholia as (or than) with hypochondria, in g.p.; and it might at least equally well be described under the head of the melancholic condition.

In this the patient believes himself to be dead; *himself*, I put it, for my experience does not enable me to say whether the condition occurs in female G.P.s. or not; it is likely it does. Believing himself to be dead, he acts the part of a corpse, and so far as psychical weakness permits, the external life and bearing are harmonized with that perverted conception; and the simulation of death is more or less complete. *Less* complete or thorough it is; and usually can be interrupted with no great difficulty; for, though silent, and lying motionless, with flaccid limbs, scarcely perceptible breathing; with eyes closed, or closed at times, and on some occasions with lower jaw rather stiffly depressed;

and giving no response to questions; the patient can usually be roused from the condition by energetic manipulation and shaking, aided by the spur of sarcastic remark. Placed upon his feet, and made to traverse the room by the united efforts of attendants, he may keep the limbs flaccid, or may stiffen them; but, in either case, must usually, at the first moment, be moved passively like a sack.

When, by manipulation, and twittingly sarcastic observation, the morbid circle of fascination is broken through, consciousness is roused to new objects, and the semi-unconscious acting is interrupted; the patient finds his tongue, declares he is dead, is indignant that the contrary opinion is expressed by those about him, and perhaps rises to his feet as the delusion recedes or vanishes. Beforehand, the patient may state that he is about to die, after which the clinical phenomenon attains development; or it may be introduced by a mild seizure of vertiginous epileptoid form (*petit mal*). In the two most marked cases I have seen it was associated with, or preceded by, left hemiparesis, or by local sinistral paresis.

One of these cases of g.p. had begun with slovenliness, neglect of duty and of orders, low spirits, discontent, the making of absurd unfounded charges against all those about him; passing into a restless, irritable, confused, sub-maniacal state; and, later on, were exalted delusions; after this mingled expansive and melancholic deliria; then necromimesis; subsequently mingled expansive and melancholic symptoms, and illusions of the muscular sense; also increased knee-jerk. Necromimesis supervened about nine months after mental disease was diagnosed, but about two years had elapsed since the beginning of the first insidious moral and emotional changes. It was as follows:—

After delusions about being starved and having bad food, at which time, also, was slight paresis of the left facial nerve, the patient, one morning in June, after saying he was about to die, refused to get up at the usual time, and kept the lower jaw held down; the limbs, however, being flaccid. Then, when he was lifted and was got out of bed, he stiffened his limbs, and alternately closed the jaws and eyes, and held the lower jaw stiffly drawn down and separated from the upper. Replaced in bed, he had a pulse, 60, regular, full; and the limbs were again flaccid. Again raised, he alternately stiffened and relaxed the limbs, but would not make the slightest effort to walk, and had to be moved like a filled sack, the eyes being kept closed. Soon a lugubrious expression on the face increased, and he began to weep silently. The pupils were now wide, sluggish, the right slightly the larger. The temp. 98·6°. Then he began to open the eyes and watch.

In the other case, before the necromimesis there had been,

successively, delusions and emotions of the melancholic type, including those concerning religious matters; and, later on, with mental confusion and defect, expansive symptoms mingled with occasional depression and causeless weeping, but expansive delirium predominating. Afterwards, were some exaggerated statements as to his muscular weakness, and, on the other hand, expansive delirium. There was also dementia. Necromimesis occurred five months after the patient had first been thought distinctly insane, but he had been "eccentric" for some time previously.

March 11th. Turned pale, then burst into a profuse perspiration, and afterwards appeared "to fall asleep." Shortly afterwards, he sat with head down, and the legs thrown out, the arms hanging limp by his sides. Skin not cold; pulse slow, regular, soft. When, by the aid of attendants, the patient was placed upon his feet and made to walk, he moved as if hemiplegic on the left side, the upper limb being quite limp, but the face not much affected. Finally, on finding his tongue, he said he had had a fit and was dead (no convulsion had occurred). Became indignant at the disbelief in his fit and death; and for a time again simulated inability to move, or death.—Subsequently, hemiparesis first on one side, then on the other; springy gait.—Later hemiplegic and contracted left limbs; increased knee-jerks; slight right ankle-clonus; apoplectiform and spasmodic seizures; modified Cheyne-Stokes's respiration.

10. "*Stupor*" *in g.p. and the form with Stupor.*—In a few cases general paralysis sets in with symptoms of the formerly so-called "acute dementia," or of *melancholia with stupor.* Here the ordinary motor and sensory signs of g.p. are either absent at first, or are masked, as it were;—when the extreme mental symptoms pass off it becomes possible to verify the existence of the physical. But not always; for a marked remission, or apparent recovery, may immediately succeed the acute symptoms, some weakness of the intellectual powers remaining. The ataxic troubles, whether of speech or of other orders, may now slowly become obvious, or, if already present, may increase, and the patient's mental alienation again become patent. The symptoms differ from those of the cataleptoid state of mental abrogation, and from *melancholia attonita,* but more nearly approach the latter.

Case 1. In one such case, afterwards under my care, a soldier, not returning home one day, was found at 4 a.m. in the street, unable to give any account of himself, but not under the influence of alcohol, or of any drug. Subsequently he ceased to speak or to take food, and, absolutely silent, sat with bowed head, taking no care of himself, and passing evacuations where he sat. When admitted, no reply could be elicited from him, his habits were still objectionable; to the necessary artificial feeding he opposed a strenuous resistance; when left alone, he sat, as if helplessly, with head for the most part hanging down, and eyes gazing in one direction; but no marked terror or emotion of any kind limned itself in his features. The pupils were sluggish, the right the larger, and somewhat dilated, the left small and slightly irregular in shape. The pulse was 114, rather small, and compressible. The viscera were healthy. Later on, diarrhœa was troublesome, but was perhaps connected with the artificial feeding. In about three weeks he began to reply in monosyllables, and after this the mental dulness gradually cleared away, the speech and writing were now noticed as being like those of g.p. at its commencement, the frame was bent, and the gait slouching and unsoldierly. Finally, he being now only somewhat childish and weakly emotional, the urgent desire of his friends for his discharge was acceded to after five months of treatment. As foretold, before very long he relapsed, but into what mental phase I am not aware.

These cases more usually take a quasi-melancholic form, and on the psychical side constitute, clinically, a variety of melancholia attonita.

Case 2. A soldier, aged 37, had suffered in India from primary syphilis, acute and chronic rheumatism, laryngitis, bronchitis, dysentery, dyspepsia, debility, hepatitis. Comparatively early, were mingled hypochondriac, melancholic and exalted delusions. Subsequently, mingled melancholic and hypochondriacal delusions, with a tendency to melancholia attonita; dilated pupils, feeble soft rapid pulse; head sometimes heated; dusky, dingy-reddish hue of face, furred tongue, costive bowels, hands blue and chilblains troublesome. Then the condition deepened. The hands were swollen, blue and chilblained; mucus ran from the nostrils, and if untouched would hang in strings upon the beard; the urine constantly dribbled away; his pupils were dilated, his gaze was fixed; at night he was sleepless. Slight indications of phthisis. To follow the case after the above symptoms ceased:—At a later period he was very incoherent, and still had delusions such as that snakes were alive inside him, that he was being killed, was being torn to pieces, and he would shriek aloud as if in agony. Later on, usually melancholic, with delusions as to being dead and as to his personal identity, he at times was laughing and chuckling or uttering pig-like grunts: got at times into dull impercipient conditions, and was for a long time affected with frequently-recurring severe epileptiform seizures, leaving behind them temporary local paralyses on the right side, often associated with impairment of speech, at times even to aphasia. Besides extraordinary perversions of general and of visceral sensibility in this case, there were auditory and tactile hallucinations, ataxy of lower limbs, absence of knee-jerk in right leg and almost so in the left. Latterly, temporary conjunctival congestion, with lessening of the usual mydriasis, and, eventually, left ulcerative corneitis.

Case 3. Another case, usually melancholic, and at times in a state rudely resembling melancholia attonita. At first, there had been mental failure; subsequently, he was silent, "wet and dirty," self-smearing with fæces, feeble, yet at times mischievous, and had bedsores. *After admission* was melancholic, rarely spoke, was "wet," helpless, vomited: tongue tremulous, speech shaky, pupils somewhat irregular unequal sluggish. Successively, epileptiform seizures, especially of right side, and right hemiparesis, right unilateral facial sweating, left pupil larger, blebs on right hand.

Trance or catalepsy in g.p. Inspector-General Dr. Wm. Macleod* described a case as g.p. with "trance or catalepsy." The patient had never been the same after being stunned, and injured in cranial nerves, five years previously, by the unexpected firing of a heavy, man of war's, gun, close to him. During a month, each day, from 6 a.m. to about 3 p.m., the patient fell into a condition of perfect stillness, apparently unconscious, with pupils sluggish, sensation and motion suspended, heightened temperature, once or twice slight twitches:—"when the stage of stillness was coming on he would break out into a most profuse perspiration, which gradually diminished as the day wore on."

It is a question whether this was not a paroxysmal variety of the condition last described, namely stupor.

11. *The form with symptoms of Circular Insanity.* Dr. Fabre† described a form of g.p. characterized by alternate excitement and depression (circular insanity). When there are only two phases, *à double phase*, they succeed each other suddenly, while in the form which is of triple phase there is an intervening period of calm, or, (1) excitement, (2) calm, (3) depression. There are, however, cases in which the phases are arranged in a manner different from that mentioned by that writer; thus, for example, the depression may precede.

I have observed the above-mentioned clinical form of g.p. of triple phase.

Case 4. After admission:—At first, were prolonged symptoms of active maniacal excitement. Then, apparent mental recovery, except for some easily-induced mental confusion. The next portion of the cycle consisted in mental dulness and confusion, sleeplessness, refusal of food under the delusion that it was poisoned, fear, delusion as to injury to him intended on the part of those about him. Acute nephritis: followed by return of active excitement and restlessness. Later on, taciturn, obstinate,

* " Journal Mental Science," July, 1879, p. 197.
† "Annales Méd.-Psych.," March, 1874, p. 197.

indifferent, apathetic, torpid, of feeble circulation. After this the alternations of expansive excitement and depression became less and less distinct and regular, and more and more blurred by an increasing state of dementia, and the symptoms more usual in advanced g.p. Cardiac disease. Apoplectiform seizures, one followed by modified Cheyne-Stokes's respiration. Albuminuria.| Choreiform movements. Finally, epileptiform seizures, acute bedsore on left natis, with some paralysis, anæsthesia, rigid contraction, and lessened reflex activity on that side. Died, aged 39; total duration of g.p. nearly 7 years. The right cerebral hemisphere was the one more diseased; the spinal cord was diseased (diffused myelitis), especially the grey matter on the left side. Slight renal disease. Atheromatous aorta, and circle of Willis: thickened left cardiac valves; gangrene of right lung, with old pleuritic adhesions.

I have, of course, frequently observed rapid and frequent alternations between excitement and depression in g.p.; sometimes the change followed one of the "seizures."

During the phase of excitement, besides maniacal symptoms, are ideas of contentment and satisfaction, expansive delirium, and sometimes delusions of persecution. During that of depression, there may be melancholia even to profound stupor, melancholic or hypochondriacal delusions, and even a suicidal tendency. Several of Fabre's cases seem to be merely marked instances of the transformation often observed in the psychical sphere in g.p., a recurring fluctuation, without the distinctness of simple circular insanity. In one case, too, the phase of depression seems to have preceded that of excitement.

MM. Renaudin and Lunier, however, had long ago reported instances of g.p. with symptoms of circular insanity, paretic signs coming on in one case in the period of depression, in another during chronic mental disorder.* Another observer† described a case in which excitement was followed by depression, and this by convalescence and discharge; then, subsequently, there appeared, in succession, high exaltation—physical signs of g.p.—depression—pronounced paralyses—and death. Twelve others mentioned by the same do not appear to be examples of a "circular" clinical variety of g.p., but only of its emotional fickleness and variability.

12. *Hallucinations and Illusions in g.p.* For the most part, the following is condensed from a series of articles by the present writer‡ on hallucinations and localization in g.p.

* "Annales Méd.-Psych.," 1858, p. 403.
† Brierre de Boismont, "Ann. Méd.-Psych.," 1859, p. 329.
‡ "Journ. Mental Science," Oct., 1881; Jan. and April, 1882.

For conciseness, the word hallucination will be used here as inclusive of both hallucination proper and illusion of any of the five fundamental special senses. Disorder of general sensibility, of muscular and other senses, will be adverted to afterwards.

Sometimes, indeed, these hallucinations are vivid, and are clearly and forcibly expressed by the sufferer, but it not seldom happens that they are by no means obvious, or even that they are only revealed by a careful research into the history and by oft-repeated observations and questioning of the patient. The sensory hallucinations and illusions must, therefore, be sought for; and their absence must be satisfactorily proved at different parts of the course of a given case before one can conclude that it has been free from hallucination throughout.

In the descriptions usually given of general paralysis, hallucination of the special senses plays but a minor part, and in some is entirely excluded. To the delusions, the exaltation, the hypochondria, and the dementia of g.p. the chief *rôles* are assigned, and on them is attention riveted. The hallucinations of special sense are comparatively neglected. Nevertheless, the sensory hallucinations are more frequent and important than is usually conceded.

To bring out clearly several points in connection with this subject I took 100 of the cases of g.p. then most recently under my care, and in which the histories and symptoms were more fully and minutely related than were those of some of the cases of g.p. previously observed by me. There is the further convenience in examining this number that the numbers of individuals under the several heads also express the percentages.

In the following numbers are included both the cases in which the existence of hallucination was indubitable, and some in which there was distinct evidence, though scarcely absolute proof, thereof. On the other hand, these hallucinations may easily exist, and yet escape being recognised or placed on record; in many cases, also, the intellectual confusion and failure preclude the patient from affording the necessary information to the observer. For these reasons, the following are *minimum* percentages.

The cases were all males. Of these 100 male cases of general paralysis, 88 were soldiers, 8 were gentlemen, and 4 were paupers.

Of the one hundred, in 55 were hallucinations or illusions

of the special senses; in 45 none were obvious, at least while the patients were under my care. But of the last admitted 50 of these 100 cases, 29 (or 58 per cent.) had hallucinations; and 21 (or 42 per cent.) had not, so far as was ascertained.

Of the total one hundred cases, there were visual hallucinations in 41; auditory hallucinations in 40; tactile hallucinations in 12; gustatory hallucinations in 12; and olfactory hallucinations in 11. More correctly speaking, many of these were illusions.

In several examples, included above as of gustatory illusion, there was scarcely proof of more than disorder of the olfactory sense, by which the flavours of food are perceived. The examples in question are also included under the head of olfactory hallucination and illusion.

Both auditory and visual (with or without other) hallucinations were present in 32 cases; the auditory predominated very decidedly in 7 of these, the visual in 5.

Of the subjects of hallucination, in about one-third there were hallucinations of one special sense only; in about two-thirds there were hallucinations of several of the special senses.

(*a*). The number with hallucination of *one* sense only was 18. These were distributed as follows:—

In 9 were hallucinations of sight only.
„ 6 „ „ „ hearing only.
„ 1 „ „ „ touch only.
„ 1 „ „ „ taste only.
„ 1 „ „ „ smell only.

(*b*). Hallucinations of *two* special senses only, were observed in 24 cases. In 21 of these the two senses affected were sight and hearing. Of the three remaining cases, hearing and touch were affected in one, and taste and smell in the other two.

(*c*). In 5 cases hallucinations of *three* special senses were noted. In 4 of these hearing, sight, and touch were affected; and hearing, taste, and smell in the remaining one.

(*d*). Hallucinations of *four* of the special senses were noted in 3 cases. In these, the sense *not* noted as being affected was either that of touch, taste, or smell.

(*e*). Hallucinations of *all the five* senses now in question were exhibited by five of the patients.

To make the subject clear at a glance most of the preceding statistics may now be recapitulated in a tabular form.

Of the *total number* of cases :—

Hallucinations were present in	55	per cent.
Or hallucinations were present in (of the last 50 cases only)	58	,,
Hallucinations were absent in	45	,,
Or including those of common sensibility, present in	63	,,

Also, of the *total number* :—

Visual hallucinations were present in	41	,,
Auditory ,, ,, ,, in	40	,,
Tactile ,, ,, ,, in	12	,,
Gustatory ,, ,, ,, in	12	,,
Olfactory ,, ,, ,, in	11	,,
Both auditory and visual (with or without other) hallucinations occurred in	32	,,
Of these, the auditory predominated very decidedly in 7, or (of the total number) in	7	,,
And the visual predominated very decidedly in 5, or (of the total number) in	5	,,
(*a*). Hallucinations of only *one* special sense were observed in	18	,,
Of these there were hallucinations of sight only in 9, or (of the total number) in	9	,,
Of these there were hallucinations of hearing only in 6, or (of the total number) in	6	,,
Of these there were hallucinations of touch only in 1, or (of the total number) in	1	,,
Of these there were hallucinations of taste only in 1, or (of the total number) in	1	,,
Of these there were hallucinations of smell only in 1, or (of the total number) in	1	,,
(*b*). Hallucinations of *two* senses only, were present in	24	,,
Of these sight and hearing were the two affected in	21	,,
(*c*). Hallucinations of *three* senses only, in	5	,,
Of these, hearing was affected in all, and vision in four.		
(*d*). Hallucinations of *four* senses in	3	,,
Sight and hearing were affected in all of these.		
(*e*). Hallucinations of *five* special senses in	5	,,
(*f*). Hallucinations, illusions, perversions of internal sensation, common, or organic, or visceral sensibility, in at least	30	,,

Then, if we take only the general paralytics actually known to be the subjects of hallucination, and estimate in them the relative frequency with which the several different

senses were affected, we find that visual hallucinations were present in 74½ per cent. ;
Auditory in 73 per cent. ;
Tactile and gustatory, each in 22 per cent. ;
Olfactory in 20 per cent.

Care must be taken not to confound with true hallucinations what are merely the relation of his dreams by the patient ; or are purely delusional.

As to the relative frequency of the several kinds of hallucinations in g.p., the leading position has sometimes been assigned to the visual, which have been considered more frequent than even the auditory. The experience summarized above scarcely supports that view, if the *absolute frequency* of each kind alone receives attention, for in this group of 100 cases there is a virtual equality between the percentage of cases with visual and that with auditory hallucinations. And, indeed, of the cases in which the two kinds of hallucinations coexisted, the auditory predominated in examples rather more numerous than were those in which the visual predominated. Nevertheless, the richness of g.p. in visual hallucinations is true, but it is more especially a *relative* richness. The *relative proportion* of the visual to the auditory is much higher in g.p. than in the other forms of insanity, taken collectively, as found in our asylums. Excluding idiocy and imbecility on the one hand, and, on the other, g.p., it is probable that auditory are more frequent than visual hallucinations in a given asylum population in this country. This is partly due to the chronicity of most of the cases, for in the acute forms of insanity visual hallucinations are relatively far more frequent than in the chronic.

From an investigation stated in summary at p. 374-5 "Journ. Mental Science," Oct., 1881, I concluded that in the military insane (mostly non-acute cases) under my care, the relative frequency of auditory to visual hallucinations was as 80 p.c. to 50 p.c. ; whereas in the G.Ps. they were, auditory to visual hallucinations as 40 p.c. to 41 p.c., or practically of the same frequency, in g.p.

Nevertheless, the hallucinations in g.p. may not only be transitory and evade notice, but also in many cases may exist before the patients come under observation, and not afterwards, and may have been unrecognised and unrecorded in the previous history obtainable.

From the above statistics of cases I observed, it may fairly be concluded that hallucination plays a more important part in the clinical phenomena of general paralysis than has hitherto been generally recognised. In several treatises on insanity, and in elaborate clinical descriptions of g.p. published by some observers, no mention whatever is made of hallucination in g.p.; others speak of hallucination as being observed occasionally, only, in it; others have considered hallucinations as extremely rare in g.p.,* or even the absence of undoubted hallucinations as characteristic of that affection; and still others assert that hallucinations of the special senses in g.p. are very rare, with the exception of those of sight, to which they concede a moderate frequency. Then, again, Magnan asserted that hallucinations were not a symptom of g.p.; that hallucinations, when observed in G.Ps., were a symptom of acute alcoholism, resulting from the previous alcoholic dissipation of the patients.

Claus † found them in about one-fourth of his cases; Brierre de Boismont in one-fourth; Mendel in 36 p.c.; Jung in 49 p.c. (over 50 p.c., as stated, by improper inclusion of cases in which they were not noted).

As for their general characters:—

In general paralysis, hallucinations and illusions are often of short, or even transitory or ephemeral duration.

Sometimes they recur irregularly, and endure for various periods of time.

They are often variable, mutable, inconsistent, being usually less fixed and systematized than the hallucinations of many of the insane of other groups.

Not seldom, those of a given sense are multiplying.

Often, too, do they harmonize in absurdity with the absurd character of the delusions present.

They are often of crude, tumultuous character, and illusions are frequent.

Coexisting hallucinations of the different senses not seldom relate to totally disconnected subjects; and one hallucination sometimes is contradictory to another.

Sometimes they are not obtruded by the patients; do not discover themselves, and are only revealed upon inquiry.

Not unfrequently they are either extremely pleasurable or extremely disagreeable in nature.

These characters of the hallucinations are not rigidly distinctive, being absent in various degrees from the hallucina-

* As Huppert, Krafft-Ebing, Simon, Hagen, Dagonet.
† "Allg. Zeitschr. für. Psych., xxxv Bd., p. 551.

tions of some G.Ps., and in various degrees present in those of some examples of other forms of mental disease.

It is not easy to separate the hallucinations of g.p. into rigidly defined clinical groups; one need only, in very broad and elastic terms, speak of them as being either neutral, and more or less free from any features indicative of emotional perturbation; or, secondly, as bearing the impress of expansive exaltation; or, thirdly, as blended in origin with, or testifying to, hypochondriacal or melancholic conditions, or at least unpleasing conceptions or internal impressions.

In the case of *visual* hallucination the first group is the most frequent, while the second occurs with a slightly greater frequency than the third.

Dividing *auditory* hallucinations, in g.p., in the same way, the three groups are more equally represented, so that the pleasing hallucinations are about as numerous as the neutral, and are slightly more frequent than the unpleasing.

The hallucinations and illusions of *touch* are most frequently of a disagreeable—less frequently, but often, of an agreeable—character.

The hallucinations and illusions of *taste* and of *smell* are usually of unpleasant or offensive character; as those relating to poison in air or food, putrefaction of food, imaginary odours from the person, &c.

13. *Hallucinations, illusions, and perversions of common, or visceral sensibility, or sensibility of organs.* These perversions of the internal sensations are frequent in g.p., and are probably important factors in the origination of its hypochondriacal delirium. The hypochondriacal delusions, in many cases, seem to be indissolubly connected with some perversion or morbidness of the visceral or common sensibility; often with anæsthesia, or hyperæsthesia, of some internal part, or of some organ. Hence, one is tempted to speak of the above as hypochondriacal hallucinations, or illusions, or sensory perversions.

These perversions appear to be of the most varied kind; but in many other cases it is impossible to do more than infer the existence of such perversions. One ought to include here all the sensations except those of sight, hearing, taste, smell, and the impressions of pressure (and of temperature) got through touch (and heat-impact) on the skin and on parts of the mucous membranes. The limits of space preclude me from inserting here the examples I had prepared; but one or two may be men-

Disorders of Muscular Sensibility. 71

tioned; *ex. gra.*:—Patient says he is "torn to pieces, has snakes inside his belly, gnawing at his ribs; is rotten inside, and injured by instruments; all his insides have been taken out in hell, his backbone was taken away, and a porcupine skin sewn inside him, and his own skin was tucked in; all sorts of rotten things run in and out of his head, and breed there."—Another "had a hole 9 inches long in the heart, which is now cemented; has disease of liver, kidney, lung and spleen; his head has been taken off, and his heart and lungs have been taken out."

14. *Disorders of muscular sensibility*, or as may be put, disorders of general sensibility as concerns the relation of the body to its environment—to the external world. This is often disordered in g.p. The patient may say he "is whirled about," or that he "flew to heaven like a bird, or all over the earth, millions of miles;" or, "can lift millions of tons weight;" or, "is riding in a railway carriage, thousands of miles in a few seconds, whirling from place to place in an instant;" or he "can't move, or walk, or cough, or put out the tongue."

The above disorders of organ-, or viscus-, or common-, sensibility, including the muscular, were noted in 30 out of the 100 cases tabulated as to the existence of hallucinations and illusions, and probably existed, at one time or other, in many more, either before or after they entered the asylum.

In short, I believe, that if all the facts could be ascertained it would be found that (about) three-fourths, or more, of G.Ps. have hallucinations or illusions of one, or more, of the special senses, at one time or other; and about one-half of them marked hallucinations, or illusions, or perversions, of internal or visceral or common sensibility.

CHAPTER IV.

SYMPTOMS PARTLY MENTAL, PARTLY PHYSICAL.

There are several phenomena which link together the psychical and somatic symptoms of g.p., and are placed here between the mental and motor (ataxic and paretic).

Speech. Although some of the alterations of speech in g.p. are of psychical and some of somatic origin, it is convenient to treat of them together, and in this place: and by speech is here meant the faculty of articulate language.

There are several affections of speech in different cases; one in one case, another in another; or several coexistent and blended. The following descriptions, therefore, are more properly applicable to an imaginary type or average of the cases.

Elision of syllables, the complete repetition once or oftener of a syllable, a species of stammering, or, again, of stuttering or of stumbling in speech, a slowness, a lingering pause or hesitation in, or even momentary arrest of, speech; all are variously combined. The disorder of speech varies considerably in different cases, and at different times in the same case, and, indeed, may be almost absent. As the lips, or tongue, are relatively the more affected so is the pronunciation of labial, or lingual, consonants relatively the more impaired. Articulate utterance is worse after sharp exercise, and in marked cases speech is attended with tremor or twitching of the labial and facial muscles. At times this tremor or twitch may be unilateral.

Later on, the articulation becomes worse, and the words are jumbled together. Let the patient attempt to speak, and often the antagonists of the muscles which commence to act are called into play: hence interference with, followed by recommencement of, the first movement; and these direct, and opposed, movements not only precede, but also follow, the purposed, and finally accomplished, movement. But the speech is often made slow and deliberate, as if better to surmount the difficulty, while the lips are sometimes brought more closely together, and the gap between the teeth lessened, whence, occasionally, a pendency or a stiffness and immobility of the upper lip. It is a somewhat strained view (as by Dr. C. Gallopain) that the hesitation, and again the slowness, of speech, and approximation of lips are always, or often, purely intentional, and are conscious stratagems on the part of the patient to obviate the lingual and labial ataxy; or, again, that, like as the early may be, so also the final muteness is sometimes sullenly willed because of the annoyance the patient feels at the speech-difficulty.

The condition of speech varies much, therefore, in different cases, and at different stages, and sometimes fluctuates from day to day. Moreover, it is usually the result not of one agency but of several; and is not a simple disorder or defect, but a complex.

It is convenient, nevertheless, to make several divisions of the speech-affection, and to mention several varieties of each, however much they may be mixed as they occur clinically. Yet it is unnecessary to speak, in connection with g.p., of all the subdivisions laid down by some as pertaining to a full classification of the known disorders and defects of articulate language in its widest aspects. Prof. Adolph Kussmaul's chief lines of division of speech-affections are adopted here.*

The disorders and defects of speech in g.p. may conveniently be arranged in three or four groups, according as they flow from derangement or defect,—of *ideation*—of *diction*—or of *articulation*;—and of *phonation*, also, as may be added for the sake of convenience, though not strictly belonging here.

1. *Ideational disorders of speech* are those from impairment or disorder of intellect. Between these and normal speech are intermediate degrees, as found in the eccentric, weak-minded, or melancholy.

The intellectual weakness, impairment of memory, and of precision of ideas, in g.p., are evinced in several ways in the speech of the patients. There is the slowness in speaking, the lingering pause, the hesitation, the search for words, the drawling in speech. There is tardy understanding of the questions put, and slow utterance of syllables and words, and delayed presentation of ideas, in reply; or, perhaps, the substitution of some commoner word for the one really required; the omission of words essential to the sense; sometimes a sudden cessation or break-off in speaking. The unwonted slowness in speaking, with, perhaps, a sudden breaking-off in the middle of sentences, transition to a new set of ideas, and entanglement of the thread of thought, is common in states of intellectual weakness. The pauses between the words or sentences may (or may not) be filled up by prolonged or repeated vowel, nasal, or diphthong sounds, a condition which is very distinctive. Here, then, are frequent halts in speech, and sentences are interrupted by drawled, or repeated, vocalic, diphthongal, or nasal, sounds, as $\bar{a}$ or *a-a*, $\bar{e}$ or *e-e, eng, ang*, &c.; and one of these is frequently suffixed, connecting the end of the word with the next. The communication the patient has to make may be concluded, or he may stop short in the course of it, lost in emotion. The hesitating speech may be confused owing to mental failure; and this condition may pass into a choreic paraphrasia.

The omission of words and of syllables, quasi-indepen-

* Consult also Dr. J. S. Bristowe's work on "The Voice and Speech," London, 1880.

dently of other decided speech-affection, mark a disordered relation between the ideation and the functional co-ordination of vocal organs.

Thus the condition is one of defective memory, impaired association of ideas, intellectual debility; of defects of will, memory, intellect.

Then, again, the frequent and irrelevant repetition of the same words or sentences, or of a few fragmentary utterances; or, again, the echo-like repetition, by him, of anything, or of many things, spoken in the patient's hearing, all, when present, testify to the mental failure and confusion. Repetition may be due, also, to a sort of spasmodic hurry in speech. Finally, the idea is forgotten before half-expressed, and sentences are abruptly discontinued.

Mutism, or voluntary aphrasia, is of mental origin, as will be more particularly stated in the chapter on pathological physiology.

Then, too, in maniacal states of g.p., we often find the patient pouring forth a flood of disconnected words and phrases, which, in the more marked examples, becomes merely a jargon, or constitutes a logorrhœa. Here the association of ideas is relaxed; words as they arise influence the train of thought very strongly, produce further confusion of the rapidly-coming ideas, so that at last sentences cannot be constructed; and the result is a senseless mob of words.

In g.p. it is not merely the vocal utterance that is affected. Often the whole bearing of the patient, his accent, emphasis, mien and gestures whilst speaking are intimately connected with the emotional and intellectual perversion, or with the dementia of the affection. The whole appearance of the patient, and his manner of speaking, may be altered, and new words are sometimes fabricated. I well remember one G.P., a sergeant of the guards, who would march up and down for hours, walking slowly; with impressive mien and attitude, in a state of emotional exaltation and expansive ideation, uttering with the utmost gusto and emphasis, and in a deep bass voice, "my most beautiful, my most righteous, my most holy, my most wise, Lord—God—Almighty—Lord —God—Almighty."

Cluttering is due, as Kussmaul states, to the combined effects of a naturally precipitate and tumultuous speech, negligence, and a faulty education. Not taking sufficient time to utter sounds and syllables clearly, the patients begin to elide or omit syllables of words, especially at the ends of sentences; and perhaps end in an unintelligible, tumultuous, jargon.

Some G.Ps. revel in hyperbole, fabricate strange words and forms of expression, inflexions and syntactical arrangements; the latter often reverting to the forms used in childhood.

Various other anomalies of speech may be observed. Thus the patients may be found, like children, holding conversation with themselves—self-question and answer—and this not in reply to hallucinatory voices, though conversation with the latter (as it were) is frequent under other conditions in g.p.

The statements, the subjects of conversation, reveal the mental disorder and the dementia. But it is not requisite to dwell here upon the loss of delicacy, of culture, tact, or right moral feeling as displayed in the language of the patients; or on the positive coarseness; or on the exaggeration and absurdity, soon giving way to childishness, silliness, stupidity, mental confusion and disintegration, and all being gradually effaced by the rising tide of mental oblivion.

2. *Disorders and defects of Diction.* These are due to defect and disorder of the process by which words as sensory signs are allied to, or combined with, ideas, and are grammatically formed and syntactically arranged to give expression to the thoughts and emotions.

In g.p., not unfrequently occur examples of aphasia both ataxic and amnesic. The ataxic form is where the patient, so to speak, has the word as a sensory image or thought-symbol, or at least recognises it on suggestion; but cannot get it out. He may retain some words or phrases, may form the sounds and syllables of these very well, even when they are difficult of articulation; but cannot group these sounds and syllables in any new way, or indeed in any way other than the one, for him, stereotyped. Though the words as motor combinations of articulate sounds fail, the same sounds are perhaps formed correctly enough in some other words. The faculty of combining words into sentences is lost.

In the amnesic form the patient has not the word in consciousness spontaneously, and never has, at least as concerns the higher, more voluntary, and more intellectual phases of speech; notwithstanding that emotional and quasi-automatic utterances may occur, and he may, perhaps, be able to transcribe or repeat the word at dictation; though he cannot repeat it if ataxic aphasia also coexists. In the amnesic form, so to speak, the internal word is wanting; in the ataxic the external. But, as Kussmaul remarks, Steinthal was right, (as long as mere derangements of simple diction are allowed to pass as aphasias) in disputing the dysphasic nature of so-called ataxic aphasia, which is cortical verbal anarthria; whereas amnesic aphasia is true aphasia, incapacity for verbal diction. G.Ps. who are aphasic often seem to have also lost the memory of the use of things, and make wrong, absurdly erroneous, use of utensils, etc.

Or, again, in g.p. may be found that heterophasia, in which the words used are, for the time at least, of a meaning different from that necessary to express the idea desired; and this in a morbid unwonted manner relatively to the education of the patient. There is incapacity to properly connect the ideas and verbal symbols, and the wrong word-images, therefore, are constantly employed, or the ones used are strange and unintelligible. So that, in severe examples of these meaningless combinations of words the statements of the patient are thus rendered confused and unintelligible.

Or, there may be that word-deafness,—a condition associated and combined as a rule with other abnormalities—in which the patient, though he can speak, write and hear, and is not extremely demented, yet is deaf to the word as a symbol of an idea, and should he respond, does so in most perverted or irrelevant words and phrases.

Or word-blindness, in which with perfect or fair sight, the patients fail to read the printed or written words which they previously could.

Or again, there may be, in g.p., other syntactic disorders; incorrect succession of words, inflexion of words, and sentence-formation; a faultily constructed expression of the ideas.

In the mentally feeble, inattentive, conditions of g.p. sufficient attention cannot be expended simultaneously on the words, their arrangements, and the ideas; and if the syntax and thoughts have not their due share of attention, confused speech results (the condition last-named and heterophasia together).

Partly here, also, is to be placed the stumbling speech, often coexisting with the elision of syllables or sounds. The sounds and syllables are, or may be, well-uttered, but in their combination into words they get out of place. In the sentence, the sounds and syllables of the word-series become intermingled; the words becoming, as it were, disjointed;—disarticulated to use an anatomical simile,—and sounds and syllables being interpolated in words to which they do not belong. There is derangement in the development of the words from sounds and syllables. The stumbler intermingles and entangles the letters and syllables of one long word, or of several consecutive words; and it is especially in uttering long or alliterative words, or in attempting to speak rapidly that the syllable-stumbling takes place. Isolated sounds can be correctly articulated without any extra effort; the words uttered inaccurately at one time may be uttered accurately at another; the more quietly and slowly the delivery is made, the less the stumbling. By the huddling of syllables together, words become strangely distorted, and even unintelligible. Here, it differs from heterophasia, inasmuch as in the latter the substituted word still retains a meaning, which the distorted word of the stumbler does not. When due to loosening of the connections of the word-pictures, it shades off into heterophasia; when occasioned by "derangements in the motor production of the words it shades off into stammering."

It occurs in other affections also; but sometimes makes an early appearance in g.p., and is a result of a diffuse, and not of a localized gross, lesion.

3. *Disorders of Articulation.* These, to follow, are truly dysarthric, that is to say are from central change, organic or functional; and not from mechanical changes in external organs of speech; which latter speech-affections are suitably termed dyslalia.

This series of speech-conditions differs from those previously mentioned, although intermediate states of speech are found, and partly to be classed here; such as the stumbling speech. With it, eventually, there may be indistinctness, mumbling, elision of syllables, and a quivering and thickness in speech.

Here also is *Stammering,* a disorder of speech depending upon defective enunciation of the literal sounds, attempts to accomplish which are repeated, perhaps several times, and eventually speech is brought forth explosively after a difficult effort of phonation and articulation, the muscles of the face, &c., spasmodically working at the same time. The stammerer replaces letters and syllables by others that do not belong to the word, inserting some definite sounds in place of other definite sounds. The difficulty is to utter this or that single literal sound; yet there is an imperfect utterance of syllables, too, inasmuch as the change of the muscles from position to position in successive transitions, and the vocalization of consonants are disordered. The muscles of speech fail, both as to individual contractions and contractions to enunciate vowels and consonants. If contraction of either kind is hindered, the hindering obstacle may be in the articulatory organs, or in the nervous centres, or may lie in an intermediate diversion or disturbance of the transmission of impulses. In g.p. the origin of stammering is internal; it is the internal part of the great articulatory mechanism that fails.

Stuttering may occur, a difficulty in the production of the separate sounds in their syllabic alliance, a recurring spasmodic inability to vocalize consonants so as to form syllables. This is a spasmodic neurosis of co-ordination, obstructing the utterance of syllables by "spastic contractions at the stop-points for vowels and consonants in the articulating tube." On the other hand, isolated sounds are correctly made. Stuttering may occur at the beginning or in the middle of speaking. In articulating a syllable, especially one beginning with an explosive consonant, there is a sudden impediment of speech, and the sound about to come, or the one just uttered, is repeated, often many times. At last the obstruction is overcome, and, for a space, speech proceeds fairly or well. The regulating centre in the central nervous system, controlling the harmonious co-ordination of muscular actions in the vocalization of sounds, is disordered.

Next is quivering, or shaky, or trembling speech; syllables being uttered with unequal pauses between them; the association of sounds being delayed by paresis of lips and of tongue, and utterance being interrupted by tongue and lip movements in the forming of some (consonantal) sounds. At last the patient may thus be rendered speechless.

The muscular movements and physiognomical play normally attending speech may be greatly exaggerated; losing in fineness and accuracy, and becoming quasi-spasmodic.

In defective articulation of sounds there is a tendency to overcome the defect by the exercise of a great effort of utterance, by a sort of explosion; uttering, thus, sounds the patient could not with the force usual in ordinary conversation.

In reading, the affection of speech is similar, but often less marked in some respects; and here it commonly is the affection of *articulation* that is obvious. In singing snatches of old song, however, the disorders of utterance are far less obvious than in ordinary speaking. There are exceptions to these rules (as, *ex. gra.*, stammering).

4. *Disorders of Phonation.* The voice may be deep, rough, hoarse. Or, the expiratory force and tension of vocal cords are lessened, the voice becoming weak, monotonous, and low in pitch. Or, the vocal expenditure of air is unregulated, the tension of vocal cords being unregulated; hence the air is soon expended; and, continuing to speak, the patient has to bring the abdominal muscles into play; and yet the voice sinks to a whisper, and the patient is soon entirely out of breath.

Dr. Wilh. Zenker* pointed to several affections of the vocal mechanism which he believed due to deterioration of motor function, and not to psychical influences, and among which he mentioned, besides the foregoing, the slow lingering way of speaking, the speaking without inflection or modulation ; the intonation, with no external occasion, in louder, raised tones, or even falsetto tones ; or the tone is maintained at an equal height, and the sound of the voice not seldom becomes hollow, hoarse and rough. This modus is chosen because the muscular adaptation necessitated by it is less complicated.

Writing. Handwriting. It is not requisite to speak here of the subject-matter of the patients' writing, such as their delusions; but of the faculty of written language only as far as concerns what might be called graphic diction, as well as manual execution.

Usually, the writing is not much affected at first in g.p. ; but sometimes, where there is early mental excitement, and where the patient is restless, sleepless, or inclined to violence, the phraseology is perhaps unintelligible; the writing is most irregular and erratic in manual execution, interrupted, or even illegible; and the paper smudged, crumpled and dirtied. Far more usually, however, in the early stages, it is merely the omissions of words and letters or of dates, of signatures or of ends of sentences; or the running together of words, which are indicative of the mental state; and, occasionally, besides these, there is a shaky and erratic condition of the handwriting which bears witness to the motor involvement. As Schüle† rightly observed, the dysgrammatographia precedes the ataxic disorders of writing.

When describing the first stage of the established disease, mention was made of the sometimes shaky and irregular handwriting, of the separation, bad formation, or hieroglyphical character of the letters ; of words left unfinished ; of

* " Allg. Zeitschrift für Psych.," xxvii. Bd., p. 673.
† " Allg. Zeitschrift für Psych.," xxxvi. Bd.

letters or words omitted; or of speedy cessation of the attempt to write. In its defective manual execution the handwriting displays ataxia, or tremulousness; in the omission of words or letters, or repetition of words, or erroneous fusion of words, it reveals fundamental intellectual failure. Yet many patients at this period are busy correspondents, and write long letters to their friends, to parliament, or to the throne. Some of the above characters are shown in Dr. Bacon's* illustrations, but this cacography usually undergoes marked remissions and exacerbations in g.p., and is only in a strained sense of the progressively degenerating nature he describes. Dr. Blandford, † also, long ago indicated the diagnostic value of the omission of words and repetition of sentences in writing, in early g.p.

In the second stage of the established disease the faculty of writing is more defective. The handwriting becomes illegible, or, at the very best, is extremely shaky, irregular and unfinished, or even fragmentary. Often, after several attempts the patient writes a word or two, or a part of a word, or several misplaced or dissociated letters thereof, and then, after a pause, makes several irregular strokes or flourishes, or puts down the pencil in momentary confusion or disgust. The last letters of words are often omitted, syllables in a word are left out, or are reduplicated; separate words are fused together, with elision of letters or of syllables, and especially when the words are somewhat alike. At worst, the words cannot be expressed in writing. In other cases words are used of wrong meaning, irrelevant to the idea to be conveyed, and this not from habitual ignorance. When written, the epistle is often full of repetitions, of disjointed or fragmentary sentences, or the purpose in writing is lost sight of.

Then as to mechanical execution; especially in the later stages the lines of the writing are irregular, pass diagonally across the paper, run into one another, or diverge irregularly; the letters and words are often made of more unequal size, as compared with the normal writing of the patient; and owing to the ataxic and spasmodic element in the muscular action, the light and heavy strokes are misplaced.

At first there is some shakiness in the formation of letters; then in making the lines and strokes there are irregular, zigzag excursions from their normal sweeping curves or folds; at last only a few illegible and jumbled strokes.

* " On the Writing of the Insane," London, 1870, p. 20; and " The Lancet," July 24, 1869.
† " Medical Times and Gazette," Nov. 3, 1866, p. 467.

Dr. Bianchi,* "on changes in handwriting in relation to pathology," says very little as to g.p., and apparently does not mention ataxic writing in it.

When commencing to write, the patient usually seeks a firm basis for the paper or book; whilst writing, his fingers press heavily on the surface inscribed. Sometimes he will readily undertake to write his name, will take book and pencil in hand confidently for that purpose, make a few shaky up and down strokes or one or two hieroglyphics; or a few irregular confused circles in one spot; and then hand the book back, sometimes with a satisfied, sometimes with a confused or impatient air. At last no attempt, even, is made to write.

If finished, the blotted, smudgy appearance of the note, written as it may be on the page of some letter received, or on a piece torn from a newspaper, is, perhaps, in striking incongruity with the address, as for example to a minister of State, or to the monarch.

In clinical fact, the several changes in the writing are usually combined in the most varied ways, in different cases. Nevertheless, in the more simple examples, conditions of writing can be made out which are analogous to one or other of the several speech-affections, also found. As the speech, so the writing may exhibit disorder or defect of thought, disorder or defect of diction, disorder or defect of muscular execution.

(*a*). As the speech, so, in a parallel way, the writing may display intellectual disorder in the statements made, the persons addressed; intellectual defect, in the use of dirty slips of paper, in the omission of letters, words, sentences; in repetitions, in slowness and painful effort, in forgetfulness of the subject in hand.

(*b*). Similarly, as in the speech, so in the writing, it may become impossible to express the idea, or to have the written symbols at command; a state of verbal agraphia analogous to some forms of aphasia.

As in speech, so in writing, too, the words used may be quite different from those proper to express the idea intended; in which case the condition is analogous to heterophasia.

There may be blindness to written as well as to printed symbols; just as with regard to speech there may be word-deafness.

* "Il Pisano Gazetta Sicula," 1882. Trans. by Dr. Jos. Workman, "Alienist" &c., Oct., 1883.

In writing, the same syntactic disorder, incorrect word-succession and sentence-formation, may be found, as in speech; a condition I would term akatagraphia.

As in the speech, so in the writing, syllables may get out of place; or may be variously reduplicated or elided.

(c). The ataxic and paretic disorders and defects of writing and of speech follow the same lines of parallelism in a laxer and less analogous manner. Thus, trembling, shaky speech finds its analogue in trembling, shaky handwriting: in which the undulating contour of curves and of up and down strokes is replaced by irregular, or, when worse, by broken lines. The ataxic disorders of speech compare with some of the spasmodic repetitions and stoppages in writing; and especially with disorderliness in contour, symmetry, and arrangement of letters, of words, and of the lines of writing.

Mumbling and indistinct paralytic speech has a similitude to the illegible and failing attempts to write; and loss of the power of writing bears a likeness to speechlessness.

For the higher elements of the faculty of writing, or graphic representation, the cortical centres form part of a wide symbolic cortical field, in which the centres of *articulate* language are the most important. But with these the cortical centres concerned in *written* language must be closely allied, and in intimate sympathy; and therefore when, and in so far as, the speech-affection in g.p. is of cortical origin, and due to a diffuse, and not to a closely circumscribed lesion, the writing must, so far as concerns the cerebral cortex, be simultaneously affected, as a rule, and in a manner somewhat proportional.

CHAPTER V.

PHYSICAL SYMPTOMS (purely physical).

1. *Motor* (ataxic and paretic and spastic).
Labial and facial muscles. Physiognomy. Facial nerve.
The tremor, twitch, or convulsive spasm, of lips and face attending speech in the marked cases, although usually bilateral, is, in some patients, occasionally unilateral. In advanced cases, this often is highly marked, the attempt to speak being preceded and accompanied by spasmodic and exaggerated repetition of movements of the muscles of face and lips, thrown into play by the voluntary impulse. In its

lesser degrees it reminds one of the facial twitches in one about to weep; in its more severe degrees it reminds one of some aspects of the stammerer, with his stoppages in speech and convulsive play of features. The twitch of the zygomaticus major and other muscles adds tumult and distortion to the broad, expansive, semi-fatuous smile of the G.P. Even in the earlier stages the eyebrows are often raised, the occipito-frontalis gathered together or twitching, the lower lines of facial expression partly obliterated; and the features either somewhat florid; or pasty; or of parchment-like appearance. While the mouth loses its firmness, and the lips and lower face become flabby, coarse and expressionless; the forehead often (and even at an early period), is corrugated; and either then, or when the patient becomes bedridden, the ribbed forehead, widely open eyes, fatuous glance, and raised head, with staring disparted coarse hair, make up a distinctive mien and cast of countenance.

The features, from being of a swollen appearance, may eventually become thin, earthy, and coarse.

In hemiparetic attacks the features are modified by the unilateral paresis, more marked in the lower part of the face; the lineaments there becoming effaced, and the lips fallen together on the side affected, the mouth drawn slightly towards the opposite side and upwards, and the tongue-tip tending towards the affected one. This temporary paresis may be limited to the facial nerve, none existing in the limbs. On the other hand, the uvula is sometimes affected in this facialis-nerve-paresis; the nerv. pet. superf. maj. supplying Meckel's ganglion, and this the levator palati and azygos uvulæ; in consequence of paresis of which the uvula-tip is turned to the side affected, on which, also, the velum palati is lowered, and the curvature and height of the palatine arches are lessened.

The frequent flushing of the visage has been mentioned.

Lip-smacking. Many G.Ps. make peculiar lip-smacking or sucking, or sniffing, or clucking, sounds, or, even, sounds as when lips disunite from kissing. And this not with an occasional, but with an oft-recurring or habitual, frequency. These movements are probably due to abnormal central irritation of branches of the fifth and facial and hypoglossal nerves; or, in some cases, simply are a reversion to coarse habits, and permitted by the failure of social refinement and of self-care, accompanying the intellectual failure and the perversion of self-consciousness.

Spasm, incited through one of the facial nerves, sometimes distorts the visage, dragging the face more awry than when the distortion is due to unilateral paresis or paralysis, or to a greater degree of the paresis upon one side than upon the other. An example of this sort of spasm was recorded by the present writer some years ago.*

In all cases, the natural conformation of the visage; and deviations due to irregularity or loss of teeth, or to natural configuration of gums, hard palate, maxillary, nasal and malar bones, must be taken into consideration before forming a conclusion as to paresis or spasm.

Paresis of the orbicularis oris animated by the facial nerve, also permits a drivelling of saliva in some advanced cases. Particles of food and drink also make easy escape, and are apt to be spilled on the beard or clothing. The saliva, escaping through loose paretic lips, and the patient breathing with half-open mouth, the buccal surface grows dry; and this excites a fresh and increased secretion of saliva.

Motor root of Fifth Nerve. In some cases, the patient grinds his teeth, even at an early stage, or champs with jaws and cheeks, making movements as of mastication. These last-named, masticatory and tasting movements, irrespective of the taking of food, are in some cases frequent or long-continued. But it is particularly in the later stages that so many G.Ps. grind the teeth frequently, or almost incessantly, and noisily, with quasi-spasmodic action; grinding by day, and perhaps discontinuing it momentarily, but only when swallowing food; and grinding through the silent watches of the night. This appears to be spasmodic in nature; whether the irritation of branches to the pterygoid, masseter and temporal muscles be of cortical, of nuclear, or of peripheral origin.

At first bolting his food, the patient at last may come to roll it about in the mouth, and between teeth and cheeks, before swallowing. Here is paresis of motor part of the fifth nerve. (See also *Eating*, below.)

The Tongue. Hypoglossal, or Ninth Nerve. It is especially in the middle and later part of g.p. that the tongue is affected. There may be fibrillar tremor of the tongue, which may be jerked in and out convulsively when the patient is desired to show it. In the advanced stages, as a rule, the tongue is only protruded, at the best, in an imperfect or momentary manner, jerkily and helplessly, moving

* "Journal of Mental Science," January, 1876.

backward and forward, or from side to side, as if the sport of varied impulses. Finally, if surviving long enough, the patient loses all power to protrude the tongue, which then lies helplessly, floating, log-like, in a salivary lake on the floor of the mouth.

The *fauces, pharynx, œsophagus* and *larynx* are sometimes relaxed and congested, of lessened sensibility, of defective motor activity. Among the effects of the anæsthesia and paresis of these parts are dysphagia and imperfect closure of the glottis. Jaccoud, Schultz, and Rauchfuss (Voisin) have described paralysis of one vocal cord, or of both vocal cords.

Dysphagia. From the condition of imperfect sensibility, and of defective reflex action, and probably of inco-ordinate action of muscles at first, and later of paresis, the act of deglutition is often imperfectly roused, or executed, and the bolus of food remains unswallowed. Dsyphagia is, broadly speaking, progressive in g.p., but, like many of the symptoms, may undergo numerous fluctuations, remissions, and exacerbations. To obviate the condition, the food of these patients is usually minced or soaked, and finally, if long surviving, they swallow fluids only. In some cases, owing to an unusual degree of bulbar lesion, or to pharnygeal and laryngeal anæsthesia, swallowing becomes almost impossible, and the fluid food is apt to be inhaled into the air ways.

Eating. Mastication is imperfectly effected; even prehension of food, and conveyance to the mouth are clumsily done; at last the food may collect in the mouth, or may be inhaled into the lung. With the greater or less abolition of the sense of smell that of taste also dwindles away, and what with this, and the intellectual failure, together with the obliteration of æsthetic feelings, of all that is self-respecting, —the eating of the patients becomes hasty and gluttonous. Seizing masses of food and hastily thrusting them into their mouths they endanger the integrity of their air-passages, and risk death by asphyxia, unless relieved by efforts of gulping, or by instrumental or digital extraction, or by an expulsive blast of air driven out by a vigorous blow on the back.

Here also must be taken into view the mental state, the failure of attention, of apprehension of surroundings, and the oft defective self-consciousness. The garbage-eating of the patients, however, is often an indication of failure of the taste-sense. As M. Voisin mentions, it is not surprising that the G.P. does not know how to eat; for, according to

Brillat-Savarin's maxim, "Il n'y a que le sage qui *sache* manger." And too often the G.P. eats with boorish coarseness, masticates imperfectly, eats too much, and with but little gustatory appreciation of what he does eat.—The muscles of the eye might well be considered here, but I will deal with the several eye-symptoms of g.p. in a separate chapter, forming the link between the motor and sensory groups of symptoms.

Limbs and Trunk. Gait. The gait may be normal in g.p. in cases of brief course, or cut short by a lethal exacerbation, or by some intercurrent malady. Omitting "spinal" cases: —in the prodromic stage the gait is ordinarily free from change, but, in some examples, is slightly irregular, awkward, and more coarsely executed than is natural to the patient.

During the course of the disease; disorder or impairment of the gait often undergoes many fluctuations; and particularly is it profoundly, though temporarily, modified by the recurring epileptiform and other seizures. Even until late in the disease the paresis is often more apparent than real; and when roused or excited the patient may throw off his inertia, and move about with great ease and celerity.

When the disease is well-established the gait is often awkward, the steps long and slightly irregular; sharp turning around may be attended with awkwardness and momentary uncertainty, or even with swaying. Trembling movements of the limbs, varying in degree and frequency, often accompany the act of walking. Later on, the gait is often slow, swerving, zigzag; the feet are widely apart, the steps short, progression is unsafe, a slight obstacle an impediment, falling-down frequent.

In different cases of g.p. are found several forms of morbid gait, and these are usually combined in different proportions, or one is modified by the coexistence of some degree of another. Elements of the ataxic and of the paralytic and of the spastic forms of gait may be found; but if the patients live long enough the tendency is for all the forms to merge at last in a state of helplessness or paralysis.

The ataxic gait. This is especially pronounced in one group of cases, those in which the spinal lesions appear simultaneously with the cephalic, or even precede the latter, or in which, although occurring later, they play an unusually effective part. Particularly in the former of these classes do we obtain a history of severe pains, said to be neuralgic or rheumatic, or supposed by the medical attendants to forecast locomotor ataxy. These, perhaps, have been protracted, have led to incapacity for work, and to tedious treatment. The gait of the patient as a rule takes the tabic or ataxiform character in these cases from the very outset, and the impairment of gait and helplessness are relatively early and prominent. Moreover, to this state, true localized paralyses, due to spinal complications, often come to join themselves, and increase the helplessness. The tabic form of g.p. is rare in females.*

But in examples running the more usual course, the gait varies with the relative amounts of ataxy and paresis present. When the special ataxy of the affection is *relatively* predominant and well-marked, the gait is more or less jerky, irregular, the steps are long, the limbs are thrown forward, the movements apt to appear hurried and made with unnecessary efforts, the lines of march swerving and crooked. In extreme cases the movements in walking are irregular and hurling, and steps hasty, spasmodic, unequal; the foot is thrown outwards and forwards, the heels are brought down more or less sharply, the leg is stiff and extended at the knee. Thus this gait has a more tottering, staggering appearance than has the paretic; and the eyes therewith are usually kept on the ground. In turning, the balance is easily lost. In its lesser degrees there is merely some inaptitude, defective dexterity, insecurity and irregularity in the muscular movements of the limbs, as in the gait; but in extreme cases static ataxy coexists with the locomotor.

As to priority, Dr. Max. Leidesdorf † in 1854 described a case in which the disease began in the spinal cord, as also did Hoffman, and Dr. Joffe.‡

Paretic gait. When paresis *relatively* predominates the gait is slow, heavy, unsteady, and the widely-separated feet readily trip at any inequality or obstacle on the ground; the fore part of the foot drags on the ground, and hangs down in progression, the sole of the foot is planted awkwardly, and usually its outer edge first, the knee is raised and pulled after in the extended position. The patient totters but little, and if he falls it is by sinking to the ground. In advanced cases, a swaying or falling follows upon any sudden attempt to turn. This slow and helpless gait, together with

* Jung.—" Allg. Zeitschrift für Psych.," xxxv Bd., 2 Heft.
† " Beiträge zur Diagnostik der Geisteskrankheiten." *Vide* " Path. u. Therapie, der Psychischen Krankheiten." Erlangen, 1860, p. 106.
‡ " Zeitsch. der K.K. Gesellsch. der Aerzte im Wien," 1857.

the bended stooping frame, imprints an aspect of premature senility.

Often the paresis is temporary and even very transitory, or at least recurring augmentations arise, and thus the impairment of gait fluctuates in degree.

The paresis is often unilateral, and when this is the case the frame is bent to the corresponding side, on which, too, the shoulder is lowered, and, in walking, the foot heavily planted.

Spastic gait. A spastic element exists in the gait in some cases. These patients usually have—exaggerated knee-jerk; often, rigidity of frame, which may be bent forward, the patients holding themselves stiffly; spasmodic twitches or jerks of the limbs, spontaneously or on passive movement; a jerky tremulous grasp of the hand; a shaky speech, with much twitching of lips and face; jerky protrusion of the tongue simultaneously with much facial and labial twitch; not infrequently paralytic and apoplectiform seizures.

Where spastic gait is pronounced it is stiff and dragging. The legs become weak stiff and heavy; the feet stick close to the ground or scrape it, and find an obstacle in every elevation. The gait becomes swaying and uncertain. At last, at each step there is a springy elevation of the whole body, giving a faintly hopping aspect to the gait, and due to muscular tension and sudden contraction. There is a tendency to get on the toes and to adduct the lower limbs strongly, so that in walking the orbits of the progressive movements of the two limbs overlap, and to compensate for this the centre of gravity of the body is thrown from side to side at each step. The patient easily stumbles and falls. The lower limbs are kept stiffly in a constrained position. At last the limbs are apt to become contracted. The wrists and fingers are at the same time often contracted, flexed; the forearms semiflexed and pronated. This description only applies to an unusually marked case of the relative prominence of the muscular tension, and increased reflex action of tendons.

For examples of the association of conditions of gait and of knee-jerk, see the section on this reflex.

Other limb-movements. In speaking of the gait, much has incidentally been stated as to the condition of the musculature of the limbs and trunk in g.p. Other limb-movements are often slow and stiff, or heavy and difficult; and muscular efforts, as the lifting of weights, readily call forth trembling of the parts. Lying in bed, the patient moves her limbs in a slow and heavy way, and finally they may become quite paralysed. Inco-ordination is succeeded by paresis, but in maniacal states the movements may be strong and free.

Arms and hands. At first, besides the changes in handwriting already mentioned, the other finer, more difficult movements, the later complex acquisitions of the hands, the manual dexterity, show some of that disorderliness, irregularity, disharmony of action, which, in this work, has been spoken of as ataxy; although this name is open to the objection that it has been so long associated with the prog. loc. ataxy described by Romberg, Todd and Duchenne, as to recall into consciousness one's idea of the latter. Thus, the movements of the hands become lessened in adroitness; in exactitude. Hence there is a failure of capability in accurately playing musical instruments, in painting, in drawing, in manual execution concerned in the various arts or trades. This condition, finally, comes home in a practical way to the patients, or to those who have the care of them, for the former cease to be able to button their clothing, or to adjust collars or other articles of apparel, break various articles they handle, let them fall, and hence may do much damage to fragile objects around them, and spoil fabrics on which they work. The manual grasp of the patient usually becomes awkward and quasi-spasmodic; he seizes one's hand, and alternately grips it hard and relaxes the pressure, and ends by pulling it towards, or pushing it from, himself, and downwards. With any muscular strain or effort of the part, the hand soon begins to tremble, except in some cases when maniacal excitement is present, or when temporary physical improvement exists.

The trunk. The body is often bent slightly or heavily forwards, or to one side or the other. With the latter is sometimes hemiparesis or hemiplegia. Rigidity may be present or not. Patients will often sit almost immovable for hours in a chair, with bent frame, and heavy ungraceful attitude, looking to the ground, or peering stupidly from under wrinkled brows.

Paresis or paralysis in general terms. Independently of real paralysis, the patients may complain of great lassitude, fatigue, weakness. The paresis or paralysis found in paretic dementia (g.p.) may come on suddenly or rapidly; or, on the other hand, slowly, *gradatim*, insidiously.

The temporary attacks of paresis or paralysis are either circumscribed and local, or unilateral, or of wider distribution than the last and almost or quite generalized. They may form part of, or follow, the special seizures in g.p., hereafter to be described.

The permanent paresis, when it exists, is usually imperfect, progressive, generalized: sometimes it is more pronounced on one side than the other: sometimes is unilateral: with it may be more pronounced paresis in one part, as for example, one side of the face. A common condition is permanent paresis; undergoing at times unilateral increase. The final tendency is to paralysis. The failures of motor power just mentioned are all referable to the ordinary lesions of g.p., are parts of its semeiology. In addition to these are, occasionally, other local or hemipareses or paralyses brought about by complications, such as local cerebral or spinal softening or hæmorrhage. But without circumscribed lesion of any kind, I have, in many G.Ps., found persistent incomplete hemiplegia during the last part of their lives. Dr. J. Christian in most cases found muscular enfeeblement as tested by the dynamometer, but the weakening of muscular power was not proportional to the marasmus in advanced cases. He maintained, however, that there is no true paralysis; that the intellectual centres are gradually destroyed, but the motor centres are only irritated secondarily; the motor symptoms being secondary also in g.p. Nevertheless, his own researches, like those of Dr. E. Chambard,* show failure of motor power in g.p.

When discussing the several "seizures" will be mentioned the incomplete, transitory, and sometimes partial, hemiplegiæ so often found in g.p. From what has been, and is yet to be, stated, I do not think it sufficiently inclusive, or entirely accurate, to say, with Dr. Ach. Foville,† that the hemiplegias found in g.p. are of two varieties, (1) the one consisting of those following apoplectiform attacks, and which are partial, incomplete, variable in intensity, mobile, and "alternative;" and due to diffuse circulatory disorder; and (2) the other variety consisting of frank or true persistent hemiplegiæ, due to a localized cerebral lesion. For, as regards those of the first variety they are sometimes, not incomplete but, complete for a greater or less length of time; and as regards the second variety, a true persistent hemiplegia in g.p. may result not only from a "local" lesion, but from a brain-atrophy predominating in one

* "Revue Scientifique," Jan., 1881.
† "Archives Générales de Médecine," Sept., 1879

cerebral hemisphere. I do not say that such hemiplegiæ from uni-hemispheral atrophy, are stationary or constantly of the same degree, or that they are absolutely complete, and, for the matter of that, hemiplegiæ from local lesions rarely are so; but I say that they may be true and permanent hemiplegiæ, and might add that with them may be "contraction" of the paralysed limbs, with descending degeneration (sclerosis) of the opposite crossed pyramidal tract in the cord, occasionally even so gross as to be visible to the naked eye, as in a case reported by me.* See also the chapter on Pathological Physiology.

Silent excitement. The motor element in "silent excitement" can scarcely be dissociated from the mental, treated of in Chapter III. Here, then, I shall merely mention the incessant, restless, rolling, pushing, thrusting activity of the patient, who seizes on objects, puts them in pocket, or, and especially if in bed, pulls them towards himself, and then thrusts them away with equal vigour, or drops them in order to seize something else within reach and pull it towards his head; grasping persons, or other objects, about him—anything, and the nearest within reach—this pulling and thrusting being made only more vigorous by any attempt to restrain it, or by resistance.

Trembling of the limbs. This has been several times mentioned; to it is sometimes partly due the loss of manual aptitude and exactitude, and the existence of other evidences of muscular unskilfulness and disharmony. Thus, tremor reinforces the ataxy of g.p. When present, the trembling is part of a more or less general condition, of which trembling speech and shaky writing are other parts. It is augmented by any muscular straining or effort. In grasping objects vigorously the trembling of the hands is replaced by irregular convulsive movements, the objects being held in a grasp which, turn by turn, squeezes spasmodically, and relaxes. The tremor of g.p., and its ataxy, are well shown graphically in Dr. E. Chambard's myographic, chronographic, and dynamometric researches. In trembling in g.p., the tracing showed equidistant, but irregular and unequal, oscillations, with now and then a series of much more ample ones; and the symptom appeared to be of passive (paralytic) nature, and due, rather, to diminished intensity of excitation of nervous centres than to lessened frequency.

* "Journal of Mental Science," Apr., 1885, p. 61.

Position of limbs. When the patient is in bed the ordinary position of the limbs is one of moderate flexion; if extended, as they easily can be, they soon resume the flexed position when interference ceases.

Contractions of the limbs. When the patient becomes helpless the limbs, as a rule, also become contracted, flexed, and rigid. Yet in many cases the limbs can be straightened and extended by the patients. In others, the contractions are permanent, and the rigidity is extreme. Some of the contractions are passive, the muscular tonus of the flexors overcoming that of the extensors when the parts are at rest, and this either in a merely helpless, or a temporarily palsied limb; but some of the contractions are spastic. Ordinarily, the legs are somewhat flexed, and the arms, also. But in extreme contraction, the knees are drawn up towards the chin, which in its turn is brought nearer to them by the co-existing flexion of the neck; the limbs are strongly flexed at the knee- and hip-joints, so that the heels tend to rest on the buttocks; the thighs are often rigidly adducted, so that the knees cross one another; the upper limbs are flexed and adducted across the chest; the wrists flexed, and the fingers either extended, or doubled up in the palms. The spine is curved, with its convexity backwards, the patient thus usually becoming like an ill-shaped ball, rolled in upon himself. One limb being often more affected than its fellow, there are corresponding deviations from their symmetrical disposition; or the legs, or one of them, may be "contracted" in the extended position, in which case, besides their more usual sites, bedsores tend to form on the heels.

Dr. Voisin * speaks of the muscular "contractions" and retractions in uncomplicated g.p. as being quite different from the late ones following cerebral hæmorrhage or localized softening, although they may be similar when cerebral or meningeal hæmorrhage complicates g.p. But, as suggested in the first edition of this work (pp. 155 and 160), the more usual contraction of the limbs in uncomplicated g.p. may at least sometimes, be attributed to the secondary, descending, systematic sclerosis of the crossed pyramidal tracts of the spinal cord, consecutive to the ordinary diffuse encephalic lesions of g.p., and independent of complicating hæmorrhages, or localized softenings; to which cause we may add changes in the spinal nerves. Contrary, therefore, to the view of M. Voisin, I find, in not a few cases of g.p.,

* *Op. cit.*, p. 149.

true "late" contractions from secondary spinal degeneration, and independent of hæmorrhagic or of other local lesion.

Contractions are often very temporary after the apoplectiform, epileptiform, and other seizures. Yet the palsied limb may be stiffly contracted, and resist firmly, without secondary spinal sclerosis having supervened. Contraction and rigidity may be due to primary changes in the muscle itself. Thus Mendel states that in two cases in g.p. he found this condition to be connected with myositis ossificans; which took its starting point in muscular hæmatoma. In neither case was the spinal cord examined. *(Op. cit.*, p. 205.)

Rigidity of the limbs. The rigidity often attending contraction of the limbs has just been described. Attempts to straighten the limbs fail in extreme cases; in less extreme cases they are successful, but, not seldom, obviously give pain or discomfort to the patient, who, though demented and inattentive, will call out, put on a pained look, and make resistance. Rigidity of limbs and resistance to passive movements are more striking after the apoplectiform and epileptiform attacks than is distortion, and are often very transitory. (See also tetaniform seizures, below).

Automatic movements. Automatic movements, as if due to some irritation of central organs, or disorder of muscular feeling, are frequent in G.Ps.; and may exhibit themselves as pulling, grasping, scratching, slapping, and rubbing movements. Some patients will continually rub one part of the head, until the hair is off; others the end of the thumb with the fingers of the same hand. It is likely that these movements sometimes depend upon direct irritation of the central nervous organs by the diffuse morbid process; sometimes upon disorder of the muscular sense, acting reflexly through some nervous centre.

The functions of bladder and rectum. These are disordered in all advanced cases of g.p.; and sometimes, indeed, at the very outset. Even in the prodromal stage may the involuntary passage of urine or of stools occur, either owing to a condition possessing apoplectiform elements; or to epileptiform seizures; or to a state of mental confusion, excitement, giddiness and hypæsthesia. Later on, these involuntary passages may occur in the apoplectiform and epileptiform attacks; attacks which also may be followed by the same, or by retention of urine; or the condition may then be one of retention combined with incontinence; the urine dribbling away constantly from an overfull and distended bladder.

The bladder at other times may become distended and require skilled interference. Independently of the above attacks, the urine and fæces in the middle and late stages may be passed into the clothing, or into the bed; and these foul habits ot G.Ps. give an enormous amount of work to nurses and attendants. The *bowels* may be too loose; they may be obstinately constipated; conditions which may recur from time to time, or may alternate. Many a patient, put on the close-stool will not pass anything; then put into bed will defile it immediately.

Sexual power and appetite. Already has this been discussed in the second chapter, so far as concerns increase or decrease of sexual power and desire in the prodromic stage, or at the onset of the affection. Sexual capacity often continues for some time; a patient under my care, removed by his friends during a remission, and against my advice, was brought back again, as bad as ever, but in the meantime had begotten a child by his previously barren wife. Sexual power usually fails progressively with the advance of g.p.

Some G.Ps. are troubled with free seminal emissions; the most marked example of this I have seen was a patient who had greatly improved, mentally, after mingled expansive, querulent, hypochondriacal, and melancholic symptoms; but to whose diminished knee-jerk an increased one had succeeded, with good plantar reflex, no ankle-clonus, and no anæsthesia in the lower limbs; whose gait was jerky and somewhat convulsive; who had had auditory, visual, and tactile hallucinations; disorder of the muscular sense, and of the feeling of equilibration; and to whose quasi-syncopal seizures there eventually succeeded desperate epileptiform attacks; who at times complained much of pains in belly, back, loins, or "all over," and whose incontinence of urine greatly improved at one period. In the exalted states of G.Ps. they often praise the other sex effusively, glow with a childish facile gushing romanticism, which, however, concerns sexual love, rather than lust, and is scarcely part of the present subject.

Electrical reaction of muscles in g.p. The utmost diversity of statement has existed on this point. M. Brierre de Boismont,* whose contributions on this subject are the

---

* "Du diagnostic différentiel des diverses espèces de paralysie générale à l'aide de la galvanisation localisée."—" Ann. Méd.-Psych.," 1850, p. 603; and " Dict. des Dicts. de Médecine. Supplément," 1851, p. 596.

earliest I have met with, stated that the irritability of the muscles to the galvanic current was conserved in cases of g.p. with mental alienation. On the other hand, Dr. J. C. Bucknill * shortly afterwards published observations to the opposite effect. Dr. v. Krafft-Ebing † found that electro-muscular contractility and sensibility were retained in g.p.: and M. Voisin ‡ declared that even in the second stage the electro-muscular sensibility is intact, as long as there is no softening of the cord.

According to some observers § the farado-contractility of the muscles of the extremities, especially of the flexors of the feet, becomes considerably and progressively lessened. Also electro-muscular contractility was found absent in 78 p.c. of G.Ps. Benedict and Svetlin ‖ found great increase of motor irritability to electricity in the lower limbs in g.p.; and the latter observed that an increase of electro-irritability on one side of the body forecasts a more diseased condition of that side, subsequently.

Speaking of the induced electric current in g.p., Dr. Tigges ¶ stated as follows:—Some of the results show increased, and some lessened, electro-muscular contractility. Strong contractions disturb the relations; and increased irritability is lessened. As regards susceptibility and reaction to pain from electricity, the extensors show increase only; the flexors, palpebraris, and zygomaticus, either increase or diminution.

The electro-muscular contractility is but little affected by the state of the general nutrition and of *embonpoint*. On the other hand, psychical (and motor) excitation and muscular tension increase the electro-muscular irritability. Paretic conditions (circumscribed, unilateral,) come on, partly with heightened, partly with lowered, contractility of the muscles concerned. Again, there is increase of the contractility, associated with recent conditions of tonic muscular tension; particularly if with these are convulsive movements.—Also, with tonic muscular rigidity of long duration is the contractility enhanced, yet often less so than in other muscle-groups presenting less muscular tension.

* Jan. 1st, 1852, "Manual of Psychol. Medicine," 3rd Ed., 1873, p. 460.
† "Allg. Zeitschr. für Psych.," xxiii. Bd., 3 Heft, p. 193.
‡ *Op. cit.*, p. 110.
§ Mr. J. Lowe, "West Rid. Asyl. Med. Reps.," Vol. iii, p. 196. Dr. Bevan Lewis, Vol. v, p. 85, and Vol. vi, p. 139.
‖ Cited by Mendel, *op. cit.*, p. 185.
¶ "Die Reaktion des nerven- und muskel-systems Geisteskranker gegen Elektricität."—"Allg. Zeitschr. für Psych.," xxx. Bd., p. 137.

In the quiet states of g.p. there is usually lessened electric irritability of the musculature; in the excited states it is increased, generally; and in the later and last stages, there is still higher enhancement of it in the spastic and paretic conditions. The increase is most in the flexors of the forearm. The facial muscles show sometimes an increased, sometimes—and possibly from over-irritation—a lessened irritability; while almost always, and throughout, the irritability of the extensors is less than that of the flexors.

Symptomatic paralysis agitans. It is not rare to find a distinct symptomatic paralysis agitans—a true tremor cöactus—which may involve the upper extremities and head or neck, and, occasionally, the lower extremities in some degree, or may be observed on one side only, or in one thoracic limb. It may last for days or weeks. In my own experience this has principally occurred in the second stage of the confirmed disease (third period), seldom in the first, and then only when the stages were ill-defined. See cases 63 and 53 at the end of this book.

Convulsive tremor, or apyrexial rigor. Nor must we omit mention of the occasional occurrence, in the first, or even a a later, stage, of a general subsultus and rapid jerk or tremor of the muscular system, somewhat like shivering or rigor, increased by the erect posture and by voluntary movement, not produced by any impression of cold, and not the rigor of any ordinary pyrexia or phlegmasia, but sometimes lasting for days together, and, perhaps, recurring. The whole frame is sometimes affected with the violent muscular tremor or shivering; or the subsultus and rapid tremor may affect one side, or one limb only. This, possibly, may be a modification and extension, as it were, of the condition last described. But it is worthy of notice that, at least after the early stages, many G.Ps. are peculiarly vulnerable by the effects of low temperature, as shown by the facile production of cutis anserina in response to comparatively slight impressions of atmospheric cold; and that the tremor in g.p. is ordinarily a *tremor a debilitate.*

Athetosis. Athetosis-movement may occasionally be seen in g.p., and especially after the apoplectiform attacks, when it occasionally lights upon the hemiparetic side. The slow peculiar flexion and extension movements and changes are sometimes, and sometimes not, associated with unusually well-localized lesions. They sometimes existed in the right upper limb in the last one of the cases related below, under "choreiform movements."

Case 5. Athetosis occurred, also, in another case which ran a course of about 7 years. Death *æt.* 39. At first, exalted delusions, etc.; then greatly improved; then depressed. Nephritis. Improved; but became taciturn, obstinate, indifferent. Erysipelas. Maniacal. Slight systolic cardiac bruit. Later, wildly maniacal: afterwards confused; repeating same phrases. Tremulous, broken speech. Albuminuria, increased arterial tension. Later, temporary apoplectiform seizures, modified Cheyne-Stokes's respiration; helplessness, more marked on right side; deviation of head and eyes to left. Later, seizure (apoplectiform) of left hemiplegia; and slight deviation of head and eyes to right. Right pupil larger, and right conjunctiva injected; bedsore on left natis, left limbs continuing palsied and rigidly contracted, sinistral epileptiform seizures. Loss of cutaneous sensibility and of superficial reflexes of left limbs. After this, bedridden, with more or less left hemiparesis, but finally about equally helpless on the two sides. Speech thick, muffled, mumbling, and slightly broken. Then, athetosis movements of right upper limb : occasional spasmodic twitch of right angle of mouth. Left pupil now the more dilated (the right *had* been so), diarrhœa. Gangrene of lung.

Necropsy (abstract). Circle of Willis atheromatous. Slight hæmorrhagic pachymeningitis over right base and convexity. Moderate changes of soft meninges, but marked thickening about vessels at base of brain. Only slight adhesion and decortication, almost confined to parts of the temporo-sphenoidal and orbital surfaces and marginal gyrus. Some cerebellar adhesion and decortication, on both surfaces near the median line. Frontal grey cortex atrophied, and chiefly so on the right side, where it was paler than elsewhere; and the white matter adjoining it slightly indurated. Cortex paler as a rule in right than in left cerebral hemisphere. Parietal cortex, less wasted than frontal. Large and granulated lateral and fifth ventricles. 4th v. much granulated. Right cerebral hemisphere 1 oz. less in weight than left. Spinal cord, softening of grey matter on the left side in the cervical region. Right lung adherent, and some gangrene in its lower anterior part. Some cardiac-valve-, and aortic-, atheroma. Kidneys, each $4\frac{3}{4}$ ozs.; pale, somewhat fatty-looking; in the right one a cicatrix, and encysted calcareous material.

Choreiform movements. Still another morbid kinesis of the second stage consists of distinct *choreiform* movements about the head and upper extremities, or more limited in their distribution. They usually last but a comparatively short time, and may readily pass unrecognized in a patient known to be restless. I have frequently noted them, and others have placed well marked cases on record.* The choreiform

* As "Journal of Mental Science," Oct. 1875, p. 421.

movements in g.p. may follow epileptiform seizures, or may be independent of the latter. Occasionally, the choreiform movements are wild, and the limbs are tossed about in the most irregular way. In several long-persisting cases published, sclerosis of the posterior-median, or of the posterior columns, or of the posterior portions of the lateral columns of the spinal cord, and even of the anterior parts of the latter; and other changes, such as spinal meningitis, were present : in one or two cases they were absent.

Case 6. Male, aged 36, subject to recurring apoplectiform seizures. The expansive mental symptoms were associated with, and succeeded by, dementia and childishness; and, later on, there were mental depression, progressive muscular paresis, tremor cöactus of forearms and legs, only slight plantar reflex; but, still later, much tremulous twitch about limbs and head, and, finally, choreiform movements of the extremities.

At the necropsy, besides the more usual lesions of g.p.—consisting in this case chiefly of general encephalic hyperæmia and softening, with adhesion and decortication of all the gyri of the frontal and parietal convexity, and affecting also, but in less degree, the orbital and temporo-sphenoidal surfaces,—there were unusually well-marked changes about the cerebellum and base; softening of fornix, hyperæmia of spinal meninges about middle of dorsal region; softening and anæmia of spinal cord in the cervical and upper half of the dorsal region, but chiefly in the cervical, where the grey-matter was almost diffluent, and the antero-lateral columns, especially, softened.

Case 7. Choreiform movements occurred from time to time during a long period; there were also easily produced shivering, and muscular tremor and much tremulous jerkiness. Trunk sometimes bent to the right. Patient often cried, or bellowed; and had both expansive and hypochondriacal emotion and idea. Later, very shaky and tremulous, bending body forwards, constant restless, fumbling, displacing movements; "foul" habits, smiling and laughing inanely; constant choreiform movements. Death after severe apoplectiform attack.

Necropsy, abstract. Pial œdema; marked meningeal thickening, especially at and about interpeduncular space. Third cranial nerves tied down by false membranes. Much adhesion and decortication of cerebral surface, chiefly on the frontal lobes, the posterior part of the parietal lobes, and the temporo-sphenoidal lobes; slight on the orbital, and the mesial cerebral, surface anteriorly. Cerebral grey cortex somewhat reddish and atrophied, especially in front, of normal or slightly softened consistence. White substance of fairly normal consistence, except the slightly indurated layer of it adjoining the frontal grey; and here the grey separated from the white with abnormal ease. Right cerebral hemisphere 1 oz. less in weight than left, and right corpus striatum somewhat wasted. Tunics of *spinal cord* adherent *inter se*, from old spinal meningitic change. Right kidney irregular cirrhosed; in its dilated hilum and calyces, a branched renal calculus.

Case 8. Both choreiform movements and athetosis. Died aged 40; g.p. of the demented form; syphilis 16 years before death. Right hemiparesis before admission. Speech stumbling, clipped, and quivering. Whilst under observation for 13 months, there were, successively, paresis of right lower limb and of left side of tongue : right othæmatoma; convulsion and spasm, usually affecting the right side, but occasionally the left; and followed by paresis. Then, twitches of right upper limb; followed by *choreiform* movements. With the latter were jerking of right leg and right side of the face, upward twitch of the right angle of the mouth, almost constant twitch of eyelids. Right conjunctiva bloodshot. Later, hand and arm twitching; and flexing and strange wandering movements were superadded. The patient made " bad shots" in attempting to seize or grasp any object. Restless, fumbling, movements also observed. Right hemiparesis promptly followed.—Afterwards, much right hemispasm, without loss of consciousness, and chiefly affecting upper limb, and hand; and therewith facial spasm, chiefly on the right side; the right upper limb occasionally rigid, and usually the less resistant to passive motion. Frequent recurring temporary right hemiparesis. Anæsthesia. Analgesia, marked. Bedridden for months.

Extreme changes of meninges, also of the enlarged ventricles. Slight induration of the cerebral grey cortex in the superoexternal fronto-parietal region, gradually diminishing thence downwards and backwards. Atrophy of frontal cortex, diminishing thence to occipital tips. Slight redness of grey cortex, paleness of white substance; white slightly indurate. Adhesion and decortication slight, and mainly about the summits of the fissures of Rolando, especially of the right. Opto-striate bodies somewhat shrunken, the left corpus striatum perhaps the more so. Left cerebral hemisphere 1 oz. less weight than right. Pons and med. obl. slightly indurated.

Reflex action. As already mentioned, even in the middle and later stages of g.p. it often is to be noticed that a patient starts back and blinks the eyes when a "dab" or rapid motion is made towards them suddenly, or he is startled into sudden lively movement by a touch. This leads to a detailed investigation of the so-called,

REFLEXES, *superficial and deep.* With regard to these the greatest possible variety obtains in g.p. The superficial or skin-reflexes, as well as the sensibility to local impressions, may be enhanced in the early and middle stages.

Late on, in chronic cases, however, the reflex activity connected with surface-impressions is often lowered or lost, and reaction to any peripheral impression is tardy. But a contrary state of things may exist.

The *plantar reflex* varies at different parts of the course of a given case, particularly in one interrupted by the special seizures. In many cases it fails, especially towards the close of life. In a large group of male G.Ps. I found the plantar reflex either much lessened or abrogated in 43½ per cent., and, on the other hand, decidedly enhanced in 17½ per cent. An enhanced plantar reflex is usually an accompaniment of exaggerated knee-jerk, if the knee-jerk be abnormal; but I have seen it associated with absent knee-jerk in the middle part of the course of a case with posterior columnar grey degeneration of the spinal cord, in g.p.

The *skin reflexes generally*, the reaction and sensibility to various species of surface-irritation made on different portions of the skin, but especially on hands and feet, were found by me to be fairly within the limits of the normal in the small majority of cases, but in 41 per cent. they appeared to be diminished or lost; in 4 per cent. exaggerated. It is not easy, however, to fix the limits of the normal with respect to this phenomenon.

Reflex excitability to heat, reaction to the application of a spoon heated to nearly 212° F., to the soles of the feet of G.Ps. was absent in 67 p.c. ("West Rid. Asyl. Med. Rep.," Vol. vi, p. 139).

The *conjunctival reflex* is lessened on the hemiparetic side, after epileptiform attacks, or when apoplectiform seizures are attended with temporary paralysis. This condition clears up more or less as the paralysis wanes. Temporarily, after severe seizures, the conjunctival reflex and sensibility may be entirely abolished. *The iridal reflexes* will be discussed when speaking of the eye. Simple direct iridoparesis, or even, as sometimes seen, absolute iridoplegia, is present in almost all cases. Of a large group of G.Ps. I examined, in only 10 per cent. was the iridal reaction to impressions of light sufficiently good to be called "fair." Whereas in none of the same group was the associated contraction of the pupils in accommodation entirely lost, although lessened in some.

The *direct quadriceps femoris contraction;* and the *Achilles tendon-reflex* are less often abolished than the patellar.—On the other hand, *front-tap-contraction* is sometimes seen: a morbid spastic element.

Patellar tendon reflex. Knee-jerk. I have condensed the results of my clinical observation on this subject in an article on the "knee-jerk in general paralysis;"* from which I quote here.

The knee-jerk may be normal; or may be increased; or may be diminished, or absent, in g.p. In my own cases, all males, those within the assumed normal range, and those with absent or almost absent or lessened knee-jerk, each outnumber those with very decidedly exaggerated knee-jerk. In some examples a little difficulty was experienced in deciding as to whether the jerk was slightly exaggerated, or was within the normal range. Upon the decision as to this point would depend some modification of the statistics.

Then, again, the knee-jerk may vary widely in the same patient at different portions of the course of g.p. For example, in cases where the knee-jerk has been absent one may find a slight return of the phenomenon months afterwards. And I have watched a knee-jerk pass, in the course of time, from a condition of complete annulment, through one of slight, and then of moderate, evincement, to one of exaggeration, although not altogether *pari passu* in the two limbs. In g.p. I have also seen the knee-jerk, examined at intervals of several months, appear on one occasion exaggerated; later, abnormally slight; and, later still, within the normal range; again to rise to exaggeration; and afterwards to sink within the normal limit. Where the fluctuations are less considerable than in this last case, it may yet be noticed that a knee-jerk, highly marked at one time, is less so on a future occasion. The jerk is by no means always alike, or nearly alike, in the two limbs. It varies much in its quickness in different cases; also in the promptitude with which it follows the stroke. I have seen a knee-jerk, in g.p., exaggerated in its extent or range, but tardy or delayed in its appearance.

The activity of the reflex movements in response to the various superficial impressions, such as tickling and pinches of the integuments of the limbs, by no means necessarily varies directly as the activity of the knee-jerk, or of other so-called deep-reflexes.

Point by point, comparing the cases having more or less exaggerated knee-jerk with those having absent or very

* "Journal of Mental Science," Oct., 1882, p. 343.

slight knee-jerk, and with those wherein it is normal—each group being taken in this collective manner—certain differences between them are found. It will only be necessary to compare the group in which knee-jerk was absent, or very slight, with that in which it was highly marked or exaggerated. Stated in abstract here, the principal differences were :—

More of early, or severe, or persistent, pain ; *more* of wet habits ; *more* of hallucinations ; and slightly more of ataxiform gait and epileptiform seizures in the group with *no knee-jerk*; and, in the same group, *less* sensibility and reaction to pinches and tickling of the feet.

And—*more* of quasi-syncopal seizures, and very slightly more of hemiparetic and apoplectiform attacks, in the group with *exaggerated knee-jerk.*

In 15 cases Dr. A. Joffroy * found knee-jerk normal in 7 ; somewhat exaggerated in 2 ; completely abolished in 4 ; diminished in 2.

With reference to the relation of the gait here to patellar tendon-reflex ; the following are the examples which happen to be the most conveniently at hand, among my notes, of several cases under my care of (*a*) absent, and of (*b*) exaggerated, patellar tendon-reflex.

Knee-jerk absent. (Besides, and purposely omitting here, cases of g.p. with marked posterior grey degeneration, and "tabic" symptoms.)

Case 9. Gait unsteady, ataxic, swerving, the heels are slightly brought down first, the feet are thrown forwards and clumsily, turning around is unsteady and unsafe. Patient looks to ground in walking. Left side paretic. At times diplopia and right external strabismus.—Spinal dura thickened irregularly over its posterior surface, and arachnoid even more so. White fibrocartilaginoid plates on the spinal arachnoid. Adhesions between spinal meninges. Myelitis of posterior columns of spinal cord ; neuritis and partial grey degeneration of right motor oculi nerve.

Case 10. Gait jerky, feet widely apart, and thrown out in walking, heels brought down a little. Bent and quasi-aged bearing and attitude ; head hangs.—Spinal vessels congested. Meninges thickened posteriorly, especially in cervical region ; adhesions between the two layers of spinal arachnoid, spinal pia thickened, rough, especially posteriorly, and beset with delicate granulations. Softening of lower portion of cord. Left anterior grey cornu smaller and paler than the right one.

Case 11. Gait at first heavy and paretic ; subsequently becom-

* " Arch. de Physiol.," No. 3, 1881, p. 474.

ing somewhat ataxic. Attitude bent and slouching. Condition modified by a suddenly-appearing and painless shortening and eversion of right lower limb, after which the right foot was planted heavily on the ground in walking.

Case 12. In another case of absent or slight knee-jerk, the posterior columns of cord unduly indurated, and turning of a somewhat reddish grey colour on section and exposure to air, especially in the cervical region.

Case 13. In another, the spinal cord pale, atrophied, the posterior columns of firmer consistence than the others. Unusual degree of brown staining of meninges over medulla oblongata.

Case 14. In another, a blod clot adhering to posterior surface of spinal cord in cervical region. Spinal veins congested. Sclerosed patches in medulla oblongata and pons. Cord fairly normal to eye, not microscopically examined.

Case 15. In another, spinal cord hyperæmic, particularly its grey matter. Posterior columns of cord slightly indurated, but not atrophied.

Knee-jerk exaggerated.

Case 16. Steps jerky; feet heavily planted; gait swerving, irregular, spasmodic; much spontaneous twitching and jerking of legs. Some of the superficial reflexes increased, *e.g.*, reaction to touches on feet; once slight ankle-clonus; grasp of hands jerky, tremulous. Frame bent forward. Bowels constipated, urine retained, owing to vesical paresis. Hemiplegic attacks recurring, and on either side, on different occasions. Only latterly any anæsthesia or analgesia of lower limbs. Pupillary changes not extreme.—Adhesion and decortication not extensive, and chiefly on first frontal g., and anterior part of first and second temporal g. Brain flabby, palish; frontal cortex atrophied. Fourth ventricle moderately rugose.

Case 17. Bent frame; slow irregular steps. Hemiparesis, affecting first one side and, later, the other, and recurring; at different times paresis, first of one, then of the other, third cranial nerve. Muscles jerk much in passive motion. Spasm in limbs, marked *tremor a debilitate*, slightly scanned speech. Later, left upper limb stiff and heavy. Final violent, jerking, quasi-shivering movement especially on the right side. Spinal dura adherent to arachnoid, &c. Spinal cord very firm in its lower, but softened in its upper, part.

Case 18. Steps long, shaky, somewhat convulsive, holds frame stiffly; urinary incontinence; pain in left arm and in legs, following severe pains over back, loins and chest. Spinal meninges opaque, thickened; spinal cord firm in dorsal region.

Case 19. Much facial and lingual twitch, frame bent forward. Recurring apoplectiform seizures and left hemipareses. Latterly, rigid, hemiplegic left limbs; impaired sensibility of feet; exaggerated left knee-jerk.—External pachymeningitis; other well-marked ordinary meningeal changes; adhesion and decortication slight; atrophy chiefly in frontal, moderate in parietal, lobes, and more in right than left hemisphere.

Case 20. Fluctuating right hemiplegia, right lower limb stiff and rigid in passive motion, occasional twitches and jerks; right arm stiff and helpless.—Atrophy, predominating in left cerebral hemisphere; spinal meninges thick and roughened; asperities on upper part of pia, lateral columns of cord indurate, with grey degeneration of right pyramidal tract.

Taking these, and other, cases I find, on the whole, that with exaggerated knee-jerk in g.p. there are :—

Often, rigidity of frame, which is bent forward, the patients holding themselves stiffly.

Often, spasmodic twitches and jerks of the limbs, occurring either spontaneously or on passive motion.

Often, the grasp of the hands jerky, tremulous.

Often, the tongue protruded jerkily, and with much twitching action of face and lips.

Often, the speech shaky and accompanied with very much twitching action of face and lips.

Often, apoplectiform, and "simple" paralytic seizures.

Often, the patients demented, depressed; often intermingled or alternating expansive symptoms.

Quasi-syncopal seizures comparatively frequent.

Atrophy of anterior two-thirds of brain, especially of frontal region; adhesion and decortication, varying from a slight to an extreme amount; in some, slightly affecting cerebellum. In some, softening of circum-ventricular part of cerebrum. Spinal meninges often thickened, adherent, granulated. Diffuse myelitis, lateral myelitis, or of pyramidal tracts only; or myelitis or degeneration causing softening of upper part of cord, induration of lower.

Dr. J. C. Shaw (U.S.) * asserted a direct connection between difficulties of speech, also hemiparetic attacks in g.p., and exaggerated tendon-reflex. It will have been seen that my observations, at the least, only partially agree with this, and Dr. E. C. Spitzka's † experience differs from Dr. Shaw's.

Dr. G. H. Savage ‡ has found lateral sclerosis of the cord (and increased knee-jerk), in g.p., chiefly in young, unmarried men who had a springy gait, very thick speech,

* "Archives of Medicine," 1881.
† "American Journ. Neurology and Psychiatry," Aug., 1883.
‡ "Journ. of Mental Science," April, 1884, p. 57.

much tremor of lips and tongue, marked restlessness, increased reflexes, marked grinding of teeth; finally, contracted limbs, bedsores, muscular wasting, &c. Comparatively little adhesion and decortication of brain; and, besides the lateral sclerosis, much atrophy of the cord.

Dissociation of spinal symptoms in g.p. It is an important feature in g.p., that owing to the complexity of the histological changes, and to their great variety of severity and distribution, in different cases, the signs usually associated in the "systematic" affections of the spinal cord, and forming a definite symptom-complex, are often dissociated in the spinal affections of g.p. I find my space will not permit of the insertion here of cases I had prepared in illustration of this point; but I may briefly refer to several examples of this analysis and recombination on new lines, in g.p., of symptoms of systematic affections of the spinal cord.

Case 21. In one example, with great exaggeration of the reflexes generally, the knee-jerk could not be spoken of as exaggerated; and during part of the time was even below the average. Another kind of dissociation occurred in a case *(Case* 22) of g.p., following tabes dorsalis ; in which, at the necropsy, were grey degeneration of posterior median columns throughout, and of posterior columns in the dorsal region, where, also, the posterior roots of the spinal nerves were grey; the grey induration becoming narrower in the cervical part; and chronic, and chiefly posterior, spinal meningitis being found. Typical brain-changes of g.p.—During the last months of life, with tabic gait, incontinence of urine and occasionally of fæces; with equal, small pupils, still acting moderately to light, the patient, seated, continually straightened and extended all the limbs, the forearms being extremely pronated, wrists flexed, and hands abducted, with the palms directed outwards, backwards, and even somewhat upwards. Twitches of the arms occurred with voluntary movements, or independently of these. Movements of the straight, extended, and somewhat rigid lower limbs were attended with twitches and jerks; and slight irritation, or passive motion, of the legs brought these on as well. The right foot was in the position of mild talipes varus. Finally, slight ankle-clonus appeared. Speech and tongue much affected, as in g.p. Some analgesia and anæsthesia, of lower limbs especially. The dissociation here is an example of some reflexes increased, and, even a degree of spinal epilepsy, on the one hand—. and, on the other, absent knee-jerk.

Or take an instance *(Case* 23) such as may be referred to in a single note : thus :—" Aug. 12, knee-jerk almost absent; the stroke is followed only by a slight twitch of the extensor digitorum, and slight jerk of great toe. There is no ankle-clonus when tested in the ordinary way, but on pinching the skin of the leg there are slight convulsive and tremulous movements of flexion and extension of the foot; tickling of the soles of the feet occasions only a slight response."

CHAPTER VI.

EYE-SYMPTOMS IN G.P. OPTIC NERVE, OCULAR SYMPATHETIC NERVE. 3RD, 4TH, AND 6TH NERVES.

Eye-symptoms in g.p. are of several orders, but it is convenient to consider them together. Inequality of the pupils is frequent. This, and their irregularity in shape, may be constant or intermittent; the conjunctiva is in some cases injected, and, later, perhaps, an irregular dilatation of the pupils may occur. But mydriasis has been far less common, at the early stages, in my practice than in that of some who have written on the subject, and the pupils may then be normal or nearly so. The average size of pupil in g.p. seems to be about $3\frac{1}{4}$ mm.

Almost always in the later stages the pupils are sluggish, irregular in shape, unequal in size, and varying from time to time in these respects. Inequality of the pupils, discovered by Baillarger, was found by Moreau in 58 per cent. of G.Ps., by Mobèche in 61 p.c., by Mendel in about 60 p.c., by Siefert (cited by Förster), and by Boy (cited by Mendel), in two-thirds of the cases; by Doutrebente* in 77 p.c.—in two-thirds of these last the left one was the wider.—Very few cases, watched closely throughout the whole course of the disease, fail to show pupillary inequality at one time or other. I do not recall a chronic case, so watched, with equal pupils continuously, throughout. In 100 patients at the Bicêtre, Moreau † observed inequality of the pupils in 58 p.c., the pupils in 24 p.c. being greater on the right; in 34 p.c. greater on the left, side; and in 42 p.c. equal. In 26 p.c. the pupils were large; in 56 p.c. of medium size; and in 18 p.c. small. Increased convexity of ocular globe in 66 p.c.: in 51 p.c., odd shape of the eyebrows, which were separated at their inner ends, or abandoned the arch at its middle, to rise on the forehead or fall like a moustache over the eyes; in one-third a bluish, sclerotic tunic; often long upper eye-lashes, and short, thin, lower ones.

* "La France Médicale," May 5th, 1880. *Vide* "Journ. Nerv. and Mental Dis.," July, 1880, p. 553.
† "L'Union Médicale," No. 78, T. vii, July 2, 1853, p. 310.

Many years ago, the sweeping generalization was made by Mr J. T. Austin * that when the patient was depressed and melancholic the right pupil mainly or almost solely was affected; and that the left was the one principally involved when the patient was maniacal and the subject of exalted delusions. Also, that the rule was confirmed by the variations of the unsymmetrical pupils. Thus :—a depressed G.P. becomes more melancholic, his slightly dilated right pupil dilates still more; he improves, the dilatation recedes:—and similarly with the exaltation of the exalted G.P. and the left pupil. Finally, that the rule was confirmed in those G.Ps. whose pupils are usually symmetrical, but who have sudden accessions of melancholy or of exaltation. I am not aware that these statements have received any corroboration, except by Mr. Thurnam. While many cases give some support to this view, there are not a few which are in direct opposition to it. Adding together the numbers at the moment before me, as given by several observers, it appears that in 101 G.Ps. the left pupil was the larger, and in 100 the right.

The contributions of Dr. Mobèche † contain a detailed consideration of the ocular, and especially of the pupillary, conditions in g.p. Deviations from the usual contracted condition of the pupils during sleep were found by Dr W. Sander ‡ in g.p. Either the contraction did not take place, or a dilated pupil dilated still more during sleep, that of the other eye contracting as much as, or less than, usual; or a pupil, contracted during the waking state, became dilated as sleep set in, as if the sphincter iridis was stimulated during the waking state.

The first edition of this work contained (at pp. 54-55) a summary of the results of an investigation I made in 1872 into the pupillary state in g.p. In different stages and conditions of g.p., the state of the pupils was minutely noted as to symmetry, size, shape, and mobility to light. It is, perhaps, unnecessary to repeat the entire summary; and here it will merely be stated that it showed that:—

(a.) Where the patients are at one time excessively elated, and at another extremely depressed, the condition of the pupils was alike, *as a rule*, under the two mental conditions.

* " A Practical Account of General Paralysis," 1859, p. 34.
† " Annales Médico-Psychologiques," Nov., 1874, p. 325; Jan., 1875, p. 19. See, also, Lasègue, " Thèse d'agregation," 1853.—Austin, *op. cit.*, 1859.— Marcé, " Traité des Maladies Mentales," 1862.—Billod, " Ann. Méd.-Psych.," 1863.—Voisin, " L'Union Médicale," 1868.
‡ " Archiv für Psych. u. Nervenkrankheiten," ix. Band, p. 129.

108 *Pupils; Changes in symmetry, size, shape, and mobility.*

In several such cases they were dilated, very sluggish, but not differing very much in size.

(β.) In the period succeeding that of acute excitement, or into which patients drift from early expansive delirium, namely, the period corresponding to the second stage of *confirmed* g.p., I found the following conditions of the pupils:—Under each head the several pupillary modifications are mentioned in their *order of relative frequency*. These modifications are described under the heads of *symmetry, size, mobility,* and *shape*.

Symmetry . Pupils equal, but one or the other *occasionally* the larger.
                Left pupil the larger.
Size . . Pupils rather small. ⎫
                „ of natural size ⎬ in equal numbers.
                „ much dilated ⎭
Mobility . Pupils very sluggish, or almost immobile, persistently.
                Pupillary sluggishness varying much in degree from time to time.
                Pupils moderately sluggish.
                „ slightly sluggish.
Shape . . Pupils,—both irregular.
                „ mutably irregular.

(γ.) In *quiet* and considerably demented general paralytics the conditions were:—

Symmetry . Pupils equal ⎫
                Right pupil *persistently* tending to become the larger ⎬ in equal numbers.
                Left pupil *persistently* tending to become the larger ⎭
                Pupils equal, or the left larger *at times*.
                „ „ „ right „ „
Size . . Pupils smallish.
                „ about natural size (relatively to light).
                „ dilated, both, or one only.
                „ small.
Mobility . Pupils, a mutable degree of sluggishness ⎫
                Pupils somewhat sluggish ⎬ in equal numbers.
                „ moderately sluggish
                „ very „

Shape . . Pupils, both irregular.
„ mutably irregular.
„ only one irregular.

(δ.) In *extremely* demented cases the conditions were very much the same as those last-mentioned (under " δ "), but, on the whole, with slightly more dilatation and sluggishness.

According to this summary, as the middle and later stages of g.p. advance towards the close of the scene, there are, in the cases taken collectively, more and more irregularity in shape, inequality in size, dilatation, and sluggishness to light. In a given eye, and under the same amount of light, the pupillary size and shape, and to some extent the degree of iridal sluggishness, may vary much from day to day, or even from hour to hour, the changes sometimes being very rapid. On the whole, and even in the later periods of g.p., I have found less pupillary dilatation than have several observers. Nevertheless, pupillary dilatation is a frequent condition. After unilateral convulsions in g.p. the pupil on the opposite side is usually, but not always, dilated for a time, or intermediately; and I have often seen the pupils dilate at the commencement of the epileptiform seizures.

Speaking in general terms, I think the course of events is usually this. The pupils become somewhat sluggish to light, this increases and some failure of their movement in accommodation often supervenes, the tendency being for the one failure to follow the other at a considerable interval. The latter failure, however, may not follow; and with immobility to light, but normal activity in accommodation, reflex iridoplegia, pure and simple, is present; or, as one might say, iridal reflex monoplegia, failure of the contraction of the sphincter iridis under the influence of light or stimulation of optic nerve. If, and when, to this reflex iridoplegia there is added cycloplegia, or failure of ciliary muscle in accommodation, internal ophthalmoplegia is present. With cycloplegia usually comes failure of the contraction of sphincter iridis during accommodation.

Skin-reflex dilatation of the iris is sometimes lost, there being, then, a failure of the normal stimulation and contraction of the dilator fibres of the iris when the skin is sharply and painfully irritated. Dr. Buccola,* of Italy, found it absent in seven out of fifteen cases, and delayed in the remaining eight; and Mr. Bevan Lewis † thinks this the

* " Journal of Mental Science," Jan., 1884, p. 59.
† Ophthalmological Sec. of United Kingdom, June 14, 1883.

most frequent of the internal ocular paralyses in g.p., having observed it complete, either in both eyes or in one eye only, in 74 p.c. of the cases, incomplete in 16 p.c. (or, respectively, in 35 p.c., and 28 p.c., in another group of cases *), and it being absent or sluggish in many cases when the reflex action was normal in the legs : loss of sympathetic or indirect light-reflex, complete in both eyes or one eye in 50 p.c., incomplete in 19 p.c.; loss of direct light-reflex, complete in 46 p.c., incomplete in 31 p.c.; loss of movements in accommodation, complete in 24 p.c., incomplete in 17 p.c.

Thus, there are irregularities in the shape of the pupil, which is often pyriform, or oval, or with notched, irregular, or dragged edges ; inequalities in size, and dissimilarities in shape, between the two pupils ; variations from the normal in size of pupils; changes in iridal and ciliary activity, of several kinds :—as impairment or loss of (*a*) skin-reflex dilatation; or (*b*) of light-reflex contraction; or (*c*) of iridal contraction associated with accommodation ; or (*d*) of ciliary muscle contraction in accommodation; or (*e*) of indirect light-reflex contraction, in opposite eye.

A comparative examination of the two eyes, to determine which is the abnormal, or more abnormal, one, should not be neglected. The one acting the more sluggishly to light, or in efforts of accommodation, is the more abnormal one. But the iridal differences between the two eyes are not always alike ; in the same patient the right pupil may sometimes be the larger, and sometimes the left; and, again, at other times the pupils of equal size. The influence of atropine, eserine, and of other drugs ; of sharp local unilateral pain; of old corneal changes ; of old iritic adhesions ; or of former operations, as iridesis or iridectomy; must all be excluded before any iridal abnormality is attributed, as a symptom, to the g.p. itself.

Myosis.—Mydriasis. Already has mention been made of pin-head pupils in the prodromal period, not infrequent in the ascending form of g.p. Narrowness of the pupils is more common in the earlier, dilatation of the pupils in the more advanced, stages. But with regard to the former (myosis), it may exist throughout, and examples have already been given of its occurrence at the end of the morbid drama : and with regard to the latter (mydriasis), it may occur at the very first, or in the prodromic stage, as in some examples

* " West Rid. Rep.," vi, p. 140.

mentioned by Mendel;* in one, the first prodromal sign of g.p. was sudden dilatation of the right pupil, followed, two years later, by dilatation of the left, and this, a year and a half still later, by mental disorder:—in another case, dilatation of the right pupil, sciatica, and rheumatism preceded the psychical disorder by one and a half year; then came an apoplectiform attack; death in a year and half; nothing unusual to g.p. at the necropsy. In another case, with right pupil dilated throughout the g.p., the necropsy, besides the ordinary changes, revealed a tumour in the upper part of an anterior central gyrus. Dr. W. A. Hammond has seen dilatation of one pupil precede g.p. by several years. Extreme myosis, followed by marked mydriasis, is of evil omen in g.p.

The pupillary condition by no means always harmonizes with that of the optic nerve. Thus, with amaurosis of both eyes, Magnan† saw an example in which only one pupil was dilated; and in an example of g.p. with tabes dorsalis the left eye was blind, but the pupil unaffected; the right eye was not blind, but its pupil was largely dilated. For the pathological basis of pupillary inequality, myosis, and mydriasis in g.p., see the chapter on pathological physiology.

External ocular palsies. Palsies of extrinsic muscles of ocular globe may also occur, giving rise to some form or other of strabismus, disorder of vision, diplopia, and, perhaps, ptosis; the symptoms of which several conditions it is unnecessary to describe, but which occur in all degrees, from the slight to the severe, and are often transitory, variable, mobile. Most frequently, I have found extrinsic ocular paralysis associated with ataxic symptoms, tabic gait, and the degeneration or myelitis of posterior columns of cord, pointing to tabes associated with g.p. In some of the G.Ps. with these palsies I have found, at the necropsy, distinct neuritis—and in some cases even naked-eye grey degeneration—of the corresponding cranial nerves. Ocular palsies may be prodromal, or may precede the symptoms of g.p. by several years. Brierre de Boismont‡ long ago observed this with regard to ptosis, diplopia, paralysis of N. abducens and N. facialis; and similar observations have been made by Falret, Billod, Esquirol, Baillarger, Foville, the present writer, and others; and in eight p.c. of his cases Mendel

* *Op. cit.*, p. 150.
† "Archives de Physiol.," 1877, p. 840.
‡ " Ann. Méd.-Psych.," 1859, p. 295. " Ann. d'Hygiène Publique," &c., T. xiv, 1860, p. 405.

obtained a history of diplopia antecedent to the g.p., and usually lasting for some months. Pierret looked on these symptoms as of sensory origin, taking rise in disorder or irritation of trigeminus nerve. Voisin* stated that in some cases, in the last stages, ptosis is caused by granulo-fatty degeneration of the levator palpebræ muscle.

Exophthalmos. Some degree of exophthalmos is occasionally observed in g.p.; the globe is more prominent, and perhaps its convexity increased. The increased convexity of the eyeball has already been spoken of. In two cases out of 90 Doutrebente† observed exophthalmos.

Eyebrows. The irregularities of the eyebrows, often observed, have, partly, been already mentioned. The eyebrows often become unsleek and irregular from separation of the hairs, which rise or fall abnormally, while the arch of the eyebrow becomes distorted owing to paresis of adjoining muscles, or because pulled upon by the spasmodic, or by the defectively-antagonized, action of the occipito-frontalis.

Narrowing of lid-chink. Narrowing of the palpebral fissure may be present, and may be due either to a paresis of the sympathetic, with which the pupil is likely to be contracted, the eye hyperæmic, and the tears and palpebral mucus to be actively secreted. Or it may result from paresis of the levator palp. superioris, owing to affection of the oculo-motor nerve, and hence ptosis, with which the pupil would usually be dilated; or it may arise from spasm of the orbicularis palpebrarum, innervated by the facial.

Conjugated deviation of eyes and rotation of head often follow the epileptiform, and occasionally follow the apoplectiform, and some other seizures of g.p. The eyes usually turn *from* the side convulsed or hemiparetic.

Amblyopia and amaurosis. There is sometimes progressive failure of vision, the outlines and colours of objects lose distinctness. Thus amblyopia or amaurosis sometimes accompanies g.p., may affect one eye or both eyes, and, if both, is often more marked in one eye than in the other. The amblyopia here meant is independent of the visual weakness and impairment due to the morbid pupillary conditions already described, and is one which remains when the latter temporarily yield to treatment or remit. But, as already mentioned, amblyopia and even amaurosis occasionally precede g.p., and this even by some years.‡ Dr. Magnan (Gowers) also observed it two to four years before. Dr. Foville §

* *Op cit.*, p. 59.
‡ Brierre de Boismont, *loc. cit.*
† "La France Médicale," *loc. cit.*
§ "Annales Médico-Psych.," 1873.

described several cases, one of them had amblyopia, and two amaurosis, the g.p. being preceded by prog. loc. ataxy; in two others locomotor ataxy was absent, but amaurosis preceded g.p.—Three cases of amaurosis preceding g.p were vouched for by Dr. Mobèche,* and examples of a like sequence have been recorded by Lélut, Lasègue, Parchappe, Calmeil, and others. Impaired, weak, indistinct vision, narrowing of the visual field, colour-blindness, betray an affection of the optic nerve, which tends to end in blindness. Atrophy of the optic nerve is found not only in cases of g.p. with degeneration of the posterior spinal cord-columns, but also with those having changes in the other columns.

Hemianopsia in g.p. points to a local lesion of the brain, or of an optic tract, or it follows some of the "seizures." See also the two next sections.

Disorder and defect of vision. Prof. C. Fürstner† described a peculiar disorder of vision in g.p., occurring in one eye only, or almost so, and especially in the right. Single letters are recognized and named, but when joined to form a word the patient fails to read it. Simple objects cannot be counted. On closing the sound eye objects placed before the affected eye are only seen confusedly; if they are objects usually coveted they are now neglected; if bright they no longer rouse attention. When this visual disorder is present, the writing is irregular, the lines and letters misplaced or jumbled together, or the patient continues writing when the pen is off the paper, or begins at incorrect places; and the gait is more affected than when the sound eye is open. In these cases the ophthalmoscope reveals nothing special; the visual defect is independent of the dementia; it is not hemiopia, nor yet ordinary amblyopia or amaurosis; it may disappear, may undergo remissions, and perhaps have exacerbations after epileptiform and apoplectiform attacks. Sight avails for the coarser actions of social life, but not for its complex processes, and there is not abrogation of the connection betwixt the defective eye and the opposite cerebral hemisphere. Some of the necropsies showed local lesion of the occipital lobes, either of both, or of the one opposite to the affected eye; the optic thalamus also being lesed in the first of the cases. Others showed no marked occipital lesions; but atrophy, chiefly frontal. A similar case was soon pub-

* *Loc. cit.*, cases 89, 90, and 91.
† "Ueber eine eigenthümliche sehstörung bei Paralytikern."—" Arch. für Psych.," viii. Bd., s. 162; u. ix. Bd., s. 90.

lished by Dr. Reinhard.* With regard to this description by Fürstner, the condition appears to be much like that mentioned in the next paragraph, being apparently a phase of

Visual Imperception. Defect of Visual Recognition. Seelenblindheit. Hirnsehschwäche. Here the patient appears to be in a condition analogous to that experimentally produced by Goltz in dogs, when, subsequent to destruction of large portions of the cerebral cortex, the animals, after the temporary primary blindness had passed off, were in a condition of defective colour- and locality-vision, as if everything to them had a grey hue or cloud-wrapt appearance; a piece of meat appearing only as a dull grey mass, and the visual images no longer united, registered, and stored away in memory. The dog avoids obstacles in his way; within narrow limits, makes appropriate use of his visual impressions. But he no longer shows any fear when threatened by fist or whip; is indifferent to the presence of persons or of strange animals; fails to recognize food thrown to him, or finds it with extreme tardiness; will eat dog's flesh, which he would not do before the brain-mutilation. For these visual defects and disorders Munk introduced the name Seelenblindheit; while Goltz, their inventor, prefers the terms Hirnblindheit, or, still better, Hirnsehschwäche. From destruction of higher visual centres, many of the intellectual visual relations are abrogated.

G.Ps., in a more or less analogous state, show a similar defective perception of their surroundings and of the use of objects. But the more diffused cortical lesion in g.p., and the demented state of the patients, make the analysis more difficult here.

Dr. Carl Stenger,† partly following Wernicke, and taking up Munk's division into seelen- and rinden-blindheit, distinguishes two visual disturbances in g.p.; (*a*) loss of power of recognition, or ideal representation:—(*b*) loss of power of perception.

(*a*). With destruction of the centre for the former power is psychical blindness, the object (dimly?) seen is no longer understood,—a condition analogous to sensory aphasia, where the correctly heard and uttered speech is no longer understood.—(*b*) With destruction of perceptive centre is cortical blindness. Here the ideal representations may or may not

---

* "Archiv. für Psychiatrie," ix, Bd., p. 147.
† "Archiv für Psych.," xiii. Bd., p. 218.

coexist; so that, according as it varies in that particular the condition is analogous in the one case to motor aphasia (speech understood but not uttered); or in the other to total aphasia (words neither understood nor uttered).—In the former " (a) "—much of the cortex, or of the occipital cortex, is destroyed: in the latter—" (b) "—the cortex of occipital tip.

In some, the mind cannot elaborate sensory impressions; in some the weakened mind cannot correct sensory defects, an uncorrected confusion of vision, as Wilbrandt thought, arising from hemiopia or partial retinal defect of sight.

In some cases of g.p. Dr. Zacher* found (1) pure soul-blindness—especially after "paralytic" attacks—bilateral, and associated with dysphasia and with dextral motor phenomena, (cortical blindness being either uni- or bilateral).—(2). Visual disturbances on both sides, which are probably true hemianopsias; and may be connected with occipital disease.—(3). A combination of (1) and (2).—(4). Pure unilateral amaurosis?

Colour-blindness, or colour-anomalies in g.p. In many cases the dementia and mental confusion of G.Ps. make it useless and misleading to attempt to test their colour-vision. Jehn and Boy, and in this country Dr. Batty Tuke,† attach importance to local colour-blindness in g.p.; and Mr. B. Lewis ‡ found colour anomalies in 23 p.c. of the cases examined with reference to this point. In many G.Ps. failure to discern colour is more apparent than real, and their replies on the subject are largely the outcome of dementia, of inattention, apathy, or downright stupidity. In many cases it is a phase of hirnsehschwäche—soul-blindness—discussed in the last section.

In 1874 Dr. Jehn wrote that incipient atrophy of the optic nerve is often indicated at its commencement by disorder and defects of colour-vision; chiefly were the colours blue and green mistaken as white and grey; more rarely, the colours crimson and red. In one case green and red were continually confounded.

Visual hyperæsthesia may occur in g.p., but must be rare. Visual hallucinations have been treated of separately in the third chapter.

Some of the affections of sight already mentioned may occur in g.p. without any ophthalmoscopic changes. Such

* "Archiv für Psych.," xiv. Bd., p. 463.
† "British Medical Journal," 1877, Vol. i, p. 744.
‡ "Trans. Ophthalmol. Soc. U.K.," Vol. iii, 1883, p. 204.

are soul-blindness, dyschromatopsy, visual hallucinations and illusions, some cases of amblyopia. We now pass to :—

Atrophy of Optic Nerve: and Ophthalmoscopic appearances of fundus oculi. Atrophy of the optic nerve occurs in many cases of advanced g.p., and when in an extreme degree explains some of the examples of dim sight and of blindness in this affection; while impairment of visual power is often the result of optic neuritis antecedent to the atrophy.

The degenerative change in the optic nerve is at least frequent in those cases of g.p. with well-marked tabic symptoms, and with posterior spinal or nerve changes; in which, indeed, may be found all the ocular and optic conditions of tabes dorsalis. It may be found, also, with changes in the other columns of the cord, as has for some years been known (Förster, 1877), and recently Dr. J. B. Lawford* has published seven cases of optic-nerve-atrophy in g.p.; in one of which there was sclerosis of the posterior columns of the spinal cord, and in four others of which there was exaggerated patella-tendon reflex; and still more recently Dr. Jos. Wiglesworth† and Mr. T. Bickerton† have published a g.p. case with well-marked typical optic-disc atrophy, with absence of spinal symptoms during life, and with perfectly healthy microscopical appearances of sections taken from all regions of the cord after death.

Slight atrophy or neuritis of optic nerve, I have often seen in g.p. Extreme atrophy of one optic nerve, I have seen following an injury to the eye with destruction of vision, and preceding g.p. In one G.P., who had become blind, or almost blind, I found the optic nerves softened, and markedly atrophied: by the ophthalmoscope had been seen, four months before death, a somewhat bright, white, atrophic pallor of the optic disc, with a somewhat irregular outline, and the vessels very much diminished in size. The angular gyri were but little affected; and not the seat of any adhesion and decortication, which were present, and considerable in extent, in many other parts.—In another case, where g.p. followed cranial injury, the blindness and optic-nerve changes were connected with unusually pronounced meningeal alterations, which, relatively to their usual distribution in g.p., were particularly well-marked at the inferior surface of the brain, and anterior part of the skull-base—about the optic nerves, and about the inter-

---

* " Trans. Ophthalmol. Soc. U.K.," Vol. iii, p. 221.
† " Brain," Apr., 1884, p. 29; July, p. 178.

peduncular space, at which latter the thickened arachnoid formed a very firm bridging membrane—; while firm strands of meningitic false-membrane tied down the optic nerves and neighbouring parts. In both of the last cases the dura was slightly thickened.

Ophthalmoscopic appearances. The reports of the ophthalmoscopic appearances in g.p. have been extremely conflicting; some asserting that the fundus oculi appears healthy in almost every case; others that in almost every case it is of morbid appearance; and every intermediate gradation of conclusion has been formed by one or by another.

In the first edition of this work I quoted (with reference) Voisin's statement that (Prof.) Koestl and (Dr.) Niemetscheck had reported marked ophthalmoscopic changes in g.p. The reference to their paper was not given by the author whom I cited, nor by Mendel who repeated his statement. Having since then come across the original contribution * of those observers I do not see any special statement as to g.p. therein; and all the cases of insanity examined by them have been spoken of as if they were G.Ps. Bouchut, in 1866, had not found marked changes, and in some relatively few cases, that is to say in 2 out of 14, ophthalmoscopic changes were observed by Westphal,† and consisted of atrophy of the optic nerve papilla; while Galezowski ‡ and Voisin ‡ but rarely found them, their numbers being, partial atrophy of disc in 2 of 40 cases; although they observed in 2 cases aneurysmal dilatation, and in many, general tortuosity and congestion, of the central artery of the retina; also occlusion of one of its branches in one instance, in which was partial atrophy of the optic disc, and the arterial wall contained fatty granulations. In most cases Magnan traced changes along the central artery of the retina.

It was stated by Dr. Clifford Allbutt § that atrophy of the optic discs occurred in 41 out of 53 cases of g.p., and in 7 more the appearances were doubtful :—the atrophy attacking them directly and not descending from the optic centres; and the morbid change often becoming apparent at first as a

* " Vierteljahrschrift für die Praktische Heilkunde," xcv. Bd., p. 134.
† " Archiv für Psych." i. Bd., p. 83.
‡ " L'Union Médicale," Aug. 4, 1868, p. 180 ; and *op. cit.*, p. 112.
§ " On the Use of the Ophthalmoscope," &c., London, 1871, p. 393. " Medico-Chirurgical Transactions," London, Vol. li, 1868, p. 97. " British and Foreign Medico-Chirurgical Rev.," Jan., 1868, p. 147. " Medical Times and Gazette," Aug. 1, 1868, p. 117.

hyperæmia, with slight exudation, and this followed by whitening of the disc, proceeding from the outer edge inwards, the nerve finally becoming white and staring, or occasionally of slate colour and its edge sharply defined. If there has been much exudation the edges are for a time uneven. Moreover, he concluded that any proportional relations of the atrophy are with the pupillary states, and not with any ataxy of the orbital muscles, the pupils contracting in the early or hyperæmic stage, and dilating as white atrophy succeeds. This, however, was denied by Jehn, who found no strict relation between pupillary width and degree of optic nerve degeneration.

Dr. Tebaldi * found more or less atrophic optic changes in 19 out of 20 cases; and he held that there was hyperæmia and inflammation of neurilemma and consecutive œdema and atrophy of the optic nerve. Yet some observers do not mention hyperæmia or inflammation as preceding the atrophy.—Dr. L. Monti (in "L'Ippocratico") described serous papillary infiltration, and congestion of the papilla and retina, as sometimes found in g.p.

That the affection of the optic discs in g.p. commences by inflammation and slight exudation, and ends in atrophy, was observed by Dr. C. Aldridge.† The atrophy of the disc was most complete in female cases, and was most advanced in the left eye, as a rule. Some discs were pink and hazy, others deep hazy-red and slightly swollen, others white with a faint capillary tint, and, lastly, some quite white and atrophic. "The disease generally ends by one side of the disc, usually the inner, becoming white and atrophic whilst the inflammatory changes are still in progress at the other. In some cases, however . . . the disc has a white rim, and a very large and shallow excavation of an extremely pearly-white tint occupies the centre, the remaining portion being of a greyish pink tinge." Yet some have held these latter cases not to be morbid, but to concern a normal disc with well-marked physiological cup.

Dr. Jehn, of Siegburg,‡ including in his paper many cases communicated to him by Prof. Saemisch and Dr. Mandelstamm, found that out of 47 cases of g.p. there were four with atrophy of optic disc in both eyes; three with the same in one eye, one of these with notable excavation of

---
* " L'Ottalmoscopie nelle Alienazione Mentale," &c., Bologna, 1870.
† " West Riding Asyl. Rep.," Vol. ii, p. 223.
‡ " Allg. Zeitschr. für Psych.," xxx. Bd., p. 519.

the papilla; two more of beginning atrophy; while in six more eyes appearances rendered degeneration or atrophy of the optic nerve likely. In two were changes from former choroiditis, with normal disc and retina. In one case, choked-disc, with head congestion, and frequent epileptiform attacks; in one, descending optic neuritis; in one, retinal cloudiness, etc. Like others, he noticed that with the above changes the patients often retained good visual power: and doubted whether the atrophy usually began as inflammation. The degeneration was ordinarily in irregularly scattered tracts of the optic nerve itself, as Virchow found usual. Thus, the atrophy of the optic papilla may be secondary only, and to a degeneration of other parts of the optic nerve. In 1871, Dr. W. A. Hammond described atrophy of optic nerve in g.p., and anæmia of retina and choroid. In a case of amaurosis with atrophy of discs, followed by g.p., Mr. E. Nettleship,* at a time when sight had been failing for six months, and nine months before the patient was found to be insane, described grey-white atrophy of discs, with diminution of the retinal vessels, the discs being uniformly pale all over, and quite clean.

Dr. W. R. Gowers † states that most of the cases he examined in various stages of g.p., presented perfectly normal conditions; in one case, only, was there simple congestion of the optic disc.

Dr. Jos. Wiglesworth ‡ and Mr. J. Bickerton,‡ in 66 cases, concluded that at least two-thirds of the optic discs were normal; that they were decidedly abnormal in about 18 p.c. (about 5 p.c. more being excluded because the changes were mainly of glaucoma, or of retinitis pigmentosa); and the remaining 12 p.c. being doubtful. They met with four abnormal conditions of the optic discs;—(1) Simple hyperæmia.—(2) Hyperæmia with some amount of softening or blurring of the edges.—(3) Simple anæmia.—(4) Distinct atrophy. These lesions fall into two main classes; the one tending in the direction of slight neuritis, the other in that of atrophy; in the former class the slight, chronic, interstitial neuritis declares itself by hyperæmia of the discs, with softened indistinct edges, tending to be replaced by atrophy and disorganization of the nerve; there being great hypertrophy of the nerve's trabeculæ, and much

* "Ophthalmic Hospital Reports," Vol. ix, p. 178.
† "A Manual and Atlas of Medical Ophthalmoscopy," 1879, p. 163.
‡ "Condition of the Fundus Oculi in Insane Individuals." *Brain*, Nos. xxv.vi.

hypertrophy and hyperplasia of its neuroglia corpuscles, and this at the expense of the nerve-elements, which subsequently undergo atrophy.—In the other class the atrophy is primary *at the disc;* the changes probably taking place in the reverse order *at the disc;* the nerve-fibres being the first to dwindle (owing to destruction of nerve-fibres higher up by a chronic inflammatory process); and the connective-tissue elements subsequently undergoing increase.—As to refraction; (of 66), in eight cases they found hypermetropia; and in one case hypermetropic astigmatism; in seven myopia; and in two myopic astigmatism.

Dr. Borrysiekiewicz,* of 28 G.Ps., observed in eight retinal opacity; in three bluish discoloration of optic nerve; in one neuro-retinitis exudat; and in some (three ?) atrophy of optic nerve; altogether, changes in 15. In the first period of g.p. Dr. Ch. Duterque † found congestion of optic disc with varicose dilatation of the veins and retinal vessels. In the second period, œdema of, and around, the disc. In the third period, the disc small, flat, atrophied. Also, there may be choroidal atrophy, retinal hæmorrhage, highly refracting granulations on retina and choroid.

CHAPTER VII.

Sensory Symptoms (purely).

Hallucinations and illusions of all the following special senses were fully discussed under the head of " Hallucinations " in Chapter III.

Vision. For other affections of the *visual* sense, see the last chapter.

Smell.—*Anosmia,* incomplete or complete. Wasting and œdematous infiltration of the olfactory bulbs and tracts is present in almost every chronic case. This atrophy of olfactory bulbs and tracts, so usually seen in necropsies on G.Ps., would give *a priori* probability to the view that the olfactory function is correspondingly defective. And so it is, in the advanced stages, when the patient takes no heed of the most loathsome smells, and may even eat the most filthy rubbish.

* " Centralblatt für Augenheilkunde," 1883, p. 494. " Alienist," Apr., 1884, p. 329.

† "Ann. Méd.-Psych.," Sept., 1882, p. 211. " L'Encéphale," Jan., 1883, p. 120.

Dr. Voisin * stated that even in the first stage of g.p. the sense of smell was, in almost all cases, diminished or lost. This symptom existing on one side or on both sides. In 1879 † he maintained and accentuated his former view, holding loss or diminution of the sense of smell on both sides, or unilaterally, to be one of the best signs of the onset of g.p., and one of the most constant, earliest, and most persistent. He stated that it is present in nearly all cases ; belongs to no other disease than g.p.,—save in exceptional conditions, such as ozœna and its sequelæ, or old fracture of ethmoid—(whereas in simple insanity the sense of smell is rather augmented than diminished); is an early sign, distinguishable even before there is trembling of the tongue, or inequality of the pupils, or enfeeblement of memory; is easily made out; and continues even should the other symptoms not. The sense of smell, gravely affected in the first stage, is abolished in the second.‡ And he added that the examination should be made by means of some common odorous substance, as pepper. At about the same date, Dr. Allbutt § had mentioned atrophy of the olfactory nerve as occurring in nearly every case of g.p.; and Westphal mentioned examples of atrophy of olfactory nerve preceding g.p. Dr. Jehn found weakening of the sense of smell in 11 out of 17 cases in which it could be tested. Dr. W. A. Hammond ‖ follows Voisin with regard to this point.

Although at the time I had specially investigated the subject, I merely stated in the first edition of this work that "my own experience differs from M. Voisin's as to this" (1st ed., p. 16).

In examining patients on this point allowance must be made for the state of dementia in which these patients are, and in consequence of which one often cannot arrive at any conclusions as to the function of smelling. Many, also, will bravely assert that they perceive " a strong smell," without being able to say what the odorous substance is. Of these, again, some, at the next moment, when questioned, will make precisely the same assertion as regards an inodorous substance, or as regards an empty bottle held to their nostrils. Others make guess-work as to the smell. In some instances there is evidence, not so much of defect as of perversion of the sense of smell. Thus one G.P.,

* "L'Union Médicale," Aug. 4, 1868, p. 180.   † *Op. cit.*, p. 39.
‡ *Ibid.*, p. 115.   § " Medico-Chirurg. Trans.," 1868, p. 97.
‖ " A Treatise on Insanity in its Medical Relations," 1883.

whom I examined with due precautions, described the strong smell of sliced raw onion as "myrrh and camphor;" and, again, as "gin and water," adding, in the manner usual to the expansive phase, that they both were "fine smells."

Many perceive an actual odour, distinguishing accurately between an odour and no odour; but cannot name or describe the odour they perceive, cannot truly *recognize* it. Some seem very rapidly to lose or exhaust their power of distinguishing smells; at the first trial or two answering correctly, but a few moments later, being unable to recognize or name any odour, even when tested with the same ones.

In some comparatively advanced cases, however, the various odorous substances used as tests are accurately perceived, recognized, and named by the patient.

With regard to pepper, it may cause lachrymation, or sneezing, or both, in those G.Ps. who cannot recognize and name any odour; and this often as well as in those who can. Yet in many cases the reflex sneezing and lachrymation are entirely absent, and no effect results from the application of an open box of strong pepper before the nostrils. With regard to this matter, one must not mistake an effect on the fifth cranial nerve for one on the first.

The substances chiefly used by me in testing G.Ps. have been onions, camphor, pepper, and some inodorous substances: whichever was used was hidden from the patients' eyes, in a wide-mouthed receptacle, applied to the patients' nostrils. For some further details concerning this sense see the next section, on taste.

Taste. The sense of smell is so closely allied with that of taste in many respects, that defect or disorder of the one is often attended by that of the other, and for the enjoyment of savoury food a healthy condition of the sense of smell is essential.

The mental failure of the patients must be kept in view; yet their garbage-eating is often an indication of a failure of the sense of taste. The manner in which they are apt to eat has already been described. Dr. Jehn, in 20 G.Ps. found the mental state prevented accurate testing: in 11 there was no alteration of taste or of smell: while in 6 there was more or less weakening of these two special senses.

The perversions of taste and of smell are of great practical importance as concerns their consequences, leading as they may, and so often do, to the refusal of food. And the failure of taste-sense, which, combined with the mental degradation of the patients, leads them to swallow filth in the latest stages,—unless this is carefully guarded against—would also be prejudicial to the patient's health.

When large portions of the cerebral cortex were removed in dogs, there was found much permanent damage to the faculty of perception through the senses of smell and taste. Whereas dogs naturally and normally turn with aversion from the eating of dog's flesh, the dog-subjects of the experiments mentioned no longer disdain to eat it. Whereas tobacco-smoke and chloroform vapour are disliked and avoided by the unoperated dog, the operated one inhales them with comparative indifference. And a somewhat analogous condition obtains in some cases of g.p., with much loss of cortical substance.

Audition.—Auditory *hyperæsthesia* is occasionally observed in g.p., and chiefly at the beginning of the mental derangement. Usually it is only one indication of an extended hyperæsthetic tendency, and of a hyperactive condition of the cerebrum, and over-lively play of its functions.

Deafness. The hearing may become obtuse in the course of g.p., or complete deafness may occur. Impaired hearing and deafness have been spoken of by Brierre de Boismont (*loc. cit.*) as prodromic in some cases. But obtuseness of hearing, or deafness, occurring thus, may, in some cases, be due to an antecedent disease unconnected with g.p., and in some G.Ps. impairment of hearing may result from plugs of wax left neglected in the ears, or from substances thrust into the meatus by the patient, the impairment of hearing being then a casual incident of the malady. In other cases the connection with g.p. is not casual but is occasioned by nerve- or centre-changes associated with the lesions of the latter.

Case 24. In one case where I observed marked auditory hallucinations, incomplete deafness came on at an early period. A convulsive seizure had first roused attention. This was followed by failure of mental powers, ending in complete dementia; by defective speech; and impaired hearing and motor power; and asthenia. Calvaria very thick and dense. Meningeal changes well-marked. Adhesion and decortication over the frontal gyri, over the anterior part of parietal lobe, and to a moderate degree over the temporo-sphenoidal gyri; not involving the occipital lobe, but strewn in patches over the mesial and inferior surfaces of the cerebral hemispheres. The brain was softened and hyperæmic, as were also the opto-striate bodies, and even the pons and med. obl. were of lessened consistence. Kidneys slightly granular.

Case 25. In another case the impairment of hearing was of a different origin. Syphilitic history;—demented form of g.p.; local pareses; hemispasm on one side, or on the other; the pareses chiefly on the right side. Later on, twitches of right upper limb, followed by choreiform movements. Much hemispasm, either on right side, or occasional partial hemispasm of left, without loss of consciousness. Right conjunctiva congested. Old right othæmatoma. Later, frequent spasms, rigidity, and pareses, of right upper limb; anæsthesia, analgesia, bedridden: injury, and gangrene, of a right finger. Pus from right ear, submaxillary swelling, deafness. Finally, drowsiness, right hemiplegia, slow and feeble pulse, very low temp. (90° F).—Calvaria osteoporotic. Brain-tissue for the most part somewhat indurated. Much meningeal and ventricular change. Some wasting of basal ganglia, chiefly of left corp. striatum. Left cerebral hemisphere the more atrophied one. Purulent fluid in right middle ear and external meatus, from the mastoid region; but no intra-cranial pus.

A peculiar case is mentioned by Mendel, (*op. cit.*, p. 154) respecting which the conclusion was drawn that the patient was unable to accurately localize the source of auditory impressions coming from the left side. With perfect freedom of movement of head and eyes in every direction; and with ready movement of them, and turning leftwards, when he was requested to turn them that way; he yet, when spoken to in the left ear, or told to look at his interlocutor, always turned the head and eyes towards the right. A local lesion of the right temporal lobe was diagnosed, in accordance with some experimental results. The necropsy revealed arachnitis, many adhesions of pia to brain-cortex, atrophy of gyri, diffuse interstitial encephalitis; also a calcareous local change, of the size of a lentil, in the right third temporal gyrus, about two inches from the tip of the lobe, and which had taken origin in a hæmorrhage. The corresponding part on the left side was highly hyperæmic.

In speaking of the speech-defects in g.p. a form of deafness was incidentally mentioned, analogous to word-blindness. This is word-deafness, or surditas verbalis; in which condition patients can speak and write, have good or sufficient general hearing-power, and a sufficiently preserved intelligence, and yet do not understand the words they hear —are deaf to the word as a symbol of the idea. Yet this seldom occurs in an isolated form: it is usually associated with other abnormal conditions, as, for example, with word-blindness, amnesic aphasia, or agraphia.

The disorders and defects of hearing in g.p. are apparently far more usually dependent upon central, and especially cortical, lesions (temporal), than upon any peripheral change in the auditory nerve itself, or peripheral auditory apparatus.

Tactile Sensibility. See also Chapter III.—In the last stages of g.p. the sense of touch is apt to be greatly lessened or abolished. The perception of slight touches, on the cutaneous surface of the hands and especially of the feet, becomes difficult to test, owing to the mental failure; but by careful comparative examination, together with check- or control-experiments elucidating the condition of the perceptive faculties of the patients, it can be distinctly made out that the cutaneous sensibility, as regards this matter, is lowered in many cases, particularly in the later stages. On analyzing my notes, I was somewhat surprised to find that this state scarcely seemed to be a symptom correlative or associated with absence or lessening of the knee-jerk, much more than with exaggeration of the same : that is to say, if one takes the cases collectively. In reference to this, any statistics must vary so much according as the cases happen to be recent and in early stages; or, on the other hand, chronic and in later stages, that it is of little use to state them; and often the same case would come under one head at one time, under the other at another.

Hyperæsthesia. Hyperalgesia. Hyperæsthesia is sometimes observed in the prodromic, or in the first or second stages of the established disease. Like anæsthesia, it may be early, and was pointed to by Michéa in the prodromic period. The hyperæsthesia may affect skin, muscles, or viscera, may be the origin of hypochondriacal ideas, and give rise to genital excitement and onanism. It may be quite local, as in the distribution of some of the cranial nerves, or it may be a general hyperæsthesia, or may be of unilateral distribution. Ordinarily, it is of short duration; sometimes occurs during remissions; in the later stages may follow unilateral epileptiform attacks affecting the same side; but may occur locally, or in widely-spread distribution, independently of special seizures. Ordinary, painless, impressions become painful; and painful ones agonizing.

As another example of that dissociation of the symptoms of systematic affections of the cord, already mentioned as often occurring in g.p., may be adduced the case of a patient, under my care, who, together with hyperalgesia, and exaggerated reflex reaction to tickling of the soles of the feet, had no knee-jerk, and no ankle-clonus; the tactile sensibility of the feet being fair.

Neuralgic and other pain. Neuralgia is sometimes a precursory sign of g.p. It may affect different parts of the head, or shoot down the spine, or the limbs; or encircle the trunk, or parts of its circumference; or be of visceral distribution. It may be present, also, in the early stages of the established disease. It may be transitory or more persistent.

Case 26. From time to time severe pains, at first about the chest and loins, but later about the head, legs, and left arm; and at times only in the left arm, which thereby was rendered nearly useless, and was kept almost immobile by the patient. A coexisting symptom was incontinence of urine. The patient, from having been excited, inclined to violence, emotional, and hallucinatory, nearly recovered mentally, but became subject to occasional frightful sets of epileptiform convulsions; and at times was depressed and irritable.—Latterly increased knee-jerk. The convulsions, in first set, were mainly of left side, then of right, with much tongue and jaw movement, and afterwards paresis of left third nerve. Later, left limbs and right face paretic. Later, fits, left hemiplegia. In fatal attack, fits and right hemiparesis, return of fits, death.—Meningeal sclerosis. Adhesion and decortication moderate, almost confined to left cerebral hemisphere; also found about the margins of great transverse fissure posteriorly; and over middle of upper surface of cerebellum. Slight cerebral atrophy. Spinal meninges thickened. Cord firm in dorsal region; lateral columns sclerosed.

Teissier of Lyons drew attention to the neuralgiæ, and also the visceral neuroses, in the prodromic stage. Especially has M. Aug. Voisin* insisted upon the import of neuralgic pain in the prodromic stage of g.p., and described eight cases of generalized neuralgiæ preceding g.p., and, in some examples preceding it by a long space of time. In some of these examples the neuralgia affected only three or four parts of the body; and was mobile, affecting now this part, now that, or a third, to return, perhaps, to its first situation. Most often, however, the neuralgiæ were much more localized, and affected this or that part of the head, sometimes the cardiac, epigastric, or other regions. Pricking or itching sensations of the skin occurred frequently; and tickling-sensation of it, has been recorded. The prodromic neuralgiæ often disappear when the delirium and special somatic signs come on. Neuralgiæ in the first stage of the established disease are often of the lumbar region, the waist, the vertex. The pains are in some cases vague and mobile, and sometimes of visceral origin. As to the manner in which generalized or severe neuralgia may bring about g.p., he attributes it to a reflex influence; citing examples of in-

* Mémoire. Prix Lefèvre, 1877

flammation of neuralgic origin—reflexly. The department of the vaso-motor concerned with the encephalic circulation, is supposed important here; though whether its disorders cause the active dilatation of the capillaries of the brain is not clear. Nor is it clear, according to Voisin,* whether the usual disappearance of the painful symptoms of neuralgia, of migraine, at the time of the onset of mental alienation in g.p. is due to a "fluxionary" state, or due to the supervention of general anæsthesia; but at all events their disappearance proves that the neuralgias were not due to any permanent spinal lesion which had preceded the appearance of the g.p. Where the spinal disorders precede the cerebral in g.p., Voisin concludes that protracted, inveterate, generalized, mobile, extremely painful neuralgia,—rendering life insupportable; and occasioned by spinal disease,—may be followed by g.p. in consequence of a reflexly produced disorder of the cerebral circulation, and the eventual results of that disorder. In other cases, the spinal lesions, although preceding the cerebral, and of painful nature, may have no causal connection with the latter. In still other cases, spinal disease attended with painful symptoms may lead to g.p. by way of "propagation," so-called.

Headache. Besides the truly neuralgic head-pains, severe headache is occasionally observed as a precursory sign; or it may be continued into the several stages of the fully-established affection. Sometimes the headache is so severe before the patient's mental state is much altered that he is forced to cease work, and take to his chair, his bed, or to hospital. It is often diffuse, of deeply-seated character. With this, the patient is apt to become sullen and taciturn; or stupid and dazed; and gradually to develop seizures of maniacal excitement; or at least of causeless outbursts of angry temper, or of dangerous violence. Another form of case in which severe pain, and, perhaps, tenderness of the head, may be found is that in which, after a history of syphilis and drink, melancholic and hypochondriacal symptoms preponderate throughout, perhaps accompanied with suicidal tendencies; the sight is impaired, rheumatic pains in the legs have long been antecedent symptoms; the knee-jerk is lessened or abolished, while hallucinations or illusions of all the special senses are found, as well as extraordinary visceral illusions. Intercurrently, is temporary numbness or incomplete anæsthesia of parts of the frame.

* Traité, *op. cit.*, p. 349

Even at a late stage the patient may rub his head, wearing an expression of pain, rubbing definite parts of the scalp so frequently and vigorously as to gradually remove the hairs growing at those situations.

Anæsthesiæ and Analgesiæ generally. In the prodromic stage (see also Chapter II), anæsthesia may occur, and preponderate on one side. Lasègue and Voisin held that anæsthesia did not occur, early, only, and clear up later, as some supposed. Voisin states that Baillarger had previously to de Crozant indicated anæsthesia at the onset of g.p., and that Brierre de Boismont did so later on.* But, in reality, both of these observers almost confined it to the second and third stages.

Reference has already been made to the frequent general and progressive impairment or loss of cutaneous and other sensibility; and the little heed apparently paid to the influence of injury, heat, or cold. Anæsthesia in g.p. may be widely-spread, or may be local; and in so far as it affects the nerves of special sense it has been described when treating of the affections of these in g.p. In one case I found anæsthesia very marked at an early date, and somewhat permanently so, on the neck, chest, abdomen and hands. In some cases observed by Dr. A. J. Linas (*op. cit.*, p. 37), sensibility failed first in the trunk. It became more obtuse on the chest than the abdomen, on the back than the front, and disappeared elsewhere in the following order of priority; —thighs, legs, arms, forearms, neck, face, and hands. It failed earlier on the extensor than on the flexor side of the limbs, and was retained longest about the mucous orifices,— as the mouth and nose—and at the peripheral ends of the limbs. Yet from my own observations, I am more inclined to agree with Dr. Mendel who found the anæsthesia most marked of all in the lower limbs, less in the upper, and least in the face; or at all events I should say, more in the limbs, less in the face.

In several instances I have known G.Ps. bite their fingers severely, and without giving any indication of being pained. In one case a patient chewed his right forefinger, taking no more notice of it, when arrested in the act, than if it had been a piece of tough food between his teeth. So terribly had he crushed it that the finger became gangrenous, and as he seemed to have no feeling in it, even in the non-gan-

---

* The reference by Brierre and Voisin to the former is erroneous; that to the latter is "Ann.| Méd.-Psych.," Oct. 1859, p. 325.

grenous parts, we proceeded to amputate it, without using any anæsthetic agent, either general or local, or taking any other measure. Whilst the incisions were being made, and partly through the sound tissues, the patient turned and gazed quietly and without flinching in the least; but also, on the other hand, without any stoical self-restraint, or voluntary repression of the indications of pain; in a word, he felt no pain from the operation; and there seemed to be complete anæsthesia and analgesia of the part. In this case, the left cerebral hemisphere was the one more atrophied: the cortical atrophy, most marked in the frontal lobes, lessened thence backwards and downwards; the white and grey were slightly indurate, this change gradually shading off from the frontal to near the occipital tip; pallor of white, faint redness of grey. Basal ganglia, especially left corpus striatum, somewhat shrunken. Meningeal and ventricular changes highly marked, comparatively little adhesion and decortication.

The same anæsthesia and analgesia may occasionally be shown in the later stages, when, under some circumstances, hot water bottles are applied to the feet of patients capable of moving their limbs; and, rarely, by mishap, the feet rest against the heated object with insufficient interposing materials, and vesication of the part follows. But here the influence of the dementia of the patients; and of the readiness with which, at slight causes, or quite spontaneously, or idiopathically, blebs form on the skin, must both be duly estimated.

In reference to cutaneous anæsthesia in the third stage, Voisin (*op. cit.* p., 144) ascribes to Baillarger a case in which amputation was performed without anæsthetics; and to Linas a case in which the patient seized a live coal, and kept it in his hand long enough to produce a burn of the fourth degree.

In relation to failure of sensory nerves in g.p., are also cases, like that reported by Westphal, where atrophy of a fifth cranial nerve and Gasserian ganglion preceded g.p.

I have seen G.Ps. who at times had to be held and prevented from violently beating their faces with their fists. Whether anæsthesia or, again, neuralgia may have co-existed I am unable to say. But the often-accompanying emotional tumult more likely effected an abstraction, or a partial one, of the consciousness from sensory impressions, and thus deadened all painful impressions at the time.

Many patients take little or no notice of pricks, pinches, painful impressions, or of the application of hot or cold articles to the skin, especially of the lower limbs. In fact, hypalgesia is more often observed, in g.p., than is tactile and cutaneous hypæsthesia of the same parts, so long, at least, as the patient's mental state does not preclude investigation of the latter point.—There is, also, an impairment of cutaneous sensibility to impressions of touch, or to painful ones, which is frequent after epileptiform or apoplectiform attacks, particularly the former; and on the affected side, the one convulsed and then paresified, so that here there is a temporary incomplete hemianæsthesia, with more or less hemiparesis.

Relevant to the present subject are Goltz's * experiments in removing large tracts of the cerebral cortex in dogs, thereby effecting, experimentally, a lesion in some respects analogous to that of advanced g.p. It was found that no sensory function is permanently extinguished by a single section, or local destruction, in cerebral cortex. But after excision of large pieces of the cerebral cortex a weakening of the powers of sensory perception occurs. Destruction of the parietal, has a greater effect in making the movements of the animal awkward, and in blunting the sensibility of the skin, than has destructive lesion of the occipital lobes. The cutaneous sensibility is not lost, but is more obtuse than is normal; to produce a given effect the irritation or excitation must be stronger; less notice than usual is taken of pain or discomfort; less easily are the animals aroused to extricate themselves from hurt and from annoy; and less guidance and slower, do they receive from impressions communicated from the periphery of the frame.

I found feeling of, and motor reaction to, pinches and painful impressions (unlike tactile perception), to be twice as often diminished or absent in cases wherein the knee-jerk was lost, as compared with cases in which it was exaggerated. In the former group, indeed, this diminution or absence was observed in the great majority of cases; in the latter group, though in some lessened, in none was sensation absent.—And with regard to another condition, in which mucous anæsthesia sometimes plays a part, and cutaneous anæsthesia may assist in a secondary way, by enhancing the effect of the mental inattention—viz., urinary incontinence,—*it* also was more frequent, early, or marked, on the average, in the group of

* "Trans. Internat. Med. Cong.," London, 1881, Vol. i, p. 228.

cases in which knee-jerk was absent than in that in which it was exaggerated.

The disorders last described refer chiefly to "common" or "general" sensibility. But there are various modifications or phases of that sensibility, in fact tactile perception —the sense of touch, proper—is a modification of it. This impression of pressure is special, has special organs, its disorders have been treated of already; and ranged nearly in rank with it is the thermal impression.—Various subdivisions are made, but here I shall merely refer to the thermal, muscular, and electrical senses, and say something of visceral sensibility, deeming it unnecessary to treat, separately, of nervous elements subservient to the sensation of pain, pricking, or pinching.

Thermal. The perceptivity for impressions of heat or cold (temperature impressions) may be increased or lessened; or morbid temperature sensations may occur, not of external objective causation, nor from heat measurable by thermometer; but of internal and subjective origin.

The *muscular* sense (see Chapter III.) is very often disordered in g.p., and is largely at fault when the patients have sensations of flying through the air, of extreme buoyancy of their frames; or, on the other hand, of having leaden limbs, difficult weary movement, inertness, and fatigue; or, again, of expansion or shrinking of the body or limbs. Here, also, is ranged a loss or disorder of perception of movements as actually made.

The *electrical* sense (if we may speak of such), when disordered, may be the source of sensations of electric currents or shocks in the head or body. But what has been supposed by some to be a disturbance of an electric sense, no doubt, in not a few cases, takes origin in neuralgic or fulgurant pain; and the description of electric shocks, causing the patients to leap from bed, are, then, merely the interpretation put upon their pains by the patients.

All the above forms of sensation may be abolished, or diminished, or exaggerated, or perverted; giving rise to abnormal subjective sensations; and these to delirious conceptions, or to hallucinations.

Visceral sensibility, another modification of common sensibility, may be disturbed. As the various parts of the nervous system and the viscera become diseased, those organs which had always previously exhibited but little sensibility, and had scarcely ever obtruded themselves by means of sensation into consciousness, become the seats of over-activity, or of perverted innervation, and hence of perversion of such sensory qualities as they are endowed with. Here is a fertile field for the generation of hypochondriacal and melancholic delirious conceptions. The sensory perversions might conveniently be spoken of as hypochondriacal illusions. Under that head, and under the hypochondria of g.p., have they been described; particularly as, owing to the mental state of the patients, these perverted sensations are difficult to examine.

Fulgurant pains: cutaneous illusions. The following concerns phenomena which I have classed elsewhere as hallucinations or illusions of touch; the patient having subjective or internal sensations, interpreted by him as cuffs, blows, or kicks, on the skin; and this when wholly unmolested. For, sometimes it has seemed to me that sharp or lightning-pain gave rise to delusions of the following kind. One G.P., when walking quite alone, and when absolutely unmeddled with, was accustomed to shriek suddenly at times, and when questioned on the subject declare that some one had that moment kicked or injured him, or that his back was broken. Another, when quite alone, would call out suddenly; and energetically grasp his perineum, or seize his coat in front with one hand and violently thrust away in the air with the other, to rid himself of the imaginary persons, or objects, who, or which, he said, squirted fluids over his legs and feet. Another, walking quietly by himself and untouched, would suddenly shriek aloud for a moment, and afterwards express delusions of personal injury. These last two G.Ps. had many symptoms as of tabes dorsalis, and one of them quasi-tabic or taboid arthropathy; and probably both had lightning-pains, which may be reputed to be the source of the symptoms on account of which I mention the cases here. These phenomena repeatedly occurred under my own observation.—If a G.P. now and then suddenly shouts or shrieks without any objective reason, and if he either gives no explanation, or evinces delusions as to corporeal injuries momentally inflicted upon him, I attribute the symptoms, in most cases, to sudden, darting pain, usually connected with spinal disease (sclerosis, myelitis, spinal meningitis). Of the three cases just mentioned, at least two had distinct spinal disease, two had a marked history of syphilis, while the previous medical history of the remaining one was defective. Most of the patients exhibiting this clinical condition are among the less communicative G.Ps.; do not readily obtrude the ills they suffer, or even are taciturn. Hence it may escape an observer who is not much with his patients; and I think I have not yet met with a description of the condition in the literature of g.p.

CHAPTER VIII.

VASO-MOTOR PHENOMENA.

Vaso-motor disorder is probably the first effect of many of the causes efficient in the production of g.p. It plays an essential part in the pathology of the disease, and is a necessary link in the series of morbid changes whose outcome is g.p. But when the disease is established, vaso-motor phenomena still occur. If the views widely held as to the influence of cerebral hyperæmia in the production of g.p. are right, it is obvious that the vaso-motor phenomena are of prime importance. (See Chap. XVII.)

In many cases, amongst the earliest symptoms noticed are a heated state of the head, a flushing of the face, and often palpitation. With these may be vertigo, or sensations of being stunned, or various sounds and bruits in the ears. The congestive symptoms may vary in intensity, in all degrees, from those just mentioned to coma, with, perhaps, paralysis and convulsion. In the later stages the graver of these symptoms may occur with increasing frequency and severity. In some cases may be observed, after local irritation of the skin, a condition like the so-called *tache cerebrale*. The immense importance of vaso-motor conditions in the production and in the course of the seizures, such as the maniacal, apoplectiform, and epileptiform, in g.p., is obvious. After death in the severe, protracted, epileptiform, or in severe apoplectiform, seizures, the encephalon is intensely engorged, passively congested with dark venous blood, as if the vessels were relaxed and palsied. But in the early stages the intense redness and turgescence of the face and head in some of the apoplectiform seizures, or in attacks of maniacal excitement; and, in the later stages, the dull-red bluish hands and feet and features, alike testify to a morbid vaso-motor condition. So, also, do states of local, and sometimes unilateral, coldness, without sufficient external cause, and often with local anæmia :—a condition which, in its turn, may make way for hyperæmia of the same parts. Thus, one often finds a G.P. with one foot or hand cold, and the other comfortably warm; or one, or an ear, heated and turgid, and the other seeming, to the observer's hand, to be of natural temperature, and presenting a natural appearance to the eye. It is particularly after the special seizures of G.Ps. that the differences in temperature between the two sides, or between symmetrically situated portions of the two sides, are to be observed, but they are frequent, even independently of these seizures. The slowly coming, and tardily disappearing, local dilatation following irritation (nail-stroke, &c.), shows a paretic vaso-motor condition. And the occurrence of urticaria may further indicate the vaso-motor disorder.

Punctiform hæmorrhage may appear in the skin after the special seizures; or may occur quite irrespectively of the latter: and now and then purpuroid blotches come on, and even bruise-like swellings about the ankles, knees, popliteal spaces, or elbows, quite independently of any traumatic cause, and reminding one forcibly of some aspects of scorbutus; and this in patients taking, throughout the time of their residence, fresh potatoes every day. Perhaps a purpuroid state, or something more profound than merely vaso-motor disorder, and having to do with changes in blood and in blood-vessel-wall, is concerned in the production of the symptoms just mentioned. In this relation, also, are to be mentioned various hæmorrhages which may occur, particularly in the later stages; hæmorrhage into the alveoli of the lungs, epistaxis, hæmatemesis, hæmaturia, intestinal hæmorrhage, menorrhagia; all occurring irrespectively of any traumatic cause, and without any ordinary local lesion of the parts affected of a kind to account for the hæmorrhage. One such of intestinal hæmorrhage I have seen in g.p.; and one is given by Dr. Mendel; and two of hæmaturia by Dr. Savage. In three cases of g.p. Dr. Jehn * found clear-red extravasations in both lungs, of irregular sharp outlines, tough and dry on section, and consisting almost entirely of red blood corpuscles; the vessels and lung tissue being healthy: and Dr. Ripping † observed a similar condition in several cases in the insane. Such lung-hæmorrhages I, also, have seen in the insane. Here also come Fleischmann and Ollivier's ‡ observations of various diseases of the pons,

---

* "Centralblatt für die Medicinischen Wissenschaften," No. 22, 1874, p. 340.
† "Allg. Zeitschr. für Psych.," xxxi. Bd., p. 600.
‡ "Arch. Générales de Médecine," Aug., 1873, p. 167.

crura cerebri, and cerebral lobes, producing lung-hæmorrhage and ecchymosis of pleura of opposite side.

In many of the trophic disorders of g.p., vaso-motor disorder no doubt plays an important part, as in its othæmatoma, or in its bedsore, while the production of rapidly-formed bullæ and eschars, in g.p., as in other diseases of the brain and cord, gives further evidence of at least some vaso-motor implication; though this perhaps is not the essential condition. It is a question whether the unilateral local sweat, sometimes met with in g.p., is a vaso-motor phenomenon (see Chapter XI). If all parts of the body are more or less represented in the encephalon, and have more or less intimate physiological alliance and pathological sympathy therewith, it is no matter of surprise that gross disease of the encephalon should be attended with lesions of the other organs; of which the most readily and directly produced may be of vaso-motor origin.

Experimental physiology has thrown some light upon this difficult subject. Thus Brown-Séquard,* by pricking and cutting some definite parts about the base of the brain caused, in the lungs, patches of anæmia, of œdema, or ecchymosis, even of large extent; or foci of congestion resembling red hepatization. And he believed the hæmorrhage and the inflammatory condition of lung, which appear to be dependent upon brain-disease, to occur more frequently with lesion of the right brain than of the left. Again, irritation of brain may alter nutrition in other parts of the cerebro-spinal nervous system; thus, for example, by some irritations of brain Brown-Séquard produced dorso-lumbar, meningo-myelitis.—This collateral effect, the production of spinal inflammation by irritation of brain, bears, possibly, on some of the spinal changes in g.p. Again, he found,† after some experimental injuries of the corpora striata, crus cerebri, or spinal cord, softening and ulceration of the mucous membrane of the stomach: and hæmorrhage usually after injury of a determinate part of the pons, although it is true he found no hæmorrhage after palsy of the vaso-motor nerves. Schiff‡ had previously found hæmorrhage and partial softening of the rabbits' stomach, after section of the optic thalami and crura cerebri. Prof. H. Nothnagel,§

---

* "Lancet," July 15, 1876, p. 77.
† "Progrès Médical," 1876, No. 8.
‡ Cited by Mendel, *op. cit.*, p. 200.
§ "Centralblatt für die Medicinisch. Wissensch.," Mar. 21, 1874, p. 209.

by injury of part of rabbits' brain-cortex, produced pulmonary hæmorrhage. Prof. Eulenberg,* also, after extensive injury of the posterior part of the brain-cortex of a dog, saw marked evacuations of blood from the bowels; which he supposed to be due to a congestive state of the vessels of the bowels, by vaso-motor change dependent on the brain-damage.—With reference, also, to irritation of cortical centres in the so-called cortical motor zone, Prof. Schiff—who believed that these were centres, not of motion but, of sensation, and that the movements attending irritation of them resemble reflex rather than direct actions—found it attended, also, with reflex secretory, cardiac, and vaso-motor phenomena.—By thermal irritation, or mechanical injury, of definite parts of the brain of rabbits and of dogs, Brown-Séquard † caused symptoms like those of paralysis of the ocular, auricular, and facial filaments of the cervical sympathetic, of the same side :—active hyperæmia of blood-vessels, elevation of temperature, contraction of pupils, and partial closure of eyelid upon the same side. And there was a still further similarity in effect; for, whereas section of the cervical sympathetic on one side is sometimes followed by atrophy of the brain and eye of the same side, thermal excitation of parts of the brain of two rabbits and of a dog was followed by atrophy of the eye of the same side. The above effects followed thermal excitation of the middle and posterior lobes more than of the anterior; and of the regions bordering on the median line, rather than of those which are placed laterally. Then, the experiments of Eulenberg and Landois ‡ show that limited destruction of the brain-cortex on one side was followed by relaxation of the arterioles and a heightened temperature in the limbs of the opposite side of the body; while, on the contrary, to stimulation of the same areas of the cortex by the induced electrical current there succeeded a brief and inconsiderable lowering of the temperature in the opposite extremities. The vaso-motor centre for the fore-limb was found to lie in front, and to the outer side, of that for the hind limb. Bochefontaine § brought about increase of the sub-maxillary secretion, as well as contractions of the intestines and of some other viscera, by electrical excitation of a definite limited part of the frontal lobe of some of the lower animals. After section of the vagi,

* " Berl. Klinische Wochenschrift," Oct. 23, 1876, p. 620.
† " Arch. de Physiol. Norm. et Path.," 1875, p. 859-60.
‡ " Centralblatt für die Med. Wissenchaften," Apr., 1876.
§ " Gazette Médicale," Aug. 28, 1875, p. 436.

Genzmer* found hyperæmia, exosmosis, œdema, and emphysema of lung:—these he attributed to damage of vaso-motor filaments in the vagus. And, indeed, intense congestion of the lungs, effusion of frothy reddish serum into air-cells and bronchi, probably with escape of the solid parts of the blood into the tissues of the lung, were long ago found after division of the vagi by Dr. Jno. Reid.

The vaso-motor system, including both its vaso-constrictor and vaso-dilator factors, no doubt also takes some, though a secondary, part in the alterations of secretion produced by emotion, by disorder of sensory nerves, by intellectual tension, and by electrical irritation of the brain or by hæmorrhages into it—most of which may have place, if not importance, in some cases of g.p.

It is not alone, however, supposed vaso-motor centres in the brain-cortex that may be disordered in g.p.; nor must the experimental and pathological evidence with regard to vaso-motor phenomena, attending various forms and situations of brain-injury, irritation, or organic change, be by any means exclusively taken into view. So far from being the sole, or even the chief, seat of vaso-motor centres it is only in recent years that the existence of cerebral vaso-motor centres has been known or established with any approach to verisimilitude, although the emotions have for countless ages revealed the play of vaso-motor activity on the human visage. For a time physiologists, founding upon the results of Schiff, von Bezold, Ludwig, and Owsjannikow, placed the great vaso-motor centre of the body in the medulla oblongata, at least in some of the lower animals, as the rabbit. Although the medulla oblongata no longer holds undisputed sway in the regulation and maintenance of arterial tonus, it, nevertheless, is still considered as of very great importance in this action. Below the medulla oblongata is also a series of vaso-motor centres throughout the whole length of the spinal cord; and, successively following these in downward gradation, a series of vaso-motor centres in the sympathetic ganglia in front of the vertebral column, or connected with the plexuses of the great cavities of the frame; and, finally, the local vaso-motor centres, which are on the walls of the vessels themselves, or in their immediate proximity, which last maintain arteriolar tonus, and upon changes in the state of which any alterations of arterial and arteriolar calibre are directly due. And one way in which modifica-

* Cited by Jehn, who gives incorrect reference to Pflüger's "Archiv."

tions of the state of these perivascular centres is brought about is by impulses coming from the higher and more widely influential centres in the spinal cord, medulla oblongata, and brain. Bearing in mind, then, the active and progressively, and successively, irritative and destructive processes in the brain, medulla oblongata, and often the cord, in g.p., there remains no room for surprise that gross lesions, due to organic changes in, or to excessive irritation of, the structural elements of the vaso-motor apparatus, should occur in g.p.; and attention has already been, and will be, directed to the functional vaso-motor disturbance at the very ground-work of g.p. (*vide* Pathology).

CHAPTER IX.

NUTRITIVE DISORDERS. TROPHIC DEFECTS. CACHEXIA.

It will be convenient to examine these in one large loosely-connected group, and in successive sections, each dealing, partly with the trophic, and partly with other, affections of different systems, or parts, in g.p.

Body-weight. Concisely stated : G.Ps., if excited, lose weight at the high tide of their restlessness, self-neglect, and excitement, as in the early—prodromic and first—stages: or at irregular points of time, later on. As a rule, also, in the cases not rapid, they gain in weight in the middle part of the course of the disease, when excitement has abated, or when the patients have become quiet. And although life may be cut short by an apoplectiform seizure, or by some other complication during the time of fatness, the more usual succession of events is that the patients emaciate more or less during the period of fluctuating decline, with its persistent down-hill tendency. In cases with this typical course, therefore, patients at first lose weight and then fatten; but it is a flabby fatness they put on, and afterwards they gradually dwindle. To this course of events, it is obvious, there are many exceptions.

Skin. The dull, earthy, parchment-like appearance sometimes assumed by the skin has already been noticed in describing the stages. The greasy appearance of the skin, the occasional clammy perspiration, and the odour often exhaled therefrom, in the later stages of g.p., are worthy of mention. Not connected by him with g.p., these conditions were noticed by Sir C. Ellis * to forecast incurability, and the finding of excess of ventricular serum. In the last stages, the skin sometimes assumes a coarse look, and is of a dull dingy hue; or else shining and "greasy" in semblance, though not in actual fact; and blebs, boils, herpetic, ecthymatous, or other eruptions, may appear about the extremities or trunk. These conditions of skin are not explained by defective washing of patients; and one who should deny the possibility of a cutaneous change in g.p. might well be charged with clinical and pathological ignorance, there being nothing unprecedented in that condition; skin-changes, even more marked, occurring in various other internal diseases, and morbid states of the viscera, or of the great anatomical systems. Nevertheless, a circumstance has been suggested which may have been of some influence in this matter; the patients from whom my description was drawn were allowed to take considerable quantities of meat; a diet which, in association with the comparative inability to take exercise, may have some effect in promoting the appearances of the skin I have mentioned;— appearances which, however, are absent, or slight, and unnoticed in many cases that are cut short, or run a rapid course, or even in some which pass through the various stages of a long-lasting case.—A glossy shining state of skin is occasionally found, especially about the hands. Also dryness, from atrophy of sweat-glands, and furfuraceous falling flakes from skin; atrophy of skin diffusely; or pigmentation of skin.—In a case recorded † as temporary nigrities of the face in g.p., the patient's age was 72, and it is not clear that he was a G.P.

Herpes zoster not infrequently occurs, affecting this or that part, and not confined to any portion of the course of the disease, though more frequently, in my experience, in the middle and last stages. A magnificent example occurred in the case of H. B., No. 68, at the end of this book.

Pemphigus blebs often appear on one or more fingers, on the forearms, or about the feet and legs or elsewhere, and when they occur usually do so at the last stage, and especially when patients are bedridden. In one case, where pemphigus blebs appeared on the forearms and legs shortly before death, Déjerine ‡ found the nerves of the subjacent

* "A Treatise on the Nature, Causes, and Treatment of Insanity," 1838, p. 133.
† Dr. Adrien Fèvre, "Ann. Méd.-Psych.," May, 1877, p. 375.
‡ "Arch. de Physiol.," 1876, p. 317.

parts undergoing a process as of atrophic breaking-up and involution. (See Chapter XVI.)

As to the appendages of the skin :—

The *nails* may become coarse, thickened, fissured, brownish or yellowish.

The *hair* often becomes coarser, duller in hue, thinned, and easily falling out still more; somewhat stiff and staring, standing on end, disparted in appearance. In describing the state of the visage in g.p., that of the hair of the head, of the eyebrows and eyelashes, has been fully mentioned. The hairs have been described as in some cases growing with abnormal rapidity and luxuriance; or turning grey rapidly and prematurely, and this sometimes on one side.

Odour from body. With regard to any peculiar odours arising from the patient in the latest stages, one must, of course, carefully and rigidly exclude everything due to the foul habits of patients; and when this has been done, and the buccal cavity also carefully attended to, in order to obviate any odorous result of neglect of the teeth, or of the retention and decomposition of particles of food about the teeth and gums, there still remains a residuum of cases in which one can only look to a perversion of secretion, or of exhalation, which is to be connected with the changes in the general nutritive activity as regards both vigour and mode; or to changes of similar kind, but existing in the blood; or to perverted action in the formation and in the absorption of ptomaines from the chylo-poietic hollow viscera; or due to a direct influence of the nervous system upon secretion; and, in any of these cases, arriving at ultimate exposition in the insensible or in the sensible perspiration, or as an exhalation from the lungs and air-passages.

Statements such as those I have just made have from time to time been denied, but there is abundant evidence that nervous disease, and even that simple nervous excitement, may cause odorous emanations from the body. The existence, occasionally, of a special odour under mental excitement, in sane, but nervous, persons, and in non-G.P. insane, is a fact of which many instances have come under my observation during the past fifteen years; and in many of which I was able to exclude the influence of any uncleanliness. An analogous condition may well exist in G.Ps.; and there is, in them, also a tendency to degradation and change in the fluids and solids of the body. As so much incredulity is apt to be expressed on the general subject, I will not adduce my own observations, nor cite, as I might do, the asserted examples of peculiar odour from the skin (or lungs) in some cases of insanity; but, merely referring to Dr. Laehr's report of several cases of special odour from the insane in puerperal and circular insanity, pass to a paper and discussion on this subject, and for the most part by general neurologists. Dr. W. A. Hammond* has placed on record the case of a young married lady, of strong hysterical tendencies; and during a paroxysm a powerful odour similar to that of violets was exhaled from the body; but only from the left lateral half of the anterior wall of the chest, at which part, also, the perspiration was much increased. From the perspiration he obtained an alcoholic distillate which had the perfume of violets, changing under the action of bicarbonate of soda to the odour of pine-apple, probably by becoming pure butyric ether, the original perspiration probably containing butyric ether, modified in odour by the presence of some other organic compound. Other cases of marked odoriferous perspiration were; one in a choreic female; another of odorous exhalation from head neck and chest of a female, when in anger, and, as in the last case, like pine-apple:—another, a male case of strong violaceous odour at times when hypochondriacal; another, a female, with odour of Limburger cheese supposed due to butyric acid, during attacks of sick-headache. Strong and disagreeable, or, on the other hand, pleasant, rosaceous, odours from women during sexual excitement and coitus, have been mentioned. In one case, communicated to Dr. Seguin by Dr. Brown-Séquard, the detection of adultery on the part of a married lady was brought about by the strong general odour she always emitted after sexual intercourse. Dr. Priestmann was alleged to have proved the existence of a peculiar odour in the breath for six hours after coitus; Dr. Weir Mitchell, a special odour of the breath in many cases of meningitis; Sir W. Gull, the odour of syphilitic persons; while many diseases, such as small-pox, measles, milk-sickness, are attended with special odours. The peculiar odour of the urine some hours after partaking of asparagus, and which I have repeatedly observed, appeared in one case in the perspiration also.

Furuncles, carbuncle, ecthyma, may make appearance.

A coarse, swollen, *spongy* state of the *gums*, which swell up around the teeth, I have seen in one or two patients who had taken sufficient fresh vegetables every day.

* "Trans. American Neurol. Assoc.," Vol. ii, p. 17.

Occasionally *diarrhœa*, in the last stages, appears to be merely an indication of cachectic break-down; or of neuro-paralytic failure of the chylo-poietic viscera; the secretions being acrid and disordered, digestion and absorption impaired, and the stools thin, fœtid, and frequent.

The *vesical inflammatory affections* in g.p. are sometimes, partly at least, of spinal origin and trophic nature. Thus, rapid inflammation and superficial necrosis of epithelial surface, with inflammatory infiltration beneath it, in g.p., is in some cases, probably, in part an indication and effect of disorder of the trophic influence of the nervous system. But the minor or more chronic affections of the urinary bladder may not be from an "irritation" of spinal centres, but from partial withdrawal of the normal influence of nerves, owing to atrophy or destructive changes in the cord and spinal nerves. They are, therefore, partly in the same category as ordinary bedsores, but sometimes acknowledge a bacterial origin.

Bedsores. If the patient survives long enough, and especially if the disease runs its full course, bedsores are apt to form, and particularly in the last stage. When they occur, their chief points of election are the skin and other tissues overlying the sacrum, great trochanters, nates, about the heels and ankles, and where the knees press together or the elbows impinge. Gradually forming, they may also affect the toes, the prominences over the vertebræ, scapula, or occiput. They sometimes form somewhat slowly, and in consequence of gradual failure of nutrition making itself felt in the parts which bear the weight of the body, or are exposed to pressure, and, perhaps, to occasional excremental irritation;—at other times they occur as acute bedsore—*decubitus acutus or ominosus*—of either the so-called cerebral, or spinal, variety.

In the former and *chronic* kind of *bedsore*, either the part becomes reddened, a few vesicles perhaps appearing on the reddened patch, the skin gives way, and ulceration ensues; or a superficial layer of integument becomes dry, brownish, or dark, and, finally, deciduous: or in other cases the integument inflames, becoming red and swollen, the epidermis is loosened, the part alternately weeps and is scabby, and a portion of the surface wears away. At lightly-affected parts, as ankles, wrists, fingers, or knees, the blebs may dry up or form scabs.

As to the latter kind of bedsore,—*acute bedsore*—but varying in degree; in the lighter cases the change is less active. On the red elevated part, blebs appear, and contain fluid of varying hue; when this is discharged the cutis beneath is seen to be purplish in parts, and in parts black, and a line of demarcation forms, within which the integument sloughs, and without which is a zone of inflammatory change. In favourable cases, this zone disappears, the dead skin and connective tissue separate from the living; and from beneath spring up healthy granulations, which may close the breach.

In cases of still more severe local change, the sore, after forming in the usual way, does not pass through the phases mentioned above as occurring with those of slight severity and favourable issue. On the contrary, the sore becomes raised by the turgescence of the subjacent tissues, its surface remains foul, and whereas, where separation has not occurred, the slough is deep and is toughly adherent, if surgical measures are not adopted the surrounding and underlying brawny parts assume a boggy feeling to palpation, and the sore is undermined by a mingled semi-purulent and sloughy material—a foul, sanious, puriform fluid, with morbose particles—which it is difficult to keep from burrowing amidst the ligaments, connective tissue, muscles, and periosteum about the hips and sacrum. Occasionally, the sloughing of skin and other tissues is so rapid as to come almost at a stroke. Thus, in wide terms, as compared with the other form, the *acute* bedsore runs a more rapid course, its blebs are large and often filled with dark fluid, these, bursting, leave a bleeding violet and purple surface, beneath whose blackened edges rapidly appear the ashen or dirty dull-white sloughs, which finally spread, or become mingled with the foul puriform fluid above described. But there is a series comprising every possible gradation between the two forms of sore; the so-called chronic, and the so-called acute or ominous form. The acute bedsore appears most frequently after one of the apoplectiform or epileptiform seizures, or in the interlude in the course of a protracted set of several of these.

I had prepared three cases of hyperacute or acute bedsore in g.p., for insertion here, but find the limit set to this work necessitates their exclusion. In these cases, all males, about 40 years of age, the acute bedsores came on towards the end of life in association with more or less of the apoplectiform order of symptoms. In two, severe convulsions occurred later, or preceded. The sores affected the lamed side if over one natis only, or else occurred over the sacrum. In all, there was a somewhat unusual course, or less ordinary association, and succession of symptoms, too long to bring out in summaries of the histories. In all, the heart showed functional, or slight organic, change. In one the temperature, with acute bedsore, rose to 102·9° F. One patient died in the epileptiform "*status*." As regards the cerebro-spinal system; in all the cases cerebro-meningeal adhesion and decortication were slight, or else absent; in two there was distinct induration of cerebral cortex; and there were decided changes in the two cords examined, particularly in the grey matter: but the precise distribution of these changes was not specified with sufficient exactitude with reference to this fine point; and the notes I have only speak in general terms of the microscopical changes.

It is in the acute form of bedsore that those extensive breaches occur from which other deadly conditions may arise, especially as found in the older literature of the subject; such as sloughing of connective tissue and ligaments; caries and necrosis of spinal processes; opening of vertebral canal; ascending ichorous spinal, or, again, ascending purulent, meningitis; pyæmia; septicæmia; embolism; or thrombosis of internal organs or about limbs or joints. Thus, the secondary lesions and their material attendants, may radiate directly from the sore, macerating the spinal cord, dyeing it of a slaty hue, and extending (or not) to the base of the brain; or absorbed bacteria, or ferments, or products of organic decomposition may be transported in the circulating fluids to distant parts, and there work their ravages in the blood or in the organs. The supposed connection of these bedsores in some cases with pulmonary infarcts or gangrene, will be adverted to in the eleventh chapter. A less lethal, but still inimical, result is the production of irregular hectic fever, which may undergo extreme and even rapid fluctuations, and appears to be a result of absorption from the part, but which, like the worse results, can often be prevented by antiseptic, and frequent cleansing, applications and syringed fluids.

The acute bedsores have long been known; were described by Dr. R. Bright and Sir B. Brodie; and their treatment and pathology were discussed in Prof. Brown-Séquard's lectures at the Royal College of Surgeons of England in 1858. Yet these referred, not to g p. but, chiefly to spinal traumatic conditions, although the acute bedsores of g.p., and their occasional lamentable results, did not pass unnoticed, and were mentioned by Foville in 1829 ; and, later on, Baillarger referred to others of the several disastrous results of deep and sloughing sacral sores in g.p.

The acute bedsores may appear in a great variety of lesions of brain or of cord. If one accepts the teaching of Prof. Charcot on this subject, in reference to the situation of bedsores in affections other than g.p., it would appear that the more severe bedsores of g.p., if of the same pathology, are far more often connected with the spinal than with the cerebral lesions of that affection. This conclusion would result from a consideration of the favourite seats of the bedsores; of the frequent contemporaneous blebs or superficial sores on knees, ankles, heels, elbows, wrists; of the occasionally coexistent or supervening turbidity, alkalinity, and decomposition of the urine, cystitis and even nephritis, or nephro-pyelo-cystitis, an almost general muco-purulent inflammation of the urinary organs and passages—and these urinary changes with or without paraplegia ;—of anæsthesia and loss of reflex action in some cases; of spasm, rigidity and " contractions " in others ;— each in its special way testifying to implication of spinal cord or meninges, to the existence of a lesion, in some irritative, in some destructive, of this or that part of the spinal mechanism and structure.—I am not quite prepared to fully accept this view as conclusive. The first of the cases I have mentioned above runs counter to it, in some respects. The bedsore came on the same side as cerebral hemiplegia, and in the situation of the bedsore of the cerebral form of Charcot, but also extending across and affecting the opposite side slightly; and yet there was marked spinal lesion consisting chiefly of softening, and partial disintegration, of the grey matter of the spinal cord, and this on the *same* side; whereas, again, according to Charcot's view, supposing the sore of spinal origin and due to this mainly unilateral spinal lesion, it would, preferably, have been on the *opposite* side. In g.p., also, the acute bedsores so very often occur in association with an exacerbation of symptoms of, or mainly of, the cerebral order, that I am inclined to think the brain-disease of more importance in regard to the production of many, and of a larger proportion, of them than would follow on the teaching I have mentioned.

As to another point, it is likely that, at least in some of the cases published, the coexisting urinary troubles were lighted-up by the introduction of catheters microscopically unclean from the presence of septic micro-organisms. The view as to this occurrence in catheterization, so much in vogue of late years, was anticipated by Prof. Traube more than twenty years ago.

Finally, it should be added that improved methods of nursing, of hardening the skin and preserving it from irritation, have immensely lessened the frequency of bedsores (see chapter on Treatment).

Muscles, (trophic changes).

Progressive muscular atrophy. Independently of the moderate wasting of the muscles in g.p. owing to their partial disuse latterly, and especially in the bedridden state, and to their participation in the general marasmus and dwindling which then take place, there is occasionally, rarely I should say from my own experience, a muscular atrophy, acute or chronic. This, of course, turns one's attention to a probable acute or chronic anterior poliomyelitis or degeneration. I have seen two cases in g.p., besides one in a borderland (of g.p.) case. One has already been placed on record in brief abstract.* In a case of more than two years' duration, the patient was far advanced in g.p., bedridden, with much restlessness and grinding of the teeth. A sharp attack of erysipelas of the left arm underwent resolution, but subsequently that member became completely paralysed, rigid, and atrophied. Here the rapid muscular atrophy was limited, at least as regards any extreme degree, to the left upper limb, which became half the size of its fellow. The other was a very chronic, slow, case, with extreme maniacal irritation and expansive symptoms, and hallucinations, at first; the physical symptoms only supervening very tardily. Later on, the patellar tendon and some other reflexes were very slight or absent, the pupils very sluggish to light and less so in accommodation, slight left ptosis, increased convexity of ocular globe. Left leg chilly. Apparently ill-defined pain. Later, clipped speech; wasting of thenar eminences, and of muscles at interosseous spaces, somewhat claw-like hand at times. Cardiac and pulmonary disease. The atrophy was apparently at a standstill, and did not progress.

Fatty degeneration occurs in some isolated muscles in some cases, but does not become general, in any very pronounced degree at least. Two cases of myositis ossificans have already been cited.

* "Journ. Mental Science," April, 1872, p. 40.

Muscular hæmatoma I have only seen as, possibly, the result of injury, not excluded with certainty; but in some cases, no doubt, it depends on a special condition of the vessel-wall and blood, and the manner in which these may become modified in the later stages.

Bruise-marks (subcutaneous ecchymoses), occur with preternatural facility in some few G.Ps. in the last stages, and, occasionally, give rise to much trouble and anxious care, when the demented patient is restless and turbulent. No doubt, here also, is a special condition of blood-vessels and of blood to be taken as the condition efficient in bringing forth this large effect from little causes.

Othæmatoma. Hæmatoma auris. Sometimes absurdly called "insane ear." *Ear-swelling* might suffice (W. J. M.). This blood-tumour of the ear usually begins in the fossa of the helix and adjoining portion of the latter, or in the hollow of the concha; it may begin elsewhere, as in the fossa of the anti-helix, or even, rarely, in the external auditory meatus. Or it may come on in a more diffuse form, affecting simultaneously, or almost so, a wide area of the external auricular surface. The affected portion of the external surface of the pinna is swollen, and, in different cases, of a bright red, dull red, or bluish red hue; in the milder cases it is only slightly discoloured; in the more severe cases its surface is tense and shiny; it is more or less heated; it gives an elastic impression to the palping finger, but a firmer impression when the distension is extreme; it is not, or but little, painful. The swelling may level the surface and conceal the natural anfractuosities of the pinna, or these may still be evident in parts. Increasing, the effusion may make the greater part of the ear swell, bulge, and assume an irregular oval form; the organ heavy, and standing out boldly from the side of the head. In severe and rapid cases, these progressive changes may occur within a few days, or even hours. In some cases, however, the advance to the *acmé* is extended over weeks.

If the surface does not give way and burst at some point, and if suppuration does not occur, the period of resorption and regressive change now succeeds. Step by step, the discolouration disappears, the swelling subsides, the abnormal local heat abates, and finally ceases. And then, during a term which lasts for some months, further changes ensue, whose result is deformity, and irregular shrivelling of the ear. The auricle becomes of a dull dead-white hue; its cartilage is distorted, irregularly folded, twisted, and bent upon itself; thickened in parts by new growth of cartilage; covered by increased, toughened and contracting areolar tissue, and this by wrinkled adherent hard skin. Its natural curves and sinuous graceful contours are lost; its concavities are partially filled up, or the projection of prominences is exaggerated, and their convex sweep is broken and made angular. The dimensions of the pinna are at last shrunk from above downwards, and from before backwards, but its diameter from the outer to the cranial surface is irregularly increased in parts; and the pinna is hard, rigid, resistant, having lost its natural resiliency and comparative pliability.

Yet in some cases, the course of events is modified by rupture of the outer surface at one point; when a thin, watery and bloody fluid is discharged for a time, with, perhaps, occasionally, clots or portions of clots undergoing organization. In other rare cases, suppuration may occur, and a free discharge of pus ensue. In either of these events, the course of the local affection is retarded, the latter usually is severe, and the eventually resulting deformity more extreme than is customary. Finally, the retrogressive course may be checked by fresh effusion, with recurrence of the heat, swelling, and discolouration; and this clinical recrudescence may be repeated.

The occurrence of the blood-tumour seems to be preceded by degenerative changes in the auricular cartilage; a partly fibrillary and partly granular degenerative change; ending in softening and potential cavitation. Or, new cartilage outgrowth from the perichondrium may precede the tumour; but more often follows it. The local blood-vessels may also be affected. If now, from any reason, the blood-vessels of the part are turgid, or become damaged by injury, blood is poured out. Usually, the clot soon assumes a semi-organized appearance, reminding one strongly of the so-called durhæmatoma, in an early stage of the latter; and of what one might call hæmorrhagic auricular perichondritis. Eventually, this clot often becomes organized, adherent to the surrounding parts, shrinks, and,—together with the fibrous tissue and cartilage, new and old,—converts into a compact, hard, distorted mass, a part or the whole of the auricle, except the lobule; which last always remains intact.

Sometimes both ears are affected, but unsymmetrically; yet in the majority of cases only one is hæmatomatous; or one may become affected singly, and the other months or years afterwards. The left is the one more frequently affected: and, of the sexes, male patients much more frequently than female. This aural affection may occur at almost any part of the course of g.p., but I have never seen it so early as to find it of diagnostic importance, as Dumesnil affirms may be the case. Its occurrence is almost always in the middle or later stages; and it particularly follows, and is associated with, exacerbations of restlessness, wildness, excitement, and cephalic congestion.

Its nature has been a much disputed point; whether it is chiefly, or solely, a part of a general disorder; or, on the other hand, of local and traumatic origin, has been the chief question. To begin with, othæmatoma is not special to g.p., nor does it occur in any large proportion of the cases of this form. It may occasionally be seen, also, in the forms of secondary dementia following the psychoses; in dementia with gross organic brain disease other than that of paretic dementia; in epileptic dementia, epileptic and other forms of idiocy, in acute or chronic mania, in melancholia, and even in those forms of hereditary psychical degeneration in which depression suspicion and sullen moroseness may be tinctured, later on, with violence and delusions of self-importance.— Nor is it exclusively limited to the insane; it, or something not very dissimilar, is sometimes seen in pugilists. In 1874, Dr. H. Sutherland showed at the Clinical Society of London an example of such in a prize-fighter; and I have read (reference mislaid) that in an ancient Greek statue of a boxer an ear is represented with a deformity of this kind. This might be called the *traumatic* form. Other cases* have been given of its occurrence in the non-insane, and without injury; but some mental disturbance or excitability seems to have existed in these cases, or in some of them, and the *spontaneous* form is probably almost exclusively found in the insane; while, as I have stated, and as Dr. E. R. Hun † observed, prior to it there is usually intense local hyperæmia, due to disorder of the sympathetic system.

Occurring in g.p., and local degenerative changes in the ear-cartilage having preceded, as described by Prof. Carl Fürstner,‡ some of the factors of the aural tumour may be stated as follows:—the degenerative changes in ear-tissues and vessels; the turgescence of the vessels; the associated

* Fischer; trans. by Dr. J. T. Arlidge, "Asylum Journ.," 1854. Gruber.
† "American Journal of Insanity," Jan., 1870.
‡ "Zur streitfrage über das Othamatom."—" Archiv f. Psych.," iii. Bd., p. 353.

mental excitement, inattention, wild restlessness, insomnia, and neglect by the patient of personal safety and comfort; and, finally, in some cases, traumatic injury—injury which is greatly facilitated by the conditions last-named, or may be inflicted by another person; and which acts as an exciting cause, as the spark that lights the prepared train.

Of 32 cases reported by Mr. Lennox Browne 8 were G.Ps.; of 10 by Dr. Dumesnil 8 were G.Ps.; and 8 of 24 by Dr. E. R. Hun. The trophic origin is admitted by Dr. Biante;* and experiments of Brown-Séquard,† have been cited by Dr. Madigan,‡ on the production of aural hæmorrhage, followed by gangrene, after section of restiform bodies of medulla oblongata in guinea-pigs; and the conclusions of Dastre and Morat that the vaso-dilator nerves of the external ear arise from the spinal cord in the upper part of the thoracic region.

The cartilages of the nose and of the ribs are said to suffer, occasionally, in the same way as the auricular (Kœppe. Mendel).

Bones. This subject is an offshoot of a wider one—the bone-condition in insanity; or cases such as bone-degeneration, with strong family-history of insanity and of dipsomania.§

General Subject. Dr. J. G. Davey seems to have been, perhaps, the first observer on the subject of bone-disease in insanity : in 1842 he wrote ‖ upon osteomalacia as observed by himself in the insane; as he also did in 1857,¶ adding that "the greater number of the patients alluded to were afflicted with general paralysis"; and in 1876 ** he asserted his apparently just claims to priority in this matter. Mollities ossium was observed in an insane patient by Dr. Lauder Lindsay.†† In 1871 Dr. Hitchman ‡‡ and I made the necropsy in a splendidly marked example of mollities ossium, which had commenced and run its course in a case of chronic insanity. Mr. Arthur Durham,§§ also, long ago indicated the history of depressing nervous influences in mollities ossium; and so did Mr. Pedler.‖‖ A peculiar form

* " Annales Méd.-Psych.," July, 1882, p. 510.
† " Arch. de Physiol.," Oct., 1882.
‡ " Alienist and Neurologist," Oct., 1883, p. 687.
§ " Lancet," May 12, 1883, p. 821. Dr. Lee's case.
‖ " Medical Times," No. 170, Vol. vii, 1842, p. 195.
¶ " Ganglionic Nervous System," p. 265.
** " British Med. Journ.," Aug. 26, 1876, p. 291.
†† " Edin. Med. Journ.," Nov., 1870, p. 414.
‡‡ " Derby Co. Asyl. Report for 1871."
§§ " Guy's Hospital Reports," 1864.
‖‖ "West Riding Asyl. Rep.," Vol. i.

of osteomalacia in the insane has also been described by Dr. Morselli * of Florence. Dr. J. W. Ogle † drew attention to experiments in which interference with the nervous supply to bone brought on hyperostosis; and to Prof. Van der Kolk's preparations, illustrating the effects of injury of nerves upon the bones of the extremities of lower animals. Another point, relating to the general subject, but of special interest in g.p., consisted of cases of simultaneous fracture of an upper and lower limb, and injury of cord, in which the union of the fracture was normal in the upper, but tardy or absent in the lower, limb, in which latter the nerve-supply was interfered with by the spinal injury. Others, in explanation, have invoked the aid of disease of periosteum; of increased activity of absorbents; or of free acid circulating in the blood; or of inflammation of the bony parts (Solly). Whether ill-health and general impairment of nutritive power lessen or disorder the nervous influence on the nutrition of the bones directly; or whether some auto-toxic material thereby engendered is the immediate agency, is not yet known. The facts of so-called tabic arthropathy, and of many other aberrant forms of bone-nutrition associated with disease or disorder of the nervous system, go far to indicate that the latter has some direct influence, more direct than has usually been supposed, upon the nutrition of the osseous system; and although centres in the nervous system which preside over bone-nutrition are not yet clearly made out, the line of discovery seems to be tending in that direction, and may end in the substitution of a more precise formulation of the influence of the nervous system, and of its depressed states, upon nutrition at large, and, in particular, upon the bones, in place of Sir J. Paget's and Mr. Durham's older views.

Alterations of bones in g.p. Dr. E. L. Ormerod‡ described the ribs in a male case of g.p., aged 46, as being "dark, singularly wet and greasy." Decomposition occurred too rapidly in the ribs. They were brittle, and snapped across readily with a clean fracture. The rib examined seemed large, "its centre was traversed by a very light open network made of the fewest possible slips of osseous tissue, all the strength of the bone lay in its outer shell of compact tissue, which yet was no thicker than cardboard." Microscopically, there was a general granular condition of the laminæ, most

* " Riv. Sper. di Frenat. e di Med. Leg."
† " Journal of Mental Science," Apr., 1873, p. 162.
‡ " Journal of Mental Science," Jan., 1871, p. 571.

marked in those immediately surrounding the Haversian canal, where the lacunæ and canaliculi were ill-marked or wanting. The Haversian systems were comparatively small, more numerous, occupied a more limited range. Here and there, one or two had disappeared, leaving holes. The Haversian canals were of large size, unequally and irregularly dilated, filled with opaque material, with a few oil-globules. Some had a distinct lining-membrane. These changes were more marked in the ribs than in the clavicle or femur. Thus, there were thickening of lining-membrane, absorption of the innermost concentric laminæ, and propagation thence of the changes tending to removal of Haversian systems. "Besides, a change seemed to have crept over the whole bone, showing itself in loosening of the mutual connections of the laminæ and in an obscure disintegration of the osseous structure itself, and this accompanied by a general infiltration of oily matter into the substance, which had intruded itself within the Haversian canal, and into whatever part of the compact structure of a bone could find room for it."—Two cases of brittle bones in g.p. were mentioned by Dr. Clouston;* a female aged 50—a male aged 46; a rib of the latter, tested, broke under far less weight than the rib of a healthy man.—Dr Laudahn † in 23 insane persons found 8 of brittleness of ribs; 2 of the 8 were G.Ps. aged 38 and 49 years. Two cases in point were mentioned by Dr. S. W. D. Williams;‡ in one, aged 38, g.p. of three years' duration, usual symptoms, force required to break rib, 9 lbs.; weight of one inch of rib, 23 grains: another, aged 48, g.p. of usual type, skull thick, ribs all very brittle; great deficiency of osseous tissue, with increase in medullary substance of bone; fractured by force of 14 lbs.; weight of one inch of rib, 18 grains. Dr. T. L. Rogers§ published three cases of g.p. in which an analysis of the ribs showed a considerable degree of that diminution of the calcic salts, and chiefly of the phosphates, in bone, relatively to the organic constituents, which, in the same, or even in a higher, degree, is found in cases of advanced malacosteon. The inorganic constituents were reduced to about 42 p.c., and the organic increased to about 58 p.c., the ratio of lime to phosphoric acid was less than in healthy bone, and it was concluded that the composition in some cases approaches that observed

* "Lancet," Feb. 5, 1870, p. 191.
† "Archiv für Psych.," iii. Bd., p. 371.
‡ "Journal Mental Science,' Apr., 1873, p. 161.
§ "Liverpool Med. and Surg. Reports," Vol. iv.

in osteomalacia. In one of these cases, and in another case of g.p. mentioned in the same paper, fractures of ribs had occurred. Dr. I. Ashe* mentioned a case of g.p. in which " in the bony structure of the ribs the Haversian canals seemed almost obliterated by a mass of degenerate deposit containing oil;" and the muscular structure of the heart and of the gastrocnemius was reported as being pale and fatty. Voisin † mentioned hyperostoses; osseous tumour of thigh in one case; hypertrophy of head of a tibia in another; and greater fragility of bones as in ataxic patients; as well as "other alterations," unspecified. Dr. Joseph Wiglesworth's ‡ cases of marked osteoporosis do not appear to have been G.Ps. Other examples of alteration of bone are in the next section, on *intra vitam* bone-fractures in g.p.

Bone-fractures during life, in g.p. A case in which fracture of five ribs, incurred by a G.P. before admission, was unheeded by the patient, was described by Dr. Workman.§ Prof. Gudden,|| whose cases were partly contributed by others, stated that in 100 cadavers of the insane (50 F. and 50 M.) there were 16 with rib-fractures, 14 of these being males, and 2, females; of the 14 males 8 were G.Ps. He admitted the occasional occurrence of a moderate degree of osteomalacia. Simultaneously with rib-fractures were often found the relics of othæmatoma; and both of these conditions are more frequent in males, and in g.p. He mentioned one case with fracture of as many as 14 ribs and osteomalacia; one with fracture of 23 ribs and not osteomalacial; but it was not stated whether these were G.Ps. Dr. T. L. Rogers¶ gave the case of a female G.P., æt. 33, who fell backward upon the grass, sustained a compound fracture of the tibia at its upper third, and attempted to walk on the protruding lower end of the upper fragment of the shaft of the bone. Dr. N. G. Mercer ** described the case of a G.P. who fell in his dormitory, and sustained a fracture of the sternum. At the necropsy, some weeks afterwards, the fractured ends were bathed in a fluid of grumous oleaginous appearance, and without any attempt at reparation. The cancellous texture of the bone was far gone in disease, "its colour

---

* "Journal Mental Science," Apr., 1876, p. 82.
† *Op. cit.*, p. 153.
‡ "British Medical Journ.," Sept. 29, 1883, p 628.
§ "American Journ. of Insanity," Apr., 1862.
|| "Archiv für Psych.," ii. Bd , p. 682.
¶ "Journal Mental Science," Apr., 1874, p. 82.
** "British Medical Journ.," Apr. 25, 1874, p. 540.

being deep-red, much resembling that of muscle. It was soft and boggy, and the knife could, with perfect ease, be made to sink into and scoop out a portion of it." The right sixth rib was removed. "On the application of very gentle violence it snapped in two, revealing a brittle exterior periosteal shell, and internally the same dark-red spongy material which had been seen in the broken sternum." A little oleo-sanguineous fluid exuded from its broken end. The cranial bones, also, were soft. In a G.P. reported by Dr. Biante * the humerus was fractured and fissured by a fall. Its medullary substance was changed to a "bouillie sanguinolente." Thinness of the compact tissue of the bones and proliferation of fatty elements in all parts of the bones were found in several cases of this kind. M. Verneuil reported an osteomalacial G.P.—The form of mental disease is not stated in most of the cases related by Dr. Laehr.†

Mr. H. Rooke Ley ‡ mentioned that in "one case, that of a G.P. in whom seven ribs were broken, and who survived the injury 15 months, no bony union was observable; the ends of the bones were strongly united by fibrous tissue. At the post-mortem examination the ribs and other bones were found to be in a very brittle condition." Dr. J. C. Shaw § observed "two or three cases of fracture of the femur in paretics due to falls while running. I have also [he says] not unfrequently observed the spongy tissue of the ribs and other bones softened, and filled with a reddish semi-fluid material." On the other hand Dr. Christian, and also Drs. Ingels and Morel of Ghent, have rarely found fractures in G.Ps.

I add brief abstracts of several cases of fracture in g.p. observed by myself. In these, the usually diminished reflex activity, and sometimes diminished sensibility in the limbs, led one to suppose that, besides anæsthesia and the disorderliness of inco-ordinate muscular contraction, a contributory factor was the inattention of the patients, and the slowness and unreadiness with which their muscles are acted on by, or respond to, the stimulus which, normally, when one trips or stumbles, causes adaptive action of a preservative nature. This is partly similar to Dr. W. H. O. Sankey's ‖ view that the slothfulness of the nerve-currents in G.Ps.

* "Annales Méd.-Psych.," Nov., 1876, p. 350.
† "Allg. Zeitschr. für Psych.," xxxvii. Bd., p. 72.
‡ "British Medical Journ.," Sept. 29, 1883, p. 630.
§ "Arch. of Medicine," 1883, p. 144.
‖ "Journ. Mental Science," Apr., 1870, p. 135.

explains the frequency of fractures in them; he, however, attributed the abnormal liability to fracture solely to this cause, associated with the sudden reckless violence of the patients: thus, at all stages of g.p., the nervous current is dull and sluggish; the excito-motory acts tardy; and, therefore, the muscular contractions too late to guard against the injurious effects of falls and blows; while the dulness of common sensation lessens the inconvenience and pain of fractures, and masks their effects.—Recently, *apropos* of a case of g.p. admitted with nine ribs fractured on one side and four on the other, he * holds that the dulness of sensibility may be at the foundation of the fractures.

Case 26a. Given in full at the end of this book (case No. 65),
Case 27. Jumped from a wall; fracture left tibia and fibula, Strange visceral sensations; disorder of muscular sense; hallucinations. Early lessened reflex action and sensibility, right limbs; left ones more affected in (apoplectiform) hemipareses.

Case 27a. Transverse fracture of patella by direct force: marked cerebral sclerosis, disease of spinal cord, and its (dorsal) posterior cornu, on same side as fracture, apparently atrophied (case 69, at end of book).

Case 27b. Rib-fracture by slight traumatic cause.

Special arthropathies in g.p. Of what nature are these? Just as we may have in some cases of g.p. the characteristic gait and the lesions of tabes dorsalis; in others an ataxic gait, but not that of tabes, nor arising from distinct lesions of the latter; in others, other forms of gait and of lesion; so, I think, with the joint-affections, they may be either tabic, or somewhat similar thereto, but not the same.

(a.) *Case* 28. Male, age 38, admitted 1878, cause "unknown." Distortion, like that of fracture of neck of femur, seemed to be part of a tabic or taboid arthropathy, and attended with erosion of head of right femur. Not long after admission, the joint-affection began to manifest itself in weaker movement of the limb, but the principal effects appeared almost suddenly and without bruise, or external injury. The left ankle and foot, also, were affected and swollen. The case ran a very chronic course. Recurring exalted delusions. Often obstinate, irritable, and at times hypochondriacal, he had many delusions as to bodily injury by imaginary persecutors. At one time or other there were hallucinations of all the special senses. He shouted suddenly, and pulled at the perineum, probably from delusions connected with fulgurant pains; knee-jerk absent; no ankle clonus. From an early date, there was urinary incontinence. Pupils usually equal and smallish,

* *Op. cit.*, 1884, p. 260.

moderately sluggish to light, and somewhat so in accommodation. Tactile anæsthesia not marked; some analgesia of feet and legs; plantar reflex almost absent. On standing with feet closely together, and eyes closed, the patient swayed, but did not fall. Before the hip-joint affection the gait was somewhat ataxic, especially as concerned the right lower limb. Speech moderately affected. No apoplectiform or epileptiform seizures; but slight dextral unilateral increase of paresis on several occasions; once there was relative paresis of the right lower face, and once of the right lower limb.

Case 29. Age 34, admitted 1878, cause "unknown." Fell, when walking quietly, no one being close to him, and incurred deformity as of fracture of the neck of the left femur. By this, he was permanently laid up, union of the bone being slight or absent, and he was very restless and excitable. Had extravagant delusions, was noisy, angry, denunciatory, insulting, and had hallucinations of sight and of hearing. Latterly, the habits were foul. Knee-jerk, throughout, was always very slight or absent. Stuttering, etc., was considerable. The only special seizures were slight apoplectiform ones towards the last. On one occasion paresis was slightly more marked in the right than in the left lower limb. The pupils were sluggish to light, equal and small, later on they were widish, and the right one slightly the larger. Before he was laid up, his gait was noted as showing inco-ordination. Feeling of, and reaction to, pinches and pricks were only slight on the hands; the reflex to tickling of the soles was slight, also. Hypæsthesia of limbs. Right othæmatoma developed. A necropsy was not permitted. The case seemed to me very like tabic arthropathy. There was, probably, a slight fracture, or merely the giving way of a spongy, eroded femur-head and neck, under the slight shock of the fall.

(*b.*) Dr. J. C. Shaw * found several cases of g.p. with bone affections as in the arthropathy of tabes dorsalis. In one case, there had been rheumatoid shooting pain in limbs, back and abdomen, incontinence of urine, anæsthesia, slightly ataxic gait, absent knee-jerk; and there was absorption of the head of the right femur, the entire upper part of the bone being more spongy than natural; also, a separate deposit of bone below the trochanters. In a second patient, redness and swelling of all the finger-joints, with distinct crepitus, but no pain on manipulation (no necropsy).—Third case: swelling of right side of lower jawbone, crepitus, necrosis fibulæ; redness, swelling and abscess, in connection with right hip-joint; anæsthesia.—Fourth case: commencing absorption of head of right femur, no joint-symptoms during life, some muscular rigidity, and increased knee-jerk.

Thus, neither of my cases ended in a necropsy; and in the

* "Seguin's Arch. of Med.," Apr., 1883.

cases I have quoted the only necroscopical information we have is of absorption of heads of femora.

Perforating ulcer of the foot, especially associated with tabes dorsalis by Duplay, Morat, Ball, and Thibierge, has been observed in g.p. by Lancereaux, Christian, and probably others.

Blood in g.p. In the last stages of g.p. there is obviously a spanæmia; and, when not congested through vaso-motor paresis, the features are often pale, or of sallow or parchment tint.

Some of the older original, practical, researches on the blood of the insane, as those by Thakrah, Hittorf,* and Erlenmeyer,† do not appear to have been separately applied to g.p. as a class apart. The second of those just named came to several broad conclusions as to the blood of the insane; one being its usual poorness in corpuscles and its hydræmia—" Das blut trägt die charaktere der hydræmie" —and others that the differences in the blood of the sexes were marked, here, as in health; and that the changes of it in recent mania were not very great.—Erlenmeyer came to similar conclusions, and to that of the frequent existence of a " dissolution " of the blood, under which head he gives, amongst others, a case, apparently g.p., with pemphigus, bedsores, &c. Erlenmeyer also found a diminution in the solids of the blood, chiefly in the fat, also well marked in the blood-corpuscles, which were about 80 p.c. of the mean given by MM. Becquerel and Rodier, and 75 p.c. of the normal according to Prof. Lehmann (" Physiolog. Chemie."); the albumen, salts, and extractive being, on the other hand, increased. Hittorf, also, found the hæmatoglobulin markedly lessened. The results of these two observers were obtained by absolute and minute laborious chemical analysis.

M. Michéa ‡ found :—1. The quantitative analyses of the blood very variable in g.p.—2. Augmentation of the blood-globules (venous crasis of Germans) in the majority of cases; while, in a strong minority they are normal in amount; and in a small minority are lessened.—3. Fibrine, normal in amount in the majority; absolutely lowered in a minority of cases, but increased in a smaller minority.—4. Solids of serum (organic and inorganic) usually normal, but in a few cases notably increased.—5. Organic solids of serum (albumen

---

* " Diss. de Sanguine Maniacorum."
† " Ueber das Blut der Irren."
‡ " De l'état du sang dans la paralysie générale des aliénés." Académie des Sciences de Paris, Nov. 29, 1847.

chiefly) notably diminished in nearly one-third of the cases.— (*Mem.* "4" and "5" are discrepant, probably from a misprint. W.J.M.)—Dr. A. J. Sutherland and Dr. Bence Jones* found the albumen in the blood diminished in g.p.

Dr. Hy. Sutherland † in 1873, examining the blood microscopically, found chiefly an augmentation of the number of leucocytes, and an absence of rouleaux-forming power in the red corpuscles, and these changes far more marked or frequent, as a rule, in *male* cases. Apparently, even some male G.Ps. examined, presented neither of these changes. Ten male G.Ps. seem to have had one or other or both (4 both), and the total number of G.Ps., male and female, to have been 29. "Absence of rouleaux, together with an increase in the colourless corpuscles, appear to be conditions almost peculiar to general paralysis, and were observed in 4 out of 29 (male and female) cases, or in 14 per cent." The two conditions rarely coexisted in other forms of insanity, (4 out of 114); either condition alone was more frequent in g.p. than in the less fatal forms of insanity.

In the final periods of g.p., Dr. Voisin ‡ describes the blood, under the microscope, as too thin, and the grouping of its red globules as too decided and too irregular; and, viewed by the naked eye, as fluid, sticky, not coagulating, or as forming clots which are defective in cohesion and float in a reddish serosity. He also noted a relative increase in the number of leucocytes. These conditions he declared to be constantly present in the third stage; and crystals of urate of soda to be sometimes observed. The *abnormal* nature of some of the appearances he describes and depicts is not clear. Dr. Daniel Brunet § examined the amount of fibrine in the blood in g.p. by the method of Andral and Gavarret. He found the fibrine to be in proportion to the inflammatory symptoms; therefore, to be not increased in cases advancing slowly and regularly, in these, indeed, it may be diminished; rising to nearly ·6 p.c. when inflammatory symptoms are intense and acute. The normal being 2·2 to 2·3 per 1000, the average in 24 cases of g.p. was 2·6. In only four cases was it above 4 per 1000; the highest proportion reached was 5·9 per 1000.—Dr. S. R. Macphail ¶ examined the blood of

* "Medico-Chirurg. Transactions," 1855.
† Roy. Med.-Chir. Soc., London, Apr., 1873. "Jl. Ment. Sci.," Apr., 1885, p. 148.
‡ *Op. cit.*, p. 160.
§ "Ann. Médico-Psych.," Jan., 1881, p. 17.
¶ "Journ. Mental Science," Jan., 1885, p. 488.

five G.Ps. on admission, of five later on, and of five in a bed-ridden and completely paralysed condition. As compared with the normal :—

1st group av. p.c. of hæmoglobin, 66·2 ; of hæmacytes, 88·7 : 1 white to 308 red b. corpuscles.

2nd group av. p.c. of hæmoglobin, 70 ; of hæmacytes 86·5 : 1 white to 176 red b. corpuscles.

3rd group av. p.c. of hæmoglobin, 60·6 ; of hæmacytes, 78·1 : 1 white to 124 red b. corpuscles.

"The percentage of hæmoglobin is. low on admission, it improves in the quiescent stage of the disease, and falls again in the paralytic stage. The red corpuscles deteriorate both in quality and quantity coincident with the progress of the disease. Small granule cells [hæmatoblasts] are not present in the blood during the last stage. The relative proportion of white to red corpuscles is increased, and this increase is coincident with the progress of the disease." In the last group the blood was dark, venous, the rouleaux-forming power slight; the red hæmacytes were crenated, some irregular in outline; in one case, many were tailed.

CHAPTER X.

SEIZURES.

Special seizures in g.p. These seizures occur in various cerebro-spinal affections, and are not limited to g.p. Some G.Ps. remain entirely free from any of them, throughout. The chief ones are the so-called "paralytic attacks," which I sub-divide into the epileptiform, apoplectiform, and simple paralytic. These cover most of the ground formerly included by some under the name of congestive seizures. Another variety is the tetanoid, another the hysteroid, seizure. The maniacal attacks are sometimes included here, but are more conveniently classed elsewhere. With regard to the several varieties of "paralytic attack," there is at least a close pathological connection between them; and it is probable that what has been called discharge, with consequent ex-haustion, according as it chiefly affects different parts of the brain may lead in one case, or in one occasion, to convulsive seizure, in another to coma and paralysis, in another to paralysis without coma. When the convulsive phenomena have passed away, the patient is often in very much the same state as regards mentality, and motility, and sensibility as he is immediately after an apoplectiform seizure. Nevertheless, they are separate branches, if of the same tree, and their clinical distinctness is sufficient.

Epileptiform seizures. The epileptiform seizures, sometimes said to occur in its advanced stages only, may in reality take place during any part of the disease, nor is it at all uncommon that the recorded case-history opens with symptoms of a convulsive nature. On the other hand, they may be entirely absent throughout. During the first few months, the seizures in question, if present, are, as a rule, comparatively rare; then after the eighth or twelfth month of the disease they are apt to become more frequent, and to continue so. Esquirol * asserted the almost invariable attendance of convulsions upon the closing days of g.p., and certainly this often occurs. Before the attack the patients often show some change, are irritable, or restless, or dull.

The utmost variety obtains in the degrees of their severity, extent, and duration. In a few cases, a severe general convulsion occurs which is precisely like the well-developed convulsion of *epilepsia gravior*. This statement is faithful to nature, although it clashes with views that have been expressed, and the characteristics sometimes described as essential points of the differential diagnosis between the convulsions of g.p. and those of true epilepsy, cannot be applied to cases such as these. Examples I have observed disprove the characteristic nature of any features of the convulsions relied on by some. M. Aug. Voisin,† also, observed the likeness between some epileptiform seizures in g.p. and the seizures of true epilepsy, and found the same high and vertical ascent, the same dicrotism, the same general form of the sphygmogram, in each, immediately after the convulsion. I have also noted in some cases the peculiar mentation, delirium, and speech-condition after the epileptiform attacks of g.p. that may be observed after some of the convulsions of some insane epileptics.

Several seizures, even when so severe as just described, may be repeated in close succession, and the *status epilepticus* be attained. These sub-intrant convulsions are often dangerous to life, or even lethal. As each violent convulsion succeeds each, the coma deepens; the face grows more turgid and livid; the respiration more laboured and noisy; the pulse more rapid and irregular, its fulness eventually making way

* " Des Maladies Mentales." T. ii, p. 264.
† " L'Union Médicale." Aug. 4, 1868, p. 87.

for smallness; swallowing more difficult, or impossible; pulmonary congestion, œdema, or lobular pneumonia more marked. All these symptoms may subside, or be relieved by treatment, and the patient be delivered from desperate straits. Or, after incomplete recovery from the effects of the first set of fits, another set may supervene, or the patient be harassed by recurring convulsions, never again quite approximating a return to the condition preceding the initial seizure.

But in the majority of cases the convulsions are neither so extensive nor so severe. They often afflict the face and arm, or the face, arm, and leg of one side; and with them there may, or may not, be slighter twitches, of less extensive range, on the opposite side. When the seizure is of a still slighter nature, twitches may involve some isolated muscles, or groups of muscles, on one side or both sides. These various convulsive and spasmodic attacks may occur in rapid succession, or there may be several in the twenty-four hours; or a set, consisting thus of a variable number of seizures, may be prolonged for several hours or days, and may recur very soon, or only after the lapse of weeks or months, or not at all. In cases such as these, consciousness is sometimes completely, at others partially, lost; but often remains apparently unaffected. This retention of consciousness is far more apt to obtain in those instances where the convulsions are resolved into local convulsive twitch of several groups of muscles, or of one group; convulsive twitch which may continue for hours or days without ceasing; may transfer itself to other parts; or may alternate with convulsive twitches elsewhere.

Then, again, it is a highly characteristic condition when the patient is seized with convulsions, at first tonic, then clonic, affecting the face mainly on one side, although the other side may also be affected in less degree;—when, with this, the muscles of the upper extremity take on marked convulsive action, which perhaps also invades the lower extremity slightly, and continues for several minutes without any, or with only slight or moderate, loss of consciousness;—when, after this, there is a calm, broken only by spasmodic jerks about the mouth or the eyes, together with twitches of the forefinger and thumb, or by jerks at the shoulder joint; —when, next, perhaps, a recurrence of the former widely-spread unilateral convulsion takes place; or a short, sharp, tonic spasm, drawing the head violently to one side, distorting the features, and stiffening the upper limb in various directions, is followed by rapid clonic succussions;—when, possibly, the spasmodic jerks now temporarily emigrate to new fields, invading fresh groups of muscles, perhaps on the opposite side of the body, and, for the time, abandoning those they formerly occupied; and, when, after a repetition of convulsions such as these, especially of the more strictly localized ones, there is more or less paralysis of the upper limb and face on the side principally convulsed, and, perhaps, in slight degree, of the lower limb also, as well as a considerably heightened axillary temperature, which, of the two, is higher on that side which was mainly convulsed at the first, and which has now become temporarily paralysed. Conjugated deviation of the head and eyes is often observed here; they usually turn *from* the paralysed side. A few years ago, this conjugated deviation was thought to be a rare condition, and I believe it was first mentioned as occurring after the seizures in g.p. by the present writer, and by M. V. Hanot. I have seen this rotation and deviation in g.p. on very many occasions. The eyes, alone, may deviate, and sometimes first to one side, then to the other; or with hemiplegia may be paralysis of third cranial nerve of same or of opposite side. During and between the actual seizures some muscles may be rigid and unyielding. Thus may the forearm be rigidly flexed across the chest, the fingers and thumb buried in the palm, the head and neck distorted.— Almost invariably the above paralysis clears up in a few hours or days. But it is not only after the most severe and extensive convulsions that a post-convulsion paralysis is seen; often a very partial and a very incomplete paralysis, essentially of the same nature as so-called epileptic hemiplegia, obviously follows the more intense of the localized and limited spasms,—spasms limited, for example, in distribution to the face, or the face and tongue, or an upper limb, or to parts of either of these. In truth, had we sufficiently delicate means of gauging it, some minute local loss of motor power would no doubt be found in the part affected after every spasm, however slight. For paralysis follows these convulsions and spasms as the shadow follows the body.

The muscular spasm or convulsion often starts from some point, as it were, becomes widely spread and severe, then ebbs away, and ceases everywhere except at the starting-point, usually the mouth, eye, or hand, where occasional jerks are seen, which may gradually die out; or, on the contrary, the preceding cycle of events may be repeated; or the renewed convulsion may chiefly affect the other side. Local convulsive jerks may affect some part or parts continuously for hours, or even days, together.

Yet often the attacks are only very slight, and analogous to the *petit mal* of epilepsy. Thus, they may merely consist of a sudden pallor, with mental confusion; or of dilatation of the pupils, with drawing of the head to one side, or of the mouth agape ; or of sudden fixation of the lineaments, or an expression as of shock, together with cold perspiration, or with the muttered automatic repetition of coherent or of incoherent phrases. No doubt these are often mistaken for simple ordinary syncopal attacks. And, indeed, if we may accept Cullen's definition of syncope—" motus cordis imminutus vel aliquamdiu quiescens "—then syncope *is* an essential element of these seizures. On several occasions, when examining the chest with monaural stethoscope, I have found the heart-sounds cease at the commencement of a slight or of a severe seizure, and, since the first occasion, have been warned thereby of the oncoming seizures before being able to see the patient's face, and thus have been able to change his position in time. Cardiac failure in epilepsy is insisted upon by Dr. W. Moxon. And I find that an older writer * has noticed the connection between syncope and epilepsy, and had " twice seen violent convulsions, quasi-epileptic, in patients not subject at all to epilepsy, attend restoration from syncope."

Sometimes with severe epileptiform seizures, chiefly unilateral, consciousness is fully retained, and the patient is the witness of his own attack, answers as well as he did just before it, and assists in the treatment of his case. As an example :—

Case 29. After left hemiparesis, numbness of left upper limb; very slight left ptosis ; occasional spasmodic jerk of right neck and left limbs ; was a prolonged sinistral epileptiform seizure, with fully retained consciousness and ability to swallow and to speak relevantly about his symptoms; the left arm was bent at a right angle at the elbow, and the patient complained of severe pains in it : face flushed, right pupil slightly the larger, both of full size. Active treatment; and the fit ceased somewhat abruptly after an expression of suicidal desire by the patient. After the seizure, left hemiplegia and numbness, particularly in left upper limb. Next day, an almost precisely similar succession of events, in a similar seizure.

* Dr. S. H. Dickson, " Elements of Medicine," 2nd edit., p. 362.

Sometimes, there is a comparatively slight convulsion, followed by relatively more pronounced and protracted apoplectiform symptoms, and forming a link between the epileptiform and apoplectiform seizures.

Case 30. Slightly convulsed "all over;" afterwards, deviation of head and eyes to left, with slight jerks of them thitherward, the iris acting with each jerk of the eyeballs: conjunctivæ injected: dusky, leaden, hue of face, partial unconsciousness, breathing easy, slight convulsive tremor. Some paresis of left limbs, especially of arm; pulse small soft 90; temp. 97·5°, feet cold; bowels costive. Soon, the eyes were turned to the right.

In some cases where the status epilepticus is established the attack is fatal, the patient dies, convulsed to the last; or, more generally, a state of profound coma supervenes, which rapidly deepens into death; or into death after a day or two. An example of the former, *i.e.* of convulsions of lethal effect continuing up to the time of death, is that of H. R., case 65 at the end of this book, in which they lasted for forty hours. An example of the latter, *i.e.*, of severe epileptiform seizure followed by profound coma for a day or two, and then by death, is one where :—

Case 31. The first "fit" was about a month before death; the second formed the onset of the fatal attack, and began as violent general convulsion, but at the close was sinistral only. The visage was then turgid, purple; the coma deep; the pupils were equal, sluggish, smallish; pulse varying from 120 to 140; respiration irregular; limbs flaccid. Under active treatment, better next day; T. left axilla 101°; p. 100, irregular; resp. 40, jerky; restless movements and spasmodic twitches of left arm, and occasionally of head and leg, and of jaws: profuse sweat; at times unconscious; unable to swallow; nutritive enemata; next day, hypostatic congestion and pneumonia. Death 44 hours after the fit.

Death in the epileptiform convulsions may appear to be directly due to the effects of the violent repeated convulsions, under which, and the circulatory and other disorders engendered thereby, the vital powers become exhausted; or it may be from asphyxia in a furious convulsion; or from pulmonary congestion, œdema, and hypostatic pneumonia; or from profound supervening coma and palsy, with or without pulmonary congestion and inflammation.

The number of fits that may occur in g.p. is sometimes very great. A soldier under my care, who for several years was subject to severe recurring epileptiform seizures, was at the last carried off in a convulsive storm, in which the fits, as they occurred, were carefully enumerated in writing by the attendants for 24 hours. During this space of time he had 245 epileptiform seizures of severe character. He lived 4½ hours longer, during which time he had still other generalised epileptiform convulsions between 30 and 40 in number.

Dr. C. F. Newcombe * found that the liability of a G.P. to epileptiform seizures did not increase in proportion with the patient's age when attacked with g.p.; that of a series of cases, the average total stated duration of the disease, and the average total length of asylum-residence until death, were longest in those with apoplectiform seizures; next, in those with epileptiform, and shortest in those not reported to have had seizures of either kind; that where epileptiform seizures *did* occur, the older the patient the shorter the average total duration of the g.p.; that while most frequently occurring towards the close, they occasionally happened shortly after the commencement, of the disease; and that, on the average, the younger the patient when attacked by g.p., the longer after its commencement did these seizures (if they occurred) first appear.

In some cases there is a broad similarity between epileptiform seizures and those produced by irritation of the epileptogenous zone (Brown-Séquard) resulting from some injuries to the spinal cord in guinea-pigs. In relation with this subject, is that of the production, or recurrence, of epileptiform convulsions in g.p., from external irritation, not seen by Westphal, but of which Calmeil † gave an instance in which irritation "of the tactile sensibility" of many parts produced, at every trial, increase of the convulsive movements of the face and four limbs. Voisin relates a case in which, after convulsive seizures, the temp. was 42·3° C.; simply the act of touching the hands or the forearms brought on slight clonic convulsion in the upper limbs, which spread to the lower limbs. Touch of the chest, neck, and face had no effect. Besides other lesions, at the necropsy, there was considerable hyperæmia of the meninges of base and of convexity, and of both grey and white substance; and in the posterior part of the spinal meninges some ecchymoses. Dr. Newcombe also mentioned a case, in which, after frequent seizures, it was noticed on one occasion, that, although they had not been convulsed, exposure of the legs brought on convulsions in them. The above condition

* "West Rid. Asyl. Med. Rep.," Vol. v, p. 198.
† *Op. cit.*, 1859, T. ii, *Obs.* 118.

is, I think, more frequent than is usually supposed; and I have so often noticed re-breaking forth, or an exacerbation, of convulsion appearing under the influence of slight forms of irritation or disturbance, that I have long made it a cardinal point in the treatment to keep the patient as undisturbed as possible, compatibly with the thorough trial of the measures available to combat the convulsive tendency.

Symptoms after epileptiform seizures. Temporarily, or permanently, after severe seizures, two conditions are usually noticed; *i.e.*, *after* the stupor, paralysis, and other symptoms about to be mentioned as immediately following the convulsions, have cleared away. These conditions are; (*a*) the mental state has further deteriorated; and (*b*) the motor disorder and failure have become more marked. Yet the slighter attacks leave no decided traces.

For the state of the *temperature* after an epileptiform attack, see Chapter XI. During the attack perspiration, and discharge of urine and fæces are common occurrences. The urine is sometimes retained, and the bladder distended, contemporaneously with coma after long-lasting convulsions. (See also the section on *urine* in g.p.).—If palsied, the limbs may be flaccid; or the arms or an arm may be rigid and half-folded. The rigid arm is usually the more helpless one. After a general convulsion, beginning at the mouth, the head and eyes may deviate to the left, the left upper limb being rigid, the right flaccid; the rigidity being, here, of the less palsied side (case). Reflex movements may be increased or lessened. Conjunctival reflex may be lessened or absent on one or both sides, and the conjunctiva injected. Dr. Claus * thought the tendon-reflex to be always heightened after "paralytic seizures" in g.p.; Dr. Zacher † found it always increased after convulsion in g.p.; but lessened when paralytic symptoms alone appear. Thus, tendon-reflexes during and after apoplectiform and epileptiform attacks, are increased on the side of the body on which motor-irritation phenomena play, even if the same parts are paretic; but are diminished or lost on the side where simple lax paralytic states have appeared; and their increase or diminution is, respectively, exaggerated or checked by any previous increase or lessening of the tendon-reflex, as the case may be; paralysing, or irritating, brain-lesions similarly affecting spinal reflex centres (?). In a case of combined posterior and

---

* ' Allgem. Zeitschr. für Psych.," xxxviii Bd.
† " Archiv für Psych.," xiv Bd., p. 464.

lateral cord-sclerosis with absent tendon-reflex, no tendon-reflex was found after the attacks. Reaction to painful impressions may be lessened in the lower limbs; as also cutaneous sensibility; but occasionally the skin is hyperæsthetic. With unilateral anæsthesia may be congestion and œdema.—Impairment of consciousness after the convulsions, when it exists, varies from deepest coma to slight obnubilation of mind, but is slight or absent with some local, or even widely-spread, spasms.—For visual disorders and defects after seizures, see Chapter VI.—Aphasia often follows the seizures, especially the dextral unilateral ones. Other impairments of speech, chiefly defects of articulation, often succeed the convulsions, of either side or of both; and gradually, sometimes rapidly, clear up.—Dysphagia is frequent; associated with this is a tendency to inhale into the lungs the fluids taken by mouth, thus causing or aggravating pulmonary congestion or inflammation.—Respiration is often heavy, even stertorous, increased in frequency, sometimes with mucous *râles*. In some cases it is irregular, even to an "up and down rhythm."—The pulse may be intermittent; or full and feeble; or soft and small; usually frequent, it may be slow. A quasi-syncopal condition may ensue.—The face may be venously congested, or flushed, or pale. As compared with their previous condition; the pupils may be dilated, one or both; or, rarely, may be irregular, equal, contracted; and dilate unequally when one attempts to rouse the patient.— The feet may be cold; the surface of the body hot and moist, or hot and dry. With recurring sinistral convulsions, the left limbs may be warm, the right limbs cold, for several days; the temperature, higher in left than in right axilla, being above normal in both (*case*).—Blebs may form on the skin; or even a rapidly-formed bedsore on the palsied side. The abdomen sometimes becomes tympanitic, distended by the gas-swollen bowels.

Apoplectiform seizures. Under this head are often included cases in which apoplectiform symptoms are associated with convulsions, and strictly speaking this is correct. But it is convenient to consider separately the apoplectiform seizures occurring without convulsions. Reference is not made here so much to the apoplectiform seizures occasionally observed at the very outset of g.p., or upon the heels of which it seems to swiftly follow, as to those which often chequer its later morbid career. Sometimes the apoplectiform attacks seem to be gradually replaced by epileptiform; so that in the earlier part of the disease are recurring apoplectiform; in the later part, recurring epileptiform, attacks. Well-marked apoplectiform seizures are less frequent in g.p. than epileptiform. Apoplectiform attacks may come suddenly and crushingly upon the patient; on the other hand, they are often preceded by the occurrence or increase of insomnia, restlessness, or excitement; by redness and heat of face and head, or by a mental heaviness, and an aggravation of the ataxic and paretic disorders. They may be ushered in by vertigo and muscæ. They are often recurrent, vary from the slightest shade to the most extreme degree of apoplectiform unconsciousness and coma, are immediately preceded, and are accompanied by some elevation of temperature, are partially expressed in turgid congestion of the face, heated head and skin, a rapidly heightened axillary temperature, sometimes dilated pupils, and involuntary passage of urine or of fæces; and are not unfrequently followed by temporary paralyses, of which the most common is a partial, or, again, an incomplete hemiplegia, and with this, conjugated rotation of the head and eyes to the opposite side, or first to one side, later to the other, may be found, while an eschar may appear on the buttock of the palsied side.* In some cases all the limbs appear to be powerless. Sometimes these attacks occur, unnoticed, in the night or, if very slight, in the daytime, yet leave recognizable traces;—local pareses, hemipareses, wide-spread motor feebleness, increased speech-disorder.

A form of alternate paralysis with, and succeeding to, apoplectiform attacks, I have met with several times, but do not recollect having seen it described as occurring *in g.p.* by any other than the present writer. It is hemiplegia of one side, with paresis or paralysis of third cranial nerve (or fibres of it) on the opposite side. In the following case there was the intrusion of some spasmodic twitches.

Case 32. · Seizure of left hemiplegia, face included; and paralysis of twigs of right third cranial nerve (dilated and immobile right pupil, and external strabismus). Occasional spasmodic twitches about left limbs and about face. Afterwards, with much tremor, were restless fidgety movements, teeth-grinding, and muttering. Temp. 97°. Next day, the left hemiplegia and right third nerve paresis were lessened; but, now, in the left eyelid ptosis had appeared, and a branch of the left third cranial nerve

---

* See cases at end of this work.—Also Dr. V. Hanot, " Comptes-rendus des Séances et Mémoires lus a la Société de Biologie," 1872, p. 61, &c.

was affected. Three days later, the left limbs were stiff and palsied. Right pupil now acted slightly but was wider than left Dysphagia. Two days later, drowsy; left limbs stiff, helpless; slight left strabismus and ptosis ; axillary temp. 99·5°. Another example is case of C. E. No. 60, at end of this book; see note of Jan. 31. Another, but connected with embolism, is case of H. L. No. 67, at end of this book. Another (see " tetaniform " seizures) is case No. 34, below.

Symptoms following apoplectiform attacks. Mental dulness, heaviness, drowsiness, or restlessness generally succeed the decided attacks, and hours, or days, or even weeks, but usually several days, elapse before the patients return to their former state; and yet the return is scarcely to their former state, but rather to a lower level on the decline due to dissolution, and after each such attack the mental and motor symptoms, as a rule, are slightly worse. Immediately after the apoplectiform seizures, or whilst they are clearing up, there may also be temporary defective-sensibility in the paralysed limbs; tremors or subsultus, occasionally; or, later on, convulsive jerks. The palsied limbs may be limp and flaccid, or this condition may alternate with rigidity ; or the upper or all the limbs may be rigid and resistant to passive motion. Dysphagia is frequent. The lower limbs are often cold, even when the temperature is above normal in both axillæ ; the temp., as a rule, is lower in the axilla corresponding to the colder foot. Sometimes the face remains flushed, or purple ; or there may be extensive venous congestion of the surface. Dr. Zacher (*loc. cit.*, p. 597) described a vaso-motor phenomenon after apoplectiform and epileptiform attacks. Lightly stroked, the skin shows a white line, this becomes intensely red, swellings follow, coalesce and form an elevation of the skin of the part, chiefly on the flexor surfaces.—Psychical blindness, when it occurs in g.p., usually follows apoplectiform and epileptiform attacks, particularly with dysphasic elements, and dextral paralysis or motor disturbance.—There may be albuminuria, with casts and hæmacytes. After severe seizures both apoplectiform and, especially, epileptiform, and partly from difficulty in food-administration, the weight of the body is sometimes found to have declined.— There may also be some or any of the following :—Sweat, hiccough, swollen tympanitic abdomen; retention of urine, frequent or laboured respiration, pulmonary congestion and œdema, moist rattles ; a pulse, usually too frequent, but varying much in other qualities. Like as at first, so in the middle and later stages, active and expansive symptoms in g.p. may follow an apoplectiform attack. For temperature-relations see *temp*. in g.p.

Death may occur, in the apoplectiform attacks, from gradually increasing coma and palsy; or from pulmonary congestion, œdema and inflammation; the latter condition sometimes a sequel to inhalation of fluid food, owing to the paralysed and anæsthetic condition of the parts, and the comatose state of the patient. While death may occur very rapidly after the patient is struck down, yet it usually occurred in the space from the fourth to the eighth day in Calmeil's cases. Space-limits preclude the insertion of illustrative cases I had prepared. In fatal seizures associated with unilateral palsy; congestion or hæmorrhage, or a predominance of the ordinary histological changes of g.p., are usually found affecting the brain or meninges on the side opposite to the paralysis.

Simple paralytic seizures. The simple paralytic seizures, as I have termed them, comparatively rare in their marked degrees, are those in which sudden motor collapse and decided local paralysis, nay, even at times a hemiplegia, occur without convulsion or spasm of any kind, as far as is known, and without any observed indications of coma. But they are usually accompanied at first by some pallor, or facial expression as of shock, or even by a momentary mental confusion and slight obnubilation, making one think of an epileptoid basis. Almost invariably, these paralyses clear up and vanish rapidly, or, at least, without any very prolonged delay. Another phase of the same is that, over and above the more or less generalized paresis from which the patients may suffer, there is sometimes found a very slight and very transient local loss of power; discoverable only by searching examination; independent, as far as known, of any epileptiform, or apoplectiform seizures; and possibly affecting now this part and now that, as, for example, one, or several muscles about the face, tongue, hand, or upper limb.

These transient paralytic seizures are widely distinct from the somewhat persistent hemiplegias occasionally observed. Yet, a latterly persistent hemiparesis may undergo sudden increases at times, without any decided convulsive or apoplectiform seizure.

Case 33. Simple paralytic seizure. When under observation, suddenly seemed to stumble, and be comparatively helpless, without any fit or marked defect of consciousness. Then he walked unsteadily to his room, spoke, but without other change became worse, aphasic, with dextral hemiplegia, flaccid right limbs, and inability to protrude the tongue; the only utterance=" no—no." Later, he muttered a few words, had cold feet, a slow pulse, and vomited once. Next day, face slightly flushed, pulse 54, T. 99·4° left axilla:—hemiplegia less; occasional jerks of right upper limb; no rigidity; patellar and plantar reflexes good: some analgesia of hands. Next day, occasional spasmodic twitches of right lower limb, which latter became rigid when tested for ankle-clonus. The hemiplegia slowly cleared up, and before the aphasia disappeared stuttering was very marked.

Tetaniform attacks. Occasionally, a G.P. is found to have recurring seizures, in which the frame, or part of it, is in tonic spasm. Usually, it is merely that the muscles of the chest and neck become temporarily rigid, the occiput boring backwards into the pillow; and these milder cases can only be called tetaniform, on the analogy of the naming of the lighter apoplectiform and epileptiform attacks.—Another condition is where tonic spasm seems very largely to replace the clonic of epileptiform seizures, giving the case clinically some tetaniform aspects. Thus, in one,

Case 34, a patient suddenly became dark in the face and passed into spasm, chiefly on the left side. Coming almost immediately, I found him nearly dead, but restored him by artificial respiration; and from time to time, the face became dark and respiration ceased in the violent tonic spasm; followed by some shaking and by rigidity, chiefly of right limbs. Pleurothotonos at times was marked; at others opisthotonos; the arms were rigid and thrust forward. Finally, the platysma and tongue were moving in spasm; teeth-grinding and champing movements of the jaws succeeded; respiration varying from 30 to 38. Pulse very rapid and feeble.—Afterwards, palsy of twigs of left third cranial nerve (external strabismus, dilated sluggish pupil), and coma. Almost universal muscular weakness followed for several days, particularly in left limbs; and during part of the time marked lower right facial paresis. Later, convulsions, left hemiplegia. Lived some time after. Spinal meninges considerably thickened and opacified; spinal columns somewhat indurate in the dorsal region. Encephalic lesions of g.p.

Dr. Voisin gives four cases; one or two of which appear to have been purer tetaniform cases than those I have mentioned; but others (certainly one) merely to have been scarcely more than a condition of great exaggeration of the ordinary superficial and deep reflexes, a condition not rare in g.p., and not deserving of the name tetaniform; another seems to have been only an example of highly-marked and immediate rigidity, accompanying some of the seizures frequent in g.p. In tetaniform attacks, he describes exaggeration of the reflex-power, rigidity of limbs and trunk, trismus and opisthotonos, and looks upon them as apparently in relation with a lesion of the anterior part of the spinal cord. The most marked case he mentions was in the last stage of g.p.; for three days the limbs were rigid; then every five minutes the head was sharply thrust backwards; there was trismus; the arms were extended and rigid; the muscles of the thorax were hard. Various forms of irritation of the trunk and limbs, and even the holding of a watch to the ears, brought on tetanic convulsions with opisthotonos.

Calmeil,* speaking of the convulsive phenomena intercurrent in g.p., says they may be accompanied with a tetanic rigidity of the muscles of the jaw and neck, a state of contraction of the arm-muscles, contraction of the wrist, stiffness of the hamstring muscles, and inability to swallow.

Hysteriform attacks. Except for emotional, hysteriform, displays, waves of uncontrollable feeling, and alternate weeping and laughter, associated, perhaps, with muscular trembling and twitch, I do not call to mind marked hysterical symptoms in g.p. In this work the description is almost entirely drawn from male cases. Voisin, whose treatise on g.p. is based almost entirely on female cases, has observed hysteriform attacks, and appearing in three cases. The briefest case is described in the following terms:— Sudden loss of consciousness, face pale, lips purple, stiffness of limbs, forcible flexion of hands, grimaces and deformity of the visage, tonic convulsions of the eyes, foam from the mouth;—after three-quarters of an hour the patient, who had recovered from the above condition, began to laugh and weep alternately. Hysterical crises in a male G.P., are also reported by Dr. Camuset.

Meningeal Hæmorrhage. Meningeal hæmorrhages, duramatral, arachnoidean, or sub-arachnoidean, are not rare in general paralysis. They give rise sometimes to no special symptoms; sometimes to apoplectic, or epileptiform symptoms and paralyses; sometimes they terminate fatally in a few hours or days; while their more usual ending (if large) is the formation of "arachnoid cysts." With reference to the symptoms immediately following considerable

* *Op. cit.*, 1859, T. i, p. 501.

effusions of blood in the arachnoid cavity, these are often attended by symptoms of the following kind.

Case 35. With hæmorrhagic clot in the left middle and posterior *fossæ basis cranii*, were; rigidity of right arm, and some paralysis of right arm and mouth and left leg; twitches of right hand and arm; then incomplete right hemiplegia—head and eyes turned to left—but in two or three days head and eyes turned to right: convulsions, especially of right side, increase of hemiplegia, frequent tonic spasm of muscles of mouth, neck and trunk. Stupor, hypostatic pneumonia.

Case 36. In another case, stertor, muscular twitching, distorted features, rolling eyes, clenched fists, flexed limbs. Later, a convulsion, followed by left hemiplegia, deviation of head and eyes to right, high temperature, drowsiness, pulmonitis. Later, widely-spread tonic convulsion, increase of hemiplegia and coma, stertor. Still later, dysphagia; sensation blunted; lessened sensori-motor activity on left side. Head and eyes were *now* turned to left; numerous epileptiform convulsions, deepening coma, death.—Clot and fluid blood in right arachnoid cavity, about base of brain, middle fossa, temporo-sphenoidal and occipital lobes.

Arachnoid Cyst. Durhæmatoma. Meninghæmatoma. Matrhæmatoma. As "arachnoid cyst" is a misnomer, and as I believe these hæmatomata not all of dural origin, but some from soft meninges, and "durhæmatoma," therefore, a misleading description, I propose to call the formation *matrhæmatoma*, or *meninghæmatoma*.

Case 37. Extremely large and thick double hæmatomata were found after death; the patient, from an early date, was more or less restless, sleepless, excited, noisy, destructive, and at times boisterous, mulishly obstinate, resistant to every form of necessary care and management, self-helpless, and neglectful, extremely demented and incoherent, and, indeed, usually uttering unintelligible sounds in monotone. Such mental action as was left to him seemed to be absorbed in the paramount activity of resistance to any and every form of care and attention. At last, severe convulsive seizures befel from time to time, and, finally, sub-intrant, and chiefly sinistral, attacks of the same lasted for 24 hours and ended in coma and death four months after admission.

Case 38. A large thick hæmatoma on the left side, somewhat flattening the gyri of the left cerebral hemisphere. Adhesion and decortication were slight, and almost confined to left hemisphere. The patient from an early date was demented; for a short time he had hallucinations and expansive ideas, but very soon was restless, excited, muttering unintelligibly, mischievous, of foul habits, and had bedsores before admission.—After admission, very demented, he scarcely spoke a word, but uttered frequent noisy inarticulate cries, was restless, mischievous, of foul habits, entirely self-helpless. Severe convulsions were followed by right hemiplegia, and, later on, by sinistral paresis. Before death, were spasmodic twitches in the right limbs, and subsultus, especially in the right lower limb. Usually, he was slavering and bending the head forward.

Case 39. In a third case, with large thick firm organized double hæmatomata, thickest over the frontal tips, with marked adhesion and decortication, congestion of brain, and the right cerebral hemisphere the one more diseased; the patient had had expansive symptoms, occasionally with hypochondriacal; but for nearly a year before death had been simply childish and demented. Partial blindness of right eye. No convulsions or hemiplegic attacks. Died with apoplectiform symptoms and high-smelling urine.— Besides the hæmatomata and lesions of g.p., and marked chronic meningitis, were right othæmatoma; aortic valvular incompetency; syphilitic (?) aortitis; extremely hypertrophied left ventricle of heart; mottled kidneys.

In two cases of g.p. with pachymeningitic neo-membranes, etc., reported by Dr. G. H. Savage,[*] the ordinary expansive symptoms were present. Incoherence was the chief symptom in another case by Drs. Savage and W. R. Wood.[†]

CHAPTER XI.

TEMPERATURE.

The Temperature in General Paralysis. A.m. temp. at 10 or 11 a.m.:—*p.m.* temp. at about 8 p.m. In the maniacal attacks of the early stage, and in apoplectiform attacks, Prof. L. Meyer found the temperature increased; and markedly so at the vertex. High temperature, and the febrile course in g.p.,—when not due to complications or intercurrent maladies,—he attributed to exacerbations of chronic meningitis. But Dr. v. Krafft-Ebing [‡] soon found cases of high temperature in g.p. without marked meningitis, and held the heightened temp., in g.p., to be a valuable diagnostic sign; as regards mental disease, occurring, besides g.p., only in some cases of acute deliria, or acute inflammations, Dr. Clouston § observed that, of the several forms of insanity, the mean temperature, of the group of patients representing each, was highest in g.p.; that in it, also, the average evening temperature was always above the average morning

[*] "Journal Mental Science," Jan., 1884, p. 512.
[†] *Ibid.*, July, 1884, p. 261.
[‡] "Allgem. Zeitschr. für Psych.," xxiii Bd., 3 Heft.
[§] "Journal of Mental Science," April, 1868.

temp., the difference between the morning and evening temperature being greater in it than in any other of the several forms of mental disease; and that "the temperature is high in the first stage of g.p., lower in the second stage, and again very high in the third. The evening temp. is most increased, as compared with the morning temp., in the third stage, and least in the second.". The convulsions were usually followed by increase of temp.

In general paralysis I have *almost* invariably found the *average evening temperature* in excess of the average morning temperature, but this not an absolute rule throughout the whole duration of every case, as may be seen by glancing at the table on page 37 of my contribution to the subject in the "Journal of Mental Science" for April, 1872. In the rare cases in which the *average* morning temperature exceeds the *average* evening temperature, its excess is but slight, lasts only a comparatively short time, and is restricted to the earlier periods of the disease. In each instance the averages in question are the averages of the axillary temperatures on thirty successive mornings and evenings, taken with every pains and precaution, and often for many months continuously so as to obtain a number of averages at different times. The following were the general conclusions at which I * arrived:—That in the middle and later stages of general paralysis of the insane with regard to *axillary* temperature:

1. A rise in the temperature often accompanies a maniacal paroxysm.

2. A rise in temperature often precedes and announces the approaching congestive or convulsive seizures, and nearly always accompanies (and follows) them.

3. When these congestive or maniacal states are prolonged the associated elevation of temp. is usually prolonged also.

4. Defervescence of temperature, after its rise with excitement or with apoplectiform attacks, often precedes the *other* indications of restoration to the usual state.

5. Moderate apoplectiform attacks, or moderate maniacal exacerbations, are, however, not invariably associated with increased heat of body.

6. A transitory rise in temperature may occur without any *apparent* change in mental or physical state to account for it.

7. The evening temperature is usually higher than the morning temperature in general paralysis, and an absolutely

* "Journ. Mental Sci.," April,1872, pp. 45, 46; words in ( ) added.

high evening temperature occurs in cases rapidly progressing towards death.

8. A *relatively* high evening temperature seems to be of evil omen, even when *not absolutely* very high.

9. Rapidly progressing cases may show temperatures above the average both in the morning and evening, for a long time before any complication exists.

10. Gradual exhaustion may pass on to death, in general paralysis, with an *average* morning temperature normal, or nearly so, (or subnormal), throughout, except when raised, temporarily, by the special attacks to which general paralytics are subject.

11. The onset, especially, of pulmonary complications, or of hectic from bedsores, is marked by much heat, and when death is accelerated by the former, the temperature and pulse rule high, often, however, sinking somewhat before death, whilst respiration then becomes very rapid.

These conclusions as to the state of the temp. in the apoplectiform and epileptiform attacks in g.p. have been confirmed by several observers, and among others by Dr. Riva * and Dr. Reinhard; † also Drs. Hanot, Magnan, and Voisin. Riva, also, has found times of excitement preceded by increase of temp. ‡ At an early date Dr. Saunders § published a highly marked case of temp. rise.—Dr. Wm. Macleod ‖ had previously stated in a Naval Medical Bluebook "that in all cases of paralysis of the insane there is a higher temperature in the evening over the morning, seldom less than 1° except in those cases where the disease is arrested, which occasionally happens; then it may be as low as ·1°."—Unless *average* temps. are meant, this statement would be too absolute, and need rectification. He found the temperature higher in the noisy sleepless and destructive G.Ps. Dr. Kroemer found the daily variations greater in the cases in which paralytic attacks occurred; and temp. higher when paralytic symptoms preponderate, or vascular stasis is considerable; lower in melancholic, tabic, and 'stupid' cases.—Now a normal temp. in the melancholic and hypochondriac, or in the quiet, demented, patient, would really be a morbidly elevated temp. for them, inas-

---

* " Rivista Sperimentale."—" Journ. Ment. Sci.," Apr., 1879, p. 125.
† " Archiv für Psych.," x Bd., 2 Heft., p. 461.
‡ " Journ. Mental Science," Jan., 1884, p. 586.
§ Dr. Maudsley, " Pathology of Mind," 1879, 3rd Ed.
‖ " Lancet," Nov. 19, 1870. His cases by Dr. Hack Tuke, " Psych. Med.," 1873, p. 325.

much as patients in those conditions, which are accompanied by defective exercise alimentation and metabolism, should naturally have subnormal temp. And, coincidently with this, moderately low temps. in G.Ps. who are in these conditions would be really normal to their mode of life and nutrition. As to the relatively higher *p.m.* temp., some of my observations on this point were first made public in a paper by Dr. J. W. Ogle.* Voisin only makes mention of the p.m. temp. being higher in those cases where febrile temps. last for several successive days.

As with regard to other conditions, so with temp. in g.p., it not seldom shows a preternatural mobility, a mercurial facility of change, as the result of relatively slight causes. Hence transitory circulatory changes may be evidenced in a rising wave of temp., and in unrest or tumult, mental and motor.

As to the final temps.; in one of 107·4° F., there were large hæmorrhages in the right hemisphere, and the temp., probably, was due to this complication (secondary rise). Mendel saw one of the same height, increasing at and after death to nearly 109°; and a case similarly ending with epileptiform attacks, and a temp. nearly as high (no necropsies mentioned). König (Mendel) and v. Krafft-Ebing † also saw cases with final temp. of about 109°. I have rarely met with extremely high temp. in g.p. The following are a few actual examples of high temp. I have seen. One case; last 5 evenings of life (hypost. pneumonia), 103·8°, 103·6°, 102·9°, 100·5°, 101·9°. Similar case, 103·5°, 102·8°, 101·9°, 103·6°, 104·8°. Eulenberg and Landois supposed thermic centres in the anterior central gyrus and anterior part of gyrus fornicatus, others in the pons; others in the upper part of the spinal cord. The question whether the fluctuations of temp. in g.p. are brought about, or partly so, by exacerbations and remissions of an inflammatory process, meningitic or encephalitic, or both; or whether they are due to affection of the above thermic centres, or, thirdly, to the pyrexiæ of intercurrent affections, is one not always easy to decide. In some cases I have felt quite able to exclude the last-named, and to distinctly say the temp. condition was due to the morbid process which terminates in the cerebro-spinal lesions found in the cadaver of the G.P. Circulatory disorders are an essential part of the affection, and either by these, or by the morbid process

---

* " Clinical Society's Trans.," London, Vol. v.
† " Allgem. Zeitschr. für Psych.," xxv Bd., p. 325.

of g.p. itself, affecting the thermic centres, are the thermometrical variations often produced; while in the cases of more active pathological changes, where adhesions are extensive, where softening has been rapid, and circulatory disorder extreme, the fluctuations depend directly on this cerebro-spinal morbid process. Of course, intercurrent diseases often do actually account for the change in temp., and it is one of the chief benefits of careful thermometrical observation in g.p., that it is precisely such variations that force us to examine closely and repeatedly for any accidental complications; or to perceive that the g.p. is in an active phase, and to adopt treatment accordingly. Here the diagnostic acumen of the skilled physician is invaluable. Sight must not be lost of the so-called "nervous" or "hysterical" pyrexia, of which M. Debove* has published many examples; nor of cases of spinal injury or shock in hysterical persons, attended with extraordinary deviations of temperature.†

It is quite true, as Voisin states, that when a patient is in a state of "stupeur," the thermometer may be of value in diagnosing as between this condition in g.p., and the simple "neuropathic" form. But I can hardly agree to his statement that the existence of fever is one of the best signs to aid in the diagnosis of g.p. at its onset. One can scarcely at this stage speak of "fever," or of "febrile temperatures." In doubtful cases I have found thermometrical observation of great value, both for purposes of diagnosis and prognosis; and, also, of great value for purposes of prognosis in marked remissions of the disease.

The local temperature of the head, and the relative height of the temp. of the head and axilla or rectum, have been investigated by L. Meyer, Albers, Voisin, Gray, Reinhard, Mendel and others. In describing the attacks, particularly the apoplectiform, I have mentioned the heat of head, perceptible to the hand, and of which there is ocular evidence not merely in the congested appearance of the face and head, but also in the rapid vaporization of cooling fluids laid or sprinkled on the head and features. Mendel failed to verify one observer's conclusion that the temp. of the head was higher than the axillary or rectal in g.p., for he usually found the temp. of the auditory meatus ·1° to ·3° C. lower than the rectal. As to the relative temp. of head and rectum in epileptiform attacks, he found, during the spasmodic attacks, a

* "Lancet," Feb. 28th, 1885.
† "Transactions Clinical Society, London," Vol. viii, p. 98.

relative increase of ear-temperature over rectum-temperature, which became equalized in the further course of the attack. But if the attacks continued long, as over an hour, the aural temperature sank 1° to 1·2° C. below the rectal, once 1·5° C. lower. One observer found head-temp., taken behind the ear, show a lowering in the beginning of the motor discharge, to rise again after a brief stage. It is an old observation* that in unilateral inflammation of the brain, the temp. of the head externally is higher on the inflamed side. In g.p., Dr. W. A. Hammond (*op. cit.*, p. 624) found the temp. at vertex, in some cases, raised 2° F.

The Temperature in Apoplectiform Seizures. It is particularly worthy of note that the axillary temp., during, and immediately after, an apoplectiform attack, is higher on the side of the body which alone, or principally, is paralysed (if there be paralysis). And that after an unilateral epileptiform seizure, the axillary temp. is higher on the side in which convulsion and paresis mainly appear. I based this general rule upon very numerous observations made during a number of years. (See 1st Ed., p. 44.) As an exception to this general rule, for apoplectiform seizure with paralysis, a case may be mentioned in which, with left hemiplegia of this nature, the thermometer rose in the left axilla to 99·4°, and in the right to 100°.

In a fatal apoplectiform seizure, in the course of g.p., the following is an illustration of the course taken by the temperature. On the first day, with extreme apoplectiform cerebral congestion, left hemiplegia also came on. The patient rallied slightly, but the symptoms returned in full force, and the pulmonary congestion began to give place to hypostatic pneumonia.

	Temperature in *left* axilla.	Temperature in *right* axilla.
1st day	103·2° F.	103·1°.
2nd ,,	103·2°.	102·9°.
3rd ,,	101·8°.	101·7°.
4th ,,	100·2°.	100·3°.
5th ,,	102·5°.	101·3°.

The difference between the temperatures of the two axillæ lasts sometimes for two or three days in the milder and non-fatal apoplectiform attacks. In these, however, the rise in temperature is not so extreme. Thus, in one case, after an

* "Lancet," March 7th, 1857, p. 236 (Mr. S. Solly).

attack of this kind with left hemiplegia, the temperature in the left axilla was 101·2°, in the right, 100·2°.

I reported many examples of the behaviour of the temperature preceding, during, and just after, apoplectiform attacks in g.p., at p. 38, *et seq.*, "Journ. Mental Science," Apr., 1872.

After these attacks the head and face are often heated; and with an axillary temp. above 100° F., the limbs, chiefly the feet and legs, may be cold, and unequally so. The rectal temp. has not been found higher by Mendel after the attack than immediately before it: but there was a relative increase of the temp. in the external auditory meatus, though only (in 2 cases) ·1° C. absolutely higher. But this observer does not appear to reckon with the fact established by my investigations that the temp. immediately before an apoplectiform attack is often a rise above the usual level, and heralding a seizure.

The Temperature in Epileptiform Seizures. After convulsive seizures in g.p., especially those that leave decided traces of palsy in the parts most convulsed; (1) the general temperature as a rule is increased, and (2) is higher on the side affected when the convulsions have been unilateral; and the temperature may remain higher for a day or two on that side than on the other.

But as to the former of these points—the general increment of body heat—it must be added that it may not occur, and again—that when the patient falls day after day into epileptiform convulsions the temperature at last is found in certain cases, to stand below the normal, thus forming exceptions to the rule.

And as to the second point also,—namely the higher relative elevation of the temperature on the side convulsed—it must be conceded that the temperature, after being higher on the side most convulsed, *may*, a day or two afterwards, sink to a lower thermometric level on that side than on the other, and this may even occur before all convulsions have ceased. Thus, in one G.P. in whom sub-intrant convulsions, chiefly dextral, produced temporary right hemiplegia with a right axillary temperature of 99·4°, and a left axillary temperature of 99°, the thermometer had sunk three days later to 97·8° in the right axilla, and to 98·3° in the left, although one or two mild convulsions still occurred each day. The right hemiplegia still persisted; yet the temperature of the right side, at first the higher, had now fallen to a point slightly below that of the left. In another case, six epileptiform seizures, chiefly sinistral, were followed by left hemiplegia, the thermometer reading 102·5° in the left, and 102° in the right axilla, or ·5° higher in the left; but three days later the left temp. (99·6°), was ·2° lower than the right (99·8°); the left hemiplegia being still highly marked, but no convulsion having occurred for nearly 24 hours.

Another exception is that in some cases the temp. is lower immediately after, than immediately before, the unilateral convulsions. Thus, on one occasion, the thermometer which had marked 95·8° in the right axilla had immediately afterwards risen to 96·6° in the left, when a fit, mainly affecting the left side, came on. The thermometer, immediately after the fit, replaced in the left axilla, now only marked 95·5°; or 1·1° F. lower than just before the convulsion.

The introductory rise of temp., Reinhard thought, might begin twenty-four hours before the attack, and rise during the attack, sometimes 1°C. With a succession of attacks a rise of temp. marked each fresh seizure. The more severe the attack, or series of attacks, the higher the temp. To this last, however, I had found many exceptions; as also to the usual rule of a slight increase of temp. with local pareses, hemipareses, and spasmodic contractions, in g.p. He also confirms my conclusions as to the occasional slight febrile rise of temperature in g.p., without obvious cause; and, like others, gives diagnostic importance to the variations in bodily heat. But Dr. Edmund Guntz* stated that the convulsive seizures were immediately preceded by, and accompanied with, a very trifling fall of temp. As a rule, this was rapidly succeeded by a rise in temp. during the continued convulsions. Or, with convulsion, there might be no temp. change; or the initial sinking alone; or a sinking followed, or not, by a temporarily stationary temp., and then by a rise; or followed at once by a rise. In either case the rise might be to the level of the former height, or might fall short of that. These fluctuations of the temp. were directly proportional to the severity of the spasms; yet this was not always true of the sinking of temp. The lowering of temp., before or at the onset of the spasms, amounted to between ·05° and ·625° C. The most rapid sinking was ·1° C in 10 seconds. An example of tardy sinking of temp. was ·1° C in 3½ minutes. Usually, the sinking of temp. followed in from 1 to 3½ minutes. Dr. Kroemer † also found temp. low one

* "Allgem. Zeitschr. für Psych.," xxv Bd., p. 165.
† *Ibid.*, xxxvi Bd., p. 137.

two or three days before the apoplectiform and epileptiform attacks, and sinking in the earlier moments of the seizure; and this supposed as an expression of the brain-irritation by which the spasms are called forth. The directly proportional relationship he claims between the severity of the attacks, and the height of the rise of temp. afterwards, is not at all true as a universal rule.

Simple paralytic seizures. In one example, a few hours after the onset of a left hemiplegia of this kind, the thermometer marked 99° in the left, and 97·8° in the right axilla (difference 1·2°). Two days later, the palsy rapidly clearing up, the left axillary temp. was 97·9°, the right 97·2° (difference ·7°). Some hours after a similar, but slight, seizure of sinistral hemiplegia the thermometer stood at 97·8° in the left axilla, at 97·3° in the right (difference ·5°).

Another patient suffered from an attack of dextral hemiplegia of this nature on the 9th, and the right temp. was 98·9°, the left 98·7°. The paralysis persisted more or less for some time, and on the 11th the temps. were: right axilla 99·3°, left 99·1°; on the 13th, right 98·8°, left 98·6°; and by the 16th, right 98·3°, left 97·4°. But frequent slight convulsive quiverings now began to play about the right side, and especially in the face, and by the 21st the right temp. was 97·7°, the left 96·7°; and by the 25th the right was 98°, the left 97°; the next phase being that dextral spasm and paralysis were augmented on the 27th, and the temp. had risen to 99·6° in the right axilla, to 99·2° in the left.

In a case of sudden right hemiplegia from embolism of the left middle cerebral artery, in a G.P., I found the temp. between one and two hours afterwards to be 101·5° in the right, and 101·3° in the left axilla.

Thermometrical Fluctuations and Pyrexia in g.p. Having thus fully spoken of the *average* temperatures throughout the course of g.p., and of the thermometrical perturbations attending its seizures, we may briefly refer to irregular fluctuations and pyrexial movements independent of special seizures. If we examine the temperature charts of G.Ps. for elevations above the normal temp., we find that they occur irregularly. From the following numbers, however, a slight deduction must be made, inasmuch as the readings at, as well as those above, 98·4° are included.

For example, in one G.P. with considerable and protracted maniacal excitement in the second stage of the confirmed disease: During one period (a) the morning temperature on fourteen of forty-six mornings was at, or above, 98·4°; then (b) during the next three months a rise above this line was observed only on the 1st and 2nd day of one month, on the 25th and 26th of the next, and on the 6th, 8th, and 9th of the next. The evening temperature, taken only during the former (a) of the above two periods, was at, or above, the same line (98·4°) on three-eighths of the evenings, and the rise occurred at irregular intervals, and on from one to three successive nights on each occasion.—Another, in the second (and early third) stage, had at one time no morning temperature above 98·4°, whereas on almost half the evenings the temperature was above that level. Three months later, the morning temperature occasionally rose to the line (98·4°). Eight months later, when the patient was much more demented and helpless, and had suffered *interim* from severe epileptiform convulsions, three-fourths of the morning, and all the evening, temperatures were above the line.— Another, in the final stage, whose morning temperature was taken regularly for four months before his death, had a temperature above the line (98·4°) on one-fourth of the mornings. His evening temperature, taken about four and three months before death, was above the line twenty-three times out of thirty-nine, remaining above it for from one to four successive nights at irregular intervals.—A somewhat rapidly advancing case of g.p. of the hypochondriacal form, at one time had nearly all the morning temperatures above the line, and *all* the evening temperatures either above, or at, it. Three months later, both morning and evening temperatures, generally above, were occasionally at or below the line. Eight months later, the same was still true of the morning temperatures, but *all* the evening temperatures were now above the line.—In another, in the second stage of typical g.p., both the morning and evening temperatures were above the line in two-thirds of the observations, remaining so during from one to four mornings or evenings in succession, at irregular but closely set intervals.

With reference to the pathological import of the greater tendency to elevation of vespertine than of matutine temperature in these cases, it must be borne in mind that, on the contrary, in health the morning temp. usually exceeds the evening.

Temperature during excitement. In the summary of the results of my thermometrical observations in g.p., it was mentioned that a rise in temp. often accompanies the paroxysmal maniacal excitement of g.p.; that if the excitement is protracted, i.e., for days, so, also, usually, is the elevation of temp.; yet that a fall of temp. here, may herald a subsidence of the excitement; and that moderate maniacal exacerbations are not invariably associated with increase of the axillary temperature. As an example of the usually associated fluctuations of mental and motor excitement and thermometrical readings, may be repeated part of a case published by me:—"When the a.m. temp. was averaging 99·2°, and the p.m. temp. 100·11°, it was noticed that on May 22nd and 23rd there was excitement, and flushed face, and a p.m. temp. rising above 101°, culminating on the 24th in a.m. temp. 101·1°, and p.m. temp. 101·6°, associated with much restless excitement. . . . Then, till the end of the month the temp. fluctuated slightly, and maintained the previous averages, quietude, or only moderate excitement and flushing, being present. In June the p.m. temp. averaged 100·53°, rising to 101·3° on the 12th, with noisy excitement; then gradually defervescing for a few nights, and shooting up suddenly to 102·4° on the 27th, when there was a paroxysm of excitement; the whole state being a strong contrast to the quietude and temp. of 98·6° on the a.m. of the same day." The flowing tide of excitement next ended in a slight apoplectiform seizure.

Nevertheless, there are exceptions; as in the case of a young, maniacal, G.P., with occasional apoplectiform and epileptiform attacks, and times of apparent exhaustion, of gastro-intestinal disturbance, or of refusal of food. At one time the temp. and waves of excitement had fluctuated in unison; but several months before death, by gradual exhaustion, the *average* a.m. temp. was 97°, the temp. was below 96° on one-sixth of the mornings, and on one was only 95·4°.

Whether as cause and effect, or whether as essentially associated coeffects of one and the same pathological state, or whether, as may frequently be the case, casually connected, certain it is that the two states, mental and motor excitement and restlessness, and a state of temperature heightened above its usual level for the patient, ordinarily go together. And they being in fact clinically associated one cannot fully agree with Prof. Westphal[*] that maniacal excitement in g.p. "does not in itself hold a relation to pathological increases of temperature;"—although there may be considerable deviations of the temp. of the body

[*] "Archiv für Psych.," i Bd., p. 204.

quite independently of excitement. The coexistence of excitement and high temp. has also been noticed by Meyer, Regis,* and Reinhard.†

Subnormal Temperature in g.p. In some examples I have found the average morning temperature (average of 30 successive mornings; 39 in the first case, below) moderately under the normal at one or at several portions of the course of an individual case. In one case the average being 96·95°; and in another, 97° one month, and 97·6° another month; in a third patient, 97·65°; in another, also 97·65°; in a fifth, 97·72°.—In the first case, with an average a.m. temp. of 96·95°, that temp. ranged between 95·1° and 99·3°, being occasionally raised during paroxysms of excitement with turgescence of head, but on eight mornings out of thirty-nine it was as low as from 95·1° to 96°. At that time the patient was bedridden, demented, and near to death. In the second case the subnormal average was obtained when the patient was bedridden and nearing death. The motor disability was extreme, so was the dementia; and many apoplectiform and epileptiform attacks had preceded. The fourth example was g.p. associated with tabes dorsalis. Nearly four and a half years after admission, and when he had been bedridden for fifteen months, the temps. in question were taken, and he was then extremely demented, noisy, destructive. He died two months later.

Thus we find comparatively low temps., especially morning temp., in not a few cases of advanced g.p., when the dementia and paresis are well-marked, excitement and maniacal restlessness absent, or not extreme, and visceral organic lesions not yet come to being. If a G.P., with febrile visceral disease, has no elevation of temp., it points to a previous subnormal state of temp., or to a peculiar inhibiting effect of the cerebro-spinal disease, or to some other influence on a heat-regulating centre. Voisin connected low temp. with uræmia in g.p. This I doubt for most cases; Mendel doubts it too.

Kroemer assigns to the slightly subnormal temp. of some G.Ps. a greater relative frequency than I do. He also, as I did previously, found the fluctuations less in quiet patients; greater in patients with seizures, or with excitement, or with increase of the motor disorders of the affection. Dr. S. W. D. Williams ‡ stated that in g.p. the body-temp. was normal at

* "Ann. Méd.-Psych.," 1879, T. ii. p. 34.
† *Loc. cit.*, p. 460.
‡ " Medical Times and Gazette," Aug. 31, 1867, p. 224.

first, then sank, and so continued "slowly to decrease as the dementia and the paralysis concurrently advance and finally extinguish life." Of nine cases of g.p., the highest temp. he gave was 98°, the lowest 95°. The inadequacy of these conclusions for many cases is obvious.

In some cases the temp. becomes quite subnormal, and even low (90°-95°) for a day or two, or longer, before death.

Examples. Case No. 70, at the end of this work: analgesia, easy gangrene, sepsis, slow pulse, low temperature—under 90°. —Another; case No. 56, at end of this work: spinal meningeal tuberculosis (and inflammation); temperature under 95°.— Another; case No. 66 : during a series of epileptiform attacks, were temps. between 95·5° and 96·6°. Several days previously, albuminuria; the kidneys looked fairly healthy.

I regret that in only one of these cases was the spinal cord examined; and the more so as out of four cases of mania, with very low temp., described by Dr. Löwenhardt, in two there were changes in the posterior or lateral columns of the spinal cord; and as Dr. H. Weber,* in a case of damage to the upper part of the spinal cord, found the temp. fall very low; and lowering of temp. in one class of spinal injuries has been emphasized by Mr. Hutchinson. None of Löwenhardt's cases were G.Ps., but in all there was exaltation increasing to maniacal excitement, in all marasmus, and final pneumonia; the sphygmograms showed the *pulsus tardus* of Wolff, a frequent condition, and found with paralysis of vaso-motors.—Dr. Wirsch † found in some cases temps. before death as low as 30° to 23° C. (37°=normal). Ulrich,‡ also, in a case of g.p., observed the temp. from 33·1° to 35·8° C. during the last twelve days; and Tiling ‡ found the temp. in one case 28·5°C.; and in another 27·7°C.—Bechterew, in one case, observed temps. varying from 32°C. to 31°C. during the last four days.

Circulation and Pulse. In the early stages of some cases of general paralysis the pulse is full and hard, the beat of the heart powerful, the first sound clear and full, the second accentuated, the arterial tension increased. This condition, however, is very far from being always present, or persistent. The pulse, indeed, has in many cases an increased frequency, a softness and unresisting feel, which, together with the characters of the heart-sounds, tell of diminished arterial

---

* "Trans. Clinical Society," London, Vol. vi, p. 75.
† "Centralblatt für Nervenheilkunde," March, 1881.
‡ Bechterew, " St. Petersburger Medicinische Wochenschr.," 1881.

tension. Finally, in some cases, or at some periods, both the pulse and heart-sounds may possess normal characters.

The above paragraph is from the first edition of this work (p. 49). It has been confirmed by the sphygmographic investigations of Mr. Bevan Lewis, who in the smaller number of cases found lessened blood-pressure; the sphygmograms presenting a nearly vertical ascent, with an acute, or a very slightly rounded summit; a tidal wave usually small, shallow, and sloping obliquely downwards; and a well-defined dicrotic wave. And in by far the larger number of cases, an increased arterial tension; the line representing the percussion impulse being short, and directed obliquely upwards; the tidal wave prominent, very convex, and prolonged; the dicrotic wave slight or absent. The latter form of tracing indicating the existence of a primary cardiac enfeeblement, which is the origin of "a torpid circulation and venous engorgement leading to the obstructed tidal wave."

As to the pulse-frequency in g.p., a mass of observations extending over many months, and epitomized under the heads of *average morning* and *average evening pulse*, are contained in my* previously mentioned contribution. There, also, are linked together simultaneous observations on the temperature, pulse, and respiration in a number of cases, and at various parts of the course, of g.p. From these it may be seen that in g.p. there is a tendency to an increase of the pulse-frequency throughout the whole nycthemeron. From these also it may be seen that the *average* pulse of the evening is almost invariably accelerated to a greater speed than that of the morning; although the patients had retired to rest before the evening pulse was counted, whereas in the morning they were usually up and about, and only assumed the recumbent posture for a few minutes.

The pulse-fluctuations only show a general tendency to harmonize, proportionally, with the fluctuations of temp. The total mean average of excess in all the cases was, *evening* pulse $4\frac{2}{3}$; *p.m.* temperature $\cdot 7°$; *p.m.* respiration $\cdot 67$ higher than the morning. In these mean total averages there is a general accord. But on closer examination of the relation of individual pulse rates and axillary temp. in g.p. we find many deviations from any approximation to a strict proportional relationship. Thus, in one case with heated head, and restlessness at night, were discrepancies between the relations of pulse and temp., comparing one a.m. with

* "Journ. Mental Science," Apr., 1872, p. 31.

the next; *ex. gra.*, pulse 12 higher, but temp. ·1° lower; and, again, pulse 10 higher, but temp. ·3° lower.

The average evening pulse exceeded the average morning pulse in every case, except during two portions of the course of one case. The *p.m.* excess varied from one to thirteen beats per minute, and the total mean average of all the p.m. excess-rates was about five and a half;—or, including the exceptional case, about four and two-thirds beats per minute. And if we exclude the figures relating to times when pyrexial complications coexisted, the total mean of the morning pulse is 78·07, of the evening pulse 82·01. The average pulse-rate, where very high, was so usually in consequence of some other affection, whether this was dependent on the g.p. or not. Each of the following figures is the average of the pulse-rate on 30 successive mornings, and on 30 evenings, respectively.

Average Morning pulse.	Average Evening pulse.	Evening excess or diminution.
68	69	+ 1
72·9	84	+11·1
73·2	84	+10·8
76·7	89·4	+12·7
94·9	100·3	+ 5·4
72	76·3	+ 4·3
73	83	+10
84	87·2	+ 3·2
80·3	84·3	+ 4
85·4	86·6	+ 1·2
111	115	+ 4
93	95	+ 2
108	121	+13
72·7	76·3	+ 3·6
71·8	75·4	+ 3·6
120	124	+ 4
84·2	84·3	+ ·1
88·7	82·6	− 6·1
82·4	79	− 3·4
67·5	73·3	+ 5·8
64	66	+ 2

Frequent pulse. Even without complications, the pulse sometimes, for weeks together, averages 80, 90, or nearly 100. And the pulse may run high in uncomplicated g.p., particularly in the acute or florid form. One source of its frequency may be a paralysing effect, on the vagus centre, of the lesion in the medulla oblongata attending g.p. Experiment, also, has shown that in the cortex are centres which influence the circulation very strongly, and which may be interfered with in g.p. Examples of rapid pulse not due to visceral disease, and occurring comparatively early in the case, were (a) pulse 114, 120 quick compressible; (b) with hypochondria in middle of course, pulse 110; (c) irritable, querulous, early stage, pulse 104; (d) exaltation, pulse 102; (e) exaltation, p. 98.—On the other hand, rapid pulse (118 to 140) for the few days before death is usually with hypostatic congestion and pneumonia.

Slow pulse. The average pulse-rates are seldom slow in g.p.: 64 average a.m. pulse, and 66 average p.m. pulse are the lowest in the adjoining table. The patient was pale, irritable, excitable. In one case, after the first apoplectiform seizure, Mendel found the pulse-rate permanently from 36 to 50 per minute. At the close of life the pulse may be very slow. Thus in case No. 63, at end of this book, two days before death, pulse slow, day before death, 44. Much intracranial serosity, large reddish kidneys, some hypertrophy of L. ventricle of heart, hypostatic pneumonia, and œdema of lungs. For another case of slow pulse before death see case 70.

Sphygmograms. Even in the early period, Voisin* finds indications of vaso-motor paralysis in the full compressible pulse, whose sphygmogram displays some elevation of the percussion-stroke, a plateau at the summit, oscillations in the line of descent, and a variable degree of dicrotism. This condition increases in proportion as the malady progresses. In the final periods the tracing is reduced to a slightly, and irregularly, wavy line. Yet the tracings given by him scarcely coincide with his interpretation of them, and moreover are lacking in detail and distinctness. On the other hand, a sphygmographic tracing similar to that found immediately after cold immersion, was thought by Dr. George Thompson,† to be characteristic of g.p., and indicative of persistent spasm of the arteries and capillaries in that disease. The absolute validity of this inference was denied by myself‡ at the time. Dr. G. H. Savage § found

* *Op. cit.*, pp. 62, 118, 153.
† West Rid. Asyl. Med. Rep., Vol. i, p. 58. "Jl. Ment. Sci.," Jan., 1875, p. 581.
‡ "British and Foreign Medico-Chirurgical Review," Jan., 1872, p. 33.
§ "Journal of Mental Science," Apr., 1875, p. 149.

great variety in the pulse-tracings in g.p., and none characteristic.

Dr. O. J. B. Wolff,* an earlier investigator than any already mentioned, and who found the pulsus tardus in the majority of the chronic incurable insane, and deemed it the expression of a chronic nervous weakness, described that of G.Ps. as like an "aged" pulse, even in those in the prime of life; adding that the normal tricrotic pulse may be present; and that heightened temperature may not affect the pulse of the G.P. as it does that of sane persons. The pulse described by Wolff is, therefore, a pulse of ventricular power defective, of arterial contraction insufficient to normally overcome the resistance, and to propel the blood freely and easily in the vessels. It is essentially a modification, and miniature representation, of the pulse and tracing of ordinary Bright's disease.

Dr. Mendel found normal pulse-tracings in the beginning of g.p., and the same even in some far advanced cases, but altered sphygmograms in the great majority of advanced cases, and consisting essentially in various modifications of the pulsus tardus.

In sphygmograms of the *carotid* pulse he† found pronounced catacrotic elevations, particularly in the earliest stages, referred to a dilatation of capillaries, and paresis of small blood-vessels of cranial cavity. Or, secondly, the carotid tracings are varieties of the pulsus tardus; occur only in the later stages, when paralyses and trophic disorders are marked; and mean vaso-motor paresis. Or, thirdly, the carotid sphygmograms show anacrotic elevations, due to defective elasticity of the arteries altered by atheroma, &c.

Respiration. Averages of 30 successive days in each example (see "Journal Mental Science," Apr., 1872). The *average* respiration in uncomplicated cases ranges between about 13 and 21. In cases of this kind :—Total mean average of all *a.m.* resps., 17·04; total mean average of all *p.m.* resps., 17·71; total mean average of *excess* of *p.m.* resps., ·67.

The average evening respiration exceeded the average morning respiration in frequency in 76 per cent. of the cases—was below the latter in 12 p.c.; and was the same in 12 p.c. The average evening excess varied from ·2 to 2·4. Taking, in the *same patients*, the average pulse corresponding to the average respirations we find the total mean average:

* Allgem. Zeitschr. für Psych., xxiii to xxvi Bd. "Beobachtungen über den puls bei Geisteskranken."

† "Arch. f. Anat. path. und Phys.," lxvi Bd.

a.m. pulse = 78·86; p.m. pulse = 82·94. Total mean average Morning *pulse respiration-ratio* (17·04 to 78·86)= 4·63; total mean average Evening *pulse respiration-ratio* (17·71 to 82·94) = 4·68.

The same applies here as with the pulse; the conditions of exercise, occupation, and rest in the two cases, favouring morning excess and disfavouring evening excess, intensify the significance of the evening excess actually found.

In general terms, the respiration is of about normal frequency in g.p., except during and immediately after the special seizures, or in connexion with pulmonary and other visceral lesions directly dependent on the cerebro-spinal, or indirectly; or, again, in association with accidental complications. Very frequent respiration usually is only at the last, and a consequence of visceral, usually pulmonary, disease. Thus; hypostatic pneumonia, wasted brain, healthy kidneys, respiration on day of death 66. In other cases the respirations on days of death were, respectively, 52, 49, 47.

In a case of which I took careful records long ago (1871), the notes state that frequently the respiration was "irregular;" or, sometimes, "very variable;" or, sometimes, "deep and irregular;" or, "deep and long." I mention this case as introductory to the following subject.

Cheyne-Stokes's respiration in g.p. Up and down respiratory rhythm in g.p. Some years ago I * described three cases of Cheyne-Stokes's respiration occurring in the insane. Two of these were perfectly typical examples; the third case was in a G.P.; and the phenomenon, although decided, was not so full and perfect in form as in the other two cases. Besides Cheyne-Stokes's respiration, I have observed several cases of g.p. in which there was what I have termed "up and down respiratory rhythm;" some of which were mentioned in a footnote to my article just cited. Clinically, this has the form of Cheyne-Stokes's respiration *minus* the period of apnœa. It consists, therefore, of the successive dyspnœal periods, which are essentially the same as those found in Cheyne-Stokes's respiration, and the several phases of each such dyspnœal period jointly constitute a cycle, and the succession of cycles constitutes the phenomenon here termed " up and down respiratory rhythm." It is, indeed, the phenomenon for which I claim that to it alone the name of "respiration of ascending and descending rhythm" is accurately applicable; but as this name has been used as synonymous with Cheyne-

* "British Medical Journal," Aug. 31, 1878, p. 308.

Stokes's respiration, I hesitate to employ it here, lest confusion should arise. This phenomenon, also, I believe, gives to Cheyne-Stokes's respiration its clinical mould; a view which I gathered from watching for hours the typical case described as "case No. 1" in my paper already referred to. On several occasions, I observed the peculiar respiration without a period of decided apnœa, pass into, or, again, succeed, either the typical Cheyne-Stokes's respiration, or what I have termed the modified form of the latter. The clinical phenomenon is this: respiration, at first light and infrequent, becomes, by an ascending scale, fuller, more forcible and frequent and exaggerated, until dyspnœa is attained; and then gradually subsides by a descending scale to the condition as at starting; after which a fresh dyspnœal period begins. In some examples the subsidence is considerably more rapid than the rise. In some cases the above conditions of respiration are probably accessory symptoms of g.p., in others independent and accidental *quoad* the g.p.

Examples of Cheyne-Stokes's respiration in g.p.; or of "modified" Cheyne-Stokes's Respiration. From the case reported in my paper as Cheyne-Stokes's respiration in g.p., I extract the following:—

Case 40. "The peculiar respiration was subsequent to the first appearance of hypostatic inflammation of the lungs; and, moreover, occurred while the patient was prostrated by epileptiform convulsions, and was being supported by nutritive enemata; departed before the cessation of the fits, or of this mode of feeding; and did not return, or at least was not observed, during the remaining ten weeks of life. . . . The aorta was healthy and the heart nearly so." And in part of the notes it is stated that, at one time, "after a severe paroxysm of coughing, the rhythm was temporarily modified, and there was for the time a regular ascending and descending rhythm of greater and less frequency, depth, fulness and loudness of respiration, without a period of true apnœa."

Case 41. Seven weeks before death; apoplectiform attack, the head and eyes turned to the left, the left forearm flexed, rigid, and hyper-pronated; the left leg rigid, somewhat flexed, the right resistant to passive motion; knee-reflex well-marked, especially in right leg, which, also, showed some ankle-clonus; ears heated, teeth-grinding, pupils equal smallish sluggish and irregular. There was a recurrent apnœal pause; a Cheyne-Stokes's respiration, but not highly pronounced.

Case 42. Five months before death, apoplectiform symptoms and right hemiparesis supervened, with some deviation of head and eyes to left. At first the pulse was 78, and the heart irregularly intermittent, as, for example, at the 4th, 8th, 20th,

24th, 30th, and 40th pulsations. Afterwards the pulse was 96, and there was an aortic systolic murmur. Temp. right axilla, 100·3°; left 99·6°. The respiration was irregular, and of a modified Cheyne-Stokes's character; with presence of the apnœal pause.

Examples in g.p. of up and down respiratory rhythm; true respiration of ascending and descending rhythm.

Case 43. Nine days before death convulsions for hours, mainly of right eyelid, and right side of face, and right upper limb. On later days, recurring convulsions, head and eyes rotated towards left side. Temp. right axilla 99·4°; left 99°. Subsequently, left axillary temp. ·5° higher than right. Right unilateral convulsions, right hemiplegia; finally, left ptosis. Two days before death, patient partially unconscious; pulse 135, feeble; respiration 50, at times noisy, but variable, and of the above-named rising and falling rhythm.

Case 44. Phthisis, pleurisy with effusion, first on right, then on left, side, the latter with lobular pneumonia. Recurring vomiting, fever, semi-coma, subsultus, marked tremulousness of movements. Finally, comatose and unable to swallow; pulse 90, full, quick, compressible; respiration varying from 44 to 54 per minute, irregular in rhythm depth and frequency. The cases are published in detail in a Journal article.*

Very recently I have become aware that Dr. Wilh. Zenker has also described Cheyne-Stokes's respiration in G.Ps., and I see from his cases that he has been content to apply the name Cheyne-Stokes's respiration to cases very similar to some of mine in the latter group above, although he points out that they were not perfectly typical examples of that phenomenon. In fact, some of his cases were neither examples of pure Cheyne-Stokes's respiration, nor even of a clinically pure respiration of ascending and descending rhythm, of the kind to which I suggest the name should be restricted; but, rather, of great irregularity of respiration, devoid of the features described above. Of the six cases he mentions, five were G.Ps., and the peculiar respiration either accompanied apoplectiform attacks, or was associated with, or alternated with, epileptiform seizures.

Although I found distinct microscopical change in the elements of the medulla oblongata in one of the cases reported by me,† I felt scarcely justified in absolutely connecting this change with the production of Cheyne-Stokes's respiration. But recently Tizzoni ‡ found in one case of

* "Journal Mental Science," April, 1886, p. 58.
† " British Medical Journal," Aug. 31, 1878, pp. 312 and 309
" Lancet," Jan. 31 1885, p. 221.

Cheyne-Stokes's respiration, chronic inflammatory changes ascending the vagi; with blood-extravasation into the lymphatic spaces of the perineurium and endoneurium. The whole length of the right nerve, the periphery only of the left, was affected. In the medulla oblongata itself were small foci, chiefly on the right side, and beneath the ependyma at the longitudinal furrow of the calamus. Similar lesion affected the upper half of the medulla oblongata in another case (uræmic), but the vagi were normal.

Pulmonary Diseases. Some pulmonary diseases, which frequently follow the apoplectiform and epileptiform seizures, at whatever stage the latter occur; and, without these seizures, may appear in the last stage of g.p.; so often modify the clinical aspects of the case, and their prevention is of such extreme practical importance that they merit attention here; although, strictly speaking, no part of the symptomatology of g.p. Owing to the mental and sensory states of the patient, these diseases are usually not complained of, their symptoms are often very latent, and physical signs inobtrusive.

Posterior pulmonary congestion and œdema, lobular and hypostatic pneumonia, whether localized or diffused, are very frequent towards the close of life, and now and then localized pulmonary gangrene occurs. The coincidence of visceral gangrene and external gangrenous affections was exemplified in a case under Prof. Charcot,* and reported by Prof. B. Ball, in which gangrenous erysipelas was accompanied by gangrenous dissecting pneumonia. Their suggestion that the pulmonary gangrene depends on the entry either of septic fluid, or of ichorous emboli, into the blood stream from the site of external sphacelus, was anticipated by the elder Foville.† I am convinced that, besides lobular pneumonia, in some cases localized pulmonary gangrene in g.p. follows the inhalation of food into the smaller bronchial passages and pulmonary vesicles. (See also Chap. XV.)

Digestion and Appetite for Food. The alimentary functions are usually active and well sustained throughout the greater part of the course of g.p., and as a rule the appetite for food remains good until the last. Often the patients eat eagerly and gluttonously, and stuff themselves if permitted an unlimited allowance ;—removed from the immediate presence of food, they show no morbid desire to gormandize,

* " L'Union Médicale," Jan. 26 and 28, 1860, pp. 162, 182.
† " Dict. de Médecine et Chirurgie Prat.," T. i, 1829.

and in some mental phases they may even refuse to eat. Their scruples, usually overcome by tact, complaisance, or persuasion, occasionally yield only to the enforcement of alimentation by mechanical means. Dr. J. Workman * deemed an increased keenness of appetite as highly significant in the diagnosis. In the later stages diarrhœa is often troublesome and at times intractable. The stools then are thin, brown or yellowish, and contain mucus and undigested particles of food. Local muco-enteritis, mucous congestion, or ileo-colonic ulceration are usually the associated pathologico-anatomical conditions. Constipation and quasi-paralytic conditions of the intestines in g.p. have already been referred to.—For manner of eating in g.p., see Chapter V.

Sweat in g.p. The state of the skin, and of its secretions, has already been mentioned in Chapter II.

Unilateral sweat in g.p., hyperidrosis unilateralis et partialis. In 1877, I † placed on record some examples of unilaterally increased perspiration, or hyperidrosis, three of the four cases described occurring in g.p. Also a fourth case of g.p., in a second paper on the subject.‡ In one of these cases of g.p. the unilateral excessive sweat, affecting one side of the face and head, was not associated with any local convulsion or paralysis, but the eyeball of the same side was sightless, and shrunk from disease following injury of old date. In the second case, convulsion and paralysis affected the same side of the body as that on which the unilateral facial sweat occurred; and in the third case, the first appearance of the sweat followed the onset of hypochondriacal symptoms, and was contemporaneous with slight, transitory, unilateral, facial paralysis of the same side, although the tendency to this sweat remained for a time after the local paresis had cleared up. In the first and third of these cases the right was the side of the face affected, and the left in the second. From the full details of the paper, a few extracts are here inserted.

Case 45. The first case, aged 40. Protracted liability, whilst bedridden for 13 months, to right unilateral sweat of face. On some occasions the beads of perspiration bedewing the right side of the face, stopped short at the zone of the upper lip, but on other occasions affected also the surface over the body of the inferior maxilla. This unilateral sweat did not appear on every day of the 13 months, nor did it usually last all day. The left side of the

---

* "The Canada Lancet," Sept., 1878, p. 3.
† "Journal of Mental Science," July, 1877, p. 196.
‡ "Journal of Mental Science," Oct., 1883, p. 396.

face appeared to be normal as regards perspiration. The shrunken, sightless right eye was occasionally bloodshot.

Case 46. In the second case, aged 43, for three weeks severe convulsions recurred nearly every day, and occasioned extreme prostration and left hemiplegia with free left unilateral cephalic sweat, which last was observed from time to time throughout the three weeks; while the temp. was higher in the *left* axilla by from ·4° to 1·1° F.

Case 47. In the third case, after a rapid increase of paretic symptoms, headache, giddiness, hypochondriasis and refusal of food, there was, for several weeks, frequently recurring profuse sweat on the right side of the face; and for one or two days this was associated with lower right facial paresis, including the nostril. The grasping power of the hands was alike, and was feeble. The equal pupils, hitherto fairly normal, were now somewhat wide and sluggish; the *left* conjunctiva was injected, and the mucus increased on one occasion : the ears were alike.

Case 48. A fourth G.P. was aged 34, emaciated, feeble, of degraded habits, melancholic and rarely spoke. Vomiting. Diarrhœa, later on. At last, local spasms, epileptiform seizures, chiefly dextral, followed by right hemiplegia. These recurred frequently, and, often, in the intervals a free sweat, confined to the right side of the face.

In that part of the above-mentioned paper which treats of the pathology of the symptom (p. 206, *loc. cit.*), I discussed the question whether the sweat was due to some disorder or lesion of the vaso-motor system, or to a morbid excitation of nerves which may be supposed to more directly control secretion, and I suggested that there may be sweat-secretory nerves, which exercise an immediate control over the perspiratory function, analogous to that which some secretory nerves were at that date believed to exercise over some other secretions, as, for example, the salivary. Since that time, views similar to those thus theoretically suggested in my paper, have resulted from the experiments of Luchsinger, Nawrocki, Adamkiewicz, and Vulpian. And as stated in my article (p. 212) "whilst, therefore, both the sympathetic and the cerebro-spinal system influence secretion, and both the vaso-motor and the so-called secretory nerves, it is mainly to the cerebro-spinal centres that we may, perhaps, look for the nervous influence on secretion, and mainly to the secretory nerves for the more immediate channel of that influence. It is only necessary to recall the cerebro-spinal morbid changes in g.p. to find the origin of a morbid influence upon secretion through the cerebro-spinal fibres supplied to the glandular system. But the simple structure and secretion, with the wide distribution of the sudoriferous glands place them under relations to the nervous system," differing from those of the complex conglomerate secretory organs. The sweat-secretory nerves of the face are, perhaps, derived from sympathetic fibres, either from the superior cervical ganglion, by the route of the vertebral artery, or from the bulbar and Varolian regions.

Bloody-sweat. In two cases of g.p. Dr. F. Servaes observed bloody sweat, limited to head and face. Probably, in at least some cases, the red coloration is of bacterial, and not of sanguineous origin.

Urinary Excretion. In 1845 Dr. Bence Jones indicated a great deficiency of phosphates in the urine of G.P.'s. The same diminution of phosphates corresponding to the progress of g.p. was one of the outcomes of some investigations published by Dr. A. J. Sutherland,* the analyses having been made by Dr. L. Beale. Speaking of the urinary excretion in g.p., Mr. A. Addison† said that " in states of excitement the quantities of chloride of sodium, urea, phosphoric and sulphuric acids are less than in the quiescent state," and less, also, than the quantities excreted in the same time in a state of health. "In the demented cases (the) quantities are about normal—some slightly above—and some below the mean" (pp. 444 and 449). According to Griesinger,‡ Sander found the excretion of urea to be small in g.p.—It was found by Dr. J. Merson,§ that in the urine of G.Ps. the quantity of urea was usually much increased ; the quantities of chlorides and of phosphoric acid notably diminished ; and the amount of sulphuric acid about normal. Also, that the *mean* specific gravity did not materially differ from that of health, and that the *absolute* quantity of urine, though slightly below that of health, was, in truth, slightly in excess of the latter, if estimated according to body-weight.

Dr. S. Rabow‖ states that the paralytic insane usually secrete in the so-called first stage, whilst there is still some mental power and psychical euphoria, an increased quantity of urine, and relatively to the greater consumption of food, more urea and chlorides than healthy individuals. As dementia advances the quantity of urine lessens, and, simul-

* " Medico-Chirurgical Transactions," London, 1855, Vol. xx, p. 261.
† " British and Foreign Medico-Chirurgical Review," April, 1865, p. 425.
‡ " Mental Pathology and Therapeutics," Trans. 1867, N.S.S.
§ " West Riding Asylum Reports," Vol. iv, p. 63.
‖ " Archiv für Psychiatrie u. Nervenkrankheiten," vii Bd., p. 72.

taneously, the absolute amount of urea and of chlorides; whilst the sp. gv. seems to be augmented; a turbidity due to urate is seldom absent, and after removal by a catheter the urine rapidly changes from an acid to an alkaline state. The most marked point about his analyses is the extreme fall in amount of urea and chlorides in the last stage with advanced dementia.—Dr. Mendel appears to agree with, or follow, Rabow. In the melancholic first stage (by first stage the latter, however, means not this, but the expansive), the quantity of urine is lessened; its phosphates are usually increased; the urea and chlorides less increased, as a rule. In the maniacal stage, his results were the same as those of Rabow. With more extreme excitement, the quantity of phosphates lessens as a rule. He found great diminution of phosphates, also, and the sp. gv. not always high in the demented stage, as Rabow thought; the solids of the urine often diminished to half the normal amount.

The urine of 36 G.Ps. was examined by Dr. v. Rabenau,* and in 20 of these he found albuminuria at some period or other, and felt justified in concluding that, in at least part of these cases, the albuminuria was dependent in some way or other on the brain-disease, and, indeed, directly so, and not on some secondary affection of the other organs, conditionated by it. Yet it appears, that of the albuminuric cases he had made necropsies in ten, and of these had found decided kidney disease in five. Dr. Richter † failed to find albuminuria in G.Ps. of either sex, at any stage, or even after the apoplectiform attacks; he excludes, however, cases of disease of the urinary organs, or other diseases, which could cause albuminuria, and the existence of which were confirmed by necropsy. Rabow agreed with Richter in opposing the view of v. Rabenau, and concluded that albumen is not frequent in the urine of G.Ps., and that if, as a matter of fact, it ever appears therein, it has no connexion with the cerebral disease. Others, also, have been unable to confirm the frequent occurrence of albuminuria in g.p.

Dr. Max Huppert ‡ found albumen, frequently with hyaline casts and blood-discs, in the urine after the epileptiform seizures in g.p.; the higher temp., the more marked albuminuria, and frequent casts and discs in urine, distinguishing these from epileptic convulsions. De Witt also found albuminuria

* " Archiv für Psych.," iv. Bd., p. 787.
† " Archiv für Psych.," vi. Bd., p. 565.
‡ " Archiv für Psych.," vii. Bd., p. 189.

after the convulsions, while König, Richter, and Mendel usually found it absent. During the attacks the amount of urine secreted is sometimes small. The amount of urine seems, in some cases, to be increased for a day or two after the seizures, and yet its sp. gv. may be high. Von Linstow gave importance to the phosphaturia after the seizures; and Mendel * observed an increase, both relative and absolute, of phosphates in the urine after both apoplectiform and long epileptiform seizures in g.p.—Sometimes comparatively early in g.p. is a rapid loss of weight, full appetite, no rise of temp., high sp. gv. of urine (1030) : here the phosphatic increase is often, but not always, proportionate to that of the other urinary solids.

Diabetes was found by Schüle † in a case of g.p.; and by Verneuil in another.

In the urine of g.p. Selmi ‡ found ptomaines—two volatile bases; one like nicotin, the other like coniin.

Salivary Secretion. The secretion of saliva is sometimes increased. A factitious appearance of the same may occur; an escape of saliva from the mouth being permitted by the paretic state of lips, face and tongue, and by the inattention and impaired consciousness of the patients; which, in some cases, may perhaps be reinforced by the influence of states of local anæsthesia. In some cases, the increase of salivary secretion in g.p. is probably of cortical origin, and due to irritation, by the lesion of g.p., of centres in the cortex which influence the secretion of saliva. In other cases, the increase may be due to irritation of chorda tympani nerve, or its central endings.

CHAPTER XII.

THE COURSE OF GENERAL PARALYSIS.

We may here resume some details as to the course of the disease.

A. *Precedence of Orders of Symptoms.* Much discussion has arisen as to which, in g.p., precede; the motor, or the mental, symptoms. Bayle asserted that the two orders of symptoms usually ran their course *pari passu,* yet that stuttering may precede any mental alienation. That the mental

* " Archiv für Psych.," iii. Bd., p. 659, 654.
† " Handb. der Geisteskr."
‡ Dr. R. N. Wolfenden, " Lancet," ii, 1883, p. 853.

usually precede the motor symptoms, but sometimes come on simultaneously, was the view of Delaye, and very similar to that of Rodriguez, Griesinger, Conolly, and Falret; and Esquirol, latterly, acknowledged the same, and even that the "paralysis" sometimes preceded. Calmeil stated that the motor symptoms sometimes followed the mental, sometimes came on at the same time, rarely preceded them; and much to the same effect were the expressions of Foville (*père*), and of Dr. Maudsley. But Daveau believed that it is because the paralytic symptoms escape observation that the mental usually seem to precede. Early in his career, Baillarger held that in g.p. the "paralysis" was the primitive and principal element; the insanity, indeed, being only secondary and accidental, and when present really a complication of the paralysis. The occasional precedence of motor (chiefly speech) symptoms was asserted by Drs. Guislain, Leidesdorf and D. Skae. Of M. Parchappe's cases, the mental and motor symptoms came on simultaneously in the majority (51); but the mental preceded the motor in a goodly number (27); while the motor were not proved to precede in any case. Compare with these the statistics of M. Brierre de Boismont;[*] mental and motor simultaneous in 34 cases; mental symptoms preceding in 42 cases; and motor in 16. Thus, while the former had not a single case in which the motor symptoms were said to be prior to the mental, the latter had 16 such; the former found mental symptoms precede in one-third of the cases, the latter in nearly half; the former found mental and motor commence together in two-thirds, the latter in scarcely more than one-third.

G.p. supervening in chronic insanity? Intimately allied with the above is the question whether g.p. ever supervenes in old cases of ordinary insanity. Falret and Guislain doubted that it ever did; although the latter admitted that g.p. might occur as an incidental termination of congestionary delirious mania. On the other hand, Delaye and Parchappe believed that the motor symptoms of g.p. sometimes arise in the course of ordinary chronic insanity, and Conolly wrote "it often happens that general paralysis supervenes after years of insanity in cases already hopeless, and it is then merely the beginning of a sure though slow decay." But no recent writer, I believe, avows the last and untenable portion of the bold generalization of Broussais that "all partial insanities tend to become general, and the

[*] Société Méd.-Psych, Dec. 27, 1848.

general tend more or less to dementia and general paralysis," although after his time similar views were put forth, as by Lunier, and the generalization was partly based on M. Georget's.

Calmeil* mentioned an insane patient, who, after 13 years' asylum-residence, and apparently about 14 years' insanity which had ended in dementia, showed physical signs of g.p., and was particularly affected with severe general tremor. The case, with its necropsy, is not an absolutely convincing one.

There is also a passage in Bayle's work, in which he speaks of some cases being quiet throughout the first period, and ordinarily dominated by a fixed ambitious delirium, capable of discussing any other subjects sensibly and coherently; yet of enfeebled faculties, and principally so as regards their memory; their pronunciation is manifestly embarrassed, or even stuttering for some words; the gait is stiff and defective in firmness, sometimes they drag one of the feet or swerve from the direct line. Nowhere does Bayle give examples corresponding to this description, but it is possible he had in mind unpublished cases, in which chronic delusional insanity finally became complicated by g.p. I have seen one or two examples of this occurrence, according to the histories of the cases; Dr. Spitzka gives an instance where symptoms of monomania had preceded g.p. by at least twenty years, and cases of still longer pre-existence of it are on record. Undoubtedly many patients are insane for some time before any motor indications can be detected, even by close examination. I have noticed the somatic signs, in different instances, supervene some weeks; some months; a year and a half; between two and three years; after the commencement of highly pronounced mental alienation. I speak now of cases in which the existence or supervention of g.p. was suspected and its physical indications carefully watched for. The primary mental alienation was of various forms: acute delirious mania;—acute mania; —expansive delirium;—intermingled hypochondria and melancholia with agitation;—dementia. In these cases, is the primary mental alienation but a phase of "general paralysis" in the psychical sphere, or is it independent of the latter? A question difficult to resolve. Voisin makes a duration of two years the criterion. For him, mental alienation, if followed by the somatic signs in less than two years,

* *Op. cit.*, 1826, Obs. liv, p. 279.

pertains to "general paralysis;" but, if not, it pertains to simple insanity; and this duration, prior to the motor signs, may be augmented by a remission of not more than two years in g.p. I should prefer to adopt the form and characteristics of the mental derangement as a criterion. Usually some of the mental traits have suggested g.p. ere its somatic signs appear.

B. *The Mode and Time of Sequence of Individual Symptoms in General Paralysis.* As to the mode and time of sequence of the individual symptoms, only the most general outlines can be indicated here. There is usually a growing intensity of the symptoms both psychical and somatic, and on the other hand irregularity in the degree and order of appearance of the different phenomena; a course progressive in its totality, irregular in its details. Moreover, there is often much fluctuation in the course of the mental symptoms, and even of the motor; nor can we agree with Esquirol when he says that the march of the paralysis is unceasing, that the latter always continues to increase.

Motor symptoms. In the motor sphere there are in general terms and in order of time:—I. ataxy; inco-ordination; II. paresis added to ataxy; and, III. increase of ataxy and of paresis to a state of helplessness.

More particularly; the motor affection of the apparatus of speech is usually that first recognized; and, together with the tremor and twitch of the tongue, lips and face, the inco-ordination of the locomotor movements of the lower limbs is sometimes noticed early in the disease. Ocular modifications, and impaired co-ordination of the finer movements of the upper extremities, may soon be observed, or occasionally may even appear to precede the labio-lingual disorder.

The word "paralysis" is immediately about to be used in deference to the writers quoted, but the motor impairment in question is chiefly an ataxia and not a paralysis, is a mingled paresis and ataxy. Foville (*père*) stated that the "paralysis" began in the tongue; Belhomme that impairment of speech preceded any other paralytic symptoms, and that the lower limbs were affected before the upper; on the contrary, Rodriguez and Brierre de Boismont, that the paralysis affected the upper before the lower extremities; Doutrebente, that when the cause is sexual excess the paralysis often begins below and travels upwards, and I have seen some examples of the association of this causation and course. Dr. D. Skae found speech alone affected in 10 cases; gait alone in 8 cases; both affected in 90.

Much discussion has arisen with regard to these points, but whether it explains the disagreement of the authors cited or not, there is no doubt that the motor disorder affects different parts of the system first in different cases. Nevertheless, it has been held that the general voluntary muscular system is uniformly affected, but that the impairment of the *speech* and of the *gait* is rendered more obvious than that of other motor activities because of the complexity and exactitude of the muscular co-ordinations and harmonious adaptations required in the one case; and the force and sustained effort necessary in the other.

Mental symptoms. The mental symptoms occur in the most varied possible time and mode of sequence. It need only be said that the exaltation and extravagant notions, if present, are, as a rule, most active near or not long after, the onset; that the more active maniacal symptoms may occur then or somewhat later, or even intervene in the period of extreme dementia; and that the hypochondriacal or melancholic symptoms may be prominent almost throughout the course of the affection, yet occur more particularly at the onset or in the middle periods, and occasionally run into, or crop up in, the later periods. Dementia, which occasionally is predominant, or even exclusive, from the outset, is present, more or less, in all cases from the early stages, is irregularly progressive, and is often very extreme towards the close of life.

After a preliminary depression, the emotional state often tends to be gay and expansive; later, to be depressed, morose or peevish; and, finally, to be reduced to the vanishing point.

The moral qualities, perturbed at first and inciting to irregular actions as if from moral turpitude, rapidly undergo disintegration and decay.

The sensory functions, both general and special, become blunted or even lost, but less so than other functions, and, as a rule, only somewhat late in the course of the disease.

C. *The Relative Degrees of Intensity of the several Symptoms.* In the motor sphere, speech and locomotion are almost invariably the functions most obviously affected. In the mental sphere, exalted delusion is a very striking phenomenon, so, too, is the hypochondriacal, or occasionally the melancholic, condition; and towards the last dementia is often so intense that mind is abolished, and vacuity reigns.

In general terms, both the mental and motor symptoms are, throughout, more intense, as a rule, than are the sensory; and the mental derangement may run riot at a time when the motor is scarcely, or is not, discernible. Yet during remissions the mental symptoms often clear up much more than do the physical.

D. *The Question of the Existence of General Paralysis without Mental Alienation.* Can g.p. exist and run its course without mental symptoms, the disease in other respects being that called "general paralysis of the insane"? A question much discussed.

The possibility of this absence of mental symptoms was at first denied by Calmeil, and a third of a century later he maintained that it only occurred with extreme rarity, if at all. The earliest recorded case supposed of g.p. without mental symptoms—namely, that by Delaye—and some, at least, of Dr. Lunier's similarly interpreted cases, were not, I believe, true instances of g.p. M. Baillarger, M. Guislain, Dr. Skae, and Dr. W. T. Gairdner had seen it with dementia or some loss of memory only, and without intellectual disorders; Dr. James Copland had asserted that he observed cases of the same import; and so did Campani in Italy.

MM. Brierre de Boismont, Duchenne, Duhamel, Prus, Requin, stated that there were cases of g.p. without mental alienation. And Requin, Lunier, Brierre de Boismont and C. Pinel erected two distinct forms of g.p.; one with, one without, alienation.

I think if cases of this kind are followed up it will usually be found that they resolve themselves into the form of g.p. with dementia only, in the psychical sphere; in some of which, perhaps, the extent of departure from the more usual and more dramatic form is widened by the relatively early, and relatively predominant, ataxy and disabling paralysis or paresis. And I believe that the greater part of this question, and of the attempts to divide g.p. into two forms, one with, one without, mental symptoms or alienation, have hinged upon the failure of physicians to recognize cases of what in this book is called the dementia form of g.p.; in which dementia is, throughout, the predominant psychical feature. Coming to hospitals or dispensaries, or met with in private practice, their true character, perhaps, is either entirely unrecognized; or, if the motor indications attract attention the dementia does not, or is neglected as of no special importance or meaning. What is not explainable in this way, is, I believe, to be accounted for by the application of the name "g.p. without mental symptoms" to examples of atrophy of brain, and to some of the cases usually classed as senile dementia and associated with organic cerebral changes; in which individual cases, also, mental symptoms exist, but are comparatively unnoticed. Next to the question of their very existence, it has principally been in dispute whether (if they exist) the cases without mental alienation constitute a distinct disease, or are really the same as "g.p.," except for the omission of the mental symptoms, which last, in this view, are said to be non-essential. If really a distinct affection it evidently constitutes a third disease; distinct from "general paralysis of the insane" on the one hand; and, on the other, distinct also from those more or less generalized *palsies* which may follow various gross changes in the nervous system. But I have already given my reasons for believing it not a distinct affection.

The Course of g.p. in Female patients. In females, as compared with males, g.p. runs, on the average, a milder and longer course; its symptoms, both psychical and somatic, are less striking. Taking the mass of cases: on the mental side, there is more often a quiet dementia: on the physical side, spinal symptoms are less marked, the ascending form of g.p. rare, the chief special seizures (apoplecti- and epilepti-form) less frequent and severe; on the contrary, the hysteriform seizures occur in women chiefly. Remissions are less frequent and less marked on the whole in females.

REMISSIONS IN GENERAL PARALYSIS. Usually the last faint physical signs that linger about the patient are some affection of speech, some change of facial expression and action; some pupillary condition; some relative increase, or higher level, of evening temperature: and the last mental trace is a psychical weakness, relatively to the age, and to the former condition normal to the particular patient, evinced in an incapacity for sustained mental attention, effort, or strain;—or to bear the struggles of life and knocks of fortune. Drs. Simon and Böttger noticed during the remissions an aversion of the patients from medical examination. This is noticeable in some cases, but not in others.

When the remission is over, and the disease resumes its course, the latter may be ushered in by an apoplectiform or epileptiform seizure (after which the mental symptoms may be different from those immediately before the remission);

or may commence with emotional dejection and depressed ideas; or with maniacal restlessness and some excitement, with expansive symptoms; or with increasing dementia. In several cases I have seen protracted remissions cut short; and, in some, the lives of the patients also; by apoplectiform or epileptiform attacks, as if the disease had silently gathered force until "discharge" with explosive violence took place.

Sometimes the remissions follow on fractures, amputations, and various diseases leading to local suppuration, or sloughing inflammation, or tubercle. There are many older examples of this; and the following are some recent ones. One case of remission following severe bone-fracture I reported in the 1st edition of this work, p. 167. One marked example I saw follow a carbuncle between the shoulders. Dr. Savage * reported a similar case; as also Dr. Pritchard Davies. † Dr. Wm. Hurd ‡ observed two very marked cases follow extensive sloughing, in one case over the heel, in the other over the sacrum. Dr. J. Christian § found two cases of marked remission after prolonged suppuration in g.p.; M. Mabille ‖ a case (or recovery?) after prolonged bath and setons; Dr. Oebeke ¶ one after erysipelas of face and head; Dr. C. B. Burr ** after sloughing.

The remissions to which the disease is occasionally subject were described by Dr. A. Sauze †† as of three kinds. Of these it is not uncommon to see two; namely, that in which there is remission of the mental and motor symptoms simultaneously; and that in which the remission pertains to the mental symptoms only, or chiefly. Clear decisive examples of the other variety are comparatively infrequent; that is to say, remission of the motor symptoms alone, the mental remaining without improvement, although a slighter degree of this is not rare. The cases I have seen for the most part coincide with the experience of Dr. Baillarger, ‡‡ who long ago spoke of a dissociation of the two orders of symptoms, of such sort that the insanity disappears—the generalized motor paresis persisting—and who referred to similar cases by Bayle, Ferrus and Rodriguez.

* "Journ. Mental Sci.," Jan., 1881, p. 566.
† Ibid., Jan., 1886, p. 508.
‡ "Alienist and Neurologist," July, 1884, p. 541.
§ "Annales Méd.-Psych.," 1880, T. i, p. 224.
‖ Ibid., Jan., 1881, p. 64.
¶ "Allgem. Zeitschr. für Psych," xxxviii. Bd., p. 301.
** "Amer. Jl. Neurology and Psych.," 1884.
†† "Ann. Méd.-Psych.," Oct., 1858, p. 493.
‡‡ Ibid., May, 1847, p. 335.

It was subsequently * asserted by Baillarger that its remissions are most frequent when the g.p. begins with a maniacal attack ;—easily explainable on the hypothesis that various deliria often complicate g.p. at its onset, of which mania is one most likely to terminate in recovery. According to this view † the form in which there is dementia only, is in reality g.p. with its essential symptoms—those which never fail—and the maniacal, melancholic, or other attacks with which the disease sometimes comes on, are not to be looked upon otherwise than as complications;—or, general paralysis, again, usually a primary affection, as sometimes being secondary, in a sense, to the above mental disorders, which he has brought within the compass of that which *he* calls *folie paralytique*, and asserts may precede or accompany paralytic dementia, but is not necessarily associated or connected with it, or dependent upon its essential lesions; and Dr. Luys ‡ declares that remissions in g.p. are merely due to the cessation of epiphenomena, the fundamental lesions, meanwhile, persisting and progressing. On the contrary, Doutrebente § argues for the complete and lasting character of some of the remissions, or, in other words, the recovery of the patients. And this most frequently in the maniacal or acute form of g.p.

That the marked remissions are relatively more frequent; or according to some occur solely; in g.p. of the chronic, sometimes called the reasoning, form, where heredity is distinctly a factor, is maintained by Drs. Legrand du Saulle, Doutrebente, and Lionet. Mendel had the same experience, heredity being a factor in 34·8 p.c. of his cases, but these hereditary cases including 62·5 p.c. of the total number of cases presenting the more marked remissions, or nearly double an equal proportional share. Baillarger, || however, collected or described several well-pronounced instances of remission, and possibly of recovery, in g.p. of the form in which the mental symptoms consist of *mélancolie avec stupeur*, or of *stupeur*. For, of the cases in which dementia predominates, the remissions occur only in those where it comes on suddenly or rapidly, and is really a pseudo-dementia, distinguished, however, from true dementia only by its rapid invasion, or by some signs of stupor. This paralytic stupor, therefore, may simulate an advanced degree of dementia; and this pseudo-dementia may be associated with some delirious conceptions. Accord-

* "Ann. Méd.-Psych.," May, 1876, p. 256.   † *Ibid.*, p. 263.
‡ *Ibid.*, July, 1877, p. 111.
§ *Ibid.*, March, 1878, p. 161; May, p. 321. || *Ibid.*, Jan., 1879, p. 4.

ing to this view true dementia does not exist from the outset; and the mobile, contradictory, absurd delirium is not an indication of it, but of a state like some cases of drunkenness, and permitting of decisive remissions. When the dementia of g.p. is slow, progressive, simple, *i.e.*, unlinked with any mental disorder, then it is real, and, like that of chronic secondary neuropathic dementia, is incompatible with the occurrence of any remissions worthy the name. Remissions in the more *usual* dementia of g.p were long ago made a principal point in its distinction from ordinary chronic mania by Dr. W. Wood.*

Case 49. Temporary maniacal symptoms in 1874. In 1875 he first came under my care with acute maniacal agitation, expansive delirium, and *masked* physical signs of g.p., the diagnosis of which affection was justified by the subsequent manifestation of slight somatic signs. Complete remission or disappearance of the mental symptoms took place, except of the incapacity for steady application to work; and almost, but not absolutely, complete disappearance of the motor. He was discharged, remained fairly well for a time, but was readmitted eight months afterwards suffering from g.p.; on this occasion with less active maniacal symptoms; again the symptoms nearly disappeared; he was again discharged for a year, and then was admitted a third time, with active maniacal and expansive symptoms, the mental powers also being shattered, and the physical signs of g.p. highly marked and characteristic. He again made considerable recovery, especially in the mental sphere. Latterly, there was persistent right hemiplegia. Death in 1882. Meningeal changes highly marked: brain soft, flabby, much adhesion and decortication, especially over left cerebral hemisphere, the more atrophied of the two. Chronic spinal meningitis, some spinal sclerosis. For other cases see *prognosis*.

In Lunier's † case of attacks and remissions or recoveries, seven in number, extended over a space of 23 years, it is doubtful whether the earlier attacks were really g.p. Died of pneumonia; no necropsy.

Remission, even temporary recovery, may follow mental depression at the onset, or follow maniacal and expansive symptoms, and generally occurs in the circular form of g.p. *Ceteris paribus*, the earlier in the disease the more complete is it possible for the remission to be. It is rare in the demented form. I have occasionally seen patients, feeble, fatuous, and bedridden, make an astonishing improvement; and if they did not take up their beds and walk, they at

* " British and Foreign Medico-Chirurgical Review," July, 1860, p. 199.
† " Annales Méd.-Psych.," March, 1879, p. 239

least resumed for a time their places amidst their comrades, and a portion of their former intelligence; Marcé, also, saw patients covered with sores, in marasmus, profound exhaustion, and complete dementia, recover to a large extent their forces, both physical and intellectual, and so remain for years.

The remission noticed by Doutrebente * to coincide with the evolution of a normal pregnancy was probably casual; and Mendel has seen the disease continually advance during pregnancy. A case of Baillarger's is of interest here: pregnancy in g.p.; a decided remission after childbirth; a rapid worsening of the state following on early weaning of the infant. Hypochondria a prominent feature.

The remissions referred to here are not mere points of sinking in the ordinary fluctuations of the disease, but are of decisive nature and persist for a time, as, for example, six months in a case reported by Dr. N. G. Mercer; † or in a case by Insp. Gen. Dr. W. Macleod; ‡ or in that of M. Bonnefous § lasting four years. Many of the cases published as recoveries of g.p. are no doubt cases of highly pronounced remission only. Of 20 cases of remission mentioned by M. Baillarger, the two longest were of about two years' duration; and of the twelve cases by Dr. Doutrebente ‖ the longest complete (?) remission endured 26 months. This followed an attack of pleurisy, but some intellectual debility remained. Of incomplete remissions, the longest two were seven years and over ten years, respectively. In Dr. T. B. Christie's ¶ case, the patient was discharged from the asylum for 27 months during a protracted remission. Some cases cited as prolonged remission, as by Ideler,** seem to have been examples of recovery. A case by Müller for ten years presented some weakness of intelligence, and nothing more. Edel reported a case of prolonged remission during which the patient went through the Franco-German war of 1870-1. In 70 cases Dr. Böttger †† found 25 with remission. Of the three forms of remission,—motor only or chiefly—mental chiefly or solely—both mental and motor—he thinks the last

---

* " Annales Méd.-Psych.," May, 1878, p. 338, Obs. xii.
† " Medical Times and Gazette," Dec., 1867, p. 617.
‡ " Journal Mental Science," July, 1879, p. 196.
§ " Ann. Méd.-Psych.," May, 1869, p. 433.
‖ *Loc. cit.*, p. 321.
¶ " Lancet," Aug. 4th, 1879.
** " Allgem. Zeitschr. für Psych.," xxxiv. Bd., 2 Heft., p. 242.
†† " Algem. Zeitschr. für Psych.," xxxiv. Bd., p. 237.

the rarest. This does not coincide with my experience. Gauster observed, of the remissions, three relapsing; and ending fatally after 8, 15, and 36 months, respectively. Legrand du Saulle reported six cases of marked remission, relapsing at points of time varying from ten months to three years; and Dr. W. H. O. Sankey two cases of remission so complete as to be apparent recoveries, one lasting for 16 months.

In the remissions, the patients should not be discharged and sent back to the conditions under which their malady took rise. And it is only under exceptional circumstances that a return to the outer world is desirable. As a rule, such patients should be kept constantly under skilled and expert observation, (not necessarily in an asylum), and, in many cases, medical treatment; so long as merely a remission lasts; or until a virtual recovery has continuously held place for two or three years. Then they should be placed in the most regular and least exciting conditions that can be arranged.

When, to superficial observation, the remission seems to have ended in recovery, there may be shamelessness, indecency of language or of act, or threatening declarations, or quiet assertions revealing an absence of moral restraint, or of appreciation of penal consequences. That is to say, although the patient appears to be recovered intellectually, there may still remain a defect of moral and æsthetic feelings; a state of moral defect or insanity. These feelings, the latest in development, and the "very bloom of culture" (Maudsley), have been the first to fail. The supreme efflorescence of the *psyche*, they have been the only faculties destroyed at the first blighting blast of disease. The tree stands, but its fruit is untimely withered.

If discharged during the remissions these patients may be the easy prey of the designing, and induced to commit illegal acts. Not being justly held responsible with respect to offences against the law; for their own sake, and for that of their families, their full civil rights and unrestrained liberty should be withheld. Neither to them, to their families, nor to society, is it fair, either on the one hand to punish them for acts against the penal law, or on the other to give them the opportunity of committing the acts in question. Certainly, if the efficacy of the penal code be the test of responsibility for crime, these persons are irresponsible.

SIMPLE ARREST OF G.P. Without any remission, the lethal course of the disease may be simply arrested for months or even years, the mental and motor symptoms, moderate in degree, being apparently stationary during that time. Sometimes this happens spontaneously, sometimes as a result of treatment. I have seen it follow the use of K.I.; the application of blisters, succeeded by Ung. Ant. Tart., to the scalp; Ferri perchlor. in the advanced stages; and Voisin gives three cases where the disease was arrested for a time by the daily use of cold baths at 12° C., or 54° F.

THE DURATION. The duration of general paralysis varies from a few weeks or months to one, two, or several years. Aside from lethal, independent and intercurrent accident or disease, death may be brought about in the early periods either by the severity and extent of the morbid process in the nervous system, or by incidental cerebro-spinal or visceral complications. In 1826, Bayle estimated the *average duration* of g.p. (including both sexes) at from one year to one year and a half; Calmeil, at thirteen months; and at one time Bayle had fixed it at ten months. On a basis of forty-two cases Parchappe* estimated the average duration at twenty-three months. He found the most rapid cases to be those in which the delirium was like that of the various acute forms of simple insanity, and also those with the "paralytic" signs from the outset; while the longest cases were those in which the "paralytic" signs appeared later than the mental disease, and those in which the latter was dementia only. At the Wakefield Asylum† the male cases of g.p. during a series of years were of the *average duration* of 20·7 months; the female, of 25·9 months; total average duration, male and female, 21·6 months; total average asylum residence 16·6 months. At the Devon Asylum‡ the average duration was: males, 15 months; females, 27 months; total average, male and female, 17 months (by error in computation stated as 21 months in original, cited).

At the date of the first edition of this work the average duration of g.p. in those soldiers on whom I had made necropsies was more than 22½ months, *i.e.*, duration of mental disease, recognized as such; although in many, symptoms noticed previously indicated that the duration was much longer than was estimated. Adding to these the other G.P. soldiers deceased; those discharged, improved (reckoning only up to date of discharge); and the then still

* "Recherches sur l'Encéphale, sa Structure," etc., 1836, p. 155.
† "West Rid. Asyl. Med. Rep.," Vol. v, p. 202. Dr. C. F. Newcombe.
‡ *Ibid.*, Vol i, p. 138.

living residue of military G.Ps.; the average total duration of known mental disease was $28\frac{1}{3}$ months. The then living residue, and those subsequently admitted, have since died, with an average duration of $40\frac{1}{2}$ months, of recognized mental disease; or of 42 months, if in some cases is added a preceding period of some known mental change or perversion. Amongst the gentlemen-patients, under my care, some examples of g.p. have been extremely protracted, as for example, more than five or six or than ten or fifteen years: and in them the *average* duration has been much longer than that found in soldiers. The duration in the last (Oct., 1885) three better-class G.Ps. who died averaged about, or nearly, five years.

Out of 95 cases, Dr. Mendel, erroneously as I think, gives four as lasting between eight and ten years, and one as lasting sixteen years; but, as he explains, all of these were of the ascending form, beginning with some disease of the spinal cord; and the time is reckoned from the first commencement of nervous symptoms. Now, as some of these cases show nervous, *i.e.*, *spinal*, symptoms for many years before the mental faculties are affected in the slightest degree, it is inaccurate to include those years in the duration of the "general paralysis." There is a slip of the pen, also, for whilst he had four cases of "galloping" g.p., the shortest duration of g.p. he gives in his table is "$\frac{1}{2}$ year to one year"—8 cases—yet the example he relates * lasted less than six months. Dr. Lunier † mentioned a case, said to be g.p., of more than 20 years' duration.

While some "ascending" cases are long, the duration, as I years ago stated elsewhere, is often short when brain and cord seem to be simultaneously attacked: and I find that Voisin has observed the same fact. The average duration is long, also, in the chronic demented form, comparatively free from special seizures (second group of this work). A few, almost from the first assume a chronic stationary form, with moderate dementia and physical signs. The shortest average duration is in those acute cases, sometimes called "galloping." Some of the unusually melancholic and hypochondriac cases run a short course.

As to sex, the relatively milder, more depressed and demented symptoms in females run a longer course *on the average* than does g.p. in the male. This fact in the natural history of the disease has been distorted from its true

* *Op. cit.*, No. 6, p. 317.　　　　† "Annales Méd-Psych.," 1877, p. 426.

meaning, and supposed to depend on differences in nursing. Sander * states that when, in statistics, the female cases are returned as of equal duration to the male, the explanation is, that owing to the differences in the domestic and pecuniary consequences of the disease in the two sexes, and its more quiet course in the females, these come later into the asylum, as a rule, than the males.

The hereditary cases, with their greater tendency to long remissions, are often of prolonged total duration. Dr. Doutrebente expressed the view—a view also espoused by Dr. Lionet †—that in those hereditarily predisposed to insanity the course of g.p. is often very chronic and remittent; or that in them g.p. is rare, and when it does occur is of the chronic remittent form. And this view he ‡ still maintained in opposition to Dr.Mirandon de Montyel § who brought forward statistics to show the short course of g.p. in cases of the kind.

Nor, with respect to duration, does the natural history of the disease include only the difference as to clinical form, sex, and heredity. Not only is g.p. a more rapid disease, on the average, in the sex in which it is more rife, but it is more rapid on the average, also, in patients of those classes of society, or following the occupations, or given to the habits or excesses, which render them prone to g.p. Hence we find the longest *average duration* of g.p. in gentlewomen; the next, in women of the lower orders; the next, in men of the upper and middle classes; and the shortest in males of the lower orders.

TERMINATIONS.

Death.—Chronic mental defect or disease.—Recovery. These are the three terminations, proper, possible to g.p.

1. *Death.* Though recovery is possible from its less advanced degrees, yet the termination of true g.p. is almost invariably in death. The patient, lying bedridden, with exhausted attenuated frame, the prominences of which lose their vitality under the mere pressure of body-weight, or parts of which melt down more rapidly or slough; may, occasionally, be cut off by the ulterior effects of these lesions,—effects such as exhaustion, septicæmia, pyæmia, spinal meningitis;—or, again, may succumb under pul-

* "Allgem. Zeitschr. für Psych.," xxxv. Bd., p. 253.
† Foville, *fils.* Art. " Nouv. Dict. de Méd. et de Chir. Prat.," 1878, p. 123.
‡ " Annales Méd.-Psych.," March, 1879, p. 201.
§ " Annales Méd.-Psych.," Nov., 1878, p. 333.

monary or abdominal diseases, sometimes accompanying g.p., particularly hypostatic pneumonia, phthisis, diarrhœa, nephritis and cystitis;—or may be cut off by such complications or incidents as intra-cranial or intra-spinal hæmorrhage, or severe epileptiform convulsions, apoplectiform stroke; or embolism; or phlegmonous erysipelas; or, the throat-power failing, may die choked by food, (or by some other substance), or, more slowly, from pulmonitis sequential to its inhalation. In some cases no new symptoms are set up; death is brought about by gradual exhaustion, apparently beginning in the cerebro-spinal system; and the patients and their powers slowly dwindle, decay, and die. Or death may be due to accident or disease entirely independent of g.p.—Suicide is rarely a cause of death in G.Ps. in asylums. The character of any suicidal attempts in the middle and later stages, as already described, sufficiently explains why this is so. In all likelihood, not a few G.Ps. commit suicide, at home, in the earliest or prodromal stage, or in the first stage, when it takes the melancholic or hypochondriac form; such persons have only been thought to be "low" or "queer," or else examples of ordinary melancholia, and are rarely understood for what they are. Many G.Ps., subsequently under my care, had had a narrow escape from successfully attempting their lives at some time before admission. In Chapter IX. several cases are quoted from medical literature in which the more immediate cause of death was rib-fracture, followed by pleurisy and pneumonia. The most dramatic among the frequent ways of dying, are by the apoplectiform and epileptiform seizures. The general exhaustion is, of course, an important factor in many cases. In a large share of cases several factors are concurrent.

Among the soldiers only, under my care during a certain term of years, of the total number of deaths, 29 per cent. were occasioned by g.p., with or without complications. A striking fact was observed in Baillarger's section of the Salpêtrière, when, in 1848, cholera raged there;—*no G.P. was affected.** The simple arrest of g.p., is not a termination.

2. *Chronic Mental Defect or Disease.* This termination of g.p. is scarcely or not at all known, even to writers on the subject. Nevertheless, in a few cases g.p. practically disappears, no decided physical signs remain, but mental health is not entirely regained. There is, per-

* "Ann. Méd.-Psych.," April, 1849, p. 315.

manently, some mental defect, a relative childishness and simplicity; perhaps with some irrelevancy and inattentiveness; or, it may be, some smiling pleased self-satisfaction, and a trace of fixed expansive idea. The patients, however, are able to work, are tractable, docile, useful, industrious; are physically strong, and, at least those I have seen, of large athletic muscular frame. So they go on for many years; no return of motor symptoms; no special sensory symptoms; no marked change in the mental state from year to year. Perhaps in the appearance of their friends, or from the family-history, we find reason to know, or to conjecture, the presence of an element of heredity. At last they die of some intercurrent malady, unconnected with g.p. At the necropsy, we find, in the meninges and brain, thickenings, opacities, and adhesions—the traces and inactive relics of former lesions of g.p.

Here, I maintain, although the patient has not recovered his full mental powers, he has recovered from his g.p.; he has recovered from that malady characterized essentially by the combination of psychic and somatic symptoms, and having a progressive course, however irregular, and however interrupted, that progression may be. A termination of the disease, therefore, has been attained. There seems to be no reason why such a patient should not still live and fulfil his limited *rôle* for twenty or thirty other years or more. The patients of this kind, under my care, died of affections which were apparently quite unconnected with their former g.p.: and if the latter in any way, even remotely, conduced to death it was merely by its unfavourable influence on the vital powers in the past; so that, finally, the organism made, so to speak, a less sturdy and strenuous fight for life than would otherwise have been the case. Like as when an adhesive pleurisy has been entirely recovered from for years past, and the patient is subsequently attacked by cardiac or pulmonary mischief; the adhesions, although in themselves harmless, may have a prejudical and hostile effect in the fight for life, by impeding the free movement of the lung, checking expulsive efforts to clear away secretions or effusions; promoting maceration, congestion, and the downward gravitation of fluids; and curtailing the recuperative forces of the parts. Yet the patient does not die of the original and long-past pleurisy, or of pleuritic adhesions; if he dies, it is of the supervening and independent pulmonary or cardiac or other disease supposed. A condition in which chronic mental defect or disorder is the only perceptible abnormality is, therefore, one of the terminations of g.p. Voisin, also, has probably seen something partially analogous; and, speaking of the pathological physiology of the remissions in g.p., although he gives no examples, says that the localization of the inflammatory lesions when the disease is not yet far advanced, must be admitted, and some necropsies had shown, at points on the brain, very old circumscribed lesions, identical with those of g.p., in individuals who long before death had shown some of its symptoms, but had returned completely or incompletely to the normal state. Yet these seem to have been, rather, examples of recovery from g.p., or of marked remission of g.p. "all round," than of the kind of case I have described above.

The cases of which I have spoken are not such as Voisin refers to under one of his so-called terminations, namely "passage à l'état chronique"; which I do not look upon as a termination, but simply—like the cases mentioned above, after the section on remissions—as an arrest of the malady; or an arrest in some directions, and the assumption of a favourable chronicity in others; a condition of no great rarity; as in the example he gives. M.B.—G.p., first year, delusions of wealth, some dementia, speech affected, convulsions, etc.:—at end of three years, no delirium, complete dementia, speech indistinct.

Nor are those to which I refer cases such as Marcé spoke of; cases with the typical expansive symptoms, passing into a condition of dementia with some isolated ambitious ideas, and an arrest of the further development of the motor disorders; and which he thought were examples of alcoholic dementia associated with ambitious ideas and transitory disorders of motility, due solely to the alcoholic poisoning; and not g.p.

Something analogous was a case by Müller in which florid g.p., with great motor weakness, came to a standstill; for ten years an unchanging mental weakness alone "testified to the paralytic disease."

The cases of which I speak are, practically, new facts in g.p., as far as I know.

3. *Recovery.* The third, and remaining, termination of g.p. is recovery. For convenience, this will be treated of in the chapter on prognosis.

CHAPTER XIII.

DIAGNOSIS.

In well-marked cases the diagnosis is unattended by difficulty, but there are many instances in which it is far from easy; or, at some periods, is impossible. The physical signs are those most relied upon, especially the characters of the speech, the fibrillary trembling and twitch of the muscles of the lips, face, and tongue when in use, the similar motor affection of the extremities, and the pupillary changes. The sensorial, spasmodic, convulsive, paretic, and paralytic phenomena are of help, when present. Another point in the diagnosis of g.p. is the clinical aspect of the mental symptoms, either a gradually progressive mental failure alone, or the same revealing itself, fundamentally, in the aspects of the several deliria, in the character of the insane conceptions exhibited, particularly when of the hypochondriacal or expansive order. The insane ideas are neither persistent, nor harmonized as to their contents, nor logically connected, nor wrought into a system dominating the individual. The third I shall mention is the coexistence of the two orders of symptoms, psychic and somatic, giving to the affection its clinical *tout ensemble*.

At the onset, when most needed, some of these points, unfortunately, often fail us. Of the above elements in diagnosis the first, namely the chief somatic signs, may for a time be absent, or ambiguous, or slight: and in such cases the third element—the union of the two orders of symptoms—necessarily fails us also, or at least becomes doubtful; since one of the orders is, practically, absent. In the conditions supposed, remains then merely the on-coming dementia, on which we cannot found a diagnosis properly so called, or more than a guess; or else there may be a hypochondria or an expansive flight of ideas, of the special features mentioned above; or, equally significant, a mingling of the two; or there may be maniacal or melancholic symptoms, or stupor, or a phase of circular insanity; in which last cases no certain diagnosis of g.p. can be made if the somatic signs are absent (as for the moment supposed to be the case). But in all such examples of insanity, occurring as a first mental attack, in males between 30 and 50 years of age, the medical attendant will act wisely if he keeps in mind the possibility of g.p., and systematically watches for any unequivocal indications thereof.

DIFFERENTIAL DIAGNOSIS. The differential diagnosis must be established between g.p. and a number of affections.

1. *Alcoholic Mental Disease.*—(a) *Chronic and subacute Alcoholism.* Alcoholic intoxication is often an exciting cause of g.p., but chronic and subacute alcoholism must be separated from g.p., by whatever cause produced.

A case admitted under my care when of one month's duration, and in which the cause of insanity was stated to be unknown, closely resembled g.p. But the diagnosis arrived at—namely, chronic alcoholism—was justified by the complete recovery of the patient, and was based upon:—(a) The general tremulous condition, and the marked tremors of the face and hands at so early a period. (b) The constant, restless, fumbling, busy movements. (c) The incomplete development of the affection of speech as compared with that usual to well-marked g.p. (d) The age of the patient (23). Nor was his admission of alcoholic excess entirely without significance. Yet was there a great resemblance to g.p. There were constant mental restlessness and confusion, and large and extravagant notions, though without much accompanying exaltation of feeling. On the negative side, again, there had been no morning vomiting, and no evidence of hallucinations, nor were there positive feelings of dread, nor red eyes, nor acne rosacea; while no tinnitus aurium, no clouds or flashes of light before the eyes, no vertigo, and no hideous nightmare were complained of. Hence a dissidence with the more ordinary features of chronic alcoholism.

But usually it is g.p. with depressed or melancholic symptoms to which some alcoholic cases bear so strong a resemblance.

In chronic simple insanity from alcoholic excess the affection of speech is usually less than in g.p., and is proportional to the amount of tremor present; (smell possibly excepted) the disorders of special, and the failure of general, sensibility, as exemplified in headache hallucinations anæsthesia hyperæsthesia and pricking sensations, are more marked in it, on the whole, than in g.p. The mental symptoms of chronic alcoholic mania, also, are ordinarily such as delusions of persecution, poisoning, suspicion, as of marital infidelity; with fear, self-abasement, sombre feeling, or tendency to suicide; while, with less inco-ordination and less frequent pupillary changes, there is more of general muscular tremulousness in it than in g.p.; and gastric catarrh, nocturnal disquietude, morning tremor, and brutish expression are frequent symptoms. The old drunkard, too, is more apt to complain bitterly of his sufferings.* Long ago Dr. Ch. Lasègue † indicated that the alcoholic trembling was universal, and without points of election as in g.p.; that at the early stages of each the muscular feebleness in alcoholism was in contrast with the spasmodic muscular incitations in g.p.; and that in the former both the contentment and the indifference of g.p. were wanting.

But when g.p. arises from alcoholic excess the diagnosis may be extremely difficult, especially if in these cases there are symptoms such as visual hallucinations, with delusions of persecution and of being poisoned.

Dr. Batty Tuke ‡ looks upon local colour-blindness, and hyperæmic retina, with contracted or irregular pupils as symptoms helping to distinguish g.p. from chronic alcoholic mania. In the latter, however, hyperæmia of the retina may also be found, but subsides when stimulants are cut off. With unaffected pupils and anæmic fundus of eye, even if there is slight atrophy of discs, he would incline to decide for chronic alcoholism. Besides some untenable points, M. Delasiauve § long ago, mentioned in the alcoholic cases, the often rapid recovery, the relapses, the more frequent hallucinations, the countenance melancholic and not expansive as in g.p., though like it in the expression of astonishment and stupidity. For alcoholic dementia, the same principles apply as in the diagnosis between secondary psychoses and g.p.

(b). *Acute Alcoholic Mania* sometimes resembles g.p. of the maniacal form, and on occasions 1 have found that final judgment must be suspended for a short time. In a case of this kind, Dr. Batty Tuke found all the symptoms of g.p., except the alterations of the pupils. Dr. Mendel, in subacute alcoholism, has observed a much greater and more rapid increase of body-weight than in the corresponding periods of g.p. Rapid cessation of symptoms, under the influence of asylum-life and of withdrawal of stimulants, may assist much in discrimination.

Exalted ideas are sometimes seen. Here the diagnostic difficulties are increased, and especially so if motor disorders and also speech-affections coexist. Dr. Bonville Fox ‖ con-

* L. Thomeuf, " Gaz. des Hôp," Jy. 19, 1859, p. 334.—" Ann. Méd.-Psych.," 1859, p. 564.
† " De la paralysie générale progressive." Thèse, 1853.
‡ " Brit. Med. Jl.," 1877, Vol. i, p. 744.—" Edinb. Med. Jl.," April, 1877.
§ " Annales Méd.-Psych.," 1851, p. 611.
‖ " Journ. Mental Science," July, 1884, p. 233.

cluded that there was no point by which the exaltation of alcoholism could be distinguished, with absolute certainty, from that of g.p.; that each separate physical symptom of g.p. might exist in chronic alcoholism; the temperature of the latter, however, being low, as a rule, and without the wide variations observed in g.p.; and its delusions usually fixed, constant, ineradicable. Now this last, alone, would form a point of strong contrast with g.p., and suffice for diagnosis; but the truth is that here, as elsewhere, there is no sure distinctive symptom which may not be found in other affections; and that where doubt arises the whole case, and its history, must be taken into full and deliberate consideration.

(c). *Delirium tremens.* *Mutatis mutandis*, similar distinctions apply.

2. *Syphilitic Disease of Brain and Meninges.* In the "British and Foreign Medico-Chir. Review," April, 1877, I treated fully of the differential diagnosis between g.p. and some cases of intra-cranial syphilis.

As regards syphilis, attention must be given to the history of the case; the coexistence of other manifestations of syphilis; the course of the affection; the indications of local straitly circumscribed lesions or growths; the results of treatment. Also, in syphilis, the absence or less marked degree of the most significant motor signs, as the affection of lips, face, tongue; the rarity of exalted delusions; the onset, rather, by indications of somatic than of psychical disorder; the more irregular course, both as regards the manifestations themselves, and their succession; the more ill-defined duration; and often the cachectic appearance. The difference between the lesions does not apply here.

(A). The g.p. most frequently simulated by intra-cranial syphilis is that in which dementia is the principal mental symptom from the first, exaltation and ideas of grandeur being absent or evanescent, while a tinge of depression, with occasional or frequent fear or terror, may exist.

On comparing its *mode of onset and earlier symptoms* with those of the form of intra-cranial syphilis which simulates it, we find, as I said in the paper just referred to, that "while there is much that is similar in the syphilitic cases, yet . . . the mental symptoms are often *preceded* by marked motor or sensory disorders. Early paralysis of some of the cranial nerves with ocular troubles, early optic neuritis, or local anæsthesiæ, often characterize the syphilitic cases. Headache, nocturnal, deeply seated, and increased by pressure or warmth, is usually a striking phenomenon, and is more urgent and persistent than is the prodromal headache observed in some cases of g.p. Convulsions and local spasms are more frequent in the first periods than in the corresponding periods of g.p.," and subsequent local motor or sensory failure, often more permanent. "Early insomnia, common in some instances to both, is, as a rule, more severe in the syphilitic cases, and the same remark applies to rheumatoid pains in the extremities and to the various neuralgiæ. A fitful appearance, and capricious association and succession of the several symptoms, and their frequent alternations, are features more evident in the syphilitic cases. Anæmia is frequent, and a sallow cachectic hue, not usual to G.Ps. at this period."* The early or rapid or even fitful decline of memory, sometimes insisted upon, is, however, common to both; but the last, or, better, an amnesic stroke, would point, rather, to syphilis; nor is there any very marked difference between the occasional early convulsive phenomena of the two diseases. The impaired articulation, and tremors and twitches of lips, face, and tongue may be absent or but imperfectly marked; but, on the other hand, they may be imperfectly marked in the demented form of g.p., also. "*Early* well-marked apoplectiform attacks, severe nocturnal cranial pain, paralysis of individual cranial nerves, local muscular contraction and rigidity, vertigo, early failure of special senses, and local anæsthesiæ, are the most important distinguishing features of intra-cranial syphilitic cases in the incipient stages;" and these, when marked, greatly facilitate discrimination between this particular class of cases of intra-cranial syphilis and the dementia form of g.p.; but each may occur early in g.p. On the other hand, the characteristic condition of the articulation, facial and labial muscles, tongue, velum and pharynx, when well developed at this early period, often suffice to establish the diagnosis of g.p. Brain syphilis is frequent under the age of 25, g.p. comparatively rare.

In the next place, the features of the same two affections when *fully developed*, are contrasted in the same article. (*a*). *Mental*. Chiefly, in the syphilitic cases, as compared with the demented form of g.p., there is at first rather an obscuration than a destruction of mind, with more frequent and marked insomnia and irritability of temper.—(*b*). *Physical*.

* "British and Foreign Medico-Chirurgical Review," Apr., 1877, p. 448.

Speech may be very similar in the two. "In the syphilitic there may be slight tremors of the tongue when it is protruded, and a speech deliberate, or slightly hesitating, but usually without any accompanying facial or labial tremor, or, again, the voice may be hoarse and low, and . . . utterance . . . slow and attended with effort. . . . But the picture is never complete, the outlines are imperfectly filled in. The tongue more often retains much of its normal condition." (*Ibid.*, p. 451.) The great distinction, however, is that in the syphilitic cases, motor affections of speech, tongue, &c., tend, rather, to be of a *paralytic* nature, than a mingled weakness and inco-ordination.

Dysphagia is sometimes sudden in syphilis, gradual and progressive in g.p. In g.p. are pupillary changes, and often atrophy of discs, after slight neuritis; in syphilis are often marked double optic neuritis, or disseminated choroiditis. Mydriasis, with paralysis of muscles in accommodation, is a frequent forerunner of psychical disorder in syphilis.

Characteristic of syphilis in this relation are palsies of individual cranial, especially ocular, nerves: these are often complete, strictly limited, and independent of convulsive action. In g.p., on the other hand, sudden local palsies or monoplegiæ usually follow local spasm or convulsion, are incomplete, and more often are transitory and recurring. The customary motor impairment in *the limbs* and other parts in g.p. is ataxic, and paretic, is general, and, for the most part, irregularly progressive: in syphilis it is usually *paralytic*, localized, and unilateral, though sometimes general, and may be stationary or retrogressive;—and palsy of one or more cranial nerves with hemispasm, convulsions, or anæsthesiæ, often coexist.

Loss of special senses occurs more frequently, and often more rapidly or suddenly, in the syphilitic cases. Cutaneous "anæsthesia is often early and local in the syphilitic; late, general, and progressive in the general paralytic;" (*loc. cit.*, p. 445); but to these there are exceptions. Severe cephalalgia, and osteocopic pains often afflict the syphilitic. In syphilis apoplectiform seizures usually leave more marked traces than in g.p.

(B). Passing now to the expansive form of g.p., and comparing it with the intra-cranial syphilitic cases most closely resembling it.*

* Dr. Silver, "Med. Times and Gazette," Oct. 26, 1872, p. 461. Dr. Batty Tuke, "Jl. Mental Science," Jan., 1874, p. 563. My cases, "Brit. and For. Med. Chir. Rev.," July and Oct., 1876. "Journ. Ment. Sci.," Oct., 1879.

When exaltation of feeling and notions of greatness arise in the syphilitic cases they are sometimes distinguished —
"*a.* By the distinct history or symptoms of syphilis.
b. By the preceding cranial pains, nocturnal and intense.
c. By the exaltation being less marked, less persistent, . . . than in most of the cases of g.p.
d. Sometimes by such complications as palsies of one or several cranial nerves, or hemiplegia, or paraplegia having the character and course of syphilitic palsies.
e. By the greater frequency of marked optic neuritis, early amaurosis, deafness, local anæsthesiæ, vertigo, or local rigid contraction.
f. By the affection of articulation being paralytic rather than paretic, and speech usually not accompanied by any facial or labial tremors or twitches.
g. By frank cerebral or spinal meningitis or pachymeningitis.
h. By the variety of the motor and sensory symptoms.
i. By the effect of anti-syphilitic treatment." *

(C). *Mutatis mutandis*, very much the same holds true of cases of g.p. taking the hypochondriacal and melancholic forms, and their separation from similar syphilitic cases.

Dr. Leidesdorf † agrees as to the importance of localized indications of disease, such as palsy of a single cranial nerve, in cerebral syphilis.‡ Headache, spots of cutaneous anæsthesia, and palsy of single cranial nerves are among the features specially relied on by Dr. Müller of Leutkirch. Hemiplegia with luetic signs; or waxy kidney; affords indications of syphilis.

Simultaneous and unilateral paralysis of the fifth and sixth cranial nerves is an almost certain indication of intra-cranial syphilis (Gräefe); the multiplicity of the nerve-trunks affected (Leudet); and the extension of the paralysis from nerve to nerve, and its sudden coming and going (Gros and Lancereaux); are striking and sufficient traits. Yet they often fail us here.

According to Heubner § the cases of intra-cranial syphilis bearing a close resemblance to g.p. are precisely those in which no very decisive or characteristic anatomical changes have been visible to the naked eye.

Nevertheless I ∥ have shown that one kind of syphilitic

* *Loc. cit.*, pp. 456, 457;—one or two slight verbal alterations.
† Imp. Roy. Med. Soc., Vienna. "Lond. Med. Rec.," Jan., 1878, p. 41.
‡ Also Wille, Müller, Gros, Lancereaux, Jackson.
§ "Ziemssen's Cyc. Trans.," Vol. xii, p. 315.
∥ "Brit. and For. Med. Chir. Rev.," July, 1876, and Apr., 1877: "British Medical Journal," July 13, 1878, p. 49.

brain disease which frequently simulates g.p. is that in which the cerebral arterioles and, usually, arteries are extensively diseased, and in which sometimes also the cortical surface of the cerebrum and the overlying meninges are the sites of gummatous infiltrations. Adhesive meningitis is another form of syphilitic disease that sometimes simulates g.p. In a case by Schüle, were gross intra-cranial lesions, also microscopic changes of cerebral cortex.

Confining attention to those published, some of my cases in the "Brit. and For. Med. Chir. Rev.," July, 1876, may be briefly referred to. I. (p. 163) Dementia, physical signs resembling g.p., headache, epileptiform attacks, transitory hemiplegia, recurrence of convulsions, death in the "status." Syphilitic disease of brain, cerebral arteries, spleen, liver, skin, and testis. Thickenings and slight adhesions of meninges, gummatous infiltration, and some irregular encephalitis.—II. (p. 174) Sudden paresis of articulation and deglutition, and right hemiparesis, followed by intellectual and moral deterioration, and by temporary attacks of dysphagia with vaso-motor disturbance. Subsequently, dementia, right hemiparesis, speechlessness, intestinal torpor, contracted limbs. Extensive disease of cerebral arteries and left corpus striatum. IV. (p. 180) Intense cranial pain, dementia, satisfaction, resemblance to g.p., symptomatic paralysis agitans, disturbed innervation of heart, dysphagia, general motor paresis, some left hemiplegia, affection of speech. Recovery under specific treatment. Also, case X., p. 193, *loc. cit.* For other cases, by present writer, of recovery from symptoms as of g.p., under specific treatment, see "Jl. Mental Science," Oct., 1879.

An apparent identity in symptoms with g.p., except for early persistent left hemiplegia, torticollis, and an unusual degree of dysphagia, was observed by Dr. Ach. Foville,[*] in a case where the cerebral lesions consisted of isolated syphilitic infiltrations and tumours in cortex, medulla, and optic thalami. In Voisin's[†] case resembling demented g.p., were syphilitic gummata in the brain-periphery. In another, with mental and motor changes, affected pupils and speech; were a syphilitic tumour in cortical motor zone, insula, and lenticular nucleus; extensive white cerebral softening; meningeal compression of a motor oculi nerve. Of cases mentioned by Zambaco,[‡] in one the convolutions were some-

---

[*] "Annales Méd.-Psych.," May, 1879, p. 355.   [†] *Op. cit.*, pp. 273, 289.
[‡] "Des Affections Nerveuses Syphilitiques."

what flattened; the peripheral grey matter was softened, spotted, and came off with the pia at several points; and the white central parts were softened. In another there was general softening of the grey substance, and opacity of the arachnoid. In another the pia was fibrous, thick, adherent, and pressed on some cranial nerves. Dr. F. Dreer* observed cases of syphilitic disease resembling general paralysis; and M. Charpentier † a female, admitted without history, and with the physical and mental symptoms as of the demented form of g.p., in an advanced stage; and going on from bad to worse, until some indications of tertiary syphilis were noticed. Under K.I. and mercurial inunction, the patient recovered.

Whether, or not, syphilis produces true g.p.,—and on this point see Chap. XIV.,—it is unquestionable that in one class of cases syphilitic disease more or less closely resembles g.p. in some of its features, and yet remains distinct therefrom in its essential nature. As between cases of this class and g.p., the distinctions drawn above will still prove serviceable, whatever be the ultimate fate of the major etiological question. There has been some misconception with regard to my writings on this subject. In laying down the above distinctions, some years ago,‡ I did not attempt to do more, at least in my later contributions, than distinguish from g.p. some cases of syphilitic brain-disease somewhat like g.p., *i.e.*, like g.p., no matter what its cause; and, whatever its cause, its symptoms will be, roughly, the same. For, although in one passage assuming syphilis not to be a cause of g.p., I left it an open question.

Is there anything that lends special features to the cases of g.p. in which there has been a distinct antecedent history of syphilitic infection, but in which, at the necropsy, no gummata, or distinctly specific lesions of syphilis, are found? The changes found being simply ordinary chronic inflammatory or degenerative ones, or both. And it being conformable with what is found in affections of analogous kind, that the manner of action of the particular causes has some influence upon the clinical aspects.

The cases of g.p. in which there is a history of antecedent syphilis may be conveniently placed in two great groups. Intermediate conditions of course exist; and not to make

* " Archivio Italiano," 1869.
† "Ann. de Dermatologie et de Syphiligraphie," Mar., 1885, p. 158.
‡ "Brit. and For. Med. Chir. Rev.," Apr., 1877; and First Ed. of this work.

distinctions that concern points too fine to be of much practical value, I shall merely mention the leading ones.

In one group are cases in no way differing as regards either symptoms or lesions from the ordinary run of cases of g.p., in their several more usual varieties. Nothing more need be added as to this group.

In another group of cases of g.p. with an antecedent history of syphilis, and without unequivocally specific lesions; several points of difference from the more usual cases of g.p. deserve some attention :—

1. The mental symptoms are apt to be predominantly either dementia or a melancholia; or the two combined. Not that other symptoms may not crop up, but such are, in that case, comparatively slight and transient, as a rule.

2. Unilateral convulsions are usually marked. Occurring, perhaps, sometimes on one side, sometimes on the other, the general tendency is to finally settle down upon one side, exclusively or chiefly.

3. The transitory post-convulsive hemipareses may affect at first one side or the other; but, finally, are solely, or mainly, of one side, which corresponds to the side convulsed; and at last that same side usually betrays evidence of a permanent incomplete hemiplegia, increased, temporarily, after the convulsions, and tending to deepen as the patient dwindles away.

4. As to lesions; (a) usually the meningeal changes are more highly marked than usual, relatively to the cerebral;— (b) and they are often more obvious on one side of the brain than on the other;—(c) adhesion and decortication predominate on one cerebral hemisphere, as compared with the other;—(d) the cerebral cortex is often the seat of a somewhat diffuse, yet incompletely circumscribed, palpable sclerosis; usually more obvious in one hemisphere than in the other;—(e) one cerebral hemisphere is often much more atrophied than its fellow.

Other points might be mentioned, but it is unnecessary to dilate on the subject in this place. When g.p. occurs after a history of syphilis, but there is no distinct evidence of syphilis therewith; and no unequivocal syphilitic lesion after death, the only cerebro-spinal lesions of any moment being the ordinary ones of g.p., as seen every day; the criterion formerly accepted was that an adhesion of the pia only, to the cortex, was never a lesion produced by syphilis alone. Therefore, when, in cases such as here supposed, I have, on very many occasions, found only the ordinary changes of g.p., I have said "there is no proof in this case that the g.p. is of syphilitic nature; and, further, if the criterion is accurate the g.p. is not syphilitic." But the criterion may not be accurate; and, if so, the lesion in question, and g.p., may, in that circumstance, be producible by syphilis, acting alone or in conjunction with other causes.

3. (A.) *Some Cases of Acute Mania.* Early in g.p. there is sometimes acute maniacal excitement, preceded, or not, by a stage of depression, of mental alteration, or of ambitious delirium, or by both. In the opinion of some, "congestive mania" often complicates or is transformed into g.p. One must, therefore, attempt to disentangle the active mania of g.p. from certain cases of similarly active but simple mania.

Here, light may be shed by the history of the case, which, in the instance of g.p., may include some transient attacks of impairment or loss of consciousness; may show the existence, and perhaps the long duration, of some of the somatic signs already mentioned as often appearing in the prodromal or first period of g.p.; and may reveal, also, some aforegoing intellectual enfeeblement. Besides the history, should that be obtainable in the cases referred to, g.p. can only be recognized by the aid of two orders of facts:—

Firstly, by the existence of some slight hesitation, or occasional pause, drawl, hemming, or stumbling, in speech; or repetition, or partial repetition, or elision of syllables; with, possibly, a faint occasional twitch of the upper lip and face during speech and independently of any emotion, or slight tremulousness and twitch of the tongue on protrusion; or some heaviness of visage. And, secondly, by the predominant expression of extravagant ideas, and emotional exaltation.

But facts of both these orders are sometimes inconspicuous or absent in mania which, without solution of continuity, subsequently merges with, or is transformed into, the psychical sphere of g.p.; so that it proves itself to have been a paralytic mania: on the other hand facts of the former, and somatic, group are not always entirely unknown in ordinary acute mania, and those of the latter, the psychic, are frequent therein.

When the above somatic signs are absent, or but doubtfully caught, a well-founded diagnosis is impossible, and that, perhaps, for weeks or months. In other cases, when the physical indications of g.p. are merely masked, as it were, the diagnosis cannot be made with certainty until the excitement subsides. As a rule this subsidence, or at least remission, in the maniacal excitement is not long delayed, but should it be tardy the diagnosis may remain doubtful. Under these circumstances of prolonged maniacal excitement, mental confusion may be more marked in the G.P. than in the maniac, and so may be any defect in writing, if the patient can be induced to write. Suspiciousness and morbid aversion or rage are more often absent, or transient and easily diverted in the G.P. of this kind as compared with the ordinary maniac; the G.P. is more easily led from his purpose or coaxed; while a clue may be afforded by the occurrence of the epileptiform or apoplectiform seizures of g.p. A history of one or more previous attacks of mental disorder would disfavour the view that the case was g.p., but would by no means preclude it. If the patient is a male between 30 and 50 years of age the likelihood of g.p. is greatly increased, and decidedly so should the most absurd and monstrous of the features of expansive delirium evolve.

(B). *Expansive Delusions.* Mania or monomania of grandeurs, wealth, or pride, often occur quite independently of g.p. In doubtful cases of this sort Esquirol * had diagnosed g.p. by an occasional slowness in pronunciation, and by the fact of the patient being calmed by a promise, and induced to forego apparently cherished projects. And it is particularly in cases such as these that his remarks apply :—that as compared with other maniacs and monomaniacs, those with g.p. have not the same energy of attention, nor the same firmly-knit association of ideas, nor the same power of will, nor the same tenacity of resolve, nor the same obstinacy of resistance. Becoming excited and flying into passion, nevertheless they yield and obey; but their acts already reveal the enfeeblement of the functions of the brain.

In both cases, there may be a mixture of exalted delusions with those of persecution, and some pupillary changes. Tremor and twitch of face and tongue, and affection of speech, may be found in the monomaniac. But the tremor and twitch is for the most part emotional, and only appearing under excitement, or obviously originating, partly, in the effects of alcoholic excess; and the same applies more or less to the affection of speech when not habitual. And the systematized, fixed, ideas, with the demeanour and conduct logically based thereupon, mark the monomania distinc-

* " Des Maladies Mentales," Paris, 1838, T. ii, p. 276.

tively; just as, on the other hand, the somatic signs, when well pronounced, mark the case as g.p.

Doubt is apt to arise in some cases of elderly patients, who, (perhaps not as the first attack), have mania with much moral perversion; changeable, inconsistent, expansive, exalted ideas; disposition to violence; and with these some irregularity in the contour of the pupils, a tongue tremulous and twitching on protrusion, when, also, the face is jerked; some facial twitches, and an occasional hesitation during speech; now and then ideas of poisoning and of injury; and the patient—often insulting, threatening, quarrelsome, destructive,—settles down in a chronic delusional state, with expansive ideas, not all fixed or systematized, mingled at times with delusions as to being poisoned and the hostility of those about him. These cases illustrate the difficulties in diagnosis between g.p. and senile mania; and, later, between g.p. and a chronic delusional state accompanied with deteriorative changes in the cortex cerebri. Patients of this kind perhaps come near ending as G.Ps., though not so ending (death by cerebral hæmorrhage in one case).

(C). *Ambitious delirium with local organic affections of brain and cord*, will chiefly be dealt with under other heads, and here we need only refer to it when supervening, for example, on local brain hæmorrhages, usually several in number, and separated by considerable intervals of time, (Baillarger, Calmeil). Softening of spinal cord was the only lesion found in a case (on which I reserve opinion) by M. Renault du Motey. In these cases, often complicated by motor and sensory signs arising from the same lesions, only the history of the case, and a comparison of the whole state, when thoroughly tested, with that usual in g.p., can prevent error. The existence, and long duration, of ambitious delirium with local lesions of brain and cord, and without chronic periencephalitis, are accepted by Baillarger as a proof that dementia paralytica, and insanity in g.p., belong to two different orders of facts; dementia paralytica, with its well-defined lesions; delirium, which has a separate existence, and an evolution of its own.

4. *Intra-cranial Tumours.* The phenomena originated by intra-cranial tumour may to some extent simulate g.p. and its epileptiform and apoplectiform attacks. But the tumour-cases are usually recognized by the existence of intense headache, vomiting, marked double optic neuritis, and failing sight; frequently, by deafness of one or both sides, or local

cutaneous anæsthesia; sometimes, by local *complete* palsy due to compression of one or more motor nerves at the base of the brain—the paralyzed muscles showing diminished farado-contractility, and increased galvano-contractility—or, also, by other localized slowly progressive paralysis; by the absence of speech-affection, or by its truly paralytic, and not ataxic and paretic, nature, when present;—by more frequent rigidity, trailing gait, or vertigo;—and, occasionally, by the indications of tumours elsewhere, or of special cachexiæ.

Yet the tumours may not be accompanied by these pronounced symptoms. Especially does the difficulty in diagnosis occur in some cases of cerebellar tumour in which there are found general impairment of muscular energy, a swaying staggering tottering unsafe gait, as well as vague and aimless movements, and, finally, perhaps, a want of control over the muscular movements—an inco-ordination—somewhat as exists in g.p. Other phases of g.p. may be simulated by certain effects of the cerebellar tumour, such as its pressure on the medulla oblongata, or intra-cerebral pressure by the ventricular effusion due to the obstructed and refluent blood-stream in the veins of Galen. Here the association of other symptoms usual in tumour—such as severe headache, vomiting, blindness—the difference in the gait and speech, and, almost always, in the mental symptoms, must guide the diagnosis, and for this purpose the most important symptoms are vertigo, occipital headache, reeling gait, vomiting, and visual failure or disorder.

In intra-cranial tumour, as compared with g.p., the mental symptoms are *comparatively* late, slight, or absent; the sensory comparatively prominent. The progressive dementia of g.p. is simulated, not the grandiose delirium. It must not be forgotten that a hydatid cyst* in the cerebrum, or other tumour, may coexist with the lesions of g.p.

5. *Other Cerebellar diseases* sometimes incompletely resemble g.p. Here the affection of the gait is often relatively very prominent; while some dementia and paresis may exist, the chief traits are often symptoms similar to those mentioned above, as the most important ones, under cerebellar tumour; some of which do not depend on the cerebellar lesion as such, but upon irritation or compression of neighbouring parts of the cerebro-spinal axis.

* Case by Baillarger, " Journ. Mental Science," Apr., 1882, p. 116.

6. *Cerebro-Spinal Disseminated Sclerosis.* Now and then a case of g.p. resembles cerebro-spinal insular sclerosis. A G.P. may have the hands, or upper extremities, or head, or even the body at large, shaken by rhythmical tremor, associated with a general constant muscular restlessness; the tremor being markedly increased during any movement, or when the patient is in the erect position; and diminishing, or perhaps ceasing momentarily, when the parts affected are placed at rest from voluntary movements. There may also be a bending forward of the head and upper part of the body; the speech may be slow, deliberate, drawling, and the words measured and alternating with pauses. In all these respects, therefore, it may closely resemble the symptoms of disseminated sclerosis; while symptoms such as the following may be common to the two cases;—increased reflexes, tremor of the tongue on protrusion, and of lips and face before and during speech, impaired and disordered lingual movements, disorder and impairment of visual power, a degree of spastic gait, increasing paresis of the limbs, their final flexed contraction, irregular fluctuations of the symptoms, and apoplectiform or epileptiform attacks. In truth, there is a secondary scattered sclerosis in some cases of g.p.

Thus, in the one direction, is a close resemblance between the two affections reached by means of an approximation of the semeiography of general paralysis to that of insular sclerosis. In the other direction, also, may the same result be arrived at, and insular sclerosis, in its turn, put on, in part, the garb of g.p.; for its mental symptoms may resemble the dementia, or even the ambitious delirium, of different cases of the latter.*

When present, this simulation of the mental symptoms of g.p., especially of the expansive, vastly enhances the difficulties in the differential diagnosis. Here we can only rely upon the presence, or upon the greater frequency in disseminated sclerosis, than in general paralysis, of such symptoms as rigidity and convulsive trembling on faradization of lower limbs, protracted existence of marked tremor

* Dr. Wilh. Valentiner, " Ueber die Sclerose des Gehirns und Rückenmarks." "Deutsche Klinik," 1856, Nos. 14, 15, and 16.
Dr. H. Liouville, " Mémoires et Comptes-Rendus," Soc. de Biol., 1868, p. 231.
Dr. S. Jaccoud, "Traité de Pathologie Interne," 1869, T. i, p. 193.
Dr. W. Leube, " Deutsches Archiv f. Klin. Med.," 8 Band, p. 1, 1871. Dr. J. M. Charcot, " Lectures on Diseases of the Nervous System." Dr. W. Moxon, " Guy's Hospital Reports," Vol. xxi. Dr. T. Buzzard, " Trans. Clinical Society of London," Vol. viii, p. 121. Dr. J. R. Gasquet, " Jl. Mental Science," Apr., 1884, pp. 74, 155-6.

during movements, temporary diplopia, nystagmus, vertigo, markedly scanned or staccato speech, bulbar paralysis—dysphagia, dyspnœa—" spinal epilepsy," formerly so-called,* as well as of paroxysms of rigid extension of the lower limbs; also, occasionally, of the upper; and ultimate permanent 'contraction,' in extension.

As for the first-mentioned class of cases, that in which g.p. counterfeits insular sclerosis, the diagnosis is aided by the fact that the usual early periods of g.p. may already have been passed through without any simulation of insular sclerosis, as well as by the character of the tremor in these cases, which does not quite correspond to that of the latter affection.

The comparative youth of many patients in insular sclerosis, its much greater *average* mean duration, and its greater frequency in females, are points of divergence from g.p.; so also is the comparative lateness of any mental disorder or decay, which, indeed, is usually preceded by spinal symptoms of some duration.

Next come several affections to be distinguished from the demented form of g.p., and from g.p. in the elderly or aged.

7. *Circumscribed Brain-lesions, with Dementia and Paralysis* (foci of softening, hæmorrhage, embolism, thrombosis). Dementia with paralysis is often confounded with g.p., and all the more readily as, in the former, speech may be lamed or stuttering, the gait impaired, the pupils unequal, and perhaps sluggish. Here, a more or less localized, usually persistent, and purely *paralytic* affection of the muscles is in contrast with the general progressive inco-ordination, and spasmodic unrest, of movement found in g.p., which, later on, is accompanied by paresis. The history of the case also aids in the differential diagnosis. A person in health is stricken with paralysis, often more or less complete, and local, and then gradually becomes demented, and, perhaps, weakly emotional. This is not the course of g.p. The condition is slowly progressive or stationary, rarely partaking of the special attacks, the crises and fluctuations, the transitory hemipareses, so common in g.p. In dementia with paralysis due to local lesion, also, there is often early contraction and pain in the paralyzed members, while the affection of speech (if any) is paralytic, and the mouth often persistently drawn awry. In it, also, the effects of apoplectiform

* Brown-Séquard, "Archives de Physiologie," 1868, T. i, p. 157;—"Journal de la Physiologie," Vol. i, 1858, p. 472.

seizures are more decisive and persistent than in g.p., and the age usually greater.

Again, as a rule, it is only g.p. *without* ambitious delirium or expansive feeling that is simulated by the condition in question. As compared with the dementia sometimes coming on primarily and exclusively in g.p., the dementia following cerebral hæmorrhage is discriminated mainly by the fact that the latter comes on after the truly paralytic symptoms have been established, and often in succession to an apoplectic comatose attack, and then sometimes progresses more rapidly than in this clinical form of g.p., and lacks its characteristic variety of symptoms, while the loss of memory may be circumscribed, relating only to a certain order of facts or events; the disorder of consciousness may be less marked than in g.p., no exalted or hypochondriac notions exist, as a rule, and motiveless weeping and laughter, transitory and frequently recurring, often separate it from the demented form of g.p. On the other hand, in the course of the dementia form of g.p. one may at times be surprised at the unwonted evincement of exalted, or of hypochondriacal delusions of the kind usual in the more typical forms.

Should the local circumscribed lesion of the brain be accompanied by delirious conceptions, hallucinations, or active excitement, marked disorder of consciousness, or delirium, the same considerations hold good, *mutatis mutandis*, as just above put forward in relation to the more usual dementia. Here, on the mental side, the comparison would be with the flight of ideas in g.p.

If numerous scattered circumscribed lesions affect the brain-cortex the symptoms may very closely resemble g.p. Marcé distinguished simple dementia due to this condition, by the absence of vermicular movements of cheeks and lips, and of ambitious delirium. But in a case of aneurysmal degeneration of capillaries of brain-cortex, following excessive emotion, reported by L. Meyer,[*] depression and excitement occurred together with paretic signs, much as in g.p.

8. *Chronic Softening of the Brain* (local). Chronic softening must be diagnosed from g.p. by the history of the case, the localization of the motor symptoms, the absence of the characteristic early irregularity and inco-ordination of movement, and of mental symptoms of the kind more usual in g.p.; as well as, often, of its more special spasmodic phenomena.

[*] "Archiv für Psych.," i Bd., p. 279.

9. (a). *Senile Dementia.* As Dr. Blandford* says, "senile dementia . . . may be characterized by loss of memory, extravagant and indecent conduct, and delusions. There will, however, be an absence of the specific delusions and the maniacal condition; neither shall we find the inequality of pupils, the stutter, nor stumbling gait. In fact the failing mind in senile dementia is manifested usually long before any symptoms of bodily paralysis, and . . . general paralysis is rare at the age of sixty, senile dementia seldom beginning so soon." Yet maniacal disturbance may be considerable in these cases, incoherence may dominate the scene, or melancholic and hypochondriac symptoms, perhaps slight and transitory, may manifest themselves. Voisin thinks some classed as senile dements should be included among the G.Ps.; I think mistake is more frequently made the other way.

(b). *Senile Dementia with Paralysis from Local Lesion,* compared with the demented, or incoherent, or depressed aspects of g.p. Here, as Dr. v. Krafft-Ebing † indicated, if there is in each a maniacal attack at the beginning, it takes place, in senile dementia, with an absence of the excessive impulse to movement and tumultuous grand delusions of g.p.; there is instead merely a childish activity and babbling restlessness; while if both begin with progressive weakening of the intelligence, there is in *dementia senilis* less of that disturbance of the consciousness which is shown by mistakes as to persons, times, or places, and more of early-coming failure of memory, than in g.p. of the kind simulated by it. Distrust, suspicion, persecutory delirium, are far more frequent in the senile dement or maniac than in the G.P., and the morbid ideas less changeable. Again, in *general paralysis* the motor affections are early, general, somewhat changeable, oscillate in their course, and are primarily of the nature of inco-ordination; while in *dementia senilis* they are usually local, are true palsies, are simply progressive, stationary or slowly improving, and often are accompanied with local contractions, even at an early date of their existence. As Marcé mentioned, under excitement much strength can be put forth in g.p., whereas the movements of the palsied senile dement may become disorderly but do not gain in power. Referring chiefly to the dysarthric forms, the disorder of speech in the two affections necessarily partakes of these relative differ-

* "Insanity and its Treatment," London, 3rd Ed., 1884, p. 310.
† "Allgemeine Zeitschrift für Psych.," xxiii Bd., p. 181.

ences: that of the palsied senile dement is the imperfect speech of true paralysis, secondary and later in appearance, and the mouth is often persistently awry from partial facial palsy, whereas in general paralysis the imperfection is primarily ataxic, and secondarily paretic, and is accompanied by tremulous, twitching and spasmodic action about the lips, tongue, and face, rarely seen in *dementia senilis*. Sometimes the age, sex, state of eyes, less impaired special senses, and grinding of teeth aid in indicating g.p.

Dr. L. V. Marcé,* the chief pioneer on this subject, relied in senile dementia, as compared with g.p., on a somewhat persistent unilateral paralysis; the absence, except rarely, of "fibrillary contractions" of the tongue and lips, simultaneously; the differences in speech; the paresis, rather than irregularity and defect of co-ordination. Ambitious delirium he valued highly, as indicating g.p.

10. *Locomotor Ataxy. Tabes Dorsalis.*† When the gait is affected very early in the disease, and the posterior columns of the spinal cord are implicated, the gait of the G.P. is somewhat like that of a subject of locomotor ataxy; and, on the other hand, the latter may have ambiguous mental symptoms. In the former, the absence or mildness of the fulgurant pains, of the early and obvious anæsthesiæ, and of the ocular troubles, of locomotor ataxy, as well as of its not infrequent *early* urinary incontinence, seminal emissions, and engirdling sensations, will help to guide the diagnosis. So also will the presence of the affection of speech of g.p., which, like its twitch of the lips and face, during articulate utterance, is scarcely simulated in locomotor ataxy. There is, also, the relationship between (1) the degree of disorderly and unsafe locomotion, and (2) the actual failure of muscular power;—the former bearing a much higher ratio to the latter in locomotor ataxy than in g.p. For in the motor sphere of neither is ataxia always everything.

The test of closing the eyes and watching its effect on equilibration and on locomotion should be employed, and the presence or absence of patellar tendon-reflex ‡ should be

* "Gazette Médicale," Aug. 1, 1863, No. 31, p. 497.
† Dr. John Hitchman, "British Medical Journal," 1871, Vol. ii.
‡ Profs. C. Westphal, Erb, O. Berger, "Arch. für Psych.," v Bd., 3 H., p. 803. "Berliner Klinische Wochenschrift," Jan. 7, 1878. Dr. T. Buzzard, "The Lancet," July 27, 1878, p. 111; Aug. 10, 1878, p. 175. Dr. W. R. Gowers, "Royal Med.-Chir. Soc.," London, Jan. 28, 1879. Muhr, "Psychiatrisches Centralblatt," 1877. "Jl. Mental Sci.," Jan., 1879, p. 680. "Eulenberg, Centralblatt für Nervenheilkunde, Psych., &c.," 1878.

verified, as its presence tends to exclude tabes dorsalis. In g.p. it may be increased, normal, lessened, or lost.

True tabes dorsalis may precede g.p., may come on simultaneously, possibly may follow. Other lesions in g.p. may occasion quasi-tabic symptoms. I have dealt with the whole subject elsewhere,* and have described an example in full.— In one class of cases it is believed that true g.p. may begin by a propagation from the lesions of true tabes dorsalis.—Of another kind were cases of tabes dorsalis with intercurrent exalted delusions or some buoyancy or optimism passing, or tending to pass, entirely away.— Of other kinds, also, were some cases in which g.p. appeared in the first or cephalic period (Duchenne) of tabes; and, as described, sometimes the former increased and the latter was arrested, sometimes the former disappeared and the latter worsened, in other cases they advanced in an equal and parallel manner. Recently the same observer reported a case of prog. locom. ataxy with ambitious delirium, but without g.p.—Another found that in a late stage of grey degeneration of the cord's posterior columns mental disease sometimes supervened; which, by its delirium, paralytic symptoms, intercurrent apoplectiform and, occasionally, convulsive seizures, bore a likeness to g.p. Amongst the clinical features viewed as separating these cases from g.p., were the absence of impairment of articulation, and the presence of staggering when the eyes were shut.

Dr. Steinthal† had long ago noticed the naïveté, carelessness, and even gaiety or serenity, in some subjects of tabes dorsalis; who, throughout their wearing painful malady, never lost courage or hope. The ataxic affection with psychic elements, described by Dr. L. Kirn,‡ also was independent of g.p.

11. *Paralysis Agitans.* A symptomatic paralysis agitans occurring in the course of g.p. may superinduce some resemblance to a case of idiopathic shaking palsy (Morbus Parkinsonii), and all the more so as the latter affection may have a speech, an impaired deglutition, a stolid expression, a slowness of movement, and even a muscular weakness, not very unlike those usually found in g.p. ;—while, on the other hand, in g.p. may occasionally be found the attitude of the hands

* "Lancet," May 21, 1881, p. 819; and May 28. "On General Paralysis of the Insane consecutive to Progressive Locomotor Ataxy."
† "Journ. der Praktischen Heilkunde," xcviii Bd., July, 1844, p. 9. "Beiträge zur Geschichte und Pathologie der Tabes Dorsualis."
‡ "Allgem. Zeitschrift für Psych." xxv Bd., p. 114.

often seen at a stage of paralysis agitans. But the diagnosis at once ceases to present any difficulty when an accurate history of the case is procurable, for the tremor cöactus occasionally seen in g.p. comes on *after* the earlier of its more customary motor and mental symptoms.

12. *Epilepsy.* Besides the convulsive seizures in epilepsy there may sometimes be a shaky, thick speech, and a jerky tremulousness of the lips and face during speech. But as far as I have seen this only occurs to any marked degree in some chronic patients, subject to frequent, severe and general convulsions—patients whose whole medical history, and whose complete return to the usual, and nearly stationary, mental and physical state in the intervals of the convulsions, differ from what is found in g.p.; while the convulsive attacks often differ from those more usual in the latter affection, the physiognomy differs, and the irritable, suspicious, surly, impulsively violent state of the epileptic is in contrast with that more usual to the general paralytic; much the same, however, may be seen in g.p. In the chronic epileptic, with all the changes, there is more of a stereotyped alternation, and small number of phases, than the variety in the paretic dement.—After severe repeated epileptic seizures may be the above affection of speech, face, and lips, also unequal pupils, a dull heavy look, and mental confusion and weakness. The history of the case, some of the points already mentioned, and the clearing up of the symptoms if the attacks keep off, make the diagnosis. But, occasionally, here the transitory mental symptoms are hallucinations, emotional exaltation, ideas of grandeur. I have seen such cases, and one is recorded by M. Foville. Dr. Huppert separates epileptiform attacks in g.p. from convulsions of true epilepsy by marked albuminuria and often casts and hæmacytes in the urine, after the former: to which we may add a more marked increase of temperature.

13. (A). *Epileptiform (and apoplectiform) attacks,* as in g.p., occur in various other cerebral affections, such as local cerebral softenings or hæmorrhages, some of which are attended by descending fasciculated sclerosis.

(B). *Cerebral hæmorrhage, or congestion.* In the initial stage, and even at a later period, the apoplectiform seizures of g.p. cannot, at the moment, well be distinguished from ordinary sudden cerebral congestion. Yet in a short time, the congestive symptoms usually pass away, leaving behind them some indications of g.p.

From the stupor or coma of actual cerebral hæmorrhage the apoplectiform seizures of g.p. may be separated by the symptoms being usually of less gravity, and by the difference in the modifications undergone by the temperature of the body under the two circumstances. In the hæmorrhagic case the temperature falls after the attack, and generally remains below the normal for some time, perhaps twenty-four hours, afterwards rising more or less above the normal, and subsequently pursuing a course which varies very greatly according to the degree of the severity and fatality of the hæmorrhagic lesion. In the apoplectiform attacks of g.p., on the other hand, the temperature often rises somewhat just before the attack, remains so immediately after it, and only subsides with the subsidence of its attendant mental phenomena, or somewhat earlier than do these. Any accompanying paralysis, also, is of shorter duration and of less strictly limited extent than that often attending hæmorrhage.

Yet meningeal and other hæmorrhages may, and sometimes, do, occur in g.p., and produce their usual symptoms, modified by the already existing disease of brain.

14. (A). *Melancholia with stupor*, distinguished from g.p. with stupor. Some of the somatic signs of g.p., the facts that the patients rarely maintain complete silence, that they show less of that rigid, fixed, contraction of the lineaments, and deep furrowing of the lines of expression, than is usually seen in the simple form of melancholia with stupor, are distinctive points. Here, Baillarger gives importance to accompanying congestions of the face in patients of middle age, not previously melancholiacs, and particularly if hypochondriacal notions are elicited, and if the pupils are unequal. Voisin deems the existence of slight fever very significant here. But, of these points, congestions of the face are of no help, and may occur in the simple form; hypochondriacal notions are ordinarily not to be elicited; and, particularly with the mental condition of these patients, slight fever may be due to some trivial, obscure, local affection, in any patient of either class, simple or paretic.

(B). *Melancholia. Mutatis mutandis*, much the same orders of facts apply, as in the case of stupor. Some fundamental psychical weakness and confusion may be revealed, gleams of expansive delirium occasionally light up the dark and sombre mental horizon of the melancholic G.P.; some of the somatic signs can usually be found, or be revealed by tests; the temperature is slightly higher than in the ordinary melancholiac, and especially at night, but this is only of value when other slightly febrile affections can be excluded with certainty.

(*C.*) *Hypochondria*, occurring in g.p., must be distinguished from simple hypochondria by the existence of some of the somatic signs of g.p., by the absurd, inco-ordinated, changeable character of the morbid ideas, and by other indications of mental failure. Difficulty is only likely to arise when the hypochondria and attendant circumstances are symptomatic of organic brain-mischief; as in some cases of cerebral syphilis; or, on the other hand, when the hypochondria, to be distinguished, occurs at a very early point in the course of g.p., and is unattended with marked somatic signs; or when it occurs later in the course of the disease, and the patients are taciturn, move but little, and do not readily reveal either the ideational or motor indications of their malady. Here, some, or all, of the points mentioned above as distinguishing the marked examples of hypochondria in g.p., may be present to a greater or less extent, must be carefully watched for, or waited for, and perhaps elicited by stratagem, or by tests; a history of slight apoplectiform or epileptiform seizures may assist (but is of no help in distinguishing from syphilis); and in the later-coming and taciturn cases the physiognomy, perhaps, is strongly indicative of g.p. The age, the pupillary and spinal symptoms, may point, preferably, to g.p.

15. *Circular Insanity* is in some cases only to be separated from the g.p. of "circular" form by the history of the case; and, occasionally, the more marked physical signs in the latter.

16. *Secondary Dementia.* The history, *i.e.* of some form of insanity gradually passing into mental decay, some indistinct outlines of the primary mental affection showing through the effacing dementia, sufficiently marks these cases off from g.p., on the psychical side; as does the absence of marked physical signs on the somatic. The only difficulty is when the history of the case is unknown, and the demented form of g.p. is present, with only slight somatic signs, and has to be separated from a secondary psychical deterioration. And the difficulty is enhanced when the latter is accompanied by some paretic symptoms. Then, only a full consideration of the whole case will clear up doubt, and long duration and sameness of symptoms are of great value in marking off the case as secondary dementia.

17. *Chronic Generalized Palsy. Double Hemiplegia, &c.* The physician may be called to a case of which there is no reliable history, in which the limbs are deprived of voluntary movement, the tongue is inert, and speech lost, and in which the intelligence is so far impaired as to prevent information being obtained from the patient in other ways. This condition, as Calmeil long ago mentioned, may arise from old double cerebral hæmorrhage, or from a general compression of the encephalon of variable origin, as well as from advanced g.p., especially if in a "state of complication," so called. Grinding of the teeth, bending forward of the head, corrugation of the forehead, and some remains of ataxic fibrillary twitch in its accustomed seats, would aid here in confirming the existence of g.p. Dr. J. C. Bucknill* proposed to found the diagnosis of such a condition, when due to the former causes and not to advanced g.p., upon "the muscular firmness and power of expression retained by the features compared with the profound palsy of the limbs, and upon the susceptibility of the limbs to excito-motory action," which latter, on the contrary, was said to be greatly lost in g.p. In many cases these proposed points of distinction would not be applicable.

18. *Acute, more or less Generalized, Palsy.* From the prostration and, frequently, paralysis of the ultimate period, a suddenly produced generalized palsy may be separated by the history of the case, and, perhaps, by the completeness of the palsy. A paralysis of this kind may arise from sudden extensive double cerebral hæmorrhage, or softening, or from similar lesions holding median, or symmetrical positions in definite basal portions of the encephalon. With the lastnamed, however, special symptoms of bulbar, or of mesocephalic, origin would assist in the diagnosis.

19. *Generalized or Diffuse Paralysis or Ataxy after Acute Affections.*† Huxham, Frederic Hoffman, and Macario were pioneers on this subject. Beau, Sée, Gubler, Maingault, Trousseau, Westphal, and Rose Cormack assisted; also others, as Faure, Jaccoud, Brenner, Eisenmann, Eulenberg, Pidoux, Leudet, Péry. M. Gubler,‡ especially, showed how

* "A Manual of Psychological Medicine," 4th Ed., p. 465.
† "Bull. de Thérapeutique," Dec., 1850. G. Sée, "L'Union Médicale," Nov. 8, 1860, p. 257. C. Westphal, "Archiv für Psych. und Nervenkrankh.," iii Bd., p. 376. A. Foville, "Ann. Méd.-Psych.," Jan., 1873, pp. 12 and 40. Sir J. Rose Cormack, "British Medical Journal," Vol. ii., 1874; Vol. i., 1875. Jaccond, "Leçons de Clin. Méd.," 1886. Drs. Whipham and A. T. Myers, Clinical Society, London, March 12, 1886.
‡ Dr. A. Gubler, "Archives Générales de Médecine," 1860, T. i, pp. 257, 402, 534, 693; 1860, T. ii, pp. 187, 718; 1861, T. i, p. 301.

various were the relations of paralysis with acute disease, and how different the pathological import of each relation. Also, he showed that *consecutive* paralysis might follow a great variety of acute affections, as pneumonia, erysipelas, cholera, dysentery, typhoid, typhus, and the exanthematous fevers, acute angina, diphtheria, a kind of erythema nodosum, miliary roseola, purpura, febrile urticaria, guttural herpes, and others.

Consisting as these consecutive paralyses or ataxies do of two varieties,—the localised and the diffuse—it must be borne in mind that it is the diffuse or generalized variety with which alone we are now concerned. It must be distinguished from g.p. by the history of the case; sometimes by the more frequent and obvious *preceding* anæsthesia, analgesia, numbness, pricking and arthritic pains; and by the circumstance that it often begins in the velum palati, almost always undergoes recovery in the space of a few weeks, and is rarely accompanied by intellectual trouble. Yet intellectual weakness, also, is sometimes observed, as in some of the instances mentioned by Westphal and Foville in relation to variola and typhus; or even excitability, altered mental state, and symptoms such as change in physiognomy, clumsy slow movements by fits and starts, shaking head, trembling limbs, disorder of speech, ataxy of limbs or of some of them; stiff gait, impaired deglutition; and, in one case, loss of smell and of power of sneezing. In the cases of the former, speech was scanned, nasal, monotonous; the letters and syllables, not misplaced but, separated by intervals and uttered jerkily, or with visible efforts yet without coexisting trembling of lips and face. The latter saw a similar case with motor signs resembling extremely advanced g.p., and with great intellectual impairment. Here also, the voice was nasal, scanned, spasmodic; but, unlike the above cases, was accompanied with marked twitches of muscles of face, the saliva often being projected also; deglutition was convulsive, and fluids inclined to return by nose.

Or, again, after acute febrile and other acute maladies there may be many of the physical signs of g.p., and even mental excitement and ideas of grandeur.* This is distin-

* Delasiauve, Christian.—Lendet, "Annales Méd.-Psych.," 1850, p. 148, reported a case; ambitious monomania temporarily followed in the period of decline of mild typhoid fever, in a female, aged 23. Dr. J. Christian, "Archives Générales de Méd.," 1873, T. ii, pp. 257 and 421, reported a case simulating g.p., and following enteric fever, and a similar one by Dr. Max Simon, *loc. cit.*, p. 269. Baillarger, "Ann. Méd.-Psych.," Jan., 1879, p. 79, a case of *délire ambi-*

guished from g.p.—and the same applies to the cases presenting a species of transitory and acute dementia—by the incompleteness of the picture, and the rapid recovery; —although true g.p. may occasionally thus take origin.

Should it be diphtheritic (and even in some other cases) the paralysis is apt to extend from the velum palati to the pharynx, thence to the lower limbs, then sight and hearing become affected, then the upper limbs, and finally the trunk and respiratory muscles, while the premonitory signs mentioned above are often present.*

It is unnecessary to refer to the anatomy and pathogeny of the paralyses after acute affections, according to Gubler, Charcot, Vulpian, Buhl, Leyden, Oertel, Bernhardt, Nothnagel, Westphal, Voisin, Ebstein, Pierret, Déjerine, Brown-Séquard, Letzerich, F. Magne, and others.

20. *Acute Ascending Paralysis.* The history and rapidity of the case, and comparative or complete absence of mental symptoms, the paralysis instead of inco-ordination with paresis, and the special lines of invasion of the musculature, separate this from g.p. Some examples follow acute febrile affections. It usually proceeds from the toes and feet to the back of the thighs and pelvis; thence, successively, to the front and inner parts of the thighs, the fingers, hands, arms, scapular region, biceps, trunk, respiratory muscles, tongue, etc.†

21. *Tremors of the Aged,* when observed in the insane, must not be confounded with those of g.p. They do not produce the speech of the latter, are not affected by its complications, nor accompanied by the same mental phenomena.

22. *Defective Speech.* Nor does the least difficult case arise when one becomes insane who for long has suffered from an impediment in speech, and especially if expansive feeling or idea displays itself. The absence of that state of pupils, tongue, gait, writing, and sensory function, both special and general,

---

tieux, of 15 days' duration, following scarlatina. Dr. Liouville, "Annales Méd.-Psych.," 1879, T. i, p. 428, a case of *délire ambitieux* during convalescence from typhoid fever in a male, aged 21. Dr. T. W. McDowall, "Jl. Ment. Sci.," July, 1881, p. 279, a case of typhoid fever with physical and mental symptoms of "typical" g.p. whilst the fever lasted.

* See also Dr. S. G. Webber, "Trans. American Neurol. Assoc.," Vol. ii.

† "Gazette des Hôpitaux," Sept. 10, 1859, p. 421; Sept. 17, 1859, p. 433. Dr. Alf. Liégard, *Ibid.*, Dec. 3, 1859, p. 562. Dr. O. Landry, "Gaz. Hebdomodaire de Méd.," July 29, 1859, p. 472; Aug. 5, p. 486; and cases by Ollivier (d'Angers) and Sandras. Dr. Geo. Harley, "The Lancet," Vol. ii, 1868, p. 451. —Some of Gubler's cases, above.

usual in g.p., and, perhaps, of its customary delirium, serve to distinguish these cases of speech-defect with mental alienation—but not always so readily as might be surmised.

23. *Plumbism.* Conditions not unlike some in g.p. have been observed in lead poisoning, as in cases by Delasiauve, Beau, Féréol (acute generalized paralysis), Tanquerel des Planches, Falret, to some extent Voisin, Dr. T. S. Dowse [*] (acute diffuse paralysis), Dr. H. Rayner,[†] and Dr. G. H. Savage.[‡]

In molybdosis, I have seen atrophy and general light sclerosis of brain, with imperfect physical signs of g.p., with marked dysphagia, and some anæsthesia; fatuity, anxiety, timidity, fear; frequently dextral convulsions, followed by right hemipareses.

The mental symptoms vary considerably in cases of this kind; incoherence and degradation, dementia, excitement, dejection, exaltation; each may be present. The motor signs have defective resemblances to those of g.p., and usually some symptoms such as marked general tremulousness, or anæsthesia of limbs, are highly pronounced. The history of the case, the sallow cachectic hue of skin, lead line on gums, and, sometimes, slight indications of drop-wrist and lead colic; or the transient nature of any accompanying acute delirium; may help in diagnosis.

M. Delasiauve,[§] writing on the diagnosis, speaks of the saturnine encephalopathy as bearing much analogy to alcoholic delirium;—the same dulness, hallucinations, expansive ideas, and muscular disorder. To these are added the cachectic appearance, and blue gum-line; while the history, and comparatively short duration, help to separate from g.p. Dr. Böttger[||] speaks of the majority of cases, which are somewhat like g.p., as often having long prodromic symptoms; the signs of metallic intoxication come either before, or simultaneously with, the mental disease; in all, memory is enfeebled or lost; tremor often is the only motor sign, but in most the gait becomes paralytic, sensibility is considerably diminished, the further course is that of dementia. Sometimes epileptiform seizure occurs; sometimes recovery.

The chief differences specified by Dr. Emmanuel Régis[¶]

[*] "Transactions Clinical Society, London," Vol. viii, 1875, p. 124.
[†] "Jl. Mental Science," July, 1880, p. 226.
[‡] *Ibid.*, p. 229.
[§] "Annales Méd.-Psych.," 1851, p. 621.
[||] "Allgem. Zeitschr. für Psych.," xxvi Bd., p. 224.
[¶] "Annales Méd.-Psych.," 1860, T. ii, p. 175.

between true g.p. and saturnine pseudo-g.p. are:—In the lead cases:—1. *Physical:* (a) blue line on gums, and earthy hue of skin; (b) headache, vertigo, heaviness of head; (c) cramps, pains, anæsthesia, hyperæsthesia, paralysis, etc.— 2. *Intellectual.* Insomnia, nightmare, hallucinations, terror, confused delusions of persecution, of poisoning, etc.—Also, points of minor difference, especially the rapid onset of the lead cases, with extreme symptoms as of advanced g.p.; and yet perhaps recovering entirely in a few months. Thus, he concludes that a true saturnine g.p. does not exist. Mendel* maintains that when lead-intoxication is somewhat like g.p. the motor symptoms are highly pronounced, but the disorder of intelligence and consciousness is not to the degree found in g.p. Symptoms closely resembling those of intra-cranial tumours, and sometimes with temporary mental derangement (mania), were usually observed by Dr. Byrom Bramwell.†

24. In *Atropine minor poisoning, or intoxication,* Michéa found hesitation of speech, staggering, disorder of co-ordination, and of consciousness; and Kowalewsky observed increased self-feeling, and rapidly changing mental symptoms. The history, condition of pupils, throat dryness, acceleration followed by retardation of pulse, put one in the way for a diagnosis.

25. *Bromide intoxication (also Iodide i.)* Voisin ‡ reported a case of bromide intoxication, mistaken for g.p., by others. After ten days of mental change and failure came a state of extreme noisy excitement, wild delirium and violence, terrifying hallucinations; and afterwards were dejection; also titubation and ataxy. The diagnosis was founded on the history of the case, the rapid course of the symptoms, the dejection and appearance of the patient, the state as of apparent drunkenness; and the equality of pupils. Séguin found in bromism: stupor, enfeebled memory, unequal pupils, disorders of articulation, tremor, uncertain movements, swaying gait. Dulness of mental faculties, loss of memory, and muscular weakness of lower limbs, occurred in cases collected by Dr. M. Clymer; § and Dr. C. H. Hughes ‖ found dementia and ataxia of speech and gait under large doses of K.Br. (and also of K.I.). Dr. Katz,¶ among other symptoms, found

* *Op. cit.,* p. 285.
† "Edinb. Med. Journ.," Dec., 1879, p. 510.
‡ "Mémoires de l'Académie de Méd.," 1871, p. 68.
§ "Medical Times and Gazette," Feb. 24, 1872, p. 238.
‖ "Alienist and Neurologist," Jan., 1885, p. 149.
¶ "Journ. Mental Sci.," Jan., 1873, p. 593.

difficulty in thinking, and motor disturbance; and Spitzka observed double ptosis, a stony stare, lax facial folds, zygomatic, labial and lingual tremor.

In some epileptics treated with full doses of K.Br., the features get a heavy, downcast, yet partially blurred look, somewhat as in hypochondriacal paretic dementia; and I have sometimes found the speech-affections highly developed. A history of bromides taken in excess; bromide rashes, if present, especially the acneform; marked faucial anæsthesia; and the effect of discontinuance of the drug, banish doubt. Yet one must not forget that a G.P. with early epileptiform seizures may be overdosed with bromides.

ACUTE GENERAL PARALYSIS. Acute g.p. must be distinguished from acute mania, acute delirium, acute melancholia, acute alcoholism, the delirium of fevers, pneumonic fever, tubercular meningitis in adults, simple meningitis, some forms of toxæmia with symptoms of cerebral congestion; and local encephalitis, whether of traumatic, or of pyæmic origin, or due to a local extension of some neighbouring inflammatory change.

CHAPTER XIV.

CAUSES OF GENERAL PARALYSIS.

A. Predisposing Causes. 1. *Sex.* The relative proportion in which g.p. attacks the two sexes is usually that of about four males to one female; * some have made the proportion more, some less, unequal; and the dictum of v. Krafft-Ebing—arrived at by a comparison of the returns of Hoffman, Duchek, Stolz, Erlenmeyer, Calmeil and others, and followed by Dr. Jung—was inaccurate, namely, that 8 males to 1 female was the usual proportion. Neumann never saw, and hence denied the existence of, g.p. in the female. Ramon found 8 male G.Ps. to 1 female; Materne 7 to 1; Linstow 10 to 1; Skae 9 to 1; Damerow 6 to 1; Sander 7½ to 1; Guttstadt,† for Prussia, 5 to 1; at the Charité, Berlin, 10⅓ to 1 (Dr. W. Sander).‡ Dr. Jung's statistics for 50 years give 23 m. to 1 f. in a public asylum; and 96 m. to 1 f. in a

* Reports xxxi to xxxviii, Commissioners in Lunacy, England.
† "Allgem. Zeitschr. für Psych.," xxxiv Bd., p. 243.
‡ "Berliner Klinische Wochenschr.," Feb. 14, 1870, p. 81.

private asylum; Spitzka's, 58 m. to 1 f.—On the other hand, of one year's admissions in France, 20 p.c. of the males and 8 p.c. of the females were G.Ps., or only 2½ m. to 1 f. ; and Dr. H. Schüle admitted the same proportion. In France also, the proportion of female to male G.P.'s. is very much higher in the lower than in the upper social classes. So also in Germany were about 4 m. to 1 f. G.P. in public asylums; and 9½ m. to 1 f. in private asylums.*—Bayle,† in 1826, reported at Charenton about 6 m. to 1 f.G.P ; Calmeil ‡ about 3½ m. to 1 f.

From the Lunacy Blue Books, § I have computed that of the *total admissions* in England and Wales into institutions and homes for those of unsound mind for the past four years —viz., 26,658 males and 27,984 females, or a total of 54,642 persons— 3,374 males, and 910 females, were G.Ps., or a total of 4,284 persons. The average proportion *per cent.* of the number of *G.Ps.* of *each sex* admitted, to the *total* number of *patients* of *each sex* admitted during these four years, was 12·65 p.c. *for the males, and* 3·25 p.c. *for the females;* or 7·8 p.c. of the total admissions of both sexes. Thus, in equal numbers of admissions the proportion is almost exactly *four male G.Ps. to one female G.P.* In the four years mentioned, the highest *relative* proportion of female G.Ps. was 3·3 p.c. of the total female admissions, as compared with 11·7 p.c., male G.Ps., of the total admissions of males ; and the lowest *relative* proportion of female G.Ps. was 3·3 p.c., as compared with a male percentage of 13·8.— Then, again, of the total number of females admitted each year, the G.P. fraction has for some years past been almost stationary, and just about $\frac{1}{30}$; whereas the proportion of male G.Ps. to the total male admissions has fluctuated considerably year by year (*e.g.* as 11·7 p.c., and 13.8 p.c.). From the 31st and 32nd Reports Dr. T. A. Chapman || constructed a table showing the number of *male* G.Ps. "that would occur at each age and condition as to marriage among the number of males living [general population] that among *females* yielding *one* G.P.":—viz., of males (to 1 female G.P.) at ages 20-30, 4·7; 30-40, 4·9; 40-50, 3·7; 50-60, 6; 60-70, 7·8; 70-80, 3.

* "Allgem. Zeitschr. für Psych.," xxxvi Bd., p. 63.
† *Op. cit.* Introduction, p. xxvii.
‡ "De la Paralysie Considérée chez les Aliénés," 1826, pp. 370, 371.
§ "Reports of Commissioners in Lunacy, England," Nos. 35-6-7-8.
|| "Journal of Mental Science," April 1879, p. 37.

Daveau * and others attributed the disproportion between the numbers of the two sexes affected with g.p. to a prophylactic influence of the menstrual discharge in women: Lunier said the sexual disproportion becomes less after the age of forty-five, owing to the occurrence of the menopause, and a consequent increase in the number of female cases; Baillarger and Doutrebente† took much the same view, a view not accepted by Duchek or Hoffman. But it is futile to strain after an explanation; for the statement (in this country) is true only (and slightly so) of widowed G.Ps. (see " condition as to marriage "), the sexual disproportion augmenting, on the other hand, among single G.Ps. at the age in question.

From the table, below, under "age," it also appears that, comparing the actual numbers, female G.Ps. relatively to male are in greater proportion at the earlier ages, under 30 years (or under 35, Rep. No. 38); that at the ages 40 to 50 (45 to 55, Rep. No. 38) the female proportion is below the average; is about the average at 50-60; but in the decade 60 to 70 is high (although in 38th Rep. low between ages 55-65). And similar general results, with variations (such as female proportion more below average at 50-60 than at 40-50), are derived from the table under " condition as to marriage " (below), showing the proportion per cent. of G.Ps. admitted to total number of patients admitted. These figures kill the statements made by many as to g.p. in the female being particularly a disease at or of the climacteric period (Voisin, Jung, *et al.*); or that the inequality in the numbers of the sexes affected lessens at that time.

Probably, the cause of the sexual disproportion mainly lies in the greater moral shocks and mental strain to which the male is subjected, as well as the greater frequency with which he indulges in excess, especially alcoholic. But the tension and effort of life among men, the more exhausting effect of sexual excess in them, the greater frequency of their intemperance in alcohol and tobacco, and greater liability to syphilis, injury and insolation, do not explain all. The male brain is innately more liable to organic disease than the female.

The main conclusions (partly erroneous, partly criticized above) of an inquiry by Dr. Jung‡ are, that g.p. is increas-

* "Diss. sur la Paralysie Générale observée à Charenton." Thèse.
† *Op. cit.* Thèse, 1870.
‡ " Allgemeine Zeitschrift für Psychiatrie," xxxv Band, pp. 235, 625.

ing in women of the lower class; is a disease of the climacteric period, whether that be normal or premature; occurs later than with men, and particularly between the ages of 35 and 45; seldom is rapid in its course; is largely favoured in women by heredity, weakness of nervous system, and predisposition for vaso-motor disturbance. Most of the women affected were barren, or had one child only, or their children were still-born or died when young. In relation to this last; although referring to either sex, Dr. Luys * had found a striking scarcity of progeny in G.Ps.—one-third of their marriages sterile, and the average progeny of the fertile marriages, 1½.

It is difficult to say how far the apparent increase of g.p. among women of late years is due to a former defective recognition of it, and faulty diagnosis, owing to the less salient features and less dramatic course of g.p. as it occurs in women, than as in men. Some older statistics, are, perhaps, unreliable : thus, for 50 years, 1 female to every 23 male G.Ps., but latterly about 1 to 4; or female G.Ps. admitted in France, increased from about 32 p.c. of male G.Ps., to 42 p.c., ten years later.† Yet Dr. A. Sauze‡ finds g.p. increasing in men, but not in women.

Again, the assertion, as by Drs. W. Sander, Jung, and others, that g.p., on the average, occurs later in life in women than in men, is disproved for this country by the above statistics from the Lunacy Blue Books. In that for 1884, while the total No. of female, to male, G.Ps. admitted was more than as 1 to 4, those aged 45 to 55 fell slightly short of 1 to 4, and those from 55 to 65 were only as 1 female to 7 males. Concerning the statement that g.p. occurs in women particularly between the ages of 35 and 45;—so it does in men, and, indeed, rather more so than in women, namely, in 45 p.c. of the male G.Ps., and in 43 p.c. of the female. Sander saw no female G.Ps. under 30 years of age, yet in this country, the fraction of *female* G.Ps. *under* the age of 30 is nearly treble that of *male* G.Ps. At a Parisian asylum Baillarger § found the average age of female G.Ps., *on admission*, one year less than of male; Burman (for Devon) 3½ years less.

2. *Age.* Found chiefly between 30 and 55, g.p. seldom

* " L'Encéphale," 1881.
† " Allgem. Zeitschrift für Psych.,"xxxvi Bd., p. 63.
‡ " Journ. Mental Science," July, 1885, p. 250.
§ " Gaz. des Hôpitaux," July 9, 1846, p. 317.

occurs except between the ages of 25 and 60: although
a recent Lunacy Blue Book returns 16 out of 1,160 G.Ps.
admitted as under the age of 25, and another returned 42
out of 1,151 as 60 years, or upwards, of age. I have seen
one case, admitted at the age of 21, the disease had come on
before the age of 20, the grandiose delusions were absent, at
least during the time I knew him. In several soldiers I have
known g.p. attack before the age of 25. M. Guislain* saw
it occur at the age of 17, Dr. Clouston† at 16, Dr. Wigles-
worth‡ at 15, and perhaps Morison§ at 19. Early-age cases
of g.p. are also reported; as, *e.g.*, by L. Meyer, Wille, v.
Krafft-Ebing, Coffin, Régis, and A. R. Turnbull (age 12).
Dr. Köhler∥ saw a case of apparent g.p. in a girl aged 6,
and Dr. Claus declared he had seen it in a young idiot
girl.

As for g.p. in the aged, most of those entered as such are,
as I stated years ago, examples of ordinary paralysis or of
feebleness, associated with senile or other mental defect or
disorder.

The statement usually made, that it is most frequent
in the decade between the ages 40 and 50,¶ is inaccurate
as regards absolute numbers, although true of the g.p. percentage of the total admissions of persons of unsound mind
in each decade-age. In the 35-6-7th Lunacy Blue Books it
is found that of the G.Ps. admitted, into the Asylums, etc., of
England and Wales in three years the number then actually
between the ages thirty to forty was larger than that of
those between the ages forty to fifty (as 1265 to 1130); but
that the proportion per cent. of the G.Ps. admitted, to
the total number of patients of all kinds, of the same decade-
age, admitted during the same years, was greater in the
decade forty to fifty than in that from thirty to forty (as 14
per cent. to 13 per cent.). And this holds good also as concerns the incidence of g.p. on the population at large, at the
same ages. The census return of 1881 is not available for
the purpose of this comparison, the age-division in *it* being
at the "5"s, but that for 1871 (England and Wales) will
give an almost absolutely correct standard for the years to
which the above figures for the G.Ps. refer.

* "Leçons Orales sur les Phrénopathies," Gand, 1852.
† "Journal of Mental Science," Oct., 1877.
‡ "Journal Mental Sci.," July, 1883, p. 241.
§ "Cases of Mental Disease," by Alex. Morison, 1828, p. 61.
∥ "Ueber Kindliches Irresein."
¶ Bayle, Calmeil, Baillarger, Boyd, Burman. But Austin, 30-40.

Gen. Pop. Age 30-40, = 2,901,348; compared with—Age 40-50, = 2,282,843.

G.Ps. admitted (3 years). Age 30-40, = 1,265; compared with—Age 40-50, = 1,130.

If the proportion of G.Ps. to total population of the same age was the same, the last number would be 994, and not 1,130; the relative excess of the proportion of G.Ps. admitted at the age of 40-50, as compared with the age 30-40, is as 1,130 to 994, or about 8 to 7; although the actual number of G.Ps. at age 40-50 is less, as already seen.

I dwell on these Lunacy Blue Book statistics, which are sufficiently large as to numbers and wide as to application, approximately settle, for a generation, several points as regards g.p. in this country, and give the quietus to some of the numerous current errors on the subject.

G.Ps. in ages by decades—on admission during 3 years.— M. 2,456; F. 668; Total 3,124.

Age 15—M. 2; F. 1. Age 20—M. 149; F. 115. Age 30 —M. 996; F. 269. Age 40 —M. 916; F. 214. Age 50—M. 312; F. 82. Age 60—M. 73; F. 26. Age 70 and upwards —M. 19; F. 8.

Reading the older literature of the subject one is led to think that formerly g.p. occurred somewhat later in life, on the average, than is nowadays the case. Of 182 cases Bayle had seen only four under the age of 30; Calmeil, with fewer cases, only two under the age of 32. It is true that in 1868 Prof. L. Meyer in over 100 male cases had found only two below 30 years of age; and König of 101 cases only two, also. And Mendel recently gives barely 4 p.c. as being 30 *years old or under*, and this even at the time of the first distinct signs. Nevertheless, there is no rarity now of cases under the age of 30, the aggregate statistics of the Commissioners' three Reports just named show 267 such cases out of 3,124; or 8·5 p.c. And some older statistics tell the same tale. Thus, those under 30 years of age, on admission, comprise about 7 p.c. of Austin's cases, and of Baillarger's; and at the Devon Asylum (1845-70) about 8·5 p.c. Finally, in the soldiers under my care nearly one-third were under that age *when attacked;* and the stated *average* age of all was then only thirty-three. This last, no doubt, is partly due to our military "short service system."

The above impression, if correct, would speak ill for the vitality of the peoples of the West of Europe, as far at least as the disease may be deemed analogous to a prodigal wasting of vital power, and premature senility; the early attainment of old age in the individual members of a race being the forerunner and prophet of its imminent decay.

3. *Temperament.* The sanguine temperament has usually been viewed as predisposing to g.p. This, perhaps, is true only of one class of the cases; certainly the sanguine temperament is very often absent.

4. *Mental Activity.* As to the state of intelligence, the same rôle has been attributed to an energetic mental life with ardent imagination, and this no doubt is an active predisponent.

5. *Character or Disposition.* The character or disposition of those who become general paralytics has often by nature been fiery, choleric, intolerant of opposition; or, on the other hand, douce, genial, and evidencing much *bonhomie;* —or sometimes proud, haughty, selfish, ambitious.

6. *Heredity* is a factor in some cases, Bayle said in nearly half; Calmeil,* König and Simon, in about one-third; Voisin, in most; Mendel, in 34·8 p.c.; Dr H. G. Stewart, in 47·6 p.c. On the other hand, Morel and his pupils held it to be rarely (some said never) hereditary, *i.e.* as a form of insanity. The apparent direct hereditary predisposition in these cases may be either to insanity, or, secondly, to some other and ordinary nervous (either neural or adneural) disease, and the occurrence of the latter in the family history has been accentuated by several writers, even by Bayle. Thus, Lunier † said it is rather that the ancestors of G.Ps. have been apoplectic, paralytic, epileptic or demented than insane; others asserted that the heredity of g.p. is very indirect, and consists in the inheritance of a tendency to cerebral congestion, which congestion produces the predisposition to g.p.; and Verga ‡ asserted that "the hereditary affinities of g.p. are not with ordinary insanity, but with paralysis, apoplexy, and other brain diseases." More recently MM. Ball and Régis § agree that g.p. does not originate in, nor engender, insanity; when hereditary it is, so to speak, the cerebral and not the insane element that is so; that the G.P. transmits to his or her offspring, not insanity, but a tendency to other cerebral affections, and this offspring may show great brilliancy. But there are

* "Traité des Maladies Inflammatoires du Cerveau," T. i, p. 272.
† "Ann. Méd.-Psych.," 1849.
‡ Abs. "Journ. Mental Sci.," Apr., 1873, p. 158.
§ "L'Encéphale," Aug., 1883.

striking exceptions to this last, and for the ills transmitted by G.Ps. see Chapter XX.

An hereditary influence is often not derived from ancestors who have been G.P. or insane, but who have possessed the character, disposition, and play of passions referred to under the last head, which, not leading to g.p. in them, may favour its occurrence in their descendants, unfortunate inheritors of the evil moral and emotional condition. As elsewhere, so here, "the fathers have eaten sour grapes, and the children's teeth are set on edge." Of 109 cases in New York;* in 39 was a family history of insanity; in 30, of other nervous diseases; and in 22, of parental intemperance.

I find, on analysing Lunacy Blue Books for England and Wales, that of the 3,374 *male* G.Ps. admitted during four years, it was stated in 490 that hereditary influence was ascertained; and in 175 of the 910 *female* G.Ps.; *i.e.* in a total of 665 G.Ps. of both sexes, out of a total of 4,284. Or, in other words, heredity was stated to be a factor in 14·52 p.c. of the males, and in 19·25 p.c. of the females; or, in 15·52 p.c. of the total G.Ps. The large inequality between the sexes as to hereditary influences, in these returns is very striking. And it is enhanced by the fact, that the causes of g.p. were, on the average, returned as being "unknown" in about 32·5 p.c. of the females, and in about 29·3 p.c. of the males. So that, with a smaller part in which any cause was *assigned*, the female G.Ps. showed hereditary influence in a larger fraction of the whole than did the male.

A second point is that the ratio per cent. of heredity in the female G.Ps. admitted each year is steady during the four years, only fluctuating between 18·3 p.c. and 20·2 p.c.; whereas it had much wider fluctuations in the males, ranging between 10·9 p.c. and 17·4 p.c.

A third point, is the frequency of assigned hereditary influences in g.p. alone, relatively to that found in the *total* admissions (all cases). The latter I find to be, for the same four years, 18·93 p.c. for the males and 21·53 p.c. for the females; or 20·27 p.c. of the total admissions. Compare with these the above much lower percentages all round for heredity in g.p. The percentage of heredity is much lower in private than in pauper female cases; and lower for private than for pauper males.

With trifling exceptions, the statistics of Ullrich, Mendel, Obersteiner, and W. H. O. Sankey also make heredity a less

---

* "American Journ. Insanity," Apr., 1877, p. 451.

frequent factor in g.p., than in the total cases of mental disease.

7. *Condition as to marriage.* See also *sex*, above. The following are *average* p.cs. (not precisely the actual p.cs. of the total no.), and compiled from Commissioners' Reports, England, Nos. 35-6-7.

Average proportion per cent. of G.Ps. to total number of all patients admitted, in each corresponding decade, in England and Wales (3 years), arranged according to their ages, and condition as to marriage.

Age.	15—			20—			30—			40—			50—			60—			70—			Total.		
	M.	F.	T.	M.	F.	T.	M.	F.	T.	M.	F.	T.	M.	F.	T.	M.	F.	T.	M.	F.	T.	M.	F.	T.
Single.	·4	·1	·2	2·3	·6	1·6	12·2	6·9	7·2	14·6	2·4	7·8	11·7	1·4	5·5	3·1	·7	1·6	1·1	·6	·8	6·	1·1	3·6
Married.	—			10·5	3·4	5·4	27·3	7·3	16·9	26·7	6·3	17·1	12·9	3·5	8·7	2·9	·7	2	2·8	·8	2·1	19·	5	12·1
Widowed.	—			2·8	3·5	3·6	25·	8·8	15	18·1	6·3	10	9·9	2·6	5·	4·2	1·9	2·7	1·5	·3	·7	9·6	3·1	5·4
Total.	·4	·1	·2	3·5	1·6	2·5	20·7	5·5	13·	21·	5·2	14·	12·2	2·9	7·3	3·4	1·3	2·3	2·2	·5	1·1	11·8	3·1	7·7

Austin, Burman, and others found that the proportion of single persons was much higher among the female than the male G.Ps. Nevertheless, larger and more recent statistics refute these, showing that in this country at large the proportion of single persons among the female G.Ps. is only equal to, or less than, that of single persons among the male. In the four years already mentioned the G.Ps. admitted were:—Single M. 740, F. 156:— married, M. 2,383, F. 612:—widowed M. 208, F. 123.—"Unknown" 62. Thus, the sexual proportions were; of the single 4·74 M. to 1 F.; of the married 3·89 M. to 1 F.: and of the widowed 1·69 M. to 1 F.—In other words, the *proportion of females to males* was lowest in single G.Ps.; and highest in widowed G.Ps., here exceeding the average at all ages except between 50 and 60, and over 70. And the proportion of single to married female G.Ps. is less than that of single to married male G.Ps., or, as 25 p.c. compared with 31 p.c.

Of the 54,642 "admissions" (all cases) during the same four years, about 3·6 p.c. of the single, about 12·1 p.c. of the married, and 5·4 p.c. of the widowed, were G.Ps. Thus there is an overwhelming majority of married G.Ps. The time of life at which g.p. is most rife partly accounts for this fact, inasmuch as it is also a time of life at which the majority of persons in this country are married. How then to arrive at a just comparison? The best method is to find the ratio of G.Ps. in the different conditions as to marriage, and in the several decades, to the entire population of the same conditions, and ages. This has been worked out from the Commissioners' tables, by Dr. T. A. Chapman,* who finds that in the general population between the ages of thirty and sixty the single are more liable to *insanity* than the married in the proportion of 2·83 to 1·,—yet that, calculated in the same way, there is only a very slight difference in the frequency of *general paralysis* according to the condition as to marriage; viz.; 16·5 single, 15·3 married, and 15·4 widowed G.Ps. *per* 100,000 of general population of the same ages (30-60), and same condition as to marriage. This is attributed to the comparative rarity of congenital defect in G.Ps.—defect on account of which so large a percentage of the population remains single:—single, because of the same factor as renders them liable to insanity and less able to earn a livelihood; and *not* that they are liable to insanity *because* single.

8. *Occupation and Social and Pecuniary position.* It is desirable to deal with these inextricably intermingled influences under one head.

(a). Military and naval life; occupations exposing the workers to great heat and sweat, or to alternate heat and cold draughts; prostitution; all favour the production of g.p. So do those which occasion emotional strain, constant worry and irritation, or intellectual overwork.

In the 31st Lunacy Blue Book the proportion of *lunatics* to *persons* in the order of persons engaged in the defence of the country is stated as being ·316 per cent., a proportion of lunatics much higher than in any other order of persons (males); the average being (males) ·063 per cent. Yet, of the military and marines admitted as insane 13·6 per cent. were G.Ps., whereas 14·1 per cent. of the *total* males of all orders were G.Ps. Thus, while mental alienation at large is far more rife in this than in any other order of the male population, yet a slightly smaller percentage of the military insane are G.Ps. than of the male insane of the country generally. The comparative youth of our soldiery is no doubt the explanation of this. But many included in these returns have been so long retired from, or so short a time in,

* "Journal of Mental Science," Apr., 1879, p. 37.

the military and naval services, as that their cases have little or no bearing on the genesis of g.p. in the army and navy.

Of the soldiers admitted under my care during a number of years, 18 p.c. were G.Ps. I found the regiments of the Guards—the flower of the army—yield the highest ratio of g.p. in the total number coming under care as of unsound mind. In soldiers there are several factors; among the officers, the tension of anxious responsibility; among all grades the violent emotions and privations of warfare; the shock of artillery-discharge, of bursting shells; but, especially, alcoholic and sexual excess, and venereal disease. G.p. was rife among the veterans of the armies of the first Napoleon; and although Dr. Lunier * found a decrease in the proportion of g.p. in France during and immediately after the Franco-German war of 1870-1, that scarcely meets the question, the element of time not being sufficiently allowed for.

(b). *Classes of Society.* A rough method is to take the ratios per cent of G.Ps. among the " private " and " pauper " patients. Of the 64,642 persons *admitted* (England and Wales), during 4 years, there were:—

5,454 private *male* patients; of these 438 were G.Ps., or 9·63 p.c.;—4,173 private *female* patients; of these 77 were G.Ps., or 1·84 p.c.—Total, 5·91 p.c. of private admissions=G.Ps.

22,113 pauper *male* patients; of these 2,936 were G.Ps., or 13·28 p.c.;—23,811 pauper *female* patients; of these 833 were G.Ps., or 3·5 p.c.—Total, 8·21 p.c. of pauper admissions=G.Ps.

Thus, of the total admissions of each class and sex, 13·28 p.c. of the pauper males, 9.63 p.c. of the private males, 3·5 p.c. of the pauper females, and 1·84 p.c. of the private females, were G.Ps. This only shows the tendency of g.p. to affect roughly-grouped classes of society, and, broadly speaking, to be more frequent in the lower; for several accidents or circumstances, affecting either large numbers or individual cases, modify the facts, and determine which class the patients in question shall happen to be in. No division of the general population can be made corresponding precisely to the sources whence the private and pauper patients, respectively, are drawn. The official return of " pauper population " of the country has no relation to

* " De l'influence des grandes commotions politiques et sociales sur le développement des maladies mentales," "Ann. Méd.-Psych.," 1872-3.

the question, since the majority of those entering asylums as pauper patients have never belonged to it. Again, the official returns take no account of cases in workhouses, or treated at home, or under " single care." Thus one cannot establish the actual *exact* proportion in which g.p. attacks the lower or higher social classes.—Whilst the percentage of pauper males who are G.Ps. is about $1\frac{1}{2}$ times that of private male G.Ps.; the percentage of pauper, is about twice that of private, female G.Ps. And it is an old observation that g.p. is rare in gentlewomen. Dr. Conolly noted it; and in 786 insane females of the better classes Dr. Laehr * found only 3 G.Ps., and in Dr. Jung's (*loc. cit.*, p. 235) experience, 31·8 p.c. of the males of the better classes were G.Ps., but only 1 out of 109 females. Is g.p. more rife among *males* of the upper or of the lower classes? Long ago, Esquirol found much more g.p. among the males of that one which, of two asylums compared, received patients who had had greater pecuniary resources, readier means of indulging their passions, and higher intellectual activity. Dr. W. H. O. Sankey's view, that g.p. occurs in a larger proportion of males of the lower than of the upper classes, broadly speaking, derives support from the figures above; although a division into upper and lower classes is not the same thing as the division into private and pauper, for the reasons already stated. And if we ascertain the relative proportions of g.p. among the insane of different " orders of persons," we must bear in mind that the statistics are vitiated, for the present purpose, by the fact that under each " order " are included persons belonging to both the upper and lower classes, those holding very inferior places being aggregated with those holding the higher. Nevertheless, some approximation to a correct notion may be gained from Table XVIII., 31st Report of Commissioners (England), containing the *Ratio per cent. of male G.Ps. to the total number of males admitted in each " order of persons,"* viz.:—Orders of persons: —Engaged in the local or general government of the country, 24·1 p.c.: in its defence, 13·6 p.c.: in learned professions, literature, art, science, with their immediate subordinates, 12·8 p.c.: in entertaining, and performing personal offices, 14·5: in commerce, and shopkeeping, etc., 18·6: in conveyance of men, animals, goods and messages, 21 p.c.: in agriculture, and about animals, 10·1 p.c.: in art and mechanic productions, 17·5 p.c.: textile fabric and dress

* " Allgem. Zeitschr. für Psych.," xxxiv Bd., p. 243.

workers and dealers, 14·2 p.c.: workers and dealers in food, drinks, animal and vegetable substances, 16·8 p.c.: mineral workers and dealers, 18 p.c.: of rank and property, or of no occupation, 7·9 p.c.

Yet in the French army, M. Colin found g.p. most frequent in the officers, three-fourths of the cases of mental unsoundness in them being due to g.p. Mental disease affected the higher grades most.

On the whole, we conclude that g.p. is more frequent in the lower classes than in the upper; that both in the upper and lower classes great differences are seen between the percentages of g.p. among the insane of the several occupations and "orders of persons": that males of the lower classes are attacked more than males of the upper, females of the lower relatively still more than females of the upper; the order in which they are most affected being, lower class males, upper class males, lower class females, upper class females.

9. *Predisposing Mental Causes.* (a). A life absorbed in ambitious projects, with all its strenuous mental efforts, its long-sustained anxieties, deferred hopes, and straining expectation; or any prolonged and violent, or sudden and frequent, play of ill-regulated passion, frequently repeated outbursts of rage; the smouldering flames of envy, of jealousy, or of unrestrained sexual passion; the disintegrating influence of prolonged anxiety, of worry, of afflictions and losses, the similar influence of exaggerated selfishness and ambition, the concussion of moral shocks; may both prepare the way for, and lead up to, g.p.

Chagrin, forced erethism of the intellectual faculties, intellectual overwork, especially if sustained by stimulants, all predispose to g.p. Its frequency among educated, highly excitable men, as poets, musicians, and literati has been noticed, amongst others, by Griesinger and Calmeil.

(b). *Previous Mental Disorder or Defect.* To a previous attack of insanity Calmeil would attribute an influence predisposing to g.p., but the correctness of this view has been impugned. Mendel only found two such cases out of 210. But these were very marked examples; there appears to have been an interval of mental soundness in one case of 23 years, in the other case (after two attacks) of 12 years. In several cases, I have known g.p. come on in those who had recovered from a previous attack of mental disorder— perhaps years before. Some of these were probably examples of recoveries from a first attack of g.p., others of extremely long and marked remissions. Dr. H. Schüle * reported some cases of the kind.

Out of 3,374 male G.Ps., 162 or 4·7 p.c. had "previous attacks" assigned as the cause, or as one of the causes; and out of 910 female G.Ps., 68 or 7·4 p.c.—Congenital defect was returned as entering into the causation of only 4 of the same 3,374 male G.Ps., and of 5 of the same 910 female G.Ps., thus showing a higher ratio for females, though an extremely small one for either sex. Köhler † and Claus, also, observed g.p. supervene in idiocy and imbecility.

(c). *Transformation of Simple Insanity into general paralysis.* One scarcely need refer to the error of many of the earlier writers who, not knowing g.p. as a separate affection, viewed its physical signs as complications or terminations of ordinary insanity. Nevertheless, g.p. may appear to supervene on simple insanity of longer or shorter duration, or the latter appear to undergo a transformation into general paralysis. Cases of such asserted transformation may be found in comparatively recent works,‡ but many of them are by no means satisfactory or convincing. I have, however, observed the physical signs of g.p. first appear as late as a year and a half, and more than two years, after the onset of uninterrupted frank mental derangement. As to this asserted transformation: of the several examples referred to in Calmeil's earlier work only one (Obs. liv, p. 279) can be accepted as being distinctly of this nature; and of the four in his later work, only one of the three given in detail, and the one given in summary, are at all satisfactory. The rest are merely examples of g.p., either with a long prodromic, or a marked melancholic stage, or period of excitement, with slight or absent motor signs; and in some cases undergoing remission. In some, only a few months or a year intervened between the first mental symptoms and the physical signs of g.p. Some of Parchappe's cases are not convincing. Voisin and Burlureaux also described cases in which they believed that simple melancholic insanity degenerated into g.p.§ The two examples given by Voisin in his treatise are not altogether convincing. He also treats of "folie con-

* " Allgem. Zeitschr. für Psych.," xxxii Bd.
† *Ibid.*, xxxvi Bd., p. 474; and Christian, "Ann. Méd.-Psych.," Jan., 1881, p. 61.
‡ Calmeil, *op. cit.*, T. i, p. 434 (4 cases); Burlureaux, Thèse, p. 41 (2 cases); Voisin, *op. cit.*, p. 361 (2 cases).
§ " De la Mélancolie dans la paralysie générale," 1875.

gestive" as a cause of g.p. What he terms "folie congestive" differs from that formerly so called by Baillarger; and the patients are without somatic disorders, have systematized delusions, expansive or depressed; and vivid hallucinations; are full of hatred and revengeful; and the cases may end in dementia, or, rarely, in g.p. To explain how any form of simple insanity may bring on g.p., reference is made to neuralgic pain producing a local determination, and to hysterical disorder ending in myelitic lesion. But it is a question whether g.p. *can* happen in chronically insane persons without the operation of a *new factor* leading to g.p. Mendel believes it cannot, and I agree with him. Dr. C. E. Hostermann observed three cases of the kind out of 160 (two were cases of verrücktheit), and founds the diagnosis on the progressive motor signs only; with which coincides also a somewhat rapidly invading mental weakness.

10. *Cessation of Discharges.* A suppression of lactation, or of the menses, seems occasionally to act as a predisponent; so does suppression of hæmorrhoidal flux, and, doubtfully, of discharge from ulcers, or, that of chronic cutaneous affections.

11. *Cranial Injuries.* In a paper on this subject I * stated that "in the cases that have come under my own observation, where cranial injury has conduced to g.p., it has, in the majority, seemed to play the part of a predisposing rather than of an exciting cause. In speaking of cranial injury as a predisposing cause of g.p., we may suppose that in consequence of latent residual results of the . . . immediate effects of trauma, either the cerebral tissues are simply less resistant to the influences of ordinary causes of the pathological process which underlies g.p., or that this process springs more fully into being by assisting in, and in its turn being assisted by, the intensification and extension of slight local inflammation or hyperplasia sequential to the brain injury; or, again, assisted by morbid vaso-motor effects of that injury."

12. *Climate, Locality, and Race.* G.p. is said to be rare in hot climates, but it is an old observation that this comparative exemption does not extend to the new-comers. It affects Saxon and Celt; but Ireland enjoys extraordinary comparative immunity from g.p., many of the large public asylums there presenting no case of the disease, and the asylums for

* "Jl. Ment. Sci.," Jan., 1883, p. 544; "Med. Press and Circ.," Jan. 10th, 1883, p. 25.

Dublin and its neighbourhood being far away the first on the list, and yet with only 2 to 3 p.c. of their admissions G.Ps.; while it has hitherto been almost unknown in the asylum for Belfast. Here, as Dr. Isaac Ashe* stated, the population is chiefly of lowland Scotch origin, and really of Saxon blood; while the Celts of Wales, Cornwall, and the Scotch highlands have a considerable share of g.p. Yet, on the whole, there are strong indications, I think, that the Celt is less liable than the Saxon to g.p., just as he exhibits less of the sustained effort, and sustained and applied cerebral energy of the Saxon. To a lesser extent, also, the north of Scotland, where the Celtic element predominates, appears to be comparatively free from g.p.† In the south of France, if one may judge by the official returns, g.p. has greatly increased during the last generation or two,‡ and at present is nearly equal to that in the north. G.p. is said not to have been recognized in America until 1843, when Dr. Luther Bell first announced its existence there. Now, at least, it is common enough on that continent. Dr. J. Workman § writes: "When I entered the Toronto Asylum in 1853, there was not a single case, as far as I could judge, in the institution, but it was not long before it began to make appearance ... in my last ten and a half years, from Jan. 1865 to July 1875, the deaths from (general) paresis amounted to seventy-two." If reliable, the asylum reports would tend to show that g.p. is much more rife in the eastern and middle than in the western and southern parts of the United States of America. Dr. A. E. Macdonald ‖ avers that statistics show an uniform order of progression in each country. 1. The appearance and recognition of the disease in males. 2. Increased frequency of occurrence in male patients, and appearance in female. 3. Increased frequency of occurrence in both sexes, and in larger proportion than the increase of ordinary forms of insanity, and increase in the proportion of females to males attacked.

Though unproven this, indeed, is what one would expect; but the numerical returns vary extremely with the varying knowledge of the disease enjoyed by different alienists and

* "Jl. Mental Sci.," Oct., 1875, p. 465; *ibid.*, April, 1876, p. 82; "Trans. International Medical Congress," London, 1881, Vol. iii, p. 650.
† 18th Report, Commissioners in Lunacy, Scotland.
‡ "Annales Méd.-Psych.," July, 1881, p. 33. (A. Sauze.)
§ "The Canada Lancet," August, 1878, p. 356.
‖ "American Journal of Insanity," April, 1877, p. 451.

medical practitioners. So that many statistics are utterly misleading as guides to a knowledge of the relative proportions of g.p. in different countries and in different localities, and are, in part, merely an index of the varying capacity of the medical men, in the past, to *recognize* the disease. To a large extent the older American, British, and French statistics in this particular relation do but limn the lines along which a practical knowledge of the disease grew in the minds of asylum physicians, the more obvious and numerous male cases being recognized first,—the less striking, and less frequent, female cases afterwards.

There are many odd and interesting local variations from the mean. Thus, in different provinces of Germany, if the returns are accurate, the percentage of g.p. in the total insanity varies enormously, or from about 4 to about 20 p.c.; in different parts of Belgium the same obtains in less degree; and the same is true of the counties of England; and, according to Verga, of the different States of Italy. In Cuba g.p. is infrequent, but affects the whites much more than the blacks. In South America, at least in several parts, it is comparatively rare. At a New York pauper lunatic asylum, among the Anglo-Saxons the proportion of G.Ps. was 13·29 per cent.; among the Celts, 11·58 p.c.; Germans, 11·13 p.c.; Hebrews, 10·29 p.c.; Negroes, 8·82 p.c. (Spitzka).

13. *Urban Life.* Town life rather than rural, seems to foster g.p. According to statistics in the 18th Scottish Lunacy Blue Book; in the *town* districts of Scotland g.p. yielded an annual death-rate of 1·9 per 100,000 of the population; but only ·8, or less than half, in the insular and mainland rural districts; so that, in equal populations, for every 100 deaths from g.p. in rural and insular parts, there were 237 in the towns. Or, taking another and subdivided classification, the annual death-rate, per 100,000, from g.p. was, in the *principal towns* 2·1, in the *large towns* 1·3, and in the *small towns and rural districts* only ·7; or as the relation 3—2—1, respectively. Moreover, "while the death-rate from g.p. is three times as great in the *principal towns* as in the *small towns and rural districts*, the death-rate from *all* causes is considerably less than twice as great."

14. Hypertrophy of the left cardiac ventricle, formerly esteemed to be an important predisposing cause, is usually, I think, rather to be taken as a collateral effect of the general condition, circulatory and renal, in which g.p. arises and pursues its course; or sometimes as, in part, a secondary

and remote result of the brain and cord changes of the paretic.

EXCITING CAUSES. Rarely does one of these act alone. As a rule, there is an alliance, a co-operation of several causes. For example, excess of one kind is joined by others, the exhausting effect of these makes work more trying, and the customary avocations to be only plied with effort and strain. From these comes speedy break-down; or else, neglect of work, from which latter, and from the reputation of a dissolute life, come loss of social esteem, friends and fortune, frustration of hopes, mental shocks, anger, jealousy. And each of these may influence the production of g.p. Dr. Verga summed up the effective causes, in broad terms, as "abuse of the moral and intellectual powers or of the cerebral." This requires development.

Some have held that the exciting causes act by producing cerebral congestion, upon which they lay emphasis as being the proximate or efficient or direct cause of g.p. So far back as 1822, Bayle had attributed this *rôle* to all the causes, and he and, later, Daveau spoke of cerebral congestion as produced by them; congestion which may be sudden and apoplectiform in its onset, or slow, and evinced by heaviness, dull weight in the head, confusion on awaking, forgetfulness, and vertigo.

I have prepared a table (see opposite page) of assigned causes of g.p.* Opposite to each cause—and subdivided according to sex—are the numbers of G.Ps., in which that cause was assigned, and the *average* percentages of these G.Ps. on the number of G.Ps. (of same sex) admitted, during four years in England and Wales.

One case I cannot quite understand: it is that of pregnancy an assigned cause in a male.

Taking the sexes together, according to this table, adverse circumstances (including business anxieties and pecuniary difficulties) : mental anxiety, worry, or overwork : and domestic trouble: are the three most efficient of the so-called "moral causes" of g.p.; whilst of the so-called "physical causes" intemperance in drink is far away the first, heredity comes next, the vague "other bodily diseases" next, then come "accident or injury," and "previous attacks."

For males only, the most efficient causes are in the following order: Intemperance in drink. Heredity. Other bodily

* Reports of Commissioners in Lunacy (England), Nos. 35-6-7-8.

Totals	Males. 3374	Females. 910	Male (p.c. of G.Ps. on admissions). 12·65 p.c.	Female (p.c. of G.Ps. on admissions). 3·25 p.c.	Both sexes (p.c. of G.Ps. on total admissions). 7·8 p.c.
CAUSES.	Males.	Females.	Males. Average p.c. on male G.P. admissions.	Females. Average p.c. on female G.P. admissions.	Total average p.c. on G.P. admissions (both sexes).
Moral.					
Domestic trouble	124	86	3·6	9·4	4·9
Adverse circumstances	373	44	10·9	4·8	9·6
Mental anxiety, worry, over-work	221	29	6·4	3·1	5·6
Religious excitement	29	7	·8	·7	·8
Love affairs (including seduction)	15	12	·4	1·3	·6
Fright, and nervous shock	15	6	·4	·6	·5
Physical.					
Intemperance in drink	795	122	23·5	13·4	21·4
Sexual excess	80	29	2·3	3·1	2·5
Venereal disease	41	8	1·2	·8	1·1
Self-abuse, sexual	10	—	·3	0	·2
Over-exertion	46	6	1·3	·6	1·1
Sunstroke	94	3	2·8	·3	2·3
Accident or injury	256	24	7·5	2·6	6·4
Pregnancy	— ?	8	—	·9	·2
Parturition and puerperal state	—	45	—	5	1
Lactation	—	15	—	1·6	·3
Utero-ovarian disease	—	12	—	1·3	·25
Change of life	—	36	—	4	·8
Fevers	12	1	·35	·1	·3
Privation and starvation	51	36	1·5	3·9	2
Old age	10	9	·25	1	·45
Other bodily disease	352	111	10·3	12·2	10·7
Previous attacks	162	68	4·7	7·4	5·3
Heredity	490	175	14·3	19·1	15·4
Congenital defect	4	5	·1	·5	·2
Other causes	38	6	1·1	·7	1
Unknown	991	297	29·3	32·6	30

disease. Adverse circumstances. For females only, the order is: Heredity. Intemperance in drink. Other bodily disease. Domestic trouble.

The above table cannot be precisely compared with the percentages in the tables concerning the *total admissions* for *mental unsoundness of all kinds* during the same four years; and this for three reasons. 1. The tables are constructed on a basis somewhat different; and in the tables of *all* cases admitted the "totals represent the entire number of instances in which the several causes, either alone or in combination with other causes, were stated to have produced the mental disorder. The aggregate of these totals (including 'unknown') of course exceeds the whole number of patients admitted; the excess is owing to the combinations."—2. A division is made into "exciting" and "predisposing" causes in the tables of *all cases*, but not in those of *g.p.*—3. Causes are not assigned in the same proportion for (a) G.Ps. admitted, and for (b) all cases admitted. Nevertheless, the returns are of value for ascertaining the relative frequency with which each etiological factor is supposed to influence the production of (a) g.p., and of (b) all cases admitted. Read on the surface, these tables of admissions would assign to g.p., as compared with the total admissions, the following relative differences as to the operation of the several causes. *For g.p.;* as compared with *all* admissions; male—female—both sexes together :—

The following causes efficient in a *larger* proportion of G.Ps.; than of *all* cases:—Adverse circumstances, (especially in aggregate of the two sexes, females being less liable to this cause, and female G.Ps. few).—Intemperance in drink, in males, by about 5 to 4; in females, by about 2 to 1; in total, by about 21 to 13.—Sexual excess, in males, by about $2\frac{1}{2}$ to 1; females, 6 to 1; total $3\frac{1}{2}$ to 1.—Venereal disease; over-exertion; sunstroke; accident or injury; each, in total, by about 2 to 1.—Privation and starvation (slightly).

The following causes efficient in a *smaller* proportion of G.Ps.; than of *all* cases :—

Domestic trouble. Mental anxiety, worry, overwork (females only; M. about same).—Religious excitement (3 to 4 times less).—Love affairs (M. and F. about $\frac{1}{2}$. Total about $\frac{1}{3}$). Fright and nervous shock. Self-abuse. Pregnancy, parturition, puerperal state, lactation, utero-ovarian disease, puberty. Fevers. Old age, (about 9 times less). Congenital defect, (very much less). Heredity. Previous attacks.

These large statistics do not harmonize with the experience of v. Krafft-Ebing * that moral causes are far more frequent in female than male cases; as 25 to 7.

I think the following efficient causes of over-excitation and exhaustion of the brain frequently bring on g.p., and of the last four each is associated in many cases with undue alcoholic stimulation. I. Alcoholic excess.—II. Excessive and prolonged intellectual labour, with undue emotional tension.—III. Protracted painful emotional strain.—IV. Exhausting heavy physical labour.—V. Sexual excess.

Exciting causes in detail. I will speak separately of (*a*) alcoholic excess; (β) sexual excess; (γ) moral causes; (δ) diet; (ε) physical labour; (ζ) injury; and more briefly of various other agencies.

(*a*). *Alcoholic excess.* In my own cases alcohol, though perhaps rarely acting alone, has appeared to be by far the most frequent and efficacious cause of g.p.

It was found to be a fertile cause of the disease by Dr. L. Thomeuf;† and Guislain,‡ previously, had stated that g.p. is the form of mental disease more particularly engendered thereby. Dr. John Hitchman § and Dr. Hack Tuke ∥ gave this cause a prominent place; Dr. L. V. Marcé observed cases after several attacks of delirium tremens, get enfeeblement of memory, shaky hands and lips, and sometimes a slight embarrassment of speech. The irregular transition from alcoholic disorder to g.p., and the greater frequency than usual of visual hallucinations, were the only special clinical features observed by him in g.p. from drink. Several cases were published by Dr. V. Magnan ¶ in which chronic alcoholism ended in g.p.; sometimes after several attacks of alcoholic insanity; between the chronic alcoholism and g.p. was often a prolonged intermediate period; or g.p. followed alcoholic insanity closely; or, intermingling with the signs of g.p., may coexist terrifying hallucinations, etc. In these cases, to fatty degeneration of the organs and vascular atheroma, interstitial diffuse sclerosis joins itself. Similar examples were recorded by MM. Gambus and Lolliot. Thus, cases were observed by the latter;** at first with alcoholic

* "Archiv für Psych.," vii Bd., p. 182.
† "Gazette des Hôpitaux," July 19, 1859.
‡ *Op. cit.*, Tom. ii, p. 55.
§ "Brit. Med. Jour.," Vol. ii, 1871.
∥ "Manual Psych. Med.," p. 334.
¶ "De l'alcöolisme, des Diverses Formes du délire alcöolique, &c.," p. 190.
** Société de Médecine de Paris, Avr. 12, 1873.

délire, perhaps several attacks and recoveries complete or incomplete, and then a period of transition, and, finally, g.p. In New York, Macdonald gave a history of alcoholic intemperance in 75 p.c. of G.Ps. Of Dr. W. Nasse's* alcoholic mental cases, one-eighth were G.Ps.; and Dr. A. Sauze makes drink the chief factor in an asserted increase of g.p.

Out of 3,374 male G.Ps., intemperance in drink was the assigned cause, or one of the assigned causes, in 795—or 23·56 p.c.; and in 122 of the 910 female G.Ps., or 13·41 p.c.; or, together, in 917 out of a total of 4,284 G.Ps., *i.e.*, in 21·4 p.c.

One caution is necessary; a G.P. in an early stage, or in the middle stages of a mild case, if at liberty, may commence to drink to excess, and this may not only modify the already existing symptoms, but lead to the whole condition being attributed to alcohol.

(β.) *Sexual Excess* (excess of coitus). Dr. J. Guislain is sometimes quoted as assigning a peculiar potency to sexual excess in the production of g.p. But he only stated that sexual excess, when conjoined with other debauchery, tends to terminate in g.p.; at the same time there is nearly always a predisposition, or the operation of a moral cause of the disease; and that we must not always assign sexual excess or spermatic emissions as the causes of g.p.; for excessive drink, fear, various reverses and disappointments, may directly bring it on, and so may intellectual overstrain. By Dr. W. H. O. Sankey† it has been deemed " remarkable how many G.Ps. had led irregular lives, and especially had been guilty of sexual impropriety of some sort." Dr. E. Sheppard‡ agrees as to the very prominent part held by sexual excess in the etiology. Dr. Maudsley § lays stress on sexual excess as a very fertile cause of g.p., and chiefly that carried on systematically by faithful married persons, "that quiet, steady continuance of excess for months or years by married people which was apt to be thought no vice or harm at all." In other quarters, this view has been most erroneously carried to a great extreme. Thus Neumann and Cavalier declared this to be the exclusive, or almost exclusive, cause of g.p.

Having sought and, though sometimes successfully, yet usually in vain for a history of antecedent sexual excess in the

* "Allgem. Zeitschr. für Psych.," xxxiv Bd., p. 171. "Ueber den Verfolgungswahnsinn der Geistesgestörten Trinker."
† *Op. cit.*, 1st Ed., p. 181; 2nd Ed., p. 284.
‡ "Medical Times and Gazette," Vol. i, 1873, p. 163.
§ "Jl. Ment. Sci.," Apr., 1873, p. 165.—"Pathology of Mind," 1879, p. 433.

cases under my own care, I do not agree with the view that excessive frequency of sexual intercourse, and especially in married life, is by far the most fertile cause of g.p.; although in some cases this is the cause, in others, one of the several causes, of the disease. And Dr. Mendel, in his work on g.p., published after the first edition of this book, goes very much further than I did, and appears to have never seen a case which he believed to be clearly due to this cause alone, or chiefly, although he admitted that in some cases it acts, or may do so, in an auxiliary manner in the causation of g.p. Of sexual excess as a symptom, an effect, of g.p., I have already spoken, but this is another matter altogether.

In only about 2½ p.c. (109 in 4,284 G.Ps.), in the statistics already given, was sexual excess even assigned as one of the causes, or the cause.

With regard to masturbation, in only 10 cases out of the same 4,284 was it even an assigned cause; and here also, as in the former case, an effect may readily be mistaken for a factor.

When concerned in the causation of g.p. sexual excess almost invariably, I think, acts in concurrence with other causes,—it then forms but part of that general sensuality and fastness which so oft incur this dread disease; or is allied with a sanguine temperament, and an overactive, protracted and exhausting output of physical and mental energy. Intense and protracted perturbation or emotion, usually of a painful nature, on the mental side; and alcoholic excess, or exhausting labour sustained by alcoholic stimulation, on the physical side, are apparently more potent and frequent causes of g.p. than is sexual excess. In the majority of cases where sexual excess has been observed in one who has become a G.P., if the research into the antecedents of the patient can happily be carried far enough, something else unusual in the demeanour, action, or mental state will be traced to the same period as the sexual excess, and will assist to establish the latter as a prodromic or early symptom, and not as the original cause of g.p., although it then comes to react unfavourably upon the disease, and hasten its progress.

Of the utmost importance in the determination of this matter is the *age* at which the disease is most prevalent. Nowhere else, save partially in the classical work of Bayle, have I met with anything similar to this view—a view previously and independently formed by myself. Most of the cases of g.p. occur between the ages of thirty and sixty years; it is particularly rife in the decade between forty and fifty; and also very rife between fifty and sixty; relatively to the number of persons in the general population alive at those ages. Allowing an appropriate interval between the incidence and action of the cause, and the commencement of the disease, it would seem that the age-period of greater frequency of g.p. is much later than it would be if sexual excess were indeed so potent a cause as is alleged by some.

For at what period of life is it that sexual excess is most frequent? Although many exceptions present themselves, this, for the most part, is the interval between the age of twenty and that of thirty years. As a rule, after the latter age, or thereabouts, the sexual instinct is less urgent and less sportive; the tempests of desire no longer perturb the being to the same degree as formerly they did. The whole domain of feeling and hence that of action, too, is sobered by the now more distinct recognition of the realities, and by the burden of the cares, anxieties, and responsibilities of life. Thus the period of life at which sexual excess, and that at which g.p., is most frequent, are far from corresponding the one to the other.

(γ.) *Moral Causes.* But if the age at which g.p. is most rife is not that at which sexual excess is most common; to the incidence of what possible causes of g.p. is the age in question most exposed? And we may specify the period from the age of thirty to that of fifty-five as the one more particularly referred to. This is especially the age of ambition, pride, selfishness, of speculation, of daring attempts to secure fame, wealth, power, prestige, and social position; the age of excessive and protracted intellectual labour done under emotional strain, of anxious and sustained strenuous efforts to provide for, and ensure future success to, a growing and exigent family; the age in which excessive physical labour is often undertaken, which, like the intellectual labour, may be sustained by too liberal potations,—and which, like it also, is no longer counteracted by the elastic power of accommodation of youth, or by its restfulness after fatigue. Then, as a necessary correlative, it is the age most liable to chagrin and mortifications of spirit, to losses, to disappointments; the age, occasionally, of sudden beggary after a life of .toil and hard-earned success; the age at which, so often, the mirage of hope vanishes and its aërial castles dissolve; at which the projects of life fail and crumble away, and its fire dies out.

No wonder that influences such as these, acting upon those whose nervous systems have lost the elasticity of youth, whose blood and blood-vessels, perhaps, are further impaired by the effects of alcohol, whose naturally hyperæsthetic brains have been exhausted by irritable reaction to every strong impression; should bring about a sudden or protracted hyperæmia, a slow irritative form of degeneration, or inflammation, in the supreme centres of the organ of mind.

Now superadd to these causes the frequently associated alcoholic excess, and their efficiency is doubled.

Already noticed as a predisposing cause, mental overstrain acts perniciously as an exciting cause also. Severe intellectual work carried on too protractedly and monotonously, and especially when carried on in an atmosphere of worry, vexation, annoyance, anxiety, may alone produce g.p. It is the long-continued dogged work, excessive in amount, lacking in variety, and trenching upon the hours of sleep, or making sleep uneasy, that kills; and especially if in alliance with emotional *malaise*. When temper becomes irritable, sleep disturbed, thought wandering and confused; and when, too, the head flushes or aches, and mental fatigue oppresses, then g.p. threatens, and it is time to stop. Emotional overstrain, whether the feelings which become passions and take possession of the life are of an expansive or depressive sort, may arrive at the same effect. Choleric, "passionate," outbursts often repeated—signs, as the word shows, of suffering, of weakness, and not of strength—prolonged play of the passions, such as rage, jealousy: thwarted ambition, disappointed and helpless pride, thirst for wealth, anxiety due to actual, to prospective, or to merely anticipated loss of means; dread of impending danger whether actual, or threatened, or imaginary; protracted strainings of ambition in all its phases; the lifelong struggles, chagrins, and heartburnings accompanying the modern conflict for existence, for place, power and prestige; domestic unhappiness, or bereavements; business anxieties or losses; all conduce to g.p. In each case the original and acquired cerebral power of resistance must be kept in view: of two persons one passes unscathed through mental trials that leave the other in a state of moral and physical wreck. Hence the immense importance of the so-called predisposing causes.

Moral causes are assigned in 22 *per cent.* of the 4,284 cases of g.p. already cited. While it is a potent cause of g.p., Mr. Austin's statement that "moral agony is the cause of the disease" was very far too sweeping and exclusive; although Dr. Conolly found moral causes in nearly two-thirds. M. Böens* seems to have little ground for his view that religious fanaticism is the most potent of the moral causes of g.p. Possibly, fright with shock, may produce g.p., as in a case described by Dr. Witkowski,† yet here were local softenings, and some final hæmorrhages of brain. Two other cases he mentioned.

Not seldom, however, does it happen that acts springing out of the intellectual enfeeblement and moral and affective perversion of the incipient disease; together with the turmoil, trouble and distress they bring into being; are mistaken as being the causes instead of, what they really are, the consequences of the disease.‡ This must be kept steadily in mind, in control and modification of the above statements.

(δ.) *Diet.* The occurrence of g.p. in connection with the pellagrous affections of Italy (Brierre de Boismont, Baillarger, Bonacossi, Girelli) due to diseased-rye bread has not, perhaps, received all the attention of which it is worthy. I believe it is unknown in this country. A toxic origin of g.p. was long ago suggested by Dr. J. Hitchman, and more recently by others. A former view by Dr. I. Ashe, that g.p. is a form of chronic phosphorus poisoning, from a diet of meat, bread and beer, over-rich in phosphates, has but a slight and insecure basis. But he has subsequently brought forward evidence that the diet has an important influence in the production of g.p.; and that differences in race, or the varying proportions, in the people, of syphilis, and of excess either venereal or alcoholic, are inadequate to explain the peculiar distribution of g.p. in the United Kingdom; and the chief place is to be given to differences in diet; the large consumption of the potato in Ireland, and of oatmeal in Scotland, being here the central points of interest and inquiry, and deserving of further investigation. In this relation, the effects of ergotism are to be borne in mind; also the spinal disease arising from the consumption of *lathyrus cicera* and described by Dr. Brunelli.

* " Gazette Hebdomadaire," Jan.. 1874, p. 91.
† " Archiv für Psych.," vii Bd., p. 87.
‡ Darde, " Du Délire des Actes dans la paralysie générale.," p. 9.

(ε.) *Exhausting Physical Work.* Heavy, exhausting, physical labour can only become a source of g.p. when it is not balanced by a legitimate and healthy exercise of the intellect, and of the affections. If the whole being is absorbed in sheer labour; and especially then if the cardiac power is jaded out by ill-timed alcoholic fillips—a present relief, a future reckoning and misery—and the call upon the power of the brain, therefore, more severe; if the sleep is disquieted; if the secretions are altered, and if the thoughts and emotions revolve monotonously around one or two centres, the condition may be g.p. in the act of forming.

(ζ) *Cranial injuries. Brain-concussion. Brain-contusion.* Where cranial injury acts as an immediate exciting cause of g.p., it does so by the direct development, into the lesion of g.p., of secondary results of the molecular, or fine, or of the gross, local damages; such as molecular perturbation of the brain, or bruises, crushes, hæmorrhages, or ruptures of its tissues; to which, vaso-motor effects of the damage may or may not give reinforcement. For injury to produce this result, the brain-tissue must, perhaps, be ready to move (as it were) in the direction of g.p.

Schläger's statistics give one-seventh of the cases of mental disease produced by this cause as being g.p.; Calmeil reported a similar case, Marcé two, Voisin one. Baillarger believed he saw acute g.p. follow *immediately* after a blow on the head; and in the discussion on my paper* on g.p. from brain-injury, read at a meeting of the Med.-Psych. Assoc. in London in Nov., 1882, Dr. H. Rayner mentioned two cases in which the g.p. followed immediately after cranial injuries in persons thought to be predisposed to it. Prof. L. Meyer† in 76 cases (two series) in which causes were clearly made out, found 15 of injury to the head. In one, g.p. followed in immediate connection with the severe head-damage; in some others it followed after a long interval. In three cases reported by Mendel the g.p. followed the injury slowly. Von Krafft-Ebing, in 80 male cases, found cranial injury the cause in 6, but in only one of 80 female G.Ps. The table above, constructed from Commissioners' Reports, gives 280 out of 4,284 cases of g.p. as from "accident or injury"; or about 7½ p.c. of the male, and 2½ p.c. of the female G.Ps., or over 6 p.c. of the total of both sexes.

---

* "Journal Mental Science," Jan., 1883, p. 544 and 647. Another paper, by present writer, "Journ. Mental Sci.," Oct., 1885, p. 375.

† "Archiv für Pysch.," iii Bd., p. 289.

But this "accident or injury," perhaps usually cranial, will in some cases not be so. Close scrutiny must be exercised before the disease is attributed to an injury. A G.P. in the early stages is apt to get his head knocked, and thus the friends may mistake an incidental and indirect effect of the disease as being its source and origin; at the most, in the case supposed, it would accelerate or precipitate the onset or march of the malady. In sailors, Dr. Wm. Macleod * not rarely found g.p. follow falls on the feet, or direct blows on the head, and, in some, from that hour a morbid change began. He described a case in which an engineer was stunned by the unexpected firing of a 25-ton gun close to him; and the same afternoon saw objects double with his left eye, and was deaf in the left ear. Shortly after this were defective vision, partial deafness; he had suddenly become irritable and exacting; and had lost his former energy and nerve. Five years afterwards he was admitted, in the expansive state, with marked somatic signs of g.p., and left descending optic neuritis.

Insolation. This cause is not uncommon among British soldiers in India. Yet in many cases, published and unpublished, occurring in different parts of the world, and attributed to sunstroke, I feel assured that early marked apoplectiform seizures or epileptiform *petit mal* have been mistaken for sunstroke; in other words a symptom of the malady has been misinterpreted as being its cause. Nevertheless, some of the most hopeless cases I have seen, and with severe extensive brain-lesions of g.p. were said to be due to sunstroke. In the returns already mentioned 94 (or 2·8 p.c.) of 3,374 male, and 3 (or ·3 p.c) of 910 female cases, were attributed, or partly so, to sunstroke. In 60 cases Meyer found 3 from sunstroke; and he attempted to distinguish clinically the cases following insolation from those following exposure to other great heat; but my experience does not lead me to the same conclusions as his about the insolation cases.

Syphilis. Syphilitic meningitis has long been known, and the affection of the brain and arteries, as in cases by Prof. W. Griesinger: and cases of syphilitic encephalitis were reported at an early date by Dr. A. Duchek. † But Virchow said little of the chronic simple inflammations of pia described by some as being syphilitic, and asserted that gumma is the

* "Journ. Mental Science," July, 1879, p. 198.
† " Vierteljahrschrift für die Practische Heilkunde," xxxvii Bd., p. 6.

proper or special syphilitic disease of the pia. Latterly, however, the special and primary syphilitic brain-lesions are acknowledged to affect meninges, or vessels, or encephalic nervous substance; and are spoken of either as cellular hyperplasiæ ending in sclerosis or in gumma; or else as sclerous or as gummatous meningitis, arteritis, or encephalitis, respectively.

It has been a much contested point whether syphilis can and does produce g.p. On the one hand, some, like Lancereaux,* while indicating that syphilitic encephalopathies may closely resemble g.p., yet maintain that they are distinct and independent in character. So also Müller, Buzzard, Fournier, Voisin, Baillarger, Delasiauve, Lasègue, Linstow, Heubner, Mauriac, Lewin, Christian and others, and to some extent Wille. Most striking of all is the experience of Lewin—20,000 cases of syphilis—one per cent. becoming insane—not a single G.P. But the predisposing effect of syphilis, as a source of constitutional weakness and of vexations, is usually not denied.

On the other hand, some have maintained that syphilis may engender true g.p. Among these are Jessen, Esmarch, Kjelberg, Steenberg, Sandberg, Jespersen, Rinecker, Pontoppidan, Rollet, Sauret, Coffin, Foville, Erlenmeyer, Oedmansson, Griesinger, Schüle, Mendel, H. C. Wood,† Legrand du Saulle,‡ Pasquale Pirocchi, § Legardelle.

G. Kjelberg (Abs. in "Jahresb.") held g.p. to be a form of cerebral syphilis; and never occurring in an organism free from both congenital and acquired syphilis. Chr. Jespersen, in 123 cases, observed 83 with constitutional syphilis, 6 with chancre, 6 with "syphilis in the highest degree probable," 13 with gonorrhœa, or other genital affection which might (?) indicate syphilis; and 15 with nothing to raise a suspicion, even, of syphilis. He found the g.p. often coming on in cases with slight luetic manifestations, after a space of 5 to 28 years from date of syphilitic infection, and with an interval of at least 4 years between the last apparent syphilitic symptoms and the g.p. In only one case were clearly syphilitic symptoms in the course of the g.p.; but in others doubtful eruptions towards the close of life. Out of 75 cases K. Pontoppidan found 25 syphilitic, and 4 more,

* Treatise on Syphilis. Trans. N.S.S., Vol. ii, p. 66.
† "New York Medical Journal," March 22, 1884.
‡ "Gaz. Hebd. des Sci. Médicales de Bordeaux."
§ "Giornale Italiano delle Mal. Ven. e della pelle," 1878, p. 258. "Ann. de Derm. et de Syph.," Jan., 1880, p. 51.

T

probably so. Rollet * summarized the symptoms and lesions of g.p. due to syphilis: the symptoms, as being, progressive motor feebleness affecting all the muscles including the tongue, inability of the lower limbs to support the frame, trembling and uncertain movements of the hands, inability to grasp small objects properly, dysphagia, difficulty in speech, mental weakness and disorder. As for the lesions, he speaks of soft meninges resistant and whitish, adherent to the softened grey substance, of which a small layer separated along with the membranes; and, in some cases, of specific tumours, and of local softenings. Mendel found a history of secondary syphilis in more than half his cases, (the estimate ⅔ is made by him by excluding all cases in which the history was not clear either way, a method almost certain to lead to inaccuracy and to exaggeration of the syphilitic element). Most of my cases who were first syphilitic and afterwards G.P. incurred the former affection from 5 to 12 years before the latter was recognized, but in some a very much shorter time intervened; and many incipient G.Ps. expose themselves freely and thoughtlessly to syphilitic infection, or in consequence of the increased sexual appetite occasionally observed as an early symptom. Some have placed the earliest date for g.p. caused by syphilis at 5, others at 2 or 3, years after primary infection.

There are several possible ways in which syphilis may conduce to g.p. 1. The syphilitic diathesis may weaken the resisting power of the organism at large, or brain particularly, to causes of g.p., and thus dispose to that affection. —2. Other influences, acting on the brain and nervous system of a syphilised person, may rouse the latent tendency to syphilitic disease, in the parts then placed under strain.—3. Syphilis may, perhaps, directly produce the inflammatory changes of brain.

Lead. M. Hippolyte Devouges † described several cases (6); in 4 at least of which he believed true g.p. was due to lead-poisoning, and that he could exclude the operation of any other cause; in two the lesions of g.p. were found on necropsy. The cases took various clinical forms of g.p. In some, were marked early melancholic symptoms; trembling, which accompanied the paralysis, or preceded it, or entirely replaced it; anæsthesia; a long duration, especially in the prodromic period; and, occasionally, beneficial effects from

---

* "Traité des Maladies Vénériennes," p. 933.
† "Annales Méd.-Psych.," Oct., 1857, p. 521.

treatment. Delasiauve finally admitted the causation of g.p. by lead; so did Dr. Kiernan.* Dr. Doutrebente † described a case of g.p., as from lead. Dr. C. v. Monakow ‡ found the cerebral changes both macro- and microscopical, almost identical with those of g.p. Sclerotic change, with atrophy of nerve-cells of cord and brain; spider-cells, and dilated pericellular spaces. Here there were not only several lead palsies and muscular atrophies, but also symptoms closely resembling, if not actually those of g.p.; and the case is, *inter alia*, really also one of g.p., probably produced by lead. Kussmaul and Mayer § found also sclerosis of the cortex and superior cervical ganglia (increase of connective tissue; deformity of cells; slight periarteritis). In g.p. from lead, Dr. Meyer ‖ found early hallucinations and terror, somewhat like delirium tremens; then, later, mental depression and physical signs; and, finally-coming, the motor and psychical of g.p. See also the chapter on Diagnosis.

Tobacco. It is an old dictum, as for example by Guislain and some of the earliest writers on the subject, that the excessive use of tobacco may produce g.p.; it has been reasserted by Lefebvre,¶ by Jolly, Voisin, Sauze, and v. Krafft-Ebing who ascribes two male cases to the excessive use of Virginia cigars (10-20 daily.)

Exposure to great furnace heat. Calmeil ** declared this to be a cause of g.p. G.p. is very frequent in labourers in iron-works, rolling-mills, and so on, but they are also " thirsty souls." Prof. L. Meyer, out of 16 cases had 4 from insolation or exposure to great furnaces or open flames. Dr. R. Victor †† has described several cases, apparently of g.p., due to exposure to excessive heat in a gun-foundry. Berstens ‡‡ published similar cases.

Sleeping after meals, especially in a hot atmosphere (Voisin, *op. cit.*, p. 320.) *Suppression of some physiological excretions, such as menses or milk; or of pathological discharges.* Non-suckling was observed by Voisin as the cause of g.p. in three cases: in one the first two children had been suckled, the third was not, and 15 days after parturition congestive

* " Journ. of Nervous and Mental Diseases," Vol. viii.
† "Ann. Méd.-Psych.," May, 1879, p. 420.
‡ " Archiv für Psych.," x Bd., p. 495.
§ Cited "American Jl. Neurol. and Psychiatry," May, 1882, p. 181.
‖ Thèse, 1881. " Journ. Nervous and Mental Dis.," Apr., 1882, p. 435.
¶ " Gazette Hebdomadaire," Jan., 1874.
** *Op. cit.*, 1826, p. 374. *Op. cit.*, 1859, T. i, p. 270.
†† "Allgem.[Zeitschr. für Psych.," xl Bd.
‡‡ " Journ.Mental Sci.," Oct., 1884, p. 442.

symptoms began, and eventually led to g.p. In some cases the suppression of lactation or of menses may have been an *effect* and not a cause of the g.p., although mistaken for the latter. Suppression of hæmorrhoidal discharges, also, apparently has not the important *rôle* formerly claimed for it, especially by Bayle.

Climacteric period in females, and cessation of menses. Although so often alleged as a fertile cause of g.p., we have seen under the heads of *sex* and *age* that this normal cessation has little or no influence on the production of g.p.

Fevers; acute inflammations; etc. Erysipelas of the face and scalp was assigned by Baillarger as the cause of g.p. in two cases observed by himself; and a similar case was reported by Bayle. The first case had had five attacks of facial erysipelas within three years. In the second case the erysipelas was followed by persistent headache, until, at the end of a year, early symptoms of g.p. appeared. In Bayle's case, the subsequent persistent headache was followed in one month by g.p. Of these patients, the first was predisposed to congestions; the second to insanity; the third drank. Voisin also observed a case. G.p. is also said to have resulted from typhus, cholera, typhoid after cholera, dysentery, diphtheria, pneumonia, enteric fever, articular rheumatism.

In 12 out of 3,374 male G.Ps., and in 1 female out of 910, " fevers " were assigned as causes. These few might perhaps be all explained away; but, probably, of those said to be caused by "other bodily diseases and disorders" some should come here. Voisin gives cases of g.p. occurring as a sequel of acute pyrexiæ or phlogoses. Thus, in pneumonia, encephalic hyperæmia may be produced by a reflex modification in the play of the vaso-motors; reflex action analogous to that which in the same disease produces the hot red cheek-patch, the differences in temperature of the two sides of the body, the abscess sometimes occurring on the inflamed side. Articular rheumatism is mentioned by Jaccoud, Contesse and Voisin as leading to g.p. The puerperal state, in which so many congestions and inflammations may occur, need not spare the encephalon; and, if not, g.p. may be the next step.

Protracted and severe neuralgia. Voisin has published cases of g.p. following severe and protracted neuralgia, in which he opines the congestion and inflammation within the cranium to be produced by a reflex influence of the neuralgic pain on the encephalic vaso-motor apparatus, causing encephalic congestion, frequently repeated or long-continued; neuralgia being known to occasionally originate congestion and inflammation in other parts or organs. The usual disappearance of these protracted neuralgic pains on the advent of g.p. was thought to prove that they were not symptomatic of a fixed spinal lesion preceding the g.p. In the cases I have seen, the painful symptoms of this kind seemed to have had origin in a chronic spinal meningitis or myelitis, or in some affection of the cranial dura; and not to have been due to a merely functional disorder of the nerves affected. But Voisin's cases were nearly all females, mine nearly all males.

Epilepsy. Epilepsy has by some been looked upon as occasionally the *cause* of general paralysis;*—so rarely, however, does the latter supervene in chronic idiopathic epilepsy that examples of this kind may better, I think, be deemed as mere coincidences.

A caution is necessary here. The epileptiform attacks of g.p. may be one of its first symptoms, and I believe they have sometimes been mistaken for the seizures of a true epilepsy preceding the g.p., and either favouring or inducing its occurrence. Of such kind appears to have been a case by Calmeil, notwithstanding his view that the early symptoms were of true epilepsy. Drs. Luys and Voisin held that the changes in meninges and brain-substance in epilepsy show how readily epilepsy, by inducing intense cerebral congestion, might cause g.p.† But the few cases of coincidence do not compel one to that view, though the possibility of epilepsy acting as a predisponent to g.p. need not be denied. Thus, Mendel states that he saw g.p. in a man, aged 35, who had been epileptic from the age of 14 to 20; and in another, aged 28, who had been epileptic from 8 to 13 years of age.

Imprisonment is not of powerful influence in the production of g.p. according to the experience of several German observers; and g.p., when it does appear in prisoners and convicts has usually existed before condemnation and imprisonment; and the crime has often been an outcome of the mental disease; and not the mental disease a result of the crime's punishment.

The much debated question whether g.p. is increasing, can scarcely be definitely answered in figures. But with the press of population into towns and cities, the increasing relative fewness of our rural populace; the tension of modern activity and debauchery; the decrease of leisure and of deliberateness; there can be little doubt that it *is* increasing.

* Parchappe, ("Traité de la Folie": various cases. Calmeil, *op. cit.*, T. i, p. 461, *et. seq.*; T. ij, p. 85. Burlureaux, "Considérations sur le Siége," &c., p. 71.
† "Arch. Générales de Méd.," Dec., 1869, p. 641.

CHAPTER XV.

MORBID ANATOMY.

A. MACROSCOPICAL. B. MICROSCOPICAL.

A.. The naked-eye appearances. Where not otherwise specified, the following description is drawn entirely from the writer's experience.

The calvaria, often large and well-made, in about one-third of the cases is more or less increased in thickness, its density also is increased in some; in a few cases it is thinner than normal, rarely is it abnormally soft. Dr. Amadei * also found it large in male G.Ps. It is often congested, in some its inner surface has a dry worm-eaten appearance. Occasionally, one or several exostoses, or a formation of new bone on the inner table, are to be seen. The brain-pan is usually flattened from left temporal to right parietal region. Occasionally, are deep grooves for meningeal arteries, extreme dryness, local depressions or attenuations of inner surface, local caries or rarefaction of internal table, rusty staining; carinate shape of vertex.

The *arteries* at the base of the encephalon and their branches in some cases present more or less opaque, whitish thickenings or patches; some of these are of yellowish cast, and some have passed into the more ordinary atheromatous state.

The dura is more or less thickened in nearly half the cases, often only slightly; in one-fourth it is unduly adherent to the calvaria, in some most tenaciously so; rarely is it adherent to the pial arachnoid except by pacchionian bodies. In more than a third of the cases the dura is congested. In the falx, or in the convexity of the dura, " limpet shell " bony plates occasionally are seen in the older patients. With the above adhesions, and wormy internal surface of calvaria, may be numerous vascular connections between dura and bone, thickened external layers of dura, thick-walled and thick-sheathed vessels. The internal surface of the dura may show evidence of internal pachymeningitis, either of the simple or of the hæmorrhagic form, being partly lined by irregular false membranes, or russet-hued hæmorrhagic relics. For, in about 18 p.c. of the cases are rusty stainings, delicate formations, or films on internal surface. These delicate filmy traces of new formation, whether of hæmorrhagic or of inflammatory origin, usually separate with ease, may occur on one or both sides over the vertex or over the fossæ of the

* " Journ. Mental Sci.," Jan., 1884, p. 590.

skull-base, especially the middle and anterior, but in nearly all the cases observed by myself have been insignificant.

Marked durhæmatomata were found in about 4 p.c. of my cases, usually with evidences of hæmorrhagic pachymeningitis. They occur on one side, or on both; vary in thickness; often extend from the tip of the frontal lobe over the convexity, almost to the occiput; and laterally to the fossæ of the skull-base; they gradually taper off at their edges to nothing; are soft and flexible; appear somewhat translucent; are more or less, and irregularly, pale-reddish or faintly yellowish; and often contain between their layers some blood or blood-clot, or serosity. They are not special to g.p.; and it is needless to take up here the debated questions as to their relations to the arachnoid or dura, or as to whether this organized formation is primarily of hæmorrhagic, or of inflammatory origin—an organized blood-clot, whose newly formed vessels may easily rupture from time to time; or an inflammatory dural product, on or in which successive hæmorrhages may occur from snap of the connecting or permeating vessels. Bayle found durhæmatoma in 18 p.c. of his cases, and sanguineous effusion in the arachnoid cavity in 13 p.c.; usually with the former. I have seen a much smaller share, per cent., than some have of these conditions, and of dural changes, in g.p. In chronic forms the dura may be flaccid anteriorly, or be pallid. In some cases no very obvious change of it is observed. In 4 p.c. I found recent blood or clot in arachnoid cavity, usually at the base. In 30 p.c. the amount of serosity in arachnoid cavity attracted attention.

Acute forms. Where death occurs early, *the brain* may even appear to be increased in volume, as well as in vascularity, thus narrowing the sulci, and distending the dura. In a case of acute g.p. in a male, aged 40, the meningeal veins were congested; the olfactory tracts, especially the left, somewhat atrophied. The arachnoid was faintly opaque, and the pia œdematous over the cerebral convexity. There were no cerebro-meningeal adhesions, the brain was slightly softened, its white substance hyperæmic, the fornix and surrounding white softened, the grey commissure indurated, the ependyma of lateral ventricles opacified, that of fourth ventricle thick, opaque, slightly granulated.

Prof. L. Meyer described the appearances in a group of cases of rapidly developing g.p. with meningitis.—The brain distends the dura, juts forth if little rents are made in the latter; if the dura is removed the brain projects, and the calvaria cannot be replaced. Marks on the gyri, corresponding to prominent fibres of the dura, also testify to the pressure that has existed. In some cases the softened brain substance presses through the pacchionian gaps into excavations in the calvaria: in these cases, on removal of the brain, small pieces of it tear off together with shreds of the soft meninges; thus superficially simulating a cohesion of all the meninges and cortex. The increased intra-cranial pressure is due to turgescence of the frontal gyri and of those next to them. The surface of the gyri, especially at the convexity, appears irregular; the anterior gyri being raised above the level of the posterior, and here and there hillocky and humpy. The sulci tend to disappear, closed up by the enlarged, widened and bulging gyri. The surface of the affected gyri is red as a rule; mottled and streaked, paler islands helping to give a marbled appearance. Here, also, the cortex glistens. On section the gorged vessel-network may readily be mistaken for capillary hæmorrhages, and the cortex appears of irregular thickness.

Chronic forms. But almost invariably the brain comes under examination at a much later period. The following are then the most marked and frequent naked-eye appearances, they are not all present in every case, and are variously grouped and associated.

Removed from the calvaria, the encephalic mass is usually more or less flaccid, and sinks under its own weight. Then, examining the parts from above downwards, the cerebral layer of the *arachnoid* is generally thickened and is always more opaque than usual, and ofttimes changed in consistence—which usually is increased—particularly over the fronto-parietal convexity and internal surface of the hemispheres, and slightly on the orbital surface: the opacity often being patchy, and generally more obvious where it covers the anfractuosities and streams along the borders of the meningeal veins. Above, on its outer surface, as first described by Bayle, in one-tenth of his cases, it is often beset with minute pearly granulations: beneath it, at non-adherent parts, lies serum in the meshes of the pia-mater and in the unduly-rounded anfractuosities of the surface. It may look milky or congested; in it are occasionally calcareous plates. The increased pacchionian bodies adhere to the dura, or excavate hollows in the skull. The interpeduncular space is often bridged by strong opaque arachnoid. At the Sylvian fissures the meninges are opaque and the lobes of the brain cohere unduly, but it is chiefly between the frontal lobes that cohesions exist. Marked interlobar adhesions are mentioned in 58 p.c. of my necropsies, but no doubt were omitted from the record in many more.

In more than three-fourths of the cases, where the point is specially noted, were the olfactory bulbs and tracts more or less wasted and softened. In 16 p.c. the optic nerves were noted as being, to the naked eye, decidedly atrophied, and either indurate or softened; and a less degree of the same changes existed in many more. In some, the cranial nerves about the front of the interior of the skull-base are tied down or compressed by white adhesion-bands or threads.

The large meningeal veins over the convexity are usually more or less congested; in some cases gorged; but may be of average fulness only, and in some are partially or incompletely exsanguinous.

The *pia mater*, usually more or less thickened, coarse, often hyperæmic either universally or in irregularly distributed patches, and—except in rare acute cases—more or less bathed in serosity, especially over the fronto-parietal regions, is also, with comparatively few exceptions, adherent to the summits of some of the gyri. The areas of adhesion are irregular, scattered, of most variable extent, and occasionally, though rarely, invade the declivities of the anfractuosities also. In not a few cases is the pia unduly tough, rarely is it too friable. In some it is pale, in some opaque. The subarachnoid fluid may be collected in lakelets, and these may drain away through the meshwork, or may be bounded by adhesive change. The pia is occasionally the seat of small bony plates, or of firm fibrous whitish knotlets, or of a few yellow tubercular nodules, or of a few patches of lymph or pus exudation. Meyer found purulent spots in 12 p.c.; Mendel saw this twice; I have seen it marked in one case, slight in two more.

The *grey cortical substance* of the cerebrum is often diminished in consistence, either in whole or in parts only; more rarely is it indurated locally, or even somewhat diffusely, and gradually shading off to the normal, or to lessened consistence, as one examines from before backwards; usually it is discoloured, is generally the site of hyperæmia of more or less irregular distribution, yet may be wan, faded, and anæmic; in hue, therefore, is usually reddish, pink, lilac, or is mottled; occasionally is pale or has a fawn colour or a yellow or a dirty-whitish or even a slaty tinge; is usually wasted in some regions, and perhaps more friable and opaque than normal, and its stratification is ofttimes indistinct. As to cortical atrophy :—In my necropsies—in 90 p.c. it is mentioned : in 4 p.c. the necropsy-note as to this point has been omitted in the final records: in only 6 p.c. is it entered as being "not marked." Scarcely a single case, therefore, was quite distinctly free from cortical atrophy. As a rule, the atrophy is chiefly in the frontal lobes, including the orbital surface; next in the parietal, though in from 5 to 10 p.c. the atrophy is chiefly here. In some cases the left or right cortex is much more atrophied than the other. Or the somewhat diffuse atrophy may be more marked in circumscribed areas.

In nearly one-third of the cases there is a marked difference and lack of bilateral symmetry between the changes, or their degree, affecting the grey cortex of the two cerebral hemispheres.

Whether as regards inflammation, degeneration, or atrophy, in g.p. the cerebral grey cortex, in my experience, is most diseased, as a rule, in the superior and external surfaces of the frontal lobe—the orbital surface, also, often suffering considerably—next in degree in those of the parietal; and next in the temporo-sphenoidal; while the occipital lobes, comparatively, escape.

Adhesion and decortication. Where adherent to the membranes, the superficial layers are stripped off from the summits of the gyri along with the meninges when the latter are removed, leaving an irregularly eroded, and usually more or less reddish, appearance of the convolutional surface, locally. This is the most characteristic of the gross lesions of g.p., but is not pathognomonic. The earlier observers, as Bayle, and, seemingly Delaye, attributed these adhesions to arachnitis, the inflammation extending here and there to the brain, though Foville in 1829 denied that they were the simple effect of an inflammation of the arachnoid, and believed that their occurrence at the *summits of the gyri, only,* was owing to the cranial compression to which the summits are exposed during inflammatory turgescence of the brain, aided by the influence of the arachnoid in increasing the thickness of the meninges at those points. Alas for the dark mantle of oblivion that falls on much good and careful work in science! Fifty years had not yet elapsed when his own son gave the whole merit of practically the same explanation to a recent writer, adding that so far as he knew no one previously to the latter had sought to explain the fact.* The special adhesive change and its effect I † have termed "ad-

.* Art. g.p. "Nouv. Dict. de Méd. et de Chir. Prat.," p. 138.
† " Journ. Mental Science," Jan., 1876, pp. 571-6. *Ibid.*, Apr., 1878, p. 29.

hesion and decortication." This adhesion of the membranes to the cerebral cortex was found by Bayle* in at least half of his cases; by Parchappe in thirty-nine out of forty-four cases; by Calmeil in seventy-four out of eighty-two cases; and by myself in ninety-two per cent. of my necropsies. Much smaller proportions than these are probably unreliable, as that given by Justin, namely, 13 out of 33; or by Dr. v. Krafft-Ebing,† 7 out of 32.

Very nearly the whole depth of the grey cortex occasionally comes off in parts, leaving the anfractuous, firm, white beneath. I saw this, extensive, in a case in 1874; Baillarger described the condition fully in 1882. But in 1824, J. B. Delaye had anticipated by some decennia the supposed discovery of Dr. Baillarger, and the contribution of Dr. Ph. Rey to the subject. My experience of this change corresponds closely to Rey's; except that I have seen it over the orbital gyri; as well as over the frontal, on the Sylvian walls and on the insula; and except that though it does generally, it does not always, correspond to adhesions. It may be found in the depths of the anfractuosities, as well as on the summits of the gyri.

Another condition I have noticed is a ready separation of the grey cortex on easy manipulation, into 2 or 3 layers; and this independently of adhesion, and most frequently met with in the insula. In the museum of the Royal College of Surgeons, England, is, I think, a specimen of similarly flaked cortex, but without any extant history, and put up by the great master—Hunter—himself.

There are two chief conditions in g.p. as regards the cerebro-meningeal connection, either a too easy separation of pia from cortex with atrophy of brain, and this chiefly of the grey cortex; or else cerebro-meningeal adhesions with disruption of the soft wine-lees-hued cortex, when the meninges are removed.

With regard to *symmetry* of disposition of the adhesions, that is to say their similar arrangement bilaterally on the two cerebral hemispheres; in my cases approximate symmetry existed in 22 p.c. of *all* cases; while in 40 p.c. symmetry was moderate or partial; and in 30 p.c. the distribution of the adhesions was decisively non-symmetrical; there being no adhesions in the remaining 8 p.c. Taking only the cases *with* adhesions the p.cs. would be about 24, 43, and 33, respectively, for symmetry, partial symmetry, and non-symmetry.

* *Op. cit.*, p. 460, 472. † "Archiv für Psych.," vii Bd., p. 182.

284 Morbid Anatomy.—Adhesion and Decortication.

In *extent*, the adhesions were marked or extreme in 30 p.c. of *all* cases; fair or moderate in 28 p.c.; comparatively slight or circumscribed in 34 p.c.

With regard to depth of the erosions left by decortication: in the cases in which it was noted, it was marked or extreme in one-third, fair or moderate in one-fifth, and comparatively shallow in nearly one-half. In *colour*, where that was noted, the erosions were red or reddish in three-fourths; of nearly ordinary cortical hue, or else pale, in one-fourth.

Table of distribution of adhesions. Notwithstanding enormous care the final records of necropsies made by me omit a few details; and where, from the description, adhesions obviously existed at a place not individually named I have counted it as ½ only; these are so few as not to make any decided difference, although the following are *minimum* percentages, in consequence.

Adhesion and decortication of some parts, present in 92 per cent.; absent in 8 p.c.

	Gyri, one or both. Adhesion and decortication—	Of these on right side only, in—	And on left side only, in—	Of the total, and whether on one side or both sides, the adhesion "slight" or "very slight," in—
Superior frontal gyrus, in	86 p.c.	8 p.c.	0 p.c.	10 p.c.
Middle „ „	76 „	10 „	12 „	14 „
Inferior „ „	64 „	4 „	4 „	10 „
Anterior central „	74 „	4 „	6 „	22 „
Posterior „ „	75 „	8 „	4 „	24 „
Superior parietal lobule	54 „	4 „	2 „	4 „
Supramarginal gyrus	71 „	4 „	10 „	16 „
Angular „	55 „	4 „	4 „	8 „
First occipital „	27 „	8 „	4 „	10 „
Second „ „	28 „	8 „	6 „	12 „
Third „ „	20 „	10 „	4 „	4 „
Descendens „ „	6 „	— „	— „	— „
Superior (1st) temporal gyrus	80 „	6 „	20 „	28 „
Middle (2nd) „ „	80 „	2 „	16 „	22 „
Inferior (3rd) „ „	64 „	6 „	12 „	22 „
G. occip. temp. lat.	28 „	8 „	8 „	14 „
„ „ medialis	20 „	4 „	2 „	6 „
„ fornicatus	14 „	0 „	0 „	2? „
Median of sup. frontal (pli de la zone ext.)	50 „	4 „	2 „	12 „
Paracentral lobule	38 „	0 „	2 „	8 „
Præcuneus	33 „	2 „	4 „	6 „
Cuneus	21 „	0 „	4 „	? „
Hippocampal g.	34 „	0 „	2 „	10 „
G. uncinatus (uncus g.h.)	50 „	2 „	8 „	24 „
Orbital; rectus	60 „	10 „	6 „	32 „
„ ; middle	46 „	4 „	6 „	18 „
„ ; external	39 „	6 „	4 „	16 „

Thus the *Left* temporal and median surfaces; in less degree, the *Right* frontal and occipital; suffer more than the corresponding parts on the opposite side.

The insula was not stripped in a number of the cases, and of it, alone of all the cortex, were the original notes defective. Adhesions were noted here in 4 p.c.; in half of which they were on the left side only, in half on both sides: in half "slight." In several cases, also, independently of adhesion, the grey of the insula showed easy cleavage in horizontal planes, or easy separation of grey from white.

The *medullary substance of cerebrum* is usually discolored, sometimes diminished, at others increased, in consistence; induration being far more frequent or decided, in the white than in the grey substance. Generally hyperæmic, reddish, lilac, or mottled, yet when increased in consistence it is often of an unnatural whiteness, and then also displays a sieve-like appearance.

In the more advanced cases there is a decided atrophy of the brain, with increase of fluid in the ventricular, subarachnoid, and arachnoid spaces; for, as the brain shrinks, the various *cerebral ventricles* become larger both relatively and absolutely, their ependyma, beset with granulations, assuming a pearly, sanded, jewelled appearance, which often is more particularly manifest in the fourth ventricle. In some cases one lateral ventricle is larger than its fellow. Usually opaque, thickened, the walls of these ventricles may be very firm, or be softened. In only 8 p.c. are the naked-eye appearances of the walls "fairly" normal. The veins coursing over their floor are often congested. Occasionally adhesions to the *velum int.* are recorded. The third and fifth ventricles are often thick-walled and granulated. The choroid plexus of the lateral ventricles often looks like a string of pale swollen bladders. Phosphate of lime crystals may affect it. Ependymal changes are still more marked in the fourth ventricle, particularly at the nib. Granulated, thickened, opaque, of swollen gelatinoid appearance, congested, tough, —these are its changes in the order of their relative frequency. In only 6 p.c. were they styled "very slight," or "absent."

Nor does the grey matter of the *ganglia at the base* of the brain escape. Changes may be mentioned in the opto-striate bodies, which are often withered and pale on the surface, are softened in half the cases; rarely indurated; are (especially the corp. striata) more or less shrunk in about one-fourth, are congested in nearly half, anæmic in fewer, occasionally adherent to the choroid fringe. Austin drew special attention to the alterations occurring here within a circumscribed limit, and principally affecting the optic thalami, soft commissure, fornix, septum lucidum, corpora albicantia, floor of third ventricle, tuber cinereum, and crura cerebri; these alterations often being accompanied by increase of the ventricular serum. In the optic thalami, he asserted that, one or other of the following changes were always to be found, (1) disorganization or softening; (2) degeneration, with or without induration; (3) atrophy (simple); and (4) mere alteration in vascularity without structural lesion.

The *pons Varolii* and *medulla oblongata* often participate in several of the general encephalic changes. In one-third or more is softening; in somewhat fewer induration; in a few distinct atrophy; in more than one-half congestion; in a number anæmia. In some the membranes were noted as being thick or adherent over the pons or med. obl., and containing dilated vessels.

The *cerebellum* is usually diminished in consistence, somewhat hyperæmic, its tunics, often moderately thickened or opaque, are sometimes adherent to parts of its surface. Cerebellar adhesion has been denied in g.p. I have found it in 44 p.c. of the cases, though ordinarily it was noted as being "slight" or "very slight." Dr. W. H. O. Sankey stated that in g.p. the cerebellum is often increased in weight, that of the cerebrum being diminished. But although, *relatively* to that of the wasted cerebrum, generally increased, I have never found a marked *absolute* increase of cerebellar weight in g.p., and this experience is corroborated by Mr. W. Crochley Clapham's * statistics. Subsequently, Dr. Sankey takes these statistics as corroborating his former view, and as showing that in g.p. "the cerebellum is not only above the average relatively to the encephalon but actually." But this interpretation is not supported by the figures. There are two fallacies. (a) The conclusion is arrived at by comparing the figures for g.p. with those for "all cases and all ages." The appropriate comparison, however, I submit, is almost exclusively with cases between 30 and 50 years of age, and here is found a slightly less weight of cerebellum, pons, and med. obl. in g.p.; viz. in grammes :—males, *all cases* (30-40, and 40-50) 180·1, and 179 : *G.Ps.*, 177·5 :— females, *all cases* (30-40, and 40-50) 160·3, and 162·7 : *G.Ps.*,

* "West Riding Asyl. Med. Rep." vi, p. 26.

159·3. The few G.Ps. above and below these age-periods will not decidedly alter this result. (*b*) The figures in debate include pons and med. obl., as well as cerebellum.—Dr. S.'s own figures (p. 295, *op. cit.*) also show a smaller cerebellar ratio to cerebrum (6·54 : *i.e.* 1 to 6·54) in g.p., than the average in ordinary insanity (6·45 : *i.e.* 1 to 6·45); and these average ratios (with which I do not agree) would have gone to negative the view cited above from their author. Dr. T. B. Peacock placed the normal ratio at 7.

Spinal Cord. Broussais, Boyd, Bucknill, and Joffe first more particularly directed attention to this in g.p. The meninges of the *spinal cord* are most often thickened, less frequently adherent *inter se* and to cord; hyperæmic, opaque, granulated; all these occurring chiefly behind. Dark distended veins often meander over the meninges; and softish dark clot from recent spinal meningeal hæmorrhage may be found, but only in about 2 p.c. In an unusual case by Drs. Williams and Savage, blood was, partly at least, between the vertebræ and spinal dura. Occasionally, are old spinal durhæmatomata, or traces of spinal hæmorrhagic pachymeningitis. In the arachnoid over the posterior surface often are numerous, minute, irregular, hard, whitish, platelets, which increase in number towards the lower part of the cord, and are composed of connective tissue, and lime salts; or the arachnoid may be patchily thickened, opaque, or semi-gelatinous in appearance; or beset with minute pearly granulations. The pia is thickened, rough, or even delicately granulated. The thickened pia forms muffs for the spinal nerves as they emerge. The meningeal, especially dural, changes may be local, confined to the parts opposite to two or three vertebræ only. The changes begin with hyperæmia; the pia and arachnoid become beset with scattered transparent granulations; then are gradually thickened; adhere to surrounding parts; and the pia may even send prolongations into the cord's interior. Later, firm white platelets may form; or the serum around the cord increase. In acute and rapid g.p. the spinal meninges may be reddened, opacified, thickened, with cloudy sub-arachnoid spinal serosity; the cord-substance reddened, or constricted, or softened (Voisin). Rarely may be spinal tubercular meningitis; or meningeal gumma, or yellow tubercle.

More or less softening of cord, or of parts of it, I found in two-thirds of the cases, more or less induration in nearly one-third, atrophy in 10 to 12 per cent. Hyperæmia of the cord-substance obtains in a few cases; pallor in some also. Postponing the microscopical and taking here only the coarse changes; one of the most striking is posterior sclerosis, or grey degeneration of the posterior and posterior median columns of the cord. This is usually most marked in the lower parts of the cord, becomes narrower and more confined to the columns of Goll in ascent, and finally ceases at the floor of the fourth ventricle. With this may be seen similar grey degeneration of the dorsal and lumbar posterior spinal nerve-roots; or, without this, the latter are enveloped in thick, rough, membraneous sheaths.

A less frequent change is a secondary, descending, grey degeneration of the posterior part of the lateral columns of the cord, an example of which I have published in the "Journal of Mental Science" for April, 1885, p. 61. But without grey degeneration, one may find, and this either in the posterior or in the lateral columns alone, or in both simultaneously, induration, with dull, faintly bluish-white hue, corresponding, microscopically, to a light sclerosis. In some cases the whole cord seems to be mildly indurate; or on the other hand to be diffusely softened, and on incising the membranes the spinal substance juts forth. In a few cases are disseminated patches in the cord, which on section and exposure to air turn of a pale reddish-grey hue. Occasionally an atrophy of part of the grey, especially of one of its horns, in a greater or less portion of its length, may be made out; even by the unaided eye. I do not recollect finding hæmatomyelia with punctiform hæmorrhage in the grey. Weiss observed three such cases in g.p. The cord may be soaking in foul septic fluid. The late Dr. Boyd, in 155 necropsies, found pus on spinal arachnoid in 3, lymph on cord in 1. Prof. Tamburini often found osteomata on spinal arachnoid.

There is not one of the above morbid alterations that may not be absent in a given case, but I have never made the necropsy of a G.P. without finding very obvious naked-eye changes in the cerebro-spinal nervous system and its protecting tunics. The view of Lélut, Aubanel, and Thore, that g.p. might occur and run its course without any alteration of the brain and meninges, was no doubt the result of faulty, imperfect, and non-microscopical observation, or of diagnostic error.

With reference to the amount of atrophy of brain in general

paralysis we may turn to the brain-weights. Mr. W. Crochley Clapham found the *average* brain-weights in g.p.: in *male* cases, entire encephalon 45·92 ozs.: cerebrum 39·66 ozs.: cerebellum, pons, and med. obl. 6·26 ozs. In *female* cases, entire encephalon 40·01 ozs.: cerebrum 34·39 ozs.: cerebellum, pons, and med. obl. 5·62 ozs.

In these cases the meninges, or portions of them, and some serum, were, probably, weighed with the nervous masses. Possibly, or probably, the "weight of the entire encephalon" was taken immediately upon its removal from the skull, in which case the pia-mater, visceral arachnoid, subarachnoid and ventricular serum, would all contribute to swell the total weight.

But it is the weight of nervous tissue that is required, and therefore in the first edition of this work I appended the average weights in forty of my cases, in which the brains were stripped before being weighed, a few shreds of meninges only remaining, and that only sometimes, and especially on the cerebellum, pons, and medulla oblongata. These general paralytics were males (soldiers), and most of them aged thirty to forty. Owing to the adhesions in g.p., the exact weight of the cerebrum usually cannot be obtained by the method of Tiedemann.

			ozs.	
Average weight of	entire encephalon...	...	44·15	
,,	,,	cerebrum	37·79	
,,	,,	right cerebral hemisphere	19·04	not ascertained
,,	,,	left... ., ,, ...	19·17	separately in all.
,,	,,	cerebellum	5·42	
,,	,,	pons Varolii and med. obl.	·945	

The above weights forming a special standard for military G.Ps.; I will now give as a general standard for males the weights in male G.Ps. of various classes of society, the necropsies in all of which I made personally. Omitting the older necropsies: of the last 90 males the two cerebral hemispheres were weighed separately in 70, together in 20. For the 90 the average weight of stripped cerebrum was 37·593 ozs.;—for the 70, 37·72 ozs. Average weight of cerebellum, only partially stripped, 5·36 ozs.; of pons. and med. obl. with membranes ·9308 oz.

					Average weight.	
90 cases, males, cerebrum stripped of meninges	...	37·593	ozs.
70 ,,	,,	,,	,,	,,	... 37·72	,,
70 ,,	,,	right hemisphere stripped of meninges	18·845	,,
70 ,,	,,	left	,,	,,	,, ... 18·875	,,

290 *Morbid Anatomy.—Weights of Cerebrum. Pons. Cerebellum.*

	Average weight.	
Cerebellum partially stripped of meninges ...	5·36 ozs.	
Pons V. and med. obl. with membranes ...	·9308 ,,	
Entire encephalon (latter group) free from serum, nearly free from meninges	44·01 ,,	
Maximum weight of cerebrum	48	,,(2 cases)
Minimum ,, ,, ...	25¾	,,
Maximum ,, ,, (right hemisphere)	24¾	,,(2 cases)
Minimum ,, ,, ,, ,, ...	12¾	,,
Maximum ,, ,, (left hemisphere)	23½	,,
Minimum ,, ,, ,, ,, ...	13	,,
Maximum ,, cerebellum	7	,,
Minimum ,, ,, ...	4¼	,,
Maximum ,, pons Var. and med. obl.	1¼	,,
Minimum ,, ,, ,, ...	¾	,,

We may compare the above average weights in my own cases, all being males, with those given by Dr. R. Boyd * for the total male insane of all kinds, including G.Ps., dying between the ages of thirty and forty at the Somerset Asylum; namely, average weight of right cerebral hemisphere, 19·82; of left 19·94, and of cerebellum 5·33. Or they may be compared with Dr. John Thurnam's † return of the *average* weights for insane males between thirty and forty years of age at the Wilts County Asylum; namely, cerebrum 39 ozs. (1,105·5 grammes): cerebellum, pons, and med. obl., 6 ozs. (170·1 grammes). Or with Mr. Crochley Clapham's figures for insane males of all kinds æt. thirty to forty; viz.: cerebrum 40·9 ozs. (1,161 grammes) : cerebellum, pons, and med. obl., 6·3 ozs. (180 grammes). Or, still better, we may compare them with the average weight of the *entire encephalon* of *sane* males of the same decade;—namely, 49 ozs. (Thurnam). The heaviest entire encephalon I met in a G.P. was almost 56 ozs. (meninges off, and therefore probably much the same weight as Dr. T. W. McDowall's case—58 ozs.—with meninges on).

The wasting would thus appear to be from one and a half to three and five ozs., but the reckoning is somewhat vitiated by fallacies; namely, the imperfect separation of meninges by other observers, and the inclusion of g.p. in the statistics quoted.

On the other hand, the average brain-weights of healthy soldiers would probably exceed the average brain-weights of healthy males of the same class, not soldiers, and if so the

* " Philosophical Transactions," 1861, p. 241
† "Journ. Mental Science," 1866, p. 34.

brain-atrophy in the military G.Ps. would be really greater than appears.

In relation to this point is the amount of serosity, or blood-stained fluid, which drains from the cranial cavity during the necropsy. In the above series of cases, I found it (collected) vary from ¾ fl. oz. to 8 fl. ozs. (10 fl. ozs. in a case not in this series); what was actually collected and measured was on the average 4·06 fl. ozs., and some more was lost by infiltration.

In the total male insane of all kinds, aged from 30 to 40, Dr. Boyd found the average weight of pons and med. obl. to be 1·05 ozs., while in g.p. I found the average weight in those of the same sex and usually of the same age to be about ·945 oz., a difference of ·6 oz.; or, in other terms, 5·9 p.c. less weight of pons and med. obl. in G.P. cases.

Thus, we find that in g.p. the wasting is chiefly of the cerebral hemispheres, of the pons and med. obl.; and probably to a less extent of the cord, though I have not, and am not aware of, weight-statistics sufficiently large and precise in proof of the cord-atrophy. The cerebellum is comparatively little affected.

Inequality in weight of the two cerebral hemispheres. In many cases I have found considerable differences between the weights of the two cerebral hemispheres in g.p., and for the purposes of this comparison I always removed and separated the hemispheres with the most scrupulous care. In one case published by me some years ago * the left hemisphere was 22¼ ozs., the right 24⅜; a difference of 2¼ ozs., or of 63·7 grammes. Other examples are at the end of this work, in both editions. Taking the second series of cases above, and counting only those in which the difference in weight between the two cerebral hemispheres was ½ oz. (14·17 grammes) and upwards, I found these differences in 49 cases out of 70, the necropsies in all being entirely made by myself. Of these 49, the right hemisphere was of lower weight and greater atrophy in 26, and the left in 23.

In 22 of these 49 the difference was 1 oz. and upwards; and in 6 of these 22 the difference was 2 ozs. and upwards. In the 49 the differences varied from ½ oz. to 3½ and 4 ozs., or from about 14 grammes to 99 or 113. (The 4 ozs.—113 grammes,—refer to a case in which some little doubt arose as to accuracy). The *average* difference in the 49 cases (excluding the case in which the difference was 4 ozs.) was 1·063

* "Journ. Mental Science," Jan., 1876, p. 572.

ozs., or about 30·15 grammes. In most, or all, of the foregoing cases the differences in weight were evidently not congenital, or normal to the individual, but were explained by more disease and atrophy of one hemisphere than of the other.

Baillarger and Baume spoke of considerable inequality in the weights of the two cerebral hemispheres in some cases of g.p.; inequality of which kind was erroneously connected by Dr. Follet* in an especial manner with epilepsy, which last he attributed to a failure of inter-hemispheral equilibrium. Dr. Baillarger† only spoke of an inequality of from 20 to 62 grammes. Dr. Baume‡ cited examples of greater disequilibrium, the greatest being 124 grammes, the least 10 grammes, the mean 35 grammes. Ten grammes ($\frac{1}{3}$ oz.), however, is too low a figure to begin at. In 27 the lower weight was of the left; in 16 of the right, hemisphere. Dr. P. Samt,§ in a case in which the left hemisphere was 50 grammes the less in weight of the two, and the several lobes were weighed separately, found the relative left atrophy about equally marked in the frontal and parietal lobes, but greater in the temporal and occipital. Dr. Boyd found inequality in a few cases, ranging from 1 oz. to $4\frac{1}{2}$ ozs.; this last being in a patient under the age of 30 (congenital?).

Specific gravity of brain in g.p. In several cases I examined this carefully some years ago, but need only quote some of the notes and remarks in a case I published ("Journ. Mental Sci.," Jan., 1876, p. 572). "Specific gravity. On the *left* side. Averages. Cortical *grey* matter of tip of first frontal, 1,039; of third frontal $1,041\frac{1}{4}$; of ascending frontal $1,041\frac{1}{4}$; of tip of occipital $1,041\frac{1}{4}$.* *White* medullary substance of first frontal, $1,040\frac{1}{2}$; of third frontal, 1,042; of ascending frontal 1,042; near the tip of occipital lobe, $1,041\frac{1}{2}$. On the *right* side the grey matter of the ascending parietal gyrus had a sp. gv. of 1,040, and the corresponding medullary tissue, $1,041\frac{1}{2}$. Portions containing about equal quantities of grey and of white substance were also tested. Their sp. gv. when taken from the tip of the right first frontal was 1,039; from upper part of right ascending frontal, 1,041; from tip of right occipital, 1,043."

In the insane collectively, the average sp. gv. of the cere-

* "Annales Méd.-Psych.," 1857, p. 512.
† Société de Médecine de la Seine, June 12, 1857.
‡ "Annales Méd.-Psych.," Oct., 1862, p. 541.
§ "Archiv für Psych.," V Bd., p. 201-14-15.

bral grey *cortex* has been variously estimated at from 1,032 to above 1,039; and of the medullary substance at from 1,039 to 1,042⅜.

It would be tedious to describe at length a number of morbid changes to which prominence has been given by various writers on g.p. Yet we may briefly glance at a few. Passing by the observations of Delaye, the elder Foville, and Legalle-Lassalle; softening of the middle layer of the cerebral grey cortex, and cerebro-meningeal adhesions, were described as the most characteristic changes by Parchappe; indurated, irradiating, tuft-like expansions in the white substance, just beneath the grey, of the cerebral convolutions, by Baillarger and Regnard;—inflammation, usually causing softening, and invading the cerebral hemispheres layer by layer from the periphery to the central parts, by Belhomme;—cortical softening and cerebro-meningeal adhesion by Bottex;—induration of the septum lucidum by Conolly;—greyish-red softening, or colouration, and partial superficial induration of the brain cortex by Griesinger;—extreme atrophic attenuation of the anterior part of the cerebral grey cortex by Erlenmeyer, Bonnet, Poincaré, and Hitzig;—induration of the grey substance by Frerichs; and, with cicatrisation, in final stage, by Salomon. Discovered by Brunner in 1694, the morbid appearance of granulations on the walls of the brain-ventricles, especially the fourth; first and fully described in g.p. by Bayle (1826), and mentioned by Calmeil (1826), and Daveau (1830); occupied the attention of Rokitansky (1844), Virchow, of L. Meyer in 1862, and in 1861 were erroneously viewed by M. Joire* as invariably found in, and special to, g.p.

Prof. Rokitansky † described four kinds (not specially in g.p.): (*a*) fine sand-like transparent or cloudy-greyish-white granulations: (*b*) larger, more projecting, and finally pedunculate nodules: (*c*) membraniform, discrete, round, quasi-fenestrated, opaque, white plaques: (*d* and *e*) pseudomembranous formations.

I have frequently seen ependymal granulations in cases other than g.p.; and with or without gross local organic brain disease. One of the most marked examples I ‡ have ever seen and published was a chronic case of insanity with widely-spread atheroma of the arterial system and renal

---
* Rep. Acad. Méd., 1861; and "Gaz. Médicale de Paris," 1864, p. 528.
† "Handb. der Speciellen Path. Anat. Wien," Bd. i, p. 748.
‡ "British. Med. Journal," Aug. 31, 1878. Case I.

disease. F. Shopfhagen, also,* has observed the granulations in maniacal, alcoholic, and even in sane, subjects.

Once, bony meningeal plates were seen pressing into left ascending gyri, by Dr. T. B. Christie.†

Other parts. For the osseous state, see Chapter IX, on bones; for the aural, see othæmatoma.

General frame and Viscera. The following particulars are entirely derived from necropsies made by myself on G.Ps.

Emaciation of the frame was well marked in 26 p.c., slight or moderate in 20 p.c. :—in 44 p.c. nutrition was fair or good; in 2 p.c. the frame was very fat; in 8 p.c. the transcribed notes are silent on this point.

Rigor mortis was marked in degree in considerably more than one-third of the cases; in more than one-third moderate, in one-sixth slight, in about 5 p.c. absent.

Rigid contraction of the limbs, or of some of them, continuing after death was observed in 9 p.c. *Cadaveric hypostatic lividity* was slight in 47 p.c., moderate in 22 p.c., marked in 31 p.c. Scarred or bronzed shins, herpetic cicatrices on trunk or thighs, blebs on limbs, are each occasionally seen. And once I met with transposition of viscera, both thoracic and abdominal.

Heart. Pericardial fluid is somewhat increased in one-third of the cases. Patchy fibroid thickening of the visceral pericardium is occasionally seen. Blood : usually, the right chambers of the heart are full, the left ventricle is nearly empty. The cardiac clots are softish, occasionally firm, rarely is the blood entirely fluid. The heart-muscle is more or less softened and unduly flabby or friable in about two-thirds of the cases. Usually of dull aspect, its hue in some cases is pale or slightly yellowish. One or both of the *valves* of the left side of the heart are altered in at least two-fifths; increased thickness, opacity, atheromatous and calcareous changes, are by far the most frequent; but vegetations, cohesions, valvular obstruction or incompetency, are occasionally seen. In 2 p.c. was marked dilatation of the heart; and in 8 p.c. marked hypertrophy, usually of left ventricle, occasionally of both left chambers. I have not included cases with slight hypertrophy here; this is frequent, and may affect either side or both sides of heart. In 80 p.c. the internal surface of the aortic arch was more or less

---

* "Jahrb. für Psych.," iii ; "Amer. Jl. Neurol. and Psych.," Feb., 1882.
† "Lancet," Aug. 2nd, 1879.

thickened, rugose, atheromatous ; or occasionally nodular, or cicatrised, or even (if calcareous) eroded. In about half the cases, one or both of the coronary arteries, especially the left, were noted as being more or less atheromatous. The *average* weight of the heart was $10\frac{1}{2}$ ozs. (male cases, most of them 30 to 40 years of age).

Heart.			
Unduly flabby, friable, or fatty		in 64	p.c.
Its muscle discoloured		,, 22	,,
Contracted		,, 6	,,
Dilated, considerably		,, 2	,,
Much hypertrophied (part or whole)		,, 8	,,
Changes of aortic valves		,, 30	,,
,, ,, mitral ,,		,, 26	,,
,, ,, left ,, (one or both)		,, 40	,,
,, ,, right ,,		,, 8	,,
,, ,, coronary arteries		,, 46	,,
Aortic arch, atheromatous or nodular, thickened, cicatrised, rugose, or eroded		,, 80	,,
Pericardial fluid considerable		,, 2	,,
Marked fibroid patch on heart		,, 8	,,
Heart and aorta healthy		,, 4	,,

In one case, granular clot in right ventricle, and in right branch pulmonary artery.

In one case, ulcerating aortic valvulitis.

Lungs. The following quotation is *verbatim* from page 117 of the first edition of this work, and is inserted here inasmuch as since its publication an erroneous statement * has been made with regard to the lung-affections in g.p. *viz.* :that it was " impossible to find on record any definite information as to the relative frequency of their different varieties."

p. 117, 1st Ed. "*Lungs.*—Old pleuritic adhesions or pleuritic thickenings in two-thirds. Hypostatic congestion, or marked congestion and œdema of bases, in more than two-thirds. Anterior emphysema in half. Some serous fluid in pleura in nearly half : much watery or frothy secretion in bronchi in one-third : occasionally, enlarged bronchial glands : calcareous nodules from former [obsolescent] caseation in one-sixth : fibroid change and cirrhosis, limited to one apex, or to both apices, in one-sixth. The foregoing changes are either chronic, quiescent, and comparatively innocuous, or connected with the mode of dying.

"But active, and even grave, disease of the lungs is not uncommon in general paralysis. In one-third of the cases there was more or less pulmonary tuberculosis, occasionally there was ordinary caseous (catarrhal) phthisis; in one-third, marked hypostatic pneumonia; and, in another one-

* " Brain," Oct., 1883, p. 317.

Morbid Anatomy.—Lungs. Stomach and Intestines.

fourth, a form of lobular pneumonia; both, occasionally, found with or passing into slight local gangrene: in one-sixth, recent pleurisy. Average weight of right lung 31·78 ozs., of left, 28·61 ozs. The right lung was the heavier in 78 per cent., the left the heavier in 22 per cent."

On comparing this with a larger, and partly new, group of cases it is only necessary to modify it for accurate application to the new group so far as to say that the anterior emphysema (found in half) was slight and recent only, in most of the cases, and was only marked in degree in one-seventh; that in fewer than the old group was there fluid in the pleural cavity; that cicatricial traces existed in only one-eighth, cirrhosis and fibroid change in one-seventh; tubercle in one-fourth; and that in one-fifth were traces, usually slight, of recent pleurisy.

Lungs.	Recorded as		
	"Lobular" pneumonia in	24	p.c.
" "	"Hypostatic" "	22	"
" "	"Pneumonia" "	12	"
	Congestion	64	"
	Œdema	26	"
	Slight gangrene	·10	"
	Tubercles, caseous masses, or cavities	26	"
	Traces of former cured tubercle	12	"
	Bronchitis or marked bronchial congestion	22	"
	Marked emphysema	14	"
	Induration or cirrhosis (in 4 p.c. with marked cicatrices)	14	"
	Recent pleurisy (mostly slight)	20	"
	Fluid collection in pleural cavity	12	"
	Old pleuritic adhesions	64	"
	Markedly pigmented and enlarged bronchial glands	2	"
	Collapse upper part of a lung	2	"
	Cartilaginoid plates in pleura	2	"
	Bronchiectasis, marked	2	"
	Splenified lung	2	"
	Healthy lungs	4	"

Abdomen, Stomach and Intestines. Peritonitis, or peritoneal fluid, in 6 p.c.; old peritoneal adhesions, 4 p.c. The gastric mucosa, occasionally pale (8 p.c.), was often (25 p.c.) the seat of patchy passive hyperæmia. The same either general or partial was seen in some (16 p.c.) in the small intestines; slight muco-enteritis in one case; intestinal hæmorrhage in one case; in a few, the mucous membrane of the colon or sigmoid flexure was either thickened or ulcerated (11 p.c.); or the colon fæces-laden; or either stomach or intestines, or both, swollen with gas (11 p.c.)

Liver. In about half the cases there was marked passive

congestion of hepatic veins; and in one-sixth of these the appearance was distinctly "nutmeggy." The colour, darker in 15 p.c., was paler than usual in 13 p.c. : occasionally it was brick-red or chocolate. The hepatic substance, unduly friable or flabby or partly decomposed in 'some cases (11 p.c.), was less often (6 p.c.) merely too firm; and was slightly cirrhotic in 11 p.c. In 8 p.c. the liver-capsule was thickened; in 6 p.c. were old perihepatitic adhesions to neighbouring parts; in 6 p.c. cicatrices on the surface of the liver invaded its parenchyma. In about 2 p.c., each of the following changes; brown discolouration of capsule: waxy liver; liver containing small cysts; numerous infarcts. Average weight 57 ozs.; maximum 84½ ozs.; minimum 44 ozs.

Spleen. In nearly one-half, the spleen was decidedly too firm: in a few (9 p.c.) unduly soft. Old perisplenitic adhesions, old cicatrices, old general capsular thickening, or a local thick hard cartilaginoid patch in the convexity of the capsule, were each observed in from 4 p.c. to 9 p.c.; in a few the spleen was unusually notched. In about 9 p.c. large, it was unduly small in about 11 p.c. Pale in 21 p.c., it was congested or dark in 19 p.c.; and in one case of slaty colour. In a few cases its capsule was extremely pigmented. The following were observed in one case each ; a hobnailed condition; an excessively distorted contour; embolism; caseous imbedded masses. Average weight 6½ ozs.; maximum 12 ozs.; minimum 2¾ ozs.

Kidneys. In 44 p.c. the kidneys showed one or several of the following changes:—markedly granular surface, adherent capsules, ordinary cystic formation; which are usually only the several parts of one and the same renal disease. But, subdividing for each of these, in 18 p.c. the kidneys were noted as being markedly cirrhotic or atrophied and granular; in 34 p.c. the capsules adherent; in 12 p.c. the ordinary cystic change.

Sub-acute or acute diffuse nephritis was recorded in 6 p.c.; and, in 8 p.c. more, a mottled condition as if of slight nephritis; suppuration, or multiple small abscesses in 8 p.c.; pyelitis in 6 p.c.; embolism in 6 p.c.; congestion of kidney 18 p.c.; anæmia 4 p.c. The following were present in from 2 p.c. to 4 p.c. each: marked lobulation; extremely thickened capsules; fatty kidney; induration (independent of "granular" change); locally cicatrised surface; old perirenal adhesions; horse-shoe kidney; renal calculus.

298 *Morbid Anatomy. Microscopical. Blood-vessels, etc.*

In one case the calculus was surrounded by pus, and the kidney was atrophied. In one case, beneath a scar was encysted calcareous matter. Average weight, right kidney 5¼ ozs.; left 6 ozs. In some, an unusual distribution, or branching, of renal arteries. In 28 p.c. the kidneys were noted as being apparently healthy.

Kidneys.		
Granular, or atrophic, cirrhotic	in	18 p.c.
Capsules adherent	„	34 „
Ordinary cystic change	„	12 „
Acute or subacute nephritis	„	6 „
"Mottled" kidney (nephritic)	„	8 „
Congestion	„	18 „
Embolism	„	6 „
Suppuration or abscess	„	8 „
Pyelitis	„	6 „
Anæmia	„	4 „
Recent perinephritis	„	2 „
Old perirenal adhesions	„	4 „
Lobulated kidney	„	4 „
Thickened capsules	„	2 „
Induration (non-granular)	„	2 „
Cicatrisation	„	2 „
Old growth or deposit	„	2 „
Renal calculus	„	4 „
Double kidney	„	2 „
Healthy kidneys	„	28 „

The *Bladder* sometimes showed traces of chronic or of sub-acute cystitis, or of slight mucous or submucous ecchymosis.

CHAPTER XVI.

Morbid Anatomy, continued.

B. The Microscopical Appearances in g.p.

The Cerebrum. The microscopical appearances in the cerebral convolutions, especially in their cortical grey matter, will be described under three heads as they affect, (1) the blood-vessels of the cortex and pia, (2) the neuroglia, and (3) the nerve-cells and fibres.

I. *The blood-vessels of the grey cerebral cortex and pia.* The microscopical condition of the minute blood-vessels of the cerebral convolutions varies very much in g.p. It is influenced by many modifying conditions which differ in each case; as, for example: (a) The age of the patients. (β) The presence or absence of degenerative vascular changes independent of any connected with the morbid process causing mental disease, and due to causes such as alcoholism or renal disease. (γ) The duration of the g.p. (δ) The mode of preparing the sections. (ε) The sites from which the specimens are taken; the vessels often being much more diseased in some cortical regions than in others, or their morbid changes, occasionally, differing in kind, also, in various regions. And what is true of the vessels in this relation applies, in part, also to the nerve-cells and to the neuroglia.

The *pia* shows dilated thick-walled vessels, is beset with numerous nuclei, its vessels, in some cases, blending with the cortex by a rich nuclear growth and connections with their walls, and cellular formation in their adventitial spaces. Obersteiner observed heaplets of fat-granules and clumps of red brown pigment in the vessels of pia, as well as of the sclerosed part of brain: in the brain, where œdematous, many lymphoid bodies, and apparently amyloid bodies, intimately coherent with the pia. Voisin sometimes observed crystals of hæmatin, or heaps of hæmatosine, necrosed pigmented dilated capillaries, abundant fibrillar tissue; and many nuclei around the vessels.

Cerebral cortex. In some cases the blood-vessels are comparatively little affected.

1. *Full of, and distended by, blood-corpuscles.* This is not a constant appearance, nor found in all parts in the cases in which it exists. I have found it sometimes very marked. Meschede, H. Schüle, Hitzig, Lubimoff and others have mentioned it. Or the vessels may contain opaque masses (Schüle), or yellow masses (Mendel), or free fat globules (Bonnet and Poincaré). I have frequently seen the gorged cerebral capillaries filled with plugs of blood-corpuscles in g.p., especially in those dying after special seizures.

2. *Increase of nuclei on walls of blood-vessels.* This, and the rich nuclear formation in the cerebral substance, seemed to me to be the most frequent of the microscopical deviations from the normal observed in g.p. The former is often to be seen when the nerve-cells still appear fairly healthy (at least under moderate powers). The proliferation (multiplication by division according to Lubimoff) may affect the nuclei of all three coats of the vessel, yet, on the whole, affects more particularly the adventitia, and is more marked at the bifurcation of the vessels. Sometimes the multiplication of nuclei (or what has been taken to be such) is so extreme that the whole structure of the wall of the vessels seems to be nuclear; these nuclei, blending with each other and the

original vessel-wall, may transform the vessel into a thick vitreous tube, which may become a formless mass; or undergo fatty change, or sclerosis. By these changes, the vessel-wall may be compressed, distorted, and its calibre here and there lessened or obliterated. This large collection, surrounding the vessel, at first, by a muff of rounded bodies, gave rise to the other chief view on this point, namely, that they are lymph-corpuscles. In the "perivascular lymph spaces" Obersteiner found a deep-tint-taking envelope, which he deemed to be lymph-corpuscles, about to organize into stellate connective tissue bodies, and form a network of fibres, which permeate, compress and destroy the brain-substance.

The truth appears to be that the one view need not exclude the other, that a collection of lymph-corpuscles in the sub-adventitial space may be associated with, may even give rise to, a nuclear proliferation of the walls of the minute vessels. Increase of nuclei of vessel-walls in the brain-cortex has been observed by Magnan, H. Schüle, Lubimoff, Hitzig, Luys, Major, Mierzejewski, Mendel; but Obersteiner could not convince himself of its existence, and L. Meyer found the new cell-growth to be independent of the nuclei of vessel-walls, which, indeed, were unchanged, or, later, underwent the changes of regressive metamorphosis only.

Here may be mentioned a spiny, prickly condition of the vessel-wall. As Mierzejewski stated, the capillaries may put on this appearance either owing to fine appendices, or to attached and ruptured bundles of fibres, or to the adhesion of cells of interstitial tissue to the vascular walls. And Rindfleisch described the wall, thickened by a fibrillar nuclear substance in which come fine, feebly-refracting, protoplasmic processes, and give a prickly appearance to the vessel. This appearance in most cases is probably the result of the attachment, to the vessels, of processes of pencil cells.

3. *Collections in the lymph spaces. Changes in sheaths of minute vessels.* The situation, the artificial or normal nature, and very existence, of lymph spaces have been matters of dispute. One space is sub-adventitial—another perivascular; the latter has been variously described, and its existence has been denied, at all events in the healthy brain. Unless he defines it, therefore, it is not always obvious what a writer means when he speaks of these spaces in g.p., or what the situation of the change he describes.

Lymph-corpuscles, collected in the lymph conduits of the blood-vessels, may find their way through the narrower paths permeating the surrounding tissues, and reach the pericellular spaces. Thus the vascular condition, independently of any engorgement or compression, may immediately modify the cortical nerve-cell in nutrition and function. And any change affecting the lymph-circulation must be of enormous importance.

Various conditions of sub-adventitial space have been described by different observers :—dissecting aneurysm ; or the spaces widened, filled with lymph corpuscles, red blood corpuscles, or these undergoing pigmentary change, pigment granules, collections of cells, homogeneous yellow masses here and there; or round opaque shining spheres, either homogeneous, or showing traces of being formed by fusion of hæmacytes. Adler observed cyst-like dilatation of sub-adventitial capillary space.

In the perivascular spaces have been described brown pigment and yellow crystals within the spaces; in the early stages an assemblage of nuclei ; in the later, leucocytes and red blood-corpuscles ; or an investing muff of lymph corpuscles (as already described) ; or hæmatosine granulations in the dilated spaces.

Here may be mentioned thickening of investing sheaths of vessels, often appearing like fusiform dilatations, (L. Clarke) ; dilated perivascular spaces or sheaths (Schüle, Marchi).

4. *Deposits in, and changes of, vessel-walls.* I have frequently observed molecular or pigmentary deposits in the wall of arteriole or capillary. In these walls, also, hæmatin crystals, hæmatoidin brownish or yellow pigment granules, yellow crystals, fatty granulations, one or other or several (Schüle, Obersteiner, Major, Bonnet, Poincaré, Voisin). A more general fatty, or atheromatous, or calcareous, or simply rigid state of vessels, whether as a final change, whether, in relation to other changes, primary or secondary, has been described, as by F. Meschede, L. Meyer, S. Wilks, Mierzejewski, Hitzig.

5. *Thickening*, or irregular thickening, of walls of minute blood-vessels, I have found not infrequent (also Schüle, J. P. Gray, Marchi). See, also, previous description of sclerosing changes, and of deposits in walls.

6. *Dilatation of blood-vessels;* irregular; fusiform; spherical aneurysmal dilatation ; dissecting aneurysm. Some or other of these are frequently seen. Capillary aneurysms being divided by Arndt into three forms; one of these, namely dilatation of the adventitia only, often,—and as Mendel says,

—represents a dissecting aneurysm, and this affecting the walls of the finest veins, and, producing fusiform, dull, reddish swellings of the vessel, causes an appearance as of punctate hæmorrhage. The mode of production of these dilatations has been variously explained. Thus, L. Meyer describes the vessels as becoming blocked here and there by some of the changes already mentioned, the circulation being thus interrupted, and under the effect of sanguineous pressure in neighbouring permeable vessels, ectasies and varices—true microscopical aneurysms—forming; or, as a later phase, the dilatation occurring in fatty or calcareous vessels. On the other hand, the degeneration or chronic inflammatory process in the aneurysmal vessel-wall itself, has, by some, as Dr. J. Luys, been made of more importance in the causation of dilatation. In the senile form of g.p., Seppilli and Riva observed minute aneurysms of arterioles of cortex.

The dilatations are fusiform, or spherical, or moniliform. They may be connected with minute dissecting aneurysm. On dilatations, consult Rokitansky, F. Meschede, L. Clarke, Arndt, L. Meyer, H. Schüle, Major, Mierzejewski, Luys, Voisin, Mendel, present writer, et al.

7. *Tortuous, varicose vessels, chiefly capillaries.* Indissolubly connected with the preceding is the varicose condition sometimes present, the capillaries chiefly being described as tortuous, or serpentine, or curved, kinked, varicose, or looped, twisted, doubled, or elongated, (W. H. O. Sankey, F. Meschede, Major, Mierzejewski, present writer).

7a. In relation to several of the preceding changes may be mentioned capillary rupture and minute ecchymoses about the vessels. Or, around the vessels, hæmacytes, or scattered hæmatin crystals, or blastemic effusion; or hyperplasia of oval, round, or fusiform embryoplastic elements (Marcé [for capillaries], Voisin); or vast collections of fatty granulations (Bonnet, Poincaré).

8. *Colloid degeneration of arterioles and capillaries.* This is not special to g.p. Colloid vessels are thickened, of glassy and nearly homogeneous, or slightly granular, appearance, sometimes yellowish, glistening, translucent; and some have spoken of it as a waxy appearance. The thick, quasi-homogeneous, shining, translucent walls in some cases show marked differences under transmitted or reflected light. In some examples a considerable proliferation of the nuclei of the vessel-walls accompanies the "colloid degeneration."

Cases and observations coming under this wide head have been published by Magnan, Arndt, Schüle, Lubimoff, Mierzejewski, Hitzig, Voisin, Mendel, *et al.*

9. *Amyloid degeneration of the arterioles*, was described by Dr. Tigges, was mentioned in some older observations by Dr. H. Schüle, and mentioned also by Hitzig. It has not been seen by Mendel, Westphal, or Simon, nor have I ever observed it.

10. *Obliteration of vessels.* Long ago described by Wedl, Marcé and by Schüle, and more recently by L. Meyer, Luys, and Voisin, but denied by Bonnet and Poincaré, this obliteration has usually been attributed to proliferation, sclerosis, and contraction, of the vessels, or of their outer tunic, or occurring external to it; the newly-formed material constricting the vessels here and there, either occluding them directly, or by the intermediation of thrombosis induced before the distortion proceeds so far; or clusters of cells piercing the internal wall of the vessel and obliterating its lumen.

11. On the contrary, the *new formation of vessels* has also been described, and variously. The formation of new capillaries may appear to begin as buds or appendices on the old capillaries; or branches of the hyperplastic and hypertrophic stellate cells of the cortical interstitial tissue, attached to the vessels, may be observed as becoming larger, and either taking the direction of the capillary and enlarging progressively so as to be confounded with it, or becoming radii as if from a centre of capillary ramification; or, again, anastomosing with branches of other cellules to form a capillary vessel. The enlarged attached process may present a conical hollow, continuous with the vessel lumen, which hollow, increasing in size and length, forms part of the new capillary. But Marchi* stated that the new capillaries originate from buds of cells on the walls (a proliferation of capillaries as it were), and not, as has been asserted by others, from connective-tissue cells. Assumed by Mettenheimer, described by L. Meyer, Hoffmann, H. Schüle (probable),† Voisin, Lubimoff, Mierzejewski, and Mendel, the occurrence of new formation of vessels was a condition of which Obersteiner and Westphal failed to convince themselves.

12. *Vacuoles in brain substance.* Found in any part, these are especially observed in the basal ganglia and pons. The

* "Rivista Sperimentale di Fren.," &c., Fasc. iv, 1883.
† "Allg. Zeitschr. f. Psych." xxxii Bd., p. 597.

condition is often artificial, produced by hardening solutions or spirit, and, under these circumstances, I have often seen gaps which did not exist in the fresh state either to the naked eye or microscopically. When not artificial, various explanations have been given of the origin of vacuoles, particularly in the form usually termed the "*état criblé;*" as that they are due to dilatation of sub-adventitial lymph-spaces; or, again, to dilatation of perivascular lymph-spaces, or of pericellular. The retention and local accumulation of nutritive or of waste-laden fluids; the compression and consequent atrophy of the surrounding cerebral tissues, easily explain the sieve-like appearance. L. Clarke, in one case, mentioned vertical fissures and oval slits containing blood-vessels, and more manifest in the white than in the grey cerebral substance. Elsewhere, he described vacuoles in the white substance of the cerebral convolutions, but occurring also in the optic thalami, pons Varolii, and anterior pyramids of the medulla oblongata—vacuoles of round, oval, crescentic, or somewhat cylindrical shape; in size from that of a grain of sand to that of a pea; having no lining membrane, and being sharply cut out of the tissue—vacuoles which are empty, or contain blood-vessels, or *débris* of blood-vessels, or hæmatoidin, and which probably represent perivascular spaces which originally contained blood-vessels surrounded by sheaths, and which subsequently became empty by the destruction and absorption of those vessels.

This seems not to be decidedly dissimilar from the condition next to be mentioned, that of:

Cystoid degeneration—a form of vacuoles in the brain described by Dr. L. H. Ripping; also by Dr. Wiesinger, and others. These the former described as larger than the vacuoles of the *état criblé*, as being confined to the grey cortex of the frontal and parietal lobes, and as differing in appearance from the last-mentioned change. He believed them to be retention-cysts, formed by local dilatations of the perivascular spaces.

Dr. G. H. Savage described two cases of g.p. in which "not only were cysts or cavities found in brain and cord, but also in lung, liver, and kidney." These were thought to have existed in the fresh state, though not so completely as in the hardened.

See, also, description of gaps in neuroglia, below.

II. *Neuroglia.* One cannot here delay to note the controversy that has taken place with regard to the nature of the so-called neuroglia, whether, as usually taught (Virchow, Kolliker, Rokitansky, Hayem, Magnan, Luys), it is a connective tissue : or whether, as Prof. Robin long ago affirmed, and as Loewe believes, it is a nervous tissue. Nor that as to whether it is finely granular and fibrillar, and bestrewn with nuclei or corpuscles; or whether it practically, if not solely, consists of stellate cells with numerous branching and communicating prolongations, which form a close meshwork of delicate fibres.

In the midst of the tissues I have sometimes seen pigment granulations, and various observers have noted hæmatoidin granules, and relics of collections of hæmacytes.

1. *Hyperplasia and hypertrophy of neuroglia.* A diffuse relative increase in the amount of neuroglia is a common occurrence, and sclerosis is the ultimate condition in which this, with several of the changes about to be mentioned, ends. Rokitansky stated that the hypertrophied connective tissue substance gradually passes from the condition of a viscid, living moisture to that of a stiff fibrous mass, disorganizing the brain-cortex, which contains some *débris* of nerve-fibres changed to colloid and amyloid bodies; and the ganglion-cells, loosened from their connections, are swollen, filled with fat-granules, transformed into colloid bodies. In the medullary brain-substance are foci of a greyish, viscid, semi-fluid material, containing imbedded nuclei; here and there are some soft, clear, nucleated cells, besides *débris* of the brain-medulla.

Broadly speaking, these views were supported by the researches of Magnan (1866), Sankey (1864), Schüle, L. Meyer, Luys. Thus Dr. Magnan described, as the constant lesion in g.p., an abundant nuclear proliferation in the interstitial tissue of the whole brain and on the walls of the minute cerebral vessels;—the morbid change in the interstitial tissue being primary, and that of the parenchymatous elements of the brain being consecutive. Dr. H. Schüle observed the same, and a dense granular fibrillar condition of neuroglia, and Dr. Luys found great increase of its fibrillæ. Dr. Joseph Wiglesworth,[*] also, believed connective tissue hyperplasia to be the primary element. On the other hand, it was denied by Westphal, and (practically) by Bonnet and Poincaré.

The diffuse increase is shown in enhanced consistence, and opacity, often proliferation of nuclei and of fusiform, or

[*] " Journ. Mental Science," Jan., 1883, p. 475.

x

of Deiter's cells forming spider-cells. The increase is very perceptible in the outermost layer of the grey cortex, as an extremely fine network of fibres. But a light diffuse sclerosis of the brain may ensue, and the glia, losing any fine-granular appearance it may still possess, is transformed into compact, thick, brightish fibres. The vessels become profoundly altered by the new growth, the blood and lymph circulation is obstructed in parts, the nerve-cells are put out of use, and atrophy of the gyri follows.

Or the change takes another mode; scattered patches of sclerosis may result,—a form of small disseminate sclerosis for which I suggest the name " islet sclerosis " (Batty Tuke, Magnan, Schüle, and observed by the present writer). Here, also must be mentioned the miliary degeneration of Dr. W. B. Kesteven—the miliary sclerosis of Drs. Batty Tuke* and Rutherford (a disputed point).

At times, are to be seen microscopic patches which take the stain badly, and have either a ground-glass-like or fibrous appearance.

Grey plaques have been described by Adler and others, consisting of compressed nuclei, cells, fine fibrils, amorphous pigment grains and amyloid bodies.

Somewhat similar is Simon's patchy vitreous degeneration of brain-cortex, consisting of roundish, or elongated, violet or lilac patches, homogeneous, glassy, shining, at the junction of cortical and medullary substance. Microscopically, are increase of granular neuroglia, transformation of its fibres into solid cords, and destruction of nerve-cells.

2. *Increase of nuclei.* This is the most readily observed of the changes of neuroglia. These nuclei increase in g.p., often to several times their usual number, or, at all events, bodies resembling them are to be seen. They are of the most varied shapes, round, oval, pyriform, baton-shaped, hemispherical, bent, horse-shoe, irregular, and varying much in size, but usually from ·005 to ·01 mm. in diameter. They may contain a nucleus, and possess a cellular form, or they may be simply nuclei with or without one or two nucleoli. They take the carmine tint well, but sometimes some take it lightly, only.

In the *white* substance of the gyri in some G.Ps., Dr. J. Luys found swelling of the neuroglia corpuscles, increase in number of their thickened and radiating processes, thus invading and destroying the nerve-fibres, and forming a

* " Brit. and For. Med.-Chir Rev.," 1873.

sclerous closely woven tissue, here and there of areolar appearance. In the superficial regions of the cortex, also, an invading sclerous tissue, which finally encroaches upon and destroys the nerve-cells, is formed by this diffuse hyperplasia of the neuroglia, which may begin in cortex, medullary white, or cord.

3. *Spider cells.* In the interstitial substance Meynert and Lubimoff found immense branched (spider) cells : and the latter observer traced them throughout all the interstitial substance of the brain, but especially in the deeper layers of the grey matter of the convolutions. Meynert, who found them in the grey of the gyri of the cortex, and marked in the gyrus fornicatus, gave them the name plasmatic cells. They are probably hypertrophied stellate cells. Another view is that they grow from effused lymph corpuscles; another that the appearance is occasioned by connective tissue nuclei fused together by coagulated fibrine. They occur chiefly in the uppermost and lowermost layers of the grey cortex, and about the vessels, with which they are connected, as shown some time ago by Hitzig and others. They are found, also, in the basal ganglia, floor of 3rd ventricle, and tuber cinereum ; their general outline is best expressed by their name—spider. They tint well, most of them contain one excentric nucleolated nucleus; some, more than one; in some a nucleus is scarcely, or not at all, distinguishable. Delicate, and perhaps non-branching, in the normal state, the processes of what become spider-cells in the brain of the G.P. are now thickened, branched, elongated, and are, perhaps, the basis for the formation of new blood-vessels. Eventually, they may undergo atrophy : their processes become a fibrous network. They have been observed in g.p. by Meynert, Obersteiner, Lubimoff, Mierzejewski, Hitzig, Mendel, Wiglesworth and others, including the present writer (1st Ed.). They are not special to g.p. I have also seen them in chronic local syphilitic cortical cerebritis ; and in connection with local destructive, and, in parts, slightly indurative, change following traumatic cranial and cerebral damage.

Mem. A word on the above three conditions, conjointly, which are intimately connected, if not indissolubly allied. Mendel, noticing the similarity of the nuclei of spider cells to the nearest free nuclei, thinks these free ones are formed in the spider cells and then extruded. The increase of nuclei is sometimes so obviously independent of spider cells that it seems to me to be quite unnecessary to adopt that view; rather, I would suggest that the relation (if any) was the reverse.

Obersteiner did not convince himself of any change in the neuroglia in g.p., all the alterations held by others to be of this nature flowing from the effused lymph-corpuscles and the changes undergone by them. For, in the fresh state, in the œdematous parts of the brain, were many lymphoid bodies; in the hardened state these parts showed deep-tint-taking bodies, enveloping the vessels, and deemed to be lymph corpuscles. In the sclerosed parts were free corpuscles; also, connective tissue corpuscles with many branching processes, forming a network, which enmeshed nerve-cells and vessels; under the pia, a layer of these fibres formed a connective tissue felt-work and was attached more deeply to connective corpuscles. At first ranged in or near the perivascular and pericellular spaces, or along the course of nerve-fibres, the lymph corpuscles next organize into connective tissue bodies, become stellate, and send out ramifications which grasp and destroy nerve-elements.

In some cases Dr. Mendel* and Dr. F. M. Cowan of Holland, have found the neuroglia unaffected, or almost so, and the changes confined to the nerve-cells and blood-vessels. Thus, the former found pericellular spaces containing flaky yellowish material and lymph bodies, and in parts dilated. Some ganglion-cells with an excentric situation, a very indistinct nucleus; or atrophy. Small veins gorged with hæmacytes, their adventitial spaces dilated, and filled with the same material as the pericellular. Endarteritis of Sylvian branchlets.—Burlureaux had previously reported a case due to arterial and capillary changes, with varicose nerve tubes, and no neurogliar lesion, though the grey was thickened in parts.

In the *second* stage of g.p., Mierzejewski described in the white substance, islands of amorphous opaque homogeneous appearance, ill-defined at the margin, of regular form, of unequal rough vacuolated surface; tinted strongly with carmine, and imprisoning nuclei, or having the latter attached to their borders. Radiating filaments pass from these and form a network in the interstitial tissue; although he thinks differently, the islands he describes are probably spider cells, agglomerated and perhaps unnucleated.

Voisin, following Robin, denies the existence of the

* Medical Society of Berlin, Feb. 14, 1883.

neuroglia as a connective tissue, and he finds in the intermedial substance of the cortex an infiltration of blastema, by transformation of which a network of nuclei appears along the course of the vessels, and, therefore, developing from the blastema at the vascular parietes. Of these, the nuclei lying in, and immediately about, the vascular walls may either remain free nuclei, or may become fibro-plastic cellules—cellules of connective tissue—or connective tissue fibrils. Others, developing in the depth of the intermedial substance, take on the characters of its normal *myélocytes*.

Thus one group of observers traces the sclerosis, etc., to an increase of the normally existing elements of neuroglia, and the effects of this upon the other tissues of the brain; another group to the exudation of lymph-corpuscles, or blastema, and the changes subsequently undergone by these. The probability is that both these factors concur in producing the ultimate result.

Two hypotheses may be put forward to explain the atrophy of brain;—one, that the connective tissue, increasing and contracting, squeezes and destroys vascular and nervous elements, and thus occasions atrophy of the brain; the other, that the interstitial change acts only in conjunction with other conditions, and atrophy partly follows as permitted by the defective blood-supply, and by degeneration and absorption of the other elements, occurring in the later stages.

4. *Adhesion of pia to cortex.* This was attributed by Wedl to penetration of the cortex by the increased grouped nuclei of the adventitial tunic of pial arterioles and tunic of the pial venules; by Mettenheimer[*] to permeation of brain by connective tissue from pia; by Besser [†] chiefly to proliferation of neuroglia and vessel-adventitia,—although dryness of tissue, defective cohesion of cortex, or losses of blood after exudative processes, may act in some cases;—by Marcé to changes in embryoplastic elements of capillaries. Bayle's original view attributed it to arachnitis, with consecutive encephalitis. Adler made it due to hypertrophy of connective tissue cells of outermost cortical layer, and increase of their processes in number and size. Mendel's view is that by pathological swelling of spider cells, increase of nuclei and of formed intercellular substance, the outermost layer of brain-cortex is swollen. Simultaneously, dilatation of vessels exists in the pia, and hindered circulation, the

---

[*] " Ueber die Verwachsung der gefässhaut des Gehirns mit der Hirnrinde."
[†] " Allgem. Zeitschr. für Psych.," xxiii Bd., p. 331.

elasticity of the walls of these vessels is lessened or lost in consequence of the nuclear increase in the adventitia, &c. From this condition of pia and cortex the space normally existing between them is obliterated, and the friction between pia and cortex leads to exudation, and nuclear proliferation in the pia, which last passes into and incorporates itself with the cortex; a condition favoured by the feltwork resulting from increase of the intercellular substance.

5. *Ependymal granulations* (also granulations on external arachnoidal surface). Ependymal granulations mainly come from hyperplasia and hypertrophy of the connective tissue, and according to Magnan, with whom Mierzejewski much agreed, result from ependymitis associated with periependymitis, forming one of the starting-points of diffuse interstitial encephalitis in g.p., and accompanied by irregular accumulations of proliferated epithelium of the ventricle. Ripping also attributed the granulations to ependymitis.

6. Final *atrophy* of so-called nuclei of neuroglia, and *of neuroglia* generally. Sometimes in the last stage the nuclei undergo atrophy, and become angular (Mierzejewski), a condition I have seen in very long-lasting cases with extreme brain-atrophy. Final atrophy of the neuroglia generally, is reported by various observers.

7. *Colloid bodies* (already mentioned) I have observed occasionally scattered amidst the elements of the cortex.

8. *Amyloid bodies* (already mentioned) are spoken of by some, as Mendel, as much increased in the ependyma ventriculorum of G.Ps.

9. *Fatty particles*, free, or in the individual tissue elements, are sometimes seen in advanced cases.

10. Schüle described bodies, not only within the subadventitial sheaths but also out-wandered amidst the elements of cortex, and here and there larger from confluence of red hæmacytes, and becoming large, opaque spheres, with separated pigment grains; and, finally, dull flat homogeneous discs.

11. *Holes.* The *géodes* described by Dr. J. Luys, and supposed due to changes in connective tissue, were stated by Prof. Robin to be artificial. In a case, Schüle found the meshes of neuroglia large, and holes in the cortex of various sizes and shapes, and supposed due to choked lymph conduits, inundation of the sponge tissue of the neuroglia, and dilatation of the normal spaces, and of the lymph spaces about the vessels.

III. *Nerve cells and fibres of cerebral cortex, and fibres of subjacent white substance.* The cortical nerve-cells are not always obviously diseased in early-dying cases of g.p. Meyer looked upon them as healthy in the early stages; Obersteiner found them healthy in œdematous brains of G.Ps., or in the œdematous parts of brain. I have seen them looking quite normal in the early stages; and Bonnet and Poincaré state that they found the nerve-*fibres* healthy in g.p. (but see below).

1. *Granular degeneration; fuscous; fatty-pigmental; granulo-fatty; pigmentary, degenerations,* may be considered under the same head. They are, usually at least, degrees and phases of one and the same morbid process. The protoplasm of the cell becomes cloudy and flecked by dim, opaque, partly glossy, strongly light-refracting granules, which, at first collected in parts, particularly about the base and nucleus, eventually affect the whole cell, and gradually obscure or destroy the nucleus and nucleolus, disintegration in the former beginning at its centre; the fibrillar and finely molecular appearance of the nerve-cells is lost; and as they become filled with granular substance, or yellow and brown pigment, they gradually cease to be stained with carmine, their nucleus disappears, and the prolongations of the cells become granular, indistinct, fragile, and at last are apparently destroyed or disappear, that of the axis-cylinder at the base proving the most resistant to change. The cells thus affected lose their normal contours, become ill-defined, somewhat rounded or pear-shaped, and may have, or not, an inflated appearance. In these podgy cells, filled with a finely granular substance, a nucleolus is often quite perceptible after the nucleus has become entirely obscured. And now comes the condition of molecular destruction of nerve-cells (Meynert), their reduction to molecular detritus (Meschede), the state described by L. Clarke: the nerve-cells fill up with granules or pigment; or, losing their sharp contours, look like a heap of particles ready to fall asunder; some are partially disintegrated; and here and there are irregular masses of fatty or other particles, scattered over areas of variable extent.

According as yellowish or brown pigment granules; or uncoloured cell-contents; or faintly yellowish fat-like molecules predominate, the change often receives different names, and some pointedly specify fat granulations unassociated with any pigmentary change or increase. Obersteiner also saw cells pigmented, yet staining too much. But it seems scarcely desirable or accurate to divide what has been described above into three morbid changes, respectively pigmental, granular, and fatty, degeneration. Yet in many cases the so-called fatty granulations are not fatty, and while, with similar appearances, some find fat, and others not, within the cells; it is well, perhaps, to retain the variety of names as pointing to some difference of detail, though probably not of essential nature.

2. *Sclerosis of the nerve-cells.* This was described by Meynert as sclerosis or sclerotic swelling, and as part of a series of changes following proliferation of the nuclei of the cortical nerve-cells. I have often found nerve-cells of dull dimmed appearance, and indistinct nuclei; which took carmine stain badly. Some describe cells of normal contour, stain-taking badly; their nuclei shrunk, and processes sharply broken off. Another condition of sclerosis is that connected with atrophy of nerve-cells.

3. *Atrophy of nerve-cells.* Particularly in advanced cases of g.p., the large cortical nerve-cells, having passed through a series of changes, are often found with dimmed protoplasm, shrunken, and encircled or not by vacuoles, or manifest pericellular spaces. The atrophy may be simple, and throughout the whole cell, or the body of the cell may shrink around the large and slightly tinted nucleus. The atrophied cell may apparently be sclerotic or not. The cells and their nuclei become condensed, reduced in size, and the nucleus indistinct, the cells assume a somewhat homogeneous, waxy appearance; at last there is a shining, angular, slender, or distorted body. Or, atrophied cells with glistening nuclei are seen (Obersteiner). Limited by some to the second layer in the type of five layers, and in which the nerve-cells may even disappear, it has by others been found chiefly in the third. Hoffmann derived the atrophied nerve-cells from fatty pigmented cells whose fat and pigment have been absorbed, and such of the protoplasm as is still remaining has contracted around the now pyriform nucleus.

It is especially with atrophied nerve-cells that are observed the pericellular spaces, whether normal, or artificially produced in preparation, or due to retraction of tissues from the pathological changes in the brain. Ordinarily clear, they may, on the contrary, in g.p., contain nuclei, or bloodcorpuscles, or coloured exudation material.

4. *Nucleus of nerve-cell.* Besides the changes already

mentioned above under "atrophy" and other heads; the nucleus, in the early stages, sometimes gets bigger and rounder, nearly filling the nerve-cell, its nucleolus indistinct or disappearing; or it becomes pyramidal, angular, or elongated, or of irregular outline, perhaps displaced, of shining appearance, or loses nucleolus.

5. *Increase of nuclei of nerve-cells.* Tigges described an increase in number by subdivision of the nuclei of the ganglionic nerve-cells of the cerebral cortex in g.p. And Meynert in some observations, not of recent date, described proliferation of these nuclei, simple or multiple nuclear division, etc. Hoffmann, Mierzejewski, and Mendel deny any such proliferation in g.p.

Lymph-corpuscles may go from the wider lymph-spaces, and passing through the narrower paths, may gain the pericellular spaces, collect about a cortical ganglionic nerve-cell, and give rise to the appearance interpreted by some as subdivision of the nucleus.

6. The *prolongations* of the nerve-cells, may become corkscrewy; unduly thin; unduly thick; rigid and glistening; or granular, fragile; or disintegrated, destroyed, invisible.

7. *Swollen or hypertrophied nerve-cells.* Meschede, with whom Hoffmann agreed, described what he called parenchymatous inflammation of the ganglionic bodies in g.p.; these becoming softened, swollen, isolated, in the early stage. Rokitansky described some of the nerve-cells as inflated; Meynert, dropsical swelling of these cells, shown by their increased size and hyaline aspect, and by the dark edges of their nuclei; Obersteiner found some in the deeper layers swollen; Batty Tuke and Rutherford stated that the nerve-cells in g.p. were sometimes hypertrophied; Mierzejewski found them occasionally increased in size and filled, or not, with a pale granular material soluble in ether; Major found immense cells in one case in the mid-depth of the cortex, in two others an irregularly inflated appearance of some cells; in the second stage, Huguenin found the nerve-cells swollen, enlarged, pyriform.

When not merely inconsiderable, and a phase in the granular degeneration described under section " 1," I think most of these were simply cells of the giant kind normally existing in the motor zone. When doing microscopic work I never felt certain of a pathological hypertrophy of a cortical nerve-cell in g.p., though some degenerate cells look big.

8. *Formation around nerve-cells and nerve fibres.* J. Henle

and Merkel's observations throw some light on this subject. The hypertrophy of connective tissue amidst, and its action on, the nervous elements; the embryoplastic nuclei permeating cortex and medullary white, and the blood-pigment and altered blood-corpuscles sometimes strewn amidst them, have been mentioned. Mierzejewski found some nerve-cells (if near blood-vessels with sub-adventitial hæmorrhage) surrounded by increased connective nuclei, which sometimes are attached to their surface; or the nerve-cells may be enwreathed by a band of fibrine. The protoplasm, also, is dimmed, and the nerve-cells are encircled by vacuoles.

9. *Calcification* of nerve-cells and similar changes. Dr. S. Wilks, in one case, found the nerve-cells altered in form and colour, and apparently containing earthy matter. In a female G.P., calcified nerve-cells were described in a local necrotic softening, together with corpora amylacea and colloid changes, by Wiedemeister.*

10. *Vacuoles* in nerve-cells were described by Adler,† with nucleoli in the protoplasm, varying in appearance.

11. *Changes in, and destruction of, nerve-fibres.* (a) Some have found the nerve-fibres tortuous and irregular in their course, or the nerve-fibres of the white medullary substance of the brain sometimes lessened and compressed by the relatively highly-stained connective tissue fibres. Dr. Franz Tuczek,‡ employing Exner's method, found that in g.p. more or less disappearance of the medullated nerve-fibres occurs in local and definite portions of the cerebral cortex. This change was artificially divided into five degrees : (1) partial disappearance of the nerve-fibres in the first layer of Meynert;—(2) disappearance of fibres in first, and partially in outer part of second layer;—(3) a more extreme condition of " (2) ; "—(4) nearly complete atrophy of nerve-fibres in the first and second layers, partial in the third;—(5) complete disappearance of nerve-fibres throughout the cortex.

The gyri most frequently affected are the rectus, fornicatus, and breves. The gyri on orbital surface, and some on the convexity, are usually affected, and of these the left third frontal the most constantly and most intensely. In the more chronic cases the first temporal g. is usually, and much, affected. The parietal lobules, in their anterior part, are affected only in some cases and slightly, and the same

---

* " Arch. f. Path. Anat. u. Phys." 1 Bd., p. 640.
† " Archiv für Psych.," v Bd., p. 374.
‡ " Beiträge zur path. Anat. und zur Path. der Dem. par," 1884.

statement applies to the central gyri and paracentral lobule. Beginning locally, often in places in the frontal cortex widely separate, the process tends to become diffuse. While leptomeningitis and interstitial encephalitis usually coexist with the nerve-fibre lesion, yet the presence or absence of the first-named is entirely independent thereof, and the second may exist without it.

With regard to the immediately *sub-cortical medullary* (white) substance of the cerebrum, Dr. Franz Meschede some years ago found grey degeneration here, in parts, in some cases of g.p. These patches and streaks were irregularly arranged, and, microscopically, exhibited disappearance of nerve-fibres, and the presence of numerous cells, nuclei, and bodies some of which were amyloid; indicating interstitial growth, and parenchymatous destruction:—or, as it seems to me, essentially the changes and the situation recently described as something new in science by Tuczek, who, in some cases, found patches of degeneration (sclerosis), transparent or greyish to the naked eye, in the medullary substance, or between it and the cortex, but not invading cortex. Microscopically, were spider-cells, increased meshwork, diminished nerve-fibres, and in parts, also, granule-cells. The change was chiefly in the frontal and in the anterior part of parietal and temporal lobes; and the ultimate result was disappearance of sub-cortical association fibres of Meynert in the anterior parts of the brain. The nerve-atrophy he thought to be primary, and not secondary to local encephalitic or other lesion. The cases were chronic, and usually had posterior spinal sclerosis.

(*b*) Amyloid bodies. Rokitansky described the growth of connective tissue as disorganizing the cortical nerve-fibres, which, in part, are changed to colloid and amyloid bodies. This formation of amyloid bodies from disintegrated nerve-tubes ("aus dem detritus der gehirnfasern") has not been always accepted; by some they are attributed to the connective tissue.

(*c*) Colloid bodies. These, also, sometimes found in g.p., have been attributed to the destructive changes occurring in nerve-fibres; also to alterations in nerve-cells (Rokitansky).

(*d*) Axis-cylinder changes. Hitzig described appearances supposed due to hypertrophy of axis-cylinders. Mierzejewski found bodies, some oval, some of riband shape, which he thought to be hypertrophied axis-cylinders, or their remains. Bodies similar to these, Hitzig thought to be produced artificially, and by cerebral injuries, and not, certainly, hypertrophied axis-cylinders. Rindfleisch looked upon them as a result of maceration, a view not accepted by Mendel for all cases.

White cerebral substance (non-gyral), corpus callosum, fornix, septum lucidum. The alterations already described in the white substance of the gyri are also more or less observed in parts of the rest of the white substance, in the corpus callosum, fornix, and septum lucidum. The last-named, often indurated in g.p., is occasionally softened. Thus, in the white cerebral substance, Voisin found changes in the vessels, and a production of numerous round or oval embryoplastic bodies, all similar to those in the cortex, but the lesions less marked in the depth of the *corona radiata* than in the superficial white.

Opto-striate bodies. Hyperæmia, œdema, clouded sections, gorged vessels, have been described in the optic thalami, but with far less infiltration and nuclear overgrowth than in the cortex: occasionally, local ampullary dilatations, or, in the later stages, sclerous change in the vessels. Many of the nerve-cells still healthy, although here, as in the corpora striata, these cells may be loaded with (fat) granulations. In the corpora striata also, but rarely, foci of hæmorrhage or softening; only slight alteration of vessels. In the geniculate bodies, endoarteritis, old yellow blood-products in and around vessel-walls: no exudation, or diffuse nuclear formation. Cells slightly altered. (Voisin.)

Ammon's horn is said by Voisin to be sometimes of a pronounced red hue. Mendel figures, in the Ammon's horn of a G.P., varicose dilatations here and there of the vessels, filled with a homogeneous yellow mass. In the tuber cinereum of a G.P., the pericellular spaces were filled with a yellowish exudation material.

Crura Cerebri. Rokitansky described the destructive affection of the cerebral cortex in g.p. as followed by an increase of connective tissue in the crura cerebri; and this, eventually, by shrinking. Rabenau traced granule-cells from the spinal cord into the crura cerebri, and Huguenin the grey degeneration of posterior spinal columns into the crura, also. Or there may be only slight vascular change; and, perhaps, moderate granulo-fatty degeneration of cell-groups.

Cerebellum. Bonnet and Poincaré found the nerve-cells of the cerebellum healthy, yet Dr. Lubimoff observed that the

Purkinje's cells were sometimes sclerosed in g.p. Also, were numerous very fine fibrils, some seeming to spring from the surface of the cells like processes, others forming a reticulum over the surface of the cell. These might be the fine processes described by Stilling; the termination of nerve-fibres in a reticulum as asserted by Gerlach and Rindfleisch; the reticulum of Golgi; or connective tissue fibrils.

In the cerebellar cortex Dr. J. Luys described marked changes in the vessels and their contents even to solidification, hyperæmia, serous infiltration, and yellow zones: and throughout the cerebellum Dr. A. Voisin found, in hyperæmic cases, extravasation of blood-globules and of embryoplastic nuclei along the blood-vessels, in their walls, and even in the nervous tissue. The corpus dentatum was little affected but, particularly in cases with epileptiform attacks, presented vascular dilatations.

Mendel as a rule had negative results here, microscopically, but he figures some ganglion-cells from the corpus dentatum cerebelli in a state of fatty pigmentary degeneration.

Pons Varolii. Some have found the nerve-cells here to be healthy: others, especially at the nuclei of origin of third cranial nerves, find fatty granulations in the misshapen notched nerve-cells; also a dense lacework of engorged vessels, the source of infiltrating embryoplastic nuclei: anteriorly, sometimes little hæmorrhagic foci, or lacunæ due to former such foci. Zones of sclerosis are sometimes seen here. I have met with patches of disseminate sclerosis. The observations of v. Rabenau and Huguenin (see crura cerebri), necessarily apply also to tracts in the pons.

Medulla Oblongata. A high interest naturally attaches to the lesions found in the medulla oblongata, in g.p.

1. Connective tissue growth. This, and consequent destruction of nerve-elements, Rokitansky first described for the medulla oblongata, as for the brain cortex. The connective growth, and fibrous transformation, in pons and bulb, may, it is said, take origin in irregularly scattered, coarse, molecular, highly-stained material. In one case, Déjerine found sclerosis of the posterior two-thirds of the pyramids of medulla oblongata. The sclerosis of med. obl. in g.p. may be in scattered patches, or it may be in zones. Round or oval nuclei and fusiform bodies may exist, especially in and alongside the walls of the vessels, and penetrating between the nerve-fibrils.

2. Ependymal changes. The connective tissue granula-

tions on the floor of the fourth ventricle have the same microscopical structure as those of the lateral ventricles. Some, as Magnan, have described the change as ependymitis with interstitial inflammation of the surrounding nervous structures. In the same locality, also, intense vascularization, and even minute hæmorrhages, may be found, and a marked cell-proliferation (Gallopain).

3. Degeneration and atrophy of nerve-cells of nuclei of origin of bulbar nerves. The fatty, or fatty-pigmentary, or granular degeneration, or an appearance as of granulations of rusty or of reddish-yellow hue, affecting the nerve-cells of the nuclei of origin of bulbar nerves, have been described, as have also their atrophy and sclerosis and sometimes complete destruction of the cell-histology. Comparatively early, the protoplasm ceases to absorb carmine well, the nuclei look spotted, the prolongations of the nerve-cells are brittle (Laufenauer). At the lower end of med. obl. and upper end of cord, W. Jessen found the nerve-fibres, ganglion cells, and their processes, all destroyed or much altered.

Investigations are discrepant as to the relative degrees in which the nuclei of the different bulbar nerves are affected. Westphal found no changes, either peripheral or bulbar, of the hypoglossal nerve. But Lubimoff observed in a few cases marked granular destruction and amyloid transformation of ganglion-cells of nuclei of facialis and hypoglossus. Voisin found the degenerative and atrophic nerve-cell changes earlier, more advanced, and extensive in the facial than hypoglossal nucleus of origin, and still more than in that of the abducens. On the contrary, it has been variously stated that the ganglionic centre of the hypoglossus is more affected than that of the facial or vagus (Gray); or than any other bulbar centre (Laufenauer); or than the facial or sixth (Mendel).

4. Atrophy is an outcome of the several changes mentioned above; atrophy of medulla oblongata; or of some of its parts (pyramids) ; or occuring in zones (Liouville). Some other changes in the med. obl. scarcely need mention.

Cranial nerves.—Olfactory nerve and bulb. The atrophied, or atrophied grey translucent condition of these, often seen in g.p., has long been known. They are often œdematous and adherent. In the bulbs, are degeneration of the cells, sanguineous effusion in the walls of the blood-vessels: in the nerves, between the fibres, and infringing upon their parallel condition, are blood-globules more or less decolorized, many connective nuclei, collections of hæmatosine in the nervous substance of the nerve, and heaps of nuclei around the vessels (Voisin).

Optic nerve. The microscopic appearances vary in different cases and stages, and with the primary or secondary nature of the atrophy of the nerve and disc. Usually, the inner sheath of the nerve and the trabeculæ are greatly thickened, a rich scattered nuclear proliferation exists amidst the nerve-bundles, some nerve-bundles are atrophied, some have almost disappeared; the coats of the vessels are slightly thickened, and around some of them is increased fibrous tissue. In more advanced conditions, the nervous elements have disappeared in parts, and, where still retained, the nerve-fibres are extremely varicose and partly granular; many granule-cells and connective tissue bundles are ranged parallel with the nerve-elements, particularly around vessels of which the bores have almost, or quite, disappeared. The fusiform and other nuclei penetrating the nerve in every direction, are here and there reinforced by islands of sclerotic tissue.

And a distinction has been made (Wiglesworth) between optic nerve atrophy in g.p. when slight chronic interstitial neuritis is succeeded by atrophy, on the one hand;—and, on the other hand, when the atrophy is primary *at the disc*, though subsequent to inflammatory destruction of nerve-fibres higher up; in the latter form, of the two, the internal sheath being more thickened, the trabeculæ less so, the nuclear and small round cell-elements much more numerous, but the nerve-fibres less degenerate. The chief microscopic investigations of optic nerve in g.p. are by Drs. Jehn, Magnan, Voisin, J. B. Lawford, J. Wiglesworth.

Third nerve (*motor. oculi comm.*). The connective tissue separating the nerve-bundles is often larger, denser, and less clear than normal. I have occasionally found this nerve sclerosed; or enlarged, reddish-grey, with thickened sheath.

Fourth nerve. In some cases sclerosed.

Fifth nerve. Atrophy of fifth nerve and Gaserian ganglion has been seen in g.p. In a case observed by Voisin, the nerve on one side had a thick red muff; in the nerve, within the pons, were vessels with nuclear proliferation in and along their walls, and, in places, amidst the fibres of the nerve.

Sixth nerve. Same changes as third nerve (*q.v.*), but less marked, and much less constant.

Seventh nerve (portio dura). For changes in its nucleus of origin, see "medulla oblongata," above.

Ninth nerve. The hypoglossal nerve may be atrophied, and present abundant fusiform bodies and embryoplastic nuclei; its roots may be atrophied, and fibrillar structure obscure. For changes in its bulbar nucleus, see above.

Sympathetic ganglia. In the sympathetic ganglia, and chiefly in the cervical, Bonnet and Poincaré describe a substitution of connective tissue and adipose cells for the nerve-cells in g.p., the other nerve-cells not destroyed are pigmented, and in the cervical ganglia the vessels may also be distended and even varicose. Pigmentation of nerve-cells of cervical ganglia; or numerous granule-cells, and a cloudy, granular state of the nerve-cells, here and in the cœliac ganglia; or thickening of cell-capsules, and multiplication of free nuclei between them; have been seen. No change appears to be constant in g.p.

Spinal Cord. Mr. Gulliver, in 1850, was perhaps the earliest to examine the spinal cord microscopically in g.p., and spoke of central softening in the lower parts of the cord, usually inflammatory, as he thought, with exudation corpuscles. Türck (1853), and Joffe (1857) found granule-cells in the cords of G.Ps.; and Rokitansky found the same connective-tissue, and other, changes in the cord as in the brain, and as already cited. Besides changes in the spinal meninges Prof. C. Westphal described others in the spinal cord itself, either (1) in the posterior columns; or (2) in the posterior sections of the lateral columns; or (3) in these two tracts simultaneously. In the *posterior* columns (grey degeneration) were atrophy of nerve-elements, growth of connective tissue in irregular plates; as a rule more marked at the periphery, especially at Goll's columns, and often more in dorso-lumbar region than in cervical. Sometimes fat-cells in fresh preparations, free, or on vessel-walls; frequently, corpora amylacea; occasionally, pale, nucleated cell-elements. In the *lateral* columns chronic myelitis, chronic interstitial connective tissue growth, formation of granule-cells. In fresh preparations, free granule-cells; in hardened preparations, the nerve-fibres surrounded by a connective network with broadened lines and knotted nucleolated points. The appearance, mainly differed from those in the posterior columns by the reticular outline, the presence of granule-cells, and absence of connective tissue plates. With reference to these changes and their import, we cannot linger to discuss the contention as to whether they are constant and characteristic in g.p., or definitely linked with some of its symptoms. Especially with regard to the presence and situation of granule-cells (or of granule-cell myelitis) in the spinal cord of G.Ps. have there been grave discrepancies of view, as to whether they were or were not characteristic or always present; as to whether, in g.p., they are in intimate relation with affection of gait, or with speech disorder, or with labial and facial tremor; as to the percentage of G.Ps. in which they could be found, as well as that of persons suffering from other forms of mental unsoundness, and of the sane; as to the ages at which they occurred; as to the coexistence with them, or not, of healthy vessels, in non-G.Ps. and G.Ps., respectively; and as to their occurrence in all inflammations, degenerations, and softenings affecting parts of the central nervous system. Connected with this, other mooted questions have been whether granule-cells are a phase through which brain-tissues pass as they undergo destruction—the visible form of physiological death—the granule-cells being the remains of cells and nuclei doomed to pass away, and granules being *débris* of protoplasm of old broken-up cells:—or whether, on the other hand, fatty granule-cells in the embryo are connected with the formation of nervous tissue, are present during regeneration of a divided nerve, and *later*, a morbid appearance; and whether they, as a phase of embryonic development, form the white substance of Schwann (Boll), or are composed of any residue of the molecular material from which the latter is developed (Jastrowitz), and are not pathological unless in large numbers. Then, with regard to the *origin* of granule-cells, spheres or masses (not necessarily in g.p.), they have variously been said to spring, in brain and cord, from fatty degeneration of walls of bloodvessels; from nuclei of neuroglia; from constituent cells forming capillaries; from nuclei of muscular coat of venules; from connective tissue nuclei about vessels, or of perivascular lymph-spaces; from spindle-cells of innermost layer of grey cortex, and probably from ganglionic cells of other layers; from connective tissue, and septa of same, between the nerve-fibres; from fatty degeneration of leucocytes; from reception of *débris* of nerve substance, chiefly nerve-medulla, into the same; from fragments of distintegrated nerve tubes, myelin making up the bulk, and axis cylinder the "nuclear" contents. (Gulliver, Türck, Joffe, Westphal, Th. Simon,

Wilh. Sander, O. Obermeier, L. Meyer, Huguenin, Jastrowitz, Boll, Cohnheim, Rabenau, Gray, Marchi.)

Westphal thought the changes in posterior columns ceased at the fourth ventricle; those in the lateral, at the foot of crus cerebri. But Rabenau found granule-cells and degeneration of the blood-vessels continued upward from the cord, through bulb and pons, into crura cerebri, radiating tract, and centrum ovale of Vieussens; and Huguenin believed that he sometimes traced grey degeneration of the posterior spinal columns into the cerebral crura; and that emigrated lymph corpuscles in the internal capsule and *corona radiata* bore some relation to the morbid process. Thus the "granule cell myelitis" would appear to follow chiefly the same track as the descending degeneration secondary to lesions of the radiations of the *crusta*, especially of the internal capsule.

Softening of cord. This may be manifest to the naked eye; but in some instances, where it was not, Dr. L. Clarke found "numerous areas of granular and fluid disintegration within and around the grey substance;" others, including the present writer, have found softening, affecting parts or tracts, or assuming a fascicular systematic form; and others, zones of granular disintegration.

Sclerosis of cord has been of the most varied distribution in different cases; sometimes symmetrically disposed bilaterally in the hinder part of the lateral columns; sometimes in the posterior columns; sometimes in zones towards the periphery and posterior roots, and chiefly affecting the white but also the grey; sometimes in patches at the deeper parts of the posterior columns, or occuring in plaques about the nerve-roots; while earlier changes are coarseness, molecular degeneration of connective substance, or abundant nuclear proliferation throughout the cord.

The nerve cells of the cord, especially of its anterior horns, have been found in various stages and phases of granulo-fatty and pigmentary degeneration and of atrophy, occasionally even to destruction and disappearance of nerve-cells. Their prolongations are sometimes rigid.

The nerve-fibres have been found atrophied and degenerate.

The blood vessels of cord often undergo fibrotic change, nuclear growth ending in the formation of connective tissue in and around their walls, sometimes almost to obliteration of their lumen; fatty and calcareous changes are occasionally seen in them.

Free fat droplets, minute hæmorrhages, have been described as occasionally in the stroma. Some have denied the existence of sclerosis, granule bodies, or fatty degeneration, in the cords of G.Ps.

The epithelial cells lining the central canal at the upper end of the cord have been found increased, heaped up in layers, or stuffing the canal. Described by Drs. W. Jessen, H. Liouville, and Ringrose Atkins, this change is not at all rare.

In the thickened pia of cord are many oval or fusiform and other nuclear elements; and hæmatin crystals. In the early stage of chronic spinal meningitis are vascularization of pia and arachnoid posteriorly; transparent granulations, composed of a fine envelope of connective tissue, connective nuclei, and thick-walled vessels. In the stellate arachnoidal bodies; fibrils of connective tissue, surrounded by carbonate of lime crystals, and, perhaps, in several concentric layers, amidst which are stellate corpuscles (Voisin). (Drs. Th. Simon, Bonnet and Poincaré, Déjerine, Ringrose Atkins, J. C. Shaw, et al., have also worked at the cord-changes.)

As concerns the changes I have seen: in speaking of the naked-eye alterations in the cord, I mentioned, as sometimes occurring in g.p., scattered patches of sclerosis; sclerosis or grey degeneration of the pyramidal tracts in the lateral columns; grey degeneration of posterior and posterior median columns; and a general diffuse change evidenced by softening or by induration. Etc.

Examining more closely, some light sclerosis is frequent; this may be general, a light chronic diffuse myelitis affecting the whole transverse section of the cord; or, falling short of this, and expending itself chiefly on the grey, may form a tract of diseased, surrounded by comparatively healthy, tissue; or some of the columns of the cord in their length may exhibit a systematic myelitis or degeneration; or the change may be at the periphery—a perimyelitis—or constituting part of a meningo-perimyelitis; or sclerosis may occur in scattered patches throughout, or in different parts of, the spinal cord.

Spinal Nerves. I have already spoken of the grey degeneration of posterior roots of spinal nerves occasionally seen in g.p., usually with grey degeneration of posterior columns. The anterior roots of the cervical nerves were found to be in advanced fatty degeneration by Dr. J. P. Gray.—In one case, with only mild spinal disease, the entire structure of some peripheral nerves was found by Déjerine to be considerably altered. Here the terminations of the nerves, subjacent to pemphigus-blebs on forearms and legs, were broken down, the axis-cylinders had disappeared, the myelin was broken into fragments; like the condition after two or three weeks in the peripheral end of a divided nerve. Mr. Bevan Lewis found the sciatic nerve atrophied in some G.Ps. There were; fasciculate atrophy of the nerve tubuli, involving both the medullary sheath and axis cylinder; defectively stained axis cylinders; increased vascularization; and connective hyperplasia. Simon and Mendel have not found these changes in the sciatic nerves.

CHAPTER XVII.

PATHOLOGY.

It is unnecessary to speak here with great fulness of the essential nature of g.p., either clinical or pathological, or of its seat.

(a). As to *the essential clinical nature,* one need only glance at the very different views maintained on this question. That g.p. is simply a complication or even a termination of insanity; or on the other hand that it is a distinct and special form of mental disease, and evidenced by symptoms both mental, motor, and sensory; or, again, that it is a distinct and special form of disease independent of, and not a form of, insanity, with which, too, was linked the view that the motor affection is the primary and essential part of the malady, the mental affection, if present at all, being secondary and accidental; or, again, that there are two distinct kinds of g.p., one with, the other without, insanity; or that it is not a distinct disease, but only an assemblage of symptoms depending on very different brain-conditions; or that it is merely a neurosis; or, finally, that g.p. is a special malady with two orders of pathognomonic symptoms —dementia and paralysis—independent of insanity, but often complicated by a special kind of insanity, which itself is neither simple insanity nor g.p.;—are views each of which has had a more or less distinguished following.

(b). Among those who hold the disease to be organic there has been a conflict of opinion as to whether the essential pathological change in g.p. is principally and primarily of a more or less frank inflammatory nature, or principally and primarily of a degenerative nature.

(c). Nor have the views as to the *seat* of the disease been less various; and some, uniting in themselves the opinions of several, have declared that g.p. arises from disease of the spinal cord, or of the brain, or of the sympathetic nerves, or from a peripheral influence on the nervous apparatus; or without any appreciable lesion of the nervous system.

1. *Starting point. G.p. by propagation.* In reference not only to the seat, or chief seat, of the disease, but, and particularly, also, to the *locality in which the disease commences* —its point of departure—it may at once be said that in the vast majority of cases this appears to be the cerebral cortex, or it and its investing meninges. Foci in the cortex are probably the starting points. I have never seen a clear case of g.p., in which, if a necropsy was made, I did not then and there find distinct evidences of disease of the cerebral cortex and meninges, even by the unaided senses. This is not true of any other portion of the nervous system in my experience; not even of the spinal cord, supposed by some to be the only, by others to be the chief, seat of organic change in g.p.

In at least the vast majority of cases of g.p., if not absolutely in all, mental defect or disorder plays an essential part. These we assign to lesions, or at least disorder, of the cerebral cortex. Many of the motor and sensory symptoms are also explainable by cerebral affection; and although some of the symptoms which go to complete the full clinical picture of typical g.p. flow from affection of parts other than the cerebrum, yet these are, rather, contributory than essential elements of that picture. See, also, " pathological physiology."

Next to the brain, the most important parts affected in g.p. are the *medulla oblongata* and *spinal cord;* which, for convenience, may be spoken of collectively, in this immediate relation, as " the cord." The brain alone may be affected; or both the brain and the cord. When both are affected :—

(*a*). The brain may be affected first, the cord subsequently; or

(*b*). The brain and cord may be attacked simultaneously (in a practical sense, not absolutely so) ; or

(*c*). The cord may be first affected, the brain secondarily (this is comparatively rare).

On the other hand, as we have said, the encephalon, or it and its tunics alone, or almost so, may be affected; the medulla spinalis being unaffected, and the medulla oblongata comparatively little.

A. Where the brain is affected first, and the cord subsequently; in different cases, the affection of the cord may be an extension of the morbid process from brain to cord, by continuity of meningeal and of nervous tissues; or irritative destruction of definite brain centres may reflexly light up inflammatory and other change in the cord, as has been produced experimentally, also; or, the brain being damaged, more and vicarious work is thrown on other parts of the nervous system, including the cord, rendering them more liable to take on morbid action. In many cases, however, none of these will afford a sufficient explanation; and for those, I think, the key lies in a law yet to be precisely formulated, but which will be ·to the general effect that organic affections of various regions of the cerebral cortex infallibly—in the sense of the circumstance-restricted or -governed infallibility of natural law—under given conditions lead to morbid changes in definite lower tracts related to them, in the brain-medulla, isthmus encephali, and cord; and this in far wider range than the known descending degenerations secondary to internal capsular lesions. So that a moderate organic cortical change throughout, say, a given cerebral lobe, brings into being parenchymatous, vascular, and connective tissue, changes, at first light and mild and not easily recognizable, in some of the lower tracts, including the cord.

I have taken, as a starting point and clue, the secondary systematic and chiefly descending degenerations or myelitic changes, which are not yet sufficiently known, even in their gross forms, and are scarcely or not at all known in their slight degrees. Like others, I have seen lesions which completely destroyed some of the convolutional surface, or some part of the corona radiata, internal capsule or crus cerebri, succeeded by descending sclerosis or degeneration, following definite anatomical tracts; or, rather, developmental tracts and regions. As the next step, I have found similar descending changes sequential to cortical cerebritis—as for example syphilitic cortical cerebritis—of moderate extent, and not comparable in grossness to the destructive lesions lastnamed. As the next step, I have found the ordinary diffuse cortical lesion of g.p. (without "focal" lesion of any kind) followed by the same sclerosis or grey degeneration from crus to cord. A notable example of this has already been published by me, in a paper on "spinal sclerosis and degeneration following brain-lesion."[*] As the last step, I

[*] "Journ. Mental Science," April, 1885. Case ix, p. 61.

have found much milder and incipient changes of similar nature in the cord in ordinary g.p., with its ordinary cerebral changes.

If the affection of cerebral cortex is light and remittent, the cord-changes secondary to it will be only a pale reflection, and promptly vanishing. But as the affection of cerebral cortex gains in intensity, and depth, and destructiveness, so the cord-change waxes.

Since the above was written I have seen the work of Dr. Franz Tuczek in which he shows the atrophy and disappearance of nerve-fibres beginning at the periphery of parts of the cerebral cortex in g.p.; and tending to involve its whole depth gradually. This seems to me to confirm my views on this subject, as pointing to a more definite lesion of nerve-fibres at the summit of the motor tracts than had hitherto been made clear; and as affording an explanation of the commencing change of a secondary and descending degeneration.

In corroboration, also, of the clinical and pathological facts I have mentioned above, are experiments, as those of Dr. Löwenthal, in which, after extensive, but not deep, removal of the so-called cortical motor area in some animals, there was secondary degeneration of the lateral spinal columns; and Prof. Schiff showed that in the ape some muscular wasting follows this secondary spinal degeneration. In this relation, also, must be mentioned the result of Pitres' investigations as to the different extent and importance of these secondary degenerations in man and in different kinds of lower animals. (See next chapter.)

B. When brain and cord are, in a practical sense, simultaneously attacked we may, I think, usually look upon it that the same pathological cause acts simultaneously on, and induces the same pathological changes in, both brain and cord.

C. The cord affected before the brain. Here the encephalic symptoms and lesions may be produced by a vasomotor disorder of cerebral circulation and nutrition, brought about by the spinal lesions.

G.p. has been considered by some as primarily and essentially a disease of the spinal cord, and others have given a most prominent place to the spinal lesions. Thus Westphal averred that the affection of the cord was often primary in g.p., preceding that of the brain, but no direct progress from one to the other being recognizable. This earlier affection

of the cord may sometimes have occurred when only a
chronic meningo-perimyelitic and meningo-periencephalitic
condition is found after death. Yet the most striking cases
are those in which there is grey degeneration, or sclerosis, of
the cord's posterior columns, together with the other changes
usually associated with that lesion; and, in this connection,
I shall, for convenience, include that which it is thought
necessary to say in this work, (besides what has been said in
the chapter on diagnosis), on the subject of g.p. by propagation, so-called; in which the pathological lesions of g.p. have
been supposed to originate in an extension, to the cerebrum,
of spinal lesions, or of disease of some cranial nerve or
nerves.

Broadly speaking, in g.p. the motor *spinal* symptoms are
either ataxiform, or paralytic, or spastic; and usually of a
mixed character, being both ataxic and paretic, or spastic
and paretic, the ataxiform are almost invariably found, the
paralytic or paretic are frequent. In cases in which the
spinal symptoms distinctly precede the cerebral, we may, in
thought, follow a morbid process which originates in the
spinal cord and meninges, and subsequently passes up to
the encephalon and its meninges. But in this group are
found the ordinary symptoms and lesions, both spinal and
cerebral, of g.p.; here the ataxy differs from the ataxy of
tabes dorsalis. Obviously, cases of this group differ from
those in which g.p. is supposed to follow upon a propagation
to the brain of disease from the cord in preceding tabes
dorsalis. Widely differing cases have been placed on record
to exemplify the relationships and coexistence of tabes
dorsalis and g.p. In some of these it is difficult to be certain,
from the description, whether or not true g.p., or, again, true
tabes dorsalis, was present; in some, either the former or
the latter was certainly absent; and in some others, where
both seem to have existed, their order of priority is by no
means clear; while, again, in some the symptoms of g.p. and
of tabes seem to have both appeared and continued, simultaneously. Thus, the subject has been obscured by the
introduction of cases that will not bear the special interpretation put upon them. Cases in which spinal disorders
precede encephalic are placed by Voisin in three groups :—
one in which, under similar causes, first a myelitis and then
the encephalic lesions of g.p. occur, successively, on a soil
prepared in advance, but without any relation between them
of causality or continuity; another in which the spinal or

spinal-nerve affection influencing vaso-motor phenomena, and affecting the cerebral circulation, is supposed to cause the encephalic lesion in a reflex manner; and, finally, in the third group there is a direct extension of disease from cord to brain, as traced in one case into the white cerebral substance.

I will briefly refer to cases supposed to be g.p. by propagation. Horn long ago described a case of tabes dorsalis, of nine years' standing, in which g.p. supervened; and H. Hoffman * a similar, but doubtful, example. Baillarger † recorded five cases; but in only one of these does it seem that g.p. may have succeeded to indubitable tabes dorsalis. Yet the history is most imperfect, and the sequence of events conjectural. In three others, grandiose mania and other symptoms of g.p. supervened in the first period (Duchenne) of tabes—the pre-ataxic stage; and it is by no means clear, therefore, that g.p. may not have existed as early as, or earlier than, the tabes. In the remaining case g.p. distinctly preceded locomotor ataxy by a year. In some of these it is said that the one affection underwent arrest or recovery on the appearance of the other; but a transitory epiphenomenal appearance of g.p. in the course of another affection, is an explanation one cannot accept. Examples of a like doubt as to the existence of g.p. occur in cases by Mr. Plaxton,‡ and apparently in some of those by Dr. Nicol.§ A case by Teissier is inconclusive. Westphal ‖ found tabes dorsalis sometimes followed in the later stages of its course by mental disease, and much resembling g.p. Yet it can scarcely be said that the latter existed. At the necropsies no cerebro-meningeal adhesions were found; there was increase of ventricular fluid, and, in some, pial œdema, or softening of central parts of cerebrum, and pallor of its grey substance. Ph. Rey ¶ described three cases supposed of tabes with consecutive g.p.; Motet, Bouchereau and Dagonet similar cases, and Falret and Dally a doubtful one, as also Magnan. Foville published four cases as examples of tabes dorsalis followed by g.p., after spaces of time varying from 2 months to 6½ years. Two of these had amaurosis. A marked example of true tabes dorsalis followed by true

* " Allg. Zeitschr. für Psych.," xliii Bd., p. 207.
† " Gaz. des Hôpitaux," Nov., 1861. " Ann. Méd.-Psych.," Jan., 1862, p. 1.
‡ " Journ. Mental Science," July, 1878, p. 274.
§ W. R. Asyl. Med. Rep., Vol. i., p. 171.
‖ " Allg. Zeitschr. f. Psych.," xx Bd., 1 Heft, p. 1.
¶ " Ann. Méd.-Psych.," Sept., 1875, p. 161.

g.p., I published,* with full details. Locomotor ataxy from diffuse sclerosis of cord, followed by g.p., was reported by Dr. C. K. Mills.†

But g.p. may be preceded for a long time by amaurosis unconnected with tabes. Three such cases were recorded by Mobèche, and similar examples by Lélut, Lasègue, Parchappe, Calmeil and others. Two such Foville adds to his four of tabes, making six, and for a seventh takes a case in which was prior paralysis of the third cranial nerve; adding, also, a case of g.p. following diphtheritic paralysis. Falret, Billod, Esquirol and Baillarger recorded, or spoke of, examples of similarly antecedent paralysis of third cranial nerve. Westphal saw g.p. follow atrophy of the fifth nerve, and Gaserian ganglion; or of the olfactory nerve; and Monakow observed apparent g.p. follow progressive amyotrophy due to lead-poisoning. I have seen g.p. follow atrophy of an optic nerve and tract. Here come some of the cases of g.p. following fevers. Doubtfully to be placed here, and in some having only an accidental connection with g.p., but in others possibly leading to the latter, are the cases in which disseminated spinal sclerosis, posterior spinal meningitis, and various forms of chronic myelitis, as a case by Dr. V. Magnan, are followed by g.p. Especially with regard to insular sclerosis are there a number of cases on record which make one inclined to admit the possibility of the origin of g.p. in that affection; not that the particular cases referred to were g.p. in themselves, but as showing how narrow the separating channel. Among the cases are those by Drs. W. Valentiner, H. Liouville, S. Jaccoud, W. Leube, and J. R. Gasquet. Indeed, cases are recorded, as by Prof. Fr. Schultze, as being insular sclerosis terminating in g.p.; or associated with g.p., as in a case by Dr. Claus. And examples have been brought forward alleged to be chorea, usually spinal chorea, ending in g.p. of the ascending form. See also, below ("2").

D. To complete this part of the subject one need only formally mention cases in which the spinal cord is unaffected; the disease being practically limited to the brain.

2. *General paralysis by extension from the vicinity; extension by contiguity.* It has been maintained that true g.p. may be produced by the inflammation occurring around cerebral tumours or around *foyers* of softening (Ch. Bur-

* "Lancet," May 28, 1881, p. 862.
† "New York Medical Record," June 23, 1883.

lureaux *), or of hæmorrhage (L. V. Marcé), and, if not starting there, eventually gaining the periphery of the brain and the meninges. Baillarger † also had seen dementia paralytica, similar to that following ambitious delirium, also follow hemiplegia from cerebral hæmorrhage. Burlureaux reported a female case (no necropsy). G.p. following these local lesions may be termed a secondary affection. For this concerns what I may term g.p. by invasion from the vicinity, the lesion of g.p. is, therefore, here a secondary one; some lesion in or near the cortex, directly or indirectly, bringing about that of g.p. Here may be mentioned g.p. occasionally following erysipelas capitis, and perhaps in some cases that occasionally following insular cerebral or cerebro-spinal sclerosis. As to this insular sclerosis and g.p., the close resemblance between them in symptoms, as already examined in detail in the chapter on diagnosis, their morbid alliances, the great similarity in nature of the pathological changes attending both; make it highly probable that the future will reveal closer relationships between them than have yet been established; and it is hoped that the present writer may have succeeded in drawing attention more pointedly to the clinical relationships of, and distinctions between, these two diseases. In a case of insular spinal sclerosis ending in g.p. were insular sclerosis of cord, diffuse sclerosis of brain; in another, insular s. of cord, diffuse s. of brain and cord (Schultze).

Here also must be named some cases of tumour, which sometimes—and perhaps by the vaso-motor disorder, and, therefore, encephalic circulatory change, they induce—give rise to expansive symptoms as in g.p., or produce dementia, as well as motor signs; as in a case with syphilitic tumours and infiltration. In relation to the present subject, it is of interest to know that this very case was the only one given in any detail out of those upon which was founded (and therefore so far erroneously) M. Colin's paper on g.p. consecutive to local lesions of the cerebrum, particularly to cerebral hæmorrhage.

Marcé, too, believed that a local hæmorrhage or softening may become the starting-point of g.p., the primary hemiparalysifying lesion—or, rather, a morbid process set up by it—advancing to the cortex, so that g.p. is implanted on the old local palsy. But here I may state that some cases are erroneously cited as examples of g.p. by propagation, or of

* *Op. cit.*, p. 76 ; and " Gaz. Hebdom.," Jan .16, 1874.
† Société Méd.-Psych., Paris, Nov. 29, 1858.

g.p. arising by what I have termed extension by contiguity—like a case thus cited by Mendel from Foville as one of g.p. coming on in connection with local softening; whereas obviously, and as believed by the reporter of the case, they were quite independent of each other: viz.—symptoms and recent lesions of g.p.—chronic persistent left hemiplegia and an old focus of gyral softening. A similar view, perhaps, must be taken of Baillarger's case,* in which progressive muscular atrophy existed for nine or ten years before the first noticed indications of g.p., which soon ended fatally. In some older alleged examples of the above it is doubtful whether g.p. ever really existed.

3. *Cerebral Congestion* and inflammation. As to the much-vexed question of the essential nature of the pathological process affecting the nervous system in general paralysis, we may briefly note the important *rôle* so often assigned to cerebral congestion of an active kind. Of the several different views entertained on this subject it will suffice to state the following.

That cerebral congestion is the proximate or determining cause of general paralysis.

That the repeated or persistent cerebral congestion causes irritation, and then a disorganizing inflammation.

That it acts by causing capillary obstruction, and consequent necrotic softening of the encephalon.

That the inflammatory cerebral congestion or the interstitial sub-phlegmasia is an epiphenomenon of degeneration.

That the cerebral congestion is secondary, and that the meningo-encephalitis, when it exists, occurs secondarily, and very late, and as a consequence of a failure of the general nutrition.

That the congestion leads to connective-tissue growth, or outwandering of leucocytes which organize; conditions which, conjointly with capillary obstruction by blood-accumulation, impair the nutrition of the brain and promote degeneration.

In relation to this I may mention the experiments of Dr. E. Mendel,† who produced a disease in dogs analogous to g.p. in man, by recurring rapid rotation of the animals fastened to a revolving table with the head at the periphery; whereby, in cases rapidly revolved, death occurred with an intense hyperæmia of the calvaria, cerebral meninges, and grey

* "Annales Méd.-Psych.," Jan. 1879, p. 76.
† "Sitzungsberichte der Königlich Preussischen Akad. der Wissenschaften zu Berlin," April 17, 1884.

cortex; and anæmia and œdema of the white cerebral substance; also punctiform hæmorrhages in the meninges and cerebral cortex, particularly in the region of the sulcus cruciatus. But when the rotation was less rapid, or of shorter duration, and was repeated from time to time; there appeared, successively, loss of muscular feeling in the posterior limbs and then in the anterior, and gradually a generalized paresis; alterations in the dog's bark, apathy, dementia, rapid loss of body-weight, although appetite for food was unaffected; and, finally, with all the appearances of a generalized paralysis, death followed. The lesions found, both macro- and micro-scopical, were very similar to those in active cases of g.p. in man. In the dogs, an intense hyperæmia is supposed to overcome the resistance of normal vessels, and lead to an acute form of the morbid process which takes place in g.p. in man; for which latter two things are necessary—a morbid change of the vessel-walls, easily permitting exudation of leucocytes and blood-plasma; and an active hyperæmia in the vessels of the brain-cortex.

4. Here also our attention is claimed by a conflict of opinion as to whether the *essential pathological change* in general paralysis is principally and primarily of a more or less frank inflammatory nature, or principally and primarily of a degenerative nature.

Thus, the observers of one group look upon an active inflammatory stage as being primary, those of the other look upon it either as secondary or as non-existent. To the minds of the former, again, the degeneration of cortico-meningeal vessels, of nerve-cells, or of neuroglia, is secondary; for the latter, these degenerations are primary.

A. *General Paralysis viewed principally and primarily as an inflammation.* Under this head, although with considerable variety in detail, must one range the views of Bayle, Calmeil, Parchappe, Belhomme, Bouillaud, A. P. Requin, Duchek (partly), A. J. Linas, L. Meyer, Lallemand, Guislain, Franz Meschede (partly), Lasègue, S. Wilks, V. Magnan, H. Schüle, A. Lubimoff, v. Krafft-Ebing, Mierzejewski, Hitzig, Burlureaux, Crichton Browne, Mendel, Blandford (fluctuatingly)* :—and *in some cases* L. Lunier, J. Baillarger, and Th. Simon. *As inflammatory or congestive.* Rokitansky, Voisin.

Well placed from the first by the weight of authority so great as that of Bayle and Calmeil, the doctrine of the inflammatory nature of general paralysis for many years ruled

* "Insanity and its Treatment," 3rd Ed., 1884, p. 320.

supreme. Not that the precise character of the inflammation or the elements primarily affected by it were agreed upon. As to what part of the encephalon or meninges was the special seat of inflammation, as to whether the latter was interstitial or parenchymatous, and as to its relations with ordinary acute inflammation of the same structures, opinion was divided, and discrepancy rife.

In fact, at an early date departures were made from the pure and simple doctrine of inflammation, and the *rôle* assigned to the latter was by no means always the same.

Baillarger described two orders of lesions as characterising different cases in g.p. I, The inflammatory—those due to peripheral meningo-encephalitis: and, II, those dependent upon hydrocephalic or serous effusion, with atrophy or softening of brain. Then, again, an identity was assumed by Lunier between g.p., hydrocephalus of adults and of the aged, and the chronic affections of meninges or of encephalon which succeed the acute inflammations of these structures.

Views such as these may be deemed as in some sort a compromise, or transitional, between the exclusively inflammatory doctrine and the views immediately to be mentioned under " B."

With reference to the opinion so commonly held that g.p. is essentially an interstitial inflammation, Dr. V. Magnan, in 1866, apparently was the first to so precisely define the lesion as a generalized diffuse interstitial encephalitis, invading the brain both from the external surface and from the ventricular.

Since Bayle, who held the disease to be a chronic meningitis, the authors who have laid most stress on *inflammation of the meninges* are perhaps; (a) L. Meyer, at one time (1858), as shown in the title of his work, according to whom chronic meningitis played the most important part, and chronic encephalitis usually followed and started from the walls of the capillary vessels of the grey cortex, and who in 12 p.c. had seen products of purulent meningitis; and (b) Th. Simon, who asserted that the only lesion present in some cases of g.p. is a hæmorrhagic pachymeningitis.

B. *General paralysis viewed principally, or primarily and principally, as a degeneration.* Here, with great differences *inter se*, are to be placed the views of Drs. L. V. Marcé, Delasiauve, Bucknill, Salomon, Bonnet and Poincaré; also Meschede and Luys partly: Baillarger in some cases:

Calmeil (at one time) perhaps in the milder cases: Obersteiner (doubtful).

At one time Luys, ascribing the changes of g.p., in and around the vessels in the nervous centres, to the effects of congestion, even if of a passive kind—and showing that the process is a complex one, in which both the over supply of blood and the proliferative reaction of the vascular walls play simultaneous and important parts—advocated that this is a special form of plastic congestion; intermediate to simple congestion, whose barriers it has overleaped; and to exudative inflammatory congestion, unto which it has not attained;—it is more than mere passive congestion, by reason of its exudative tendencies, yet it is not inflammation, at the outset (at least), by reason of the absence of primordial erethism and of purulent formation.

Duchek long ago made atrophy one of the essential processes in g.p. Partly in continuance of this subject of the lesion in g.p. as a degeneration must here be added,

5. *The vaso-motor theory of g.p.*

Bonnet and Poincaré held that inflammation is not necessarily present in g.p., and that when it does occur it is late in the course of the disease, and is the result of defective nutrition. They agree with the earlier writers as to the existence of cerebral congestion, and the *rôle* that it plays; but not as to the *primum movens* of that congestion. Especially do they agree with Bayle as to the unity of the disease and the congestive element; while they differ from him, essentially, as to the interpretation of the symptoms and the point of departure. They believe that the alterations found in the encephalon are the consequences of disorders of the cerebral circulation following impairment of the function of the sympathetic ganglia, owing to their diseased condition, as described in a preceding chapter. Also, that in these ganglia they find the anatomical point of departure of the disease; and that the changes in the encephalon are only the consequences of disorders produced in the cerebral circulation by the sclerosis and fatty substitution in the cervical sympathetic ganglia,—changes which have a paralyzing effect upon the vessels, induce congestive disorder of cerebral circulation, and alterations in the brain itself. " All the alterations we have described bring on disorders of nutrition in most of the organs—disorders which tend to fatty degeneration, or other modifications of their elements, and betray themselves on the physiological side, at first by ataxy, and then by prostration of the functions both of the life of relation and the vegetative." Thus, in their view, vaso-motor lesions and disorders are of the very essence of the disease.

Or there is vaso-motor paralysis in the brain, followed by exudation, in which new tissue forms and compresses the nervous elements (Obersteiner, Lubimoff). Spitzka concludes that the essential and primitive anomaly is a vaso-motor disturbance and that g.p. " is a progressive deterioration of the central nervous system, chiefly affecting the brain, and the result of a chronic inflammatory process of an angio-paralytic nature, whose essential element, the vaso-motor weakening, is due to overstrain of the encephalic vaso-motor centre."

The insufficiency of the theory that g.p. is due to disease of cervical sympathetic ganglia is shown by the fact that similar changes have been seen in these ganglia in other forms of insanity and in mentally healthy persons.

6. *General paralysis viewed as a meningeal affection.* The insufficiency, also, of the view that g.p. is simply a meningitis; still more, is simply a pachymeningitis; is shown by rare cases in which the meningeal changes are absent, or in which they are so slight, or so similar to what obtains in other mental disease, or in some persons not insane, that they cannot be deemed the organic basis of g.p. in these cases; while the cerebral or cerebro-spinal tissues will here show more distinct alterations.

7. *General paralysis viewed as chiefly or primarily a spinal affection.* This was the view chiefly accepted in this country and Germany some years ago. What has already been stated shows its inaccuracy.

8. *General paralysis viewed as a neurosis.* Since the discovery of g.p., there have rarely been lacking those according to whom it is not an organic affection of the cerebro-spinal nervous system, or of any part thereof; but is a neurosis, a vesania, a neuropathic affection. A modification of this is the view that some organic affection of the spinal cord exists in g.p., but, (unless casually), no organic affection of the brain. And a cognate view is that it is merely an assemblage of symptoms arising from various conditions, and not a definite disease, although in the later stages there may be pachymeningitis as well as disorders of nutrition and innervation.*

* Th. Simon, " Arch. für l'sych.," i Bd p. 624.

Thus, at an early date the views of Bayle were attacked even by Georget; and later on Lélut, Aubanel, and Thore denied that organic change necessarily exists in the brain or meninges in g.p.; even when it runs its full course. Others, as Brierre de Boismont, held that the brain was not affected in some cases, and that the only alterations in g.p. might be in the cord; or in the sympathetic; or in peripheral nerves; or that it might occur without any appreciable lesion of nervous system. Of late years, Dr. Th. Simon declared g.p. to be merely a group of symptoms arising sympathetically or reflexly from very different conditions, and that in the present state of our knowledge no naked-eye or microscopical changes were to be found in the brain in many cases of g.p. And Dr. v. Rabenau stated that disease of the brain existed in only a few cases of g.p.; that the changes were so little constant that no mental image of the disease could be set up; and that the attempt to find changes in the cadaver to explain the clinical phenomena was fruitless. One would say, rather, that to deny, as he did, the changes and their validity was a fruitless attempt.

9. *Obstacles to return of intra-cranial blood.* As bearing upon some of the pathological changes in g.p. it is well to bear in mind how, in active states of the cerebral circulation, the natural obstacles to the local circulatory freedom bring about chronic thickening of the pia-arachnoid, increase in size of Pacchionian bodies, increase of sub-arachnoid fluid, thickening of the skull inwards, and some increase in the firmness of the brain.*

10. *Durhæmatoma. Matrhæmatoma. Meninghæmatoma. Arachnoid cyst.* This, sometimes found in g.p., results, either from the transformation of blood effused into the arachnoid cavity, whether by primary hæmorrhagic extravasation, or, probably, by rupture of fattily degenerated vessels of pachymeningitic neo-membranes; or perhaps, occasionally, from fibrinous effusion. Occurring in various cerebral affections, g.p. is one of the forms in which this is most frequent (3 out of 10, Dr. H. Sutherland). It attracted attention at a comparatively early date, and later on F. Lélut† published a dozen cases mostly occurring in G.Ps. Ferrus mentioned it as of hæmorrhagic origin. W. Schuberg gave an account

* Wilks and Moxon, "Path. Anat.," 2nd Ed., p. 205.
† "Gazette Médicale de Paris," Jan. 2, 1836, p. 1. *And*, "Sammlung zur Kenntniss der Gehirn und Rückenmarks Krankheiten."—Ed. F. Nasse. 3 Heft. Stuttgart, 1840. "Ueber die pseudomembranen der Spinnenwebehaut des Gehirns."

of the several different situations of the effusion. Most of the earlier observers in France believed the extravasation of blood to be primary, a view supported by Sir (then Mr.) Prescott Hewitt,* and opposed by Heschl and Virchow. Bayle said it might result from inflammation of inner surface of dura; but Baillarger and Boudet denied this. In some of the cases I have seen I thought the origin rather from the soft meninges than from the dura.

11. *General remarks and summary.*

Inflammatory doctrine of g.p. As to which part is first or mainly inflamed; some placed the inflammatory action chiefly or primarily in the cerebral meninges; others in the cortex generally; some in both; others in the middle layer of the cortex; others in its deeper layers; some in the anterior lobes; while some described it as invading the brain layer by layer from periphery to central regions; others, again, as beginning in the grey, or in the white, or in the cord; some gave cerebellar inflammation a share; still others asserted that the morbid process attacks the whole encephalon; or even the whole cerebro-spinal nervous system. Many held the lesions, whatever their nature, to be essentially spinal, or spinal and bulbar.

Then, as to which of the histological elements entering into the formation of the nervous organs were those first or chiefly affected with inflammation in g.p.; according to one view the inflammation is parenchymatous; another makes it start from the walls of the blood vessels; and a third is that it is interstitial.

And something of the same remarks applies to those who have looked on g.p. as being primarily and chiefly of a degenerative nature; although, on the whole they have attempted far less limitation and localization of the lesions they described, and described in terms often more or less vague. But, for the most part, even they have admitted a congestive element in the affection or in its causation, and have not regarded it as a pure and simple degeneration of structure *ab initio*. Chronic or frequently repeated congestion of the nervous system, degenerative changes in the vessel-walls of its blood-supply, passive blood-stasis, are conditions which have played an important part in their views. Nay, further, it has by some of them been held that inflammation may, and sometimes does, supervene as a result of the coexisting general disorder of nutrition.

* Roy. Med.-Chir. Soc., Feb. 11, 1845.—" Med. Times," Feb. 22, 1845, p. 452.

Two principal reasons may be assigned for this diversity of view:—

1. The too great inclusiveness, in pathology, and hence the vagueness and width of meaning, of the word *inflammation*.
2. The actual occurrence of *degeneration* as an ultimate result of inflammation.

The causes of inflammation induce a depression in the vital powers of the parts affected, which manifests itself as deficient normal formative activity, and increased tendency to degeneration.

In inflammation there is question of damage. As Dr. Burdon-Sanderson says, inflammation is the physiological effect of injury; *i.e.*, the inflammatory objective changes are the direct physiological effects of injury.

Thus, the inflammatory phenomena are the expression of loss and failure. On the true tissue of the part the *direct* effect is necrosis; as to its total amount there is virtually atrophy of the true tissue; and any newly-formed texture readily undergoes degeneration and absorption.

The older hæmatic and vascular view as to the condition in inflammation, although not fully tenable, yet does embody large elements of truth, and the evidences of vascular turgescence exist during life and even to some extent after death.

While the *acute* inflammations of a nervous organ reveal themselves by changes that are easily recognizable this is not the case with the *chronic* inflammations, and the difficulty in distinguishing these from simple degenerative states is at times not inconsiderable. The nomenclature of many morbid changes in the nervous system shows this clearly. The limits of inflammation here, are not yet definitely fixed, and no exact parallels of pathological latitude bound its territories.

In fact the products of an inflammatory process are either exudation or new growth, the former scarcely concerning us so much at this moment, the latter being a vascular connective tissue framework with leucocytes. And it is claimed that in the interstitial tissue of solid organs chronic inflammation produces a connective-tissue growth, which compresses and causes atrophy of the special elements of the part. But this is often called "fibroid degeneration." To avert this discrepancy, and for purposes of discrimination, there is something to be said in favour of a tri-partite division into: inflammation—simple degeneration—connective overgrowth.

Therefore, much of the discrepancy we have referred to is more apparent than real; and if, for a moment, we assume that g.p. is of inflammatory nature, we may say that the results of inflammation in the parts affected in this disease are practically of the nature of degeneration; and that a chronic inflammatory process may supervene in a brain or cord already damaged by a degenerative condition. For the chronic inflammation supervenes in a brain which has been subjected to some persistent, or frequently repeated, active hyperæmia, whose ulterior result has been to mechanically affect, and impair the nutrition of, the vessel-walls, rouse connective-tissue overgrowth, degrade the nutrition of the delicate nerve-elements, both directly by watery exudation and out-wandered blood-corpuscles, and indirectly by the changes in other elements just named. At first, we may suppose that the causes which induce g.p. act on the encephalic vaso-motor centres, or that these are disordered, secondarily, by the operation, on the ganglionic nervous apparatus, of the causes of g.p. The vaso-motor disorders occurring in the course of the *established* disease may be induced by implication of vaso-motor centres in brain, bulb, and cord, in the ordinary encephalic, bulbar, and spinal lesions of g.p.

And, indeed, it must be admitted that in many cases of g.p. all one sees may fairly be explained without calling in inflammation as a factor; and that, besides hyperæmia or its traces, and atrophy and degradation of true elements, some substitution of a texture or substance of altered and lowered, type, vital character, degree of development, and differentiation; for, and in place of, the usual texture of the part—which may then be termed transformed or replaced by substitution—is, in some cases, practically what we see. It also seems to me that here far too little has been thought of the embryonic-tissue nature of much that is seen in the brains of G.Ps., in the less advanced stages. It is not merely that the enormous nuclear growth—or (as some prefer to view it) nuclear formation in exudation material—or modification of out-wandered corpuscles—is so like what is normal to the brain of the fœtus and new-born child. It is not merely this. For even the morbid state which so much strikes attention in most cases of g.p., namely the adhesion and decortication at parts of the brain-surface, is, after all, in some respects, very much like a textural reversion or retrogression in type, and incidentally due to the substitution of a lower and embryonic type of texture for a higher. And this textural reversion and retrogression carries its retracing steps along the lines of development both of the animal kingdom and of the human individual.

The condition of the brain in acute cases of g.p., with its red, violaceous, blood-injected and-distended, or even ecchymotic, membranes—sometimes with fibrino-albuminous flecks—and its soft, swollen, hyperæmic cortical, and hyperæmic medullary, substance, may remind one of the "congested kidney," a first stage of one form of Bright's disease; or, again, of that congestion of kidney, enlargement of its cortical substance, and swelling of its glandular elements, occurring early in some cases of that disease.

But it is particularly when the patient dies after a long-lasting g.p. that the adherent meninges, the atrophied cortex, the indurated medulla, and the great shrinking of the brain (together with the microscopic changes), have led observers, (Dr. E. Salomon perhaps the first), to liken it to the contracted kidney of Bright's, or to a cirrhotic liver. With this is allied the view as to the toxic origin of g.p., as by Dr. J. Hitchman, and others (see also p. 141, 1st Ed.). Now, it might be said that the homology between these two conditions enables one to bring g.p. into line with diseases of interstitial tissue, that being the more commonly accepted view as to a contracted Bright's. But I think the evidence is now in favour of the view that Bright's disease is essentially a diffuse nephritis, which in some cases may affect the interstitial tissue mainly or primarily; in others the glandular elements mainly or primarily; and that some cases of so-called interstitial nephritis are chiefly and essentially examples of arterial sclerosis; viz., the small red kidney. Be this as it may, one must not press the homology too far; yet the same broad laws of pathology hold sway in both organs.

Then, if we enquire where the starting-point of the change is in g.p., our histological investigations would, on the whole, refer us to the vessels and interstitial tissue, or at least intercellular elements. Nevertheless, it may well be that the starting-point of the trouble is often in the ganglionic nerve-cells, and that the excessive action, over-strain, and morbid stimulation or irritation by which the nerve-cells are brought into an invisible state of damage, injury, lowered vitality,—or however else one may express it—superinduces frequent or persistent active afflux of blood, whence the various ulterior changes in sequence. It may be said; *then* one would expect *here* the commencement of material organic change evidenced as parenchymatous cerebritis—as inflammation of the ganglionic nerve-cells. But the nerve-cells by virtue of minute molecular changes and disturbances, not recognizable by us, will set in action enormous gross muscular activity and vascular alteration, in health and in disease;—and, an equal degree of departure from a healthy condition existing in nerve-cell, in vessel, and in intercellular substance, the last two would give vastly more obvious histological indications of it than the first would. In the nerve-cell is an excessively delicate organization, the field of enormously active dynamic perturbations, at present imperceptible and inscrutable by our means of observation;—although eventually even coarse change affects the nerve-cell;—while, on the other hand, the blood-vessel—and similar—textures have a comparatively coarse and extensive exhibition of motion, or of gross material reaction and change, in health or in disease. This, indeed, is an inherent quality, a necessary result of their essential constitution.

Thus, then, the effects of severe disturbance, and the resulting, comparatively gross, morbid change will first be visible in the vessels and interstitial (intercellular) substance. Hence we see and hear much of interstitial inflammation or change; and correctly so from one, and the ordinary, point of view.

Summary. On the whole we may view g.p. as essentially commencing with hyperæmia, and ending with chronic cortical degenerative cerebritis, and, usually, embryonic- and connective-tissue substitution, the change, fundamentally parenchymatous, affecting all the elements of the part, but usually, under the methods of examination hitherto chiefly in use, presenting a more obvious and more marked affection of the blood-vessel walls, and of the interstitial elements; and this for the reasons already named. The cerebral, or even encephalic, are almost invariably associated with lesions of other parts of the cerebro-spinal nervous system; and the morbid action begins, exceptionally, in parts other than the cortex cerebri. Obviously, this view allows of several varieties of g.p.

Mem. The above views were expressed in my paper introductory to a discussion on g.p. at the annual meeting, Brit. Med. Assoc., Liverpool, 1883. As this sheet is passing through the press, I see Dr. Wm. Leah's interesting paper in the "Birmingham Med. Rev.," June, 1886, in which he very ably argues for a parallelism between g.p. and systemic diseases of the cord; and that the disease in g.p. extends in the direction of nervous impulse, and from its starting-point in one, to the corresponding part in the opposite cerebral hemisphere, by sympathetic inflammation, or a process akin thereto.

CHAPTER XVIII.

PATHOLOGICAL PHYSIOLOGY.

1. *Introductory*. To trace the many symptoms of general paralysis to the various effects upon physiological function of the several stages of the morbid process, or to its several kinds, or to the several localities mainly affected, is a task which has varied in its execution with each advance, or supposed advance, or each caprice of fashion, in physiological and in pathological teaching.

Broadly speaking, the motor signs and the mental symptoms more usually proceed, concurrently, in somewhat parallel lines, and in reference to this association, " not only does the morbid state of the motor centre lead to a difficulty of expression by the appropriate movements, but the diseased motor intuition enters into the intellectual life, and in conjunction with morbid ideas there, gives rise to all sorts of extravagant and outrageous delusions as to personal power " (Dr. Maudsley); and more than usually impossible is the correction of the delusions, on account of the failure of the muscular sense, partially closing the avenue whereby a knowledge of the qualities of objects is acquired, and leaving the patient a prey to internal disorder.

I shall not tarry in order to relate the several hypotheses that have been put forward on this subject, such as the more or less hypothetical views of Bayle, Foville *(père)*, Bouillaud, Linas, Marcé, Dagonet, Obersteiner, Luys, Lasègue, and other more recent observers. But the historic interest of one of these justifies its insertion here. In the first two of the three stages described by Bayle, the "paralytic" symptoms were attributed by him to the effects of congestion of the pia-mater (and brain), and of thickening of the meninges; and in the third and last stage, partly to the same congestion, but especially to the pressure exerted by the now excessive arachnoid, subarachnoid, and ventricular serosity. The delirium, exaltation, agitation, and furor, he assigned to the irritation exercised by the inflamed meninges on the cortical substance of the convexity and internal surface of the cerebral hemispheres, and thence on the entire encephalon :—the ambitious ideas to the indirect effect on the cortical substance, and successively, on the whole cerebrum, of the sanguineous congestion of the pia-mater and of the inflammation or irritation of the internal surface of the cerebral arachnoid:—the dementia to compression of the brain by serosity :—the apoplectiform attacks to sudden sanguineous congestion in the vessels of the pia-mater and brain :—and the convulsions to consecutive inflammation of the grey cortex of the cerebrum ; or, rarely, to sudden serous effusion upon its surface, or into its ventricles. Nor need I forbear to add the practical experience of Prof. Tamburini and Dr. Ricci *:—in 55 cases, weakness of voluntary motion, some part of motor area of cortex affected : in 21, unilateral motor affections, and unilateral and opposite cerebral lesions : in most, the foot of third frontal gyrus affected: in 20, either hallucinations, or sensory weakness or loss, and affection of parietal, occipital and temporal regions : in 10, unilateral motor disorder with some anæsthesia and analgesia, and lesion limited to central gyri : of 60 G.Ps. somewhat demented, lesion of frontal gyri in 56.

2. *General aspects. Groups of symptoms.*

Proposed View. In the interpretation of the various symptoms of g.p., it may be broadly stated that the morbid process in the nervous system first deranges, and then destroys, or tends to destroy, the functions of the parts affected. Yet is the interpretation surrounded by grave difficulties. For: —1, The morbid process is modified in different cases.—2, Its extent varies likewise.—3, The part or parts at which it begins, also vary in different instances.—4, In some cases the greater portion of the cerebro-spinal system may be more or less involved in a general disturbance of nutrition and disorder of function, while, in others, the diseased action may long be *comparatively* localized.—5, As different nervous districts are successively and progressively implicated, it usually happens that the function of one part is merely exaggerated or disordered at the same time as that of another is lessened or destroyed.—6, The destruction of an inhibitory centre, under these circumstances, will permit comparative

* "Riv. Sper. di Fren. e di Med. Leg.," F. iv, 1883. Abs. Dr. Huggard.

over-action of functionally related centres, which have escaped from its regulative influence.

Motor and mental nervous centres are markedly implicated in the morbid process. The motor disorders cannot be fully explained by any secondary disease of the fibres which merely transmit motor impulses, although this disease plays its share in the later stages. On the one hand, as the principal lesions of g.p. apparently are situate in the cerebral cortex, and, on the other, as investigations tend to show that in the cortex are situated not only the mental centres, as ordinarily believed, but the highest motor centres also; there is an inclination to adopt a well-knit theory that the mental and motor symptoms in g.p. are due to a widely-spread cortical lesion, including within its area of action both the motor and the mental centres, to which, indeed, we may add the sensory. And here I would bring to the notice of modern experimenters that practically this physiological view as to the existence of cortical psycho-motor centres is not so new as they deem it to be. It is constantly implied by Bayle. Parchappe, again, speaks of softening of the cerebral cortex in general paralysis as producing the diminution of the muscular forces, and difficulty in the movements co-ordinated by the will. On this point also may be consulted the *Revue Médicale* for Sept., 1846, in reference to lesions of superficial parts of the brain producing paralysis; and many cases in Lallemand's works.

As to the particular parts of the cortex necessarily involved, under this theory, they vary as one adopts the conclusions of this or that physiologist. But one may utilize the actual outcome of experimental interference with the brains of lower animals, and yet be animated by a feeling of caution and reserve in accepting many of the rigid conclusions drawn therefrom. If the true higher motor centres are not *in* the cerebral cortex they are at least in evident intimate functional connection therewith; a functional connection which undoubtedly involves a pathological alliance—a community in pathological suffering—a true sympathy—also. Be the functional relation what it may between centres in the cerebral cortex on the one hand, and basal and spinal motor centres on the other, it must not be forgotten that in one group of G.Ps. the first apparent symptoms are spinal, and only the later are mental. Believing that almost invariably the disease is primarily and mainly of the cerebral cortex and its tunics, I think it desirable that minute observations of the cadaver, and the comparison of these with the results of physiological experiments, should be continued. Guided by that scheme of localization (Ferrier's) which is best known in this country one would be led to assign the *mental* disorder and failure in general paralysis to the morbid process or change, of the greater part of the cortex of the frontal lobes—morbid stimulation of the motor and sensory centres also modifying the mental condition and conducing to mental disorder:—the *motor* troubles mainly to that of the central and paracentral districts;—the *sensory* affections to that of the temporo-sphenoidal, and part of the parietal, lobe; and any perversion or destruction of *organic feeling*, to that of the occipital.

Restricting one's attention to the clinical aspects and to the morbid histology as actually observed, the course of g.p. is probably somewhat as follows.

In the vast majority of cases the cerebral cortex is primarily affected, the meninges usually being more or less involved almost simultaneously.

In many cases the morbid process apparently is most active, and at first active only, in circumscribed regions of the cerebral cortex. In others the morbid action is more diffused.

Taking the mass of cases, the convolutions of the frontal and of the parietal lobes suffer more than those of other parts of the brain.

The pia-matral adhesions to the cortex often form a valuable index to the localities principally diseased in general paralysis. These adhesions vary very much in their extent, degree, and situation, in various cases, in correspondence with the greater activity of the morbid process in different circumscribed regions of the cortex.

The morbid process in general paralysis is primarily set up by excessive, irregular, protracted, activity and overstrain of a larger or smaller number of the active functional elements of the cerebral cortex, which subserve the higher faculties of the organism. It is usually admitted that these elements are the so-called ganglionic nerve-cells of the cortex. Of the most potent and frequent causes of g.p., each, in its own way, brings about the primary step to which we refer, namely, the excessive, irregular, protracted, activity or over-strain of a larger or smaller number of the nerve-cells. Partly in consequence of the different modes of action of the several causes, partly owing to the varieties in habits, work, and circumstances of the patients, producing habitual relative over-activity, activity, or disuse, of this or that part of the brain, and, hence, relative higher development in the one case, and relative feebleness of action and simplicity of inter-communication in the other; and partly owing to hereditary and to diathetic causes and, possibly, temperament as well; it comes about that the cortical regions or centres most severely affected vary, as we have said, in different cases.

Over-activity of the nerve-cells and their over-strain induce contemporaneous hyperæmia of the part, and this hyperæmia tends to keep in action its own causes. From frequent repetition of this condition the normal tonus of the arterioles is gradually lost, and not only in the cortex but in the over-lying meninges also. Hereby is prepared the way for sudden or protracted meningeal and cerebral hyperæmias, which embarrass the brain-circulation and brain-nutrition in a more or less protracted manner, and leave behind them more or less permanent effects or traces.

It may be that the exhaustion from over-excitation of the cerebral nerve-cells induces paralysis, ever-renewed, of the vaso-constrictor filaments, usually assigned to an origin in the cerebro-spinal system through the sympathetic. But it is by no means clear that the vascular dilatation and sanguineous over-fulness may not be kept up by frequent, or ever-repeated, direct, or reflected irritation of the vaso-dilator or vaso-inhibitory filaments supplied to the local peri-vascular vaso-motor centres which maintain the normal arteriolar tonus—centres whose stimulation, direct or propagated, causes contraction of the blood-vessels animated thereby; and the inhibition of which causes vascular dilatation; the inhibiting influence being transmitted to the peri-vascular ganglia by the vaso-dilator or vaso-inhibitory filaments. Furthermore, we know that vaso-motor centres are situate in the cerebral-cortex, as well as in other parts of the nervous system, and that the irritation or destruction of these may influence the blood-supply of the brain, of other parts, and glandular secretion.

Be this as it may, we resume the consideration of the repeated, and more or less persistent, cerebro-meningeal hyperæmia. In consequence of this there is distension of the vessels, circulatory impediment, irritative overgrowth of the connective nuclei of the walls of the vessels, and probably also of the neuroglia, while others of the nuclei and cells often termed embryoplastic, or their materials, are, perhaps, directly effused. Other changes occur in both, also, as a result of the lowered standard of the local processes of nutrition. Moreover, out-wandering of white blood-corpuscles, and escape or extravasation of red blood-corpuscles, may further choke the parts. There is a constant tendency to diffusion of all the morbid processes, and among the macroscopic changes are a thickening, opacity, and œdema of the superjacent meninges. In the meanwhile, the nerve-cells of the parts diseased, under the influence of morbid and excessive activity, have failed in their nutrition, while *they* also feel the effects of the surrounding vascular and neurogliar changes;—hence their swelling, cloudiness, and final degeneration. Then, if a chronic, mild, adhesive form of inflammation sets in, fibrinous effusion, and the subsequent production of connective tissue, assist in more completely involving the nerve-cells, and in tying down the membranes to the cortex. These changes proceed in the usual degenerative coruse, and finally as a result of them we find the processes of the nerve-cells cut off by the way—the cells themselves atrophic and degenerate—the blood-vessels fatty, calcareous, pigmented, and misshapen,—and the, formerly hyperplastic, neuroglia now atrophied.

The morbid process having once set in it must produce a series of effects in any given centre or region. Let us see for a moment what occurs in such centre or region, (1) when considered theoretically as if detached, and (2) when the total effects of the lesions of all the centres diseased are considered.

In the simplest case of all there is merely progressive impairment of the function of mental, motor, and sensory centres; any over-action observed being the result of partial withdrawal of the normal inhibitory influence of higher centres, and hence freed action of lower centres. In these cases hyperæmia is either less, or is negatived in some of its effects by the predominance of parenchymatous, or of interstitial, changes, or of both; which, as it were, jugulate at its very birth the functional energy of the active elements.

But in the majority of cases there is more or less over-activity of an irregular and inharmonious kind. The nerve-cells, embarrassed by the surrounding neurogliar and vascular changes, and flooded with superabundance of blood, discharge fitfully and irregularly. Their activity, excessive in amount, is irregular in rhythm and inharmoniously adjusted, or inco-ordinate, as to normal functional relationships.

A. *Mental Centres. Dementia.* (a). From disorder of this kind affecting the centres which are concerned with the anatomical substrata of purely mental function there is embarrassment in mental operation—inability for prolonged definite effort directed to any distinct purpose—inattention —a want of fixity of thought and of determination—a return to childish ways, and childish and more simple modes of thinking—a forgetfulness—and a failure to register and retain impressions—a dimming of the lustre of the higher virtues, of the general moral and æsthetic culture. Still later, incoherence and feebleness of thought, and a disappearance of all sense of shame, of all moral feeling, are observed:—while mental confusion, and failure of memory, of perception, and of judgment are painfully obvious. All these psychical troubles may be assigned to the morbid process in the frontal lobes of the cerebrum more particularly, and the diffuse character of the lesion renders the mental defect general, blurs registration of impressions, restricts and confuses reproduction, obscures judgment, and (after morbid excitement of it, or not) destroys imagination.

The most important question in this matter is whether the dementia is due solely to the lesions and wasting of the frontal lobes, or of some other part, or of the cerebrum generally. Voisin thinks that the demented form of g.p., besides the same coarse lesion as the other forms, has atheroma of the cerebral arterioles; the lesion is for a long time vascular only, yet bears more on brain than on membranes, and there are few apoplectiform or epileptiform attacks. In the second group of cases at the end of this work—very chronic cases, characterized on the mental side chiefly by dementia, as a rule; and singularly free from apoplectiform or epileptiform seizures—there is atrophy of the brain, and increase of intra-cranial serum. The gyri are wasted, especially on the upper surface and at the frontal region, their grey cortex being either softened, or, occasionally, of about normal consistence, pale, watery, sodden, or mottled. The meningeal changes are very marked, and extend to the base, and, like the other changes, are symmetrical. Microscopically, some of the cases in this group showed the atrophied and granular nerve-cells to be apparently more affected than the other elements of the cerebral grey cortex, especially in the frontal lobes. And in the third group, in which dementia is frequently early and predominant, the left cerebral hemisphere is the one chiefly diseased, both as regards its cortex and its general atrophy; the meningeal changes, as a rule, are well-marked, and, like the adhesion and decortication, are often well seen about the lower parts or base of the brain. In some other cases in which dementia predominated throughout, there has been marked diffuse induration of the cerebral white, or both grey and white; in others, the same in much slighter degree; in others, old durhæmatoma, or encysted collections of serum between frontal pia and arachnoid; in still others, large soft brains.

Case 50. In another case, microscopically, the nerve-cells of the frontal gyri were considerably granular, &c., the disease of vessels, also, was marked there, and the growth of nuclei; whereas in the temporal gyri the vessels were more healthy, although the nerve-cells were even more affected there than in the frontal; in the medulla oblongata the vessels were comparatively healthy, the nerve-cells somewhat affected.

Case 51. In another, my notes say "right frontal lobe; vessels, some with thick, coarse, irregular walls; many big spider-cells; numerous neuroglia nuclei of all sizes and shapes, and much wavy connective tissue; nerve-cells granular, fatty-like; but some only obscured, and defective in taking the stain" (aniline-magenta). Adhesion, chiefly on parietal and temporal g.; fronto-parietal cortex atrophied; slight induration of it and of subjacent white. Spinal meninges, thick, granulated.

On the other hand, Goltz found dogs with destroyed posterior lobes more stupid than those with destroyed parietal; and very stupid with destroyed parietal and occipital lobes, but spared frontal. Dogs with extensive destruction of both cerebral cortices showed disorder and defect of psychical perception of special sensory impressions. The temporary blindness and paralysis are supposed to be due to inhibition of base of brain and cord by the cerebral damage; they clear up when that inhibition ceases; the permanent results are then to be perceived as coarse movement, dulled sensibility of skin, defects of all special sense-perceptions, dementia.

Tuczek inclines partly to the view of Henle that the nerve-fibres, and not the nerve-cells, are the seats of mental activity —of the nervous process which expresses itself in motion and sensation—and in relation to diversity of function present differences *inter se* as much as, or more than, the nerve-cells: and while in g.p. there are no characteristic changes of the nerve-cells, there is always destruction of the delicate medullated nerve-fibres so abundant in the cortex; and his researches tend to show that the mental failure, as well as the motor disorder, in g.p., depends more immediately on wasting and disappearance of the subcortical association fibres of Meynert in the frontal lobes.

These failing first, there is at first defect of the higher moral and intellectual activities, of the highest, latest-acquired, ideal representations gained through speech:—reason, judgment, fail; and the acts cease to be guided, as formerly, by altruism, religion, morality, duty, patriotism, love of family, of truth, of friendship, of beauty; there is defect also of the close alliance of conception with motor impulses. Thence come change of character, and of the highest and most complicated activities; and incapacity for storing up and combining new mental acquisitions.

(β.) But mingled with the above in the early stages there is often *Exaltation or ambitious delirium* in some of its phases. Partly does this occur as a result of the unwonted and morbid stimulation of the ideational centres by hyperæmia, which rouses old paths of ratiocination, and brings vividly into consciousness the old day-dreams, the relics of ambitious thought and of long and fondly cherished visions of the mirage of hope—the whole taking a gay colouring, either from the excitation of the brain generally, or of definite parts of it possibly connected with the representation of the organic and other feelings. Partly from the above, then, comes the ambitious mania or expansive delirium, and partly as a result of the morbid excitation of cortical motor centres, giving rise to the subjective impression of an enormous outflow of energy —thence to a pleasurable feeling of power and force, and thence extending beyond its primary sphere to tincture all the thoughts, and to swell the happy patient with exuberant spirits, with exultation, and with extravagant notions as to everything in relation with himself, his honours, wealth, position—extravagant notions which the failure of perception and of the intellect generally no longer permits him to doubt, much less to appreciate in their true character. The hallucinations and illusions, especially perhaps illusions of the muscular sense, from which the patients may suffer can also originate or foment their expansive state of feeling and idea.

Again, what with hyperæmia, protracted excitation, and the *relative* embarrassment of the highest centres, there may be excessive and discordant uncontrolled activity of the lower; and, in consequence, protracted or paroxysmal mental and motor excitement.

Expansive delirium, attributed by Bayle to the effect on the cortical substance of congestion of pia, and inflammation of arachnoid, and by others supposed to be connected with the abnormal nutritive activity attending hyperæmia and inflammatory change, or to be the psychical response to over-irritation or over-heating of nerve-cells, is by Voisin confined to cases where the disease is simple and of cerebrum, pia, and arachnoid only; and by Mendel held to be inexplicable.

(γ). But what as to the cases in which *hypochondriacal or melancholic symptoms* are found, either sharing the throne with ambitious delirium, or excluding it temporarily, or, more rarely, excluding it permanently? If early activity of the disease-process, with its hyperæmia, leads to gaiety of feeling and optimistic delusions, there would be little wonder if, after its force was largely expended, the centres of motion and sensation, now exhausted and partially destroyed, gave origin to organic feelings mutilated and modified, to subtle impressions of constraint, limitation of power, and prostration of energy,—thence to depressing emotion, and thence to ideas of the hypochondriacal or melancholic order. Yet a recrudescence of the primary morbid process in the same centres, or its activity elsewhere, may traverse these secondary effects, and restore the exaltation and grandiose ideas. Not chimerical is this view that hypochondriacal delirium may be due to lesions of the cerebral cortex; the various visceral organs as well as the musculature are represented therein; by its experimental local irritation or destruction are these viscera affected; in the brain-cortex do organic sensations, as well as special sensations, find intimate alliance and highest representation. The early and rapid —or late and slow—destruction of motor centres, also, is adapted to call forth hypochondriacal or melancholic conceptions, the very antitheses of those former grandiose ideas the outcome of the wild whirling tide of life, in the motor ganglionic cells, begotten of hyperæmia.

Again, it is probable that temporary vaso-motor disturbance in parts of the encephalon may not unfrequently bring about a temporary mental change, and hence a transitory hypochondria or melancholia.

So, too, may alterations in the constitution of the blood

account for some of the mental phases, as the hypochondriacal or melancholic. In relation to this we may bear in mind the mental colouring usually associated with states of toxæmia; as, for example, the depression attending cholæmia; the ill temper, anxiety, and depression of chronic lithæmia; and the apathy and unconcern of pyæmia. So also in phthisis; bright, swift, and lively as are often the mental powers in the early periods, yet later,—and when the blood is probably much altered—caprice, fickleness, variability, and impatience are observed far more often than the so-called *spes phthisica*.

Again, the mental depression may be merely an ultimate reflex effect of impressions starting from some distant organ. The hypochondria of general paralysis may, like simple hypochondria, have its rootlets in some visceral or some peripheral affection or organic disease; in some disorder or loss of sensorial function; whether of the special, as evinced by hallucinations, illusions, or failures or losses of the special senses; or in failure or perversion of common, or organic, or visceral sensation, or in various neuralgiæ, hyperæsthesiæ, and anæsthesiæ, or in spasm of tubular or hollow organs, or of voluntary muscles. Now these may be present from the outset, and the case may take the melancholic or hypochondriacal form from the outset, also; or they may occur late in the course of general paralysis, and produce an intercurrent and temporary hypochondria. Vivid hallucinations of a distressing nature may call forth melancholic conceptions. Or when the patient finds that his imaginary untold wealth and power avail him nought *quoad* his actual possessions, liberty, and enjoyments, he oftentimes, in the early stages, becomes wrathful, or weeps and sobs like a grief-stricken child.

Finally, when melancholic symptoms in general paralysis are more permanent, or when, throughout, they are the most striking mental features, next to the failure of mental powers, I would refer to the evidence given below that in many cases the phenomena, perhaps, owe their explanation to the predominance of the morbid changes in the left cerebral hemisphere as compared with the right.

Other assigned sources of the melancholic delirium are lesions of dura (Meschede, Voisin); of cranial nerves, or of cord; anæmia (Gubler); cystoid degeneration of brain (Ripping).—Mairet has averred that in melancholic g.p., the sphenoidal region is chiefly affected. In some cases there is ground for this statement; a conclusion which, indeed, had previously been implicitly contained and partially expressed in my paper in the "Jl. of Mental Science," April, 1878, where, in group III, the cases of g.p. with melancholic symptoms are particularly concerned, and where I stated in italics (p. 36) that in these cases the meningeal changes *are often well seen at the base;* and one of the conclusions regarding this group, on p. 45, is as follows: "Adhesion and decortication usually more on left side, with equal frequency on frontal and parietal lobes, while the temporo-sphenoidal suffer very considerably, and the changes in question may be well marked on the inferior surfaces." The left hemisphere is more diseased than the right. Though the gross changes are more marked in the frontal lobes, yet the temporo-sphenoidal, especially the *left*, are comparatively much more affected than in the run of cases of g.p. (see Chap. XXI.) Spitzka explains the exhibition by G.Ps. of emotional manifestations opposed to their true emotional state by affection of medulla oblongata and pons. Luys * connects emotional displays with lesion of right first temporal g. (not g.p.)

Hypochondria in g.p., Voisin would deduce, in galloping g.p., from very rapid and extensive softening of all the periphery of the encephalon. Anæmia (Gubler), cystoid brain-degeneration, and colloid vessel-degeneration, were connected by some (Ripping, Schüle) with hypochondriacal g.p.; Mendel deems it to be unexplainable.

B. *Motor Centres. The widely spread ataxia and paresis in g.p.* If the centre under consideration should be a motor centre, besides the motor intuitions and the motor element in ideas involved, as already described, there will be the production of motor symptoms. But here the cortex is by no means alone affected. The vast disease at the periphery of the brain may be followed by secondary disease travelling through the white substance to the basal ganglia, and thence downwards; often to break forth, so to speak, with renewed energy in the medulla oblongata and spinal cord, and in the nerves thence derived. Nor, from this point of view, should the experiments of Lussana and Lemoigne be left unnoticed; experiments in which apparent foci of various special movements were found about the base of the encephalon, in the pons, medulla oblongata and other parts. The considerable amount of lesion often discovered at the floor of the fourth

* "L'Encéphale," 1881, p. 378.

ventricle in general paralysis, and the degeneration of cerebral and spinal nerves, warn us against too ready an indictment of motor centres in the cerebral cortex as answerable for the most frequent and characteristic motor impairment; that of the lips, tongue, face, and articulatory organs generally. But even in the production of this, the cortical lesion is at the very least an important factor, nor can I agree with those who attribute this motor disorder and impairment to bulbar lesions exclusively. Here, too, the function of the centres is disordered, and voluntary movements that require accurate and complex co-ordination can no longer be deftly or properly performed. For not only is *transmission* of impulses to movement embarrassed, obstructed, and here and there absolutely blocked, by the products of hyperæmia, nuclear overgrowth, or inflammatory effusion, but by the same means, also, is the energizing of the brain as an organ of volition made irregular, fitful, and disharmonious — characters which are immediately imprinted upon the voluntary movements, the general activity. The speech and writing often display this to perfection. Intimately associated with the disorder in the motor centres —if not produced by the same cause—is that perversion of the muscular sense, which plays so important a part in general paralysis.

Yet as the disease progresses, and as the cortical centres become more and more dilapidated, as their efferent fibres become more involved in the lesion, and especially as the centres of the peripheral nerves in the medulla oblongata and cord become affected, the symptoms partly flow therefrom. Sometimes the spinal and bulbar affection sets in with, or precedes, the cerebral. Further, as regards ataxic states, while there is disease of a system of subcortical nerve-fibres, and destruction of many fibres in the medullary substance, the intimate connection of the intellectual and motor functions and centres is shown by secondary descending degeneration from focal frontal lesions—the implication of the radial projection fibres in cases of extreme fibre-destruction in frontal lobes—and the circumstance that every act of thought involves a centrifugal activity.

C. Sensory Centres. Should one of the *sensory* centres, suppose the visual, be affected, there is primarily the rousing of ocular spectra, and even of hallucinations or illusional forms in accordance with the predominant nature of the thoughts and feelings at the time. The intellectual centres also react upon the sensory, and the excitement of cerebral circulation being general one would expect sensory disorder, even without *special* involvement of the cortical centres subservient to sensation. From the intimate association of sensation and thought in the genesis and edification of mind, the only matter of surprise is that sensory disorders are not more frequent and marked in general paralysis. The partial distribution of the more extreme morbid changes is the plausible explanation of the fact. Finally, like all the animal functions, that of the sensory ganglia fails as the brain sinks in progressive degeneration, and not only the brain itself, but also the peripheral nerves and nerve expansions subservient to sensory functions.

Other symptom-groups. The extensive relations of *speech* and of *gait* justify their insertion here.

Speech. For the several varieties of speech-affection in g.p., see Chapter IV. *Firstly.* The ideational disorders of speech, in g.p., are to be referred to affections of brain-cortex. In some of them the normal process of thought, of ideation, as exhibited in speech is deranged or defective; for disturbance or failure of the mental operations has its correlative expression in secondary disorders or defects of speech, such as slowness, pauses, drawling, verbal substitution, omissions, cluttering, repetitions, parrot-speech, fragmentary utterance, some forms of aphrasia, logorrhœa, self-conversation.

Interruption of speech may, in different instances, arise from intellectual weakness, from emotional tumult or distraction; or from hallucinations; or from the sudden invasion of a recurring delusion; or from the influence of words themselves upon the train of thought, which produces ideal confusion when the association of ideas is lax. Cessation of speech, in g.p., may have a mental origin; it may be due to absence of ideas, owing to the almost complete obliteration of mind; or may depend on the influence of vivid hallucinations and delusions in the earlier stages, as when a state like that of melancholia with stupor is present, or a hypochondriac exacerbation; and where the hallucinatory voices command the patient to keep silence, or threaten with punishment for speaking; or silence may be sullenly willed by the patient under the influence of morose feeling and delusions.

Secondly, are disorders of speech due to absence or defect of inward speech, symbolic expression, or diction. Of these the chief are ataxic and amnesic aphasia, heterophasia, word-deafness, asyntaxy (akataphasia), syllable-stumbling. They are of cerebro-cortical origin, the cortical action concerns itself with both the articulation of the word as a motor sign, and its recollection as an acoustic sign. And lesions of fibres between cortex and med. obl. may assist.

In reference to the temporary aphasic attacks in g.p., which are attended sometimes by abolition, sometimes only by cloudiness, of intellect, Kussmaul attributes them either to disorder of blood-circulation, or to swelling of spindle-shaped neuroglia-cells, in the cortex (Lubimoff). In my experience, attacks of temporary aphasia in g.p. usually follow distinct epileptiform or apoplectiform seizures, they are of the utmost variety in character, and degree, and duration, and I cannot linger to describe the varieties.* Thus, when a G.P. is smitten with aphasia it is usually † temporary, and after one of the several seizures already mentioned; and, as a rule, no localized destructive lesions are found. But permanent true aphasia may occur in g.p. with gross localized cortical lesions. Symmetrical lesions of the posterior and inferior parts of the third frontal convolution on both sides, and a slight lesion on the anterior and inferior part of the left one were observed in one case. ‡ The lesions were old, localized softening almost limited to the grey cortex.

Thirdly. Disorders of articulation; such as mumbling; stammering; stuttering; quivering or trembling speech. There is reason to believe that these are largely due to bulbar lesion; partly, and in the early stages perhaps solely, due to cortical lesions, or to those affecting the basal ganglia. The cortex has to do with the pronunciation of words not only as acoustic symbols of ideas, but also as the motor aggregates of sounds—organized motor units—which, as it were, it fashions and makes over to the lower articulatory centres for outward expression.

Whether fibrillar or coarse, the twitching movements of the face, lips, and tongue, during speech; and with these the shaky or quivering form of speech, may be explained as under "tremor" (see below).

Phonation may be fundamentally affected. From paresis

---

* Cases by present writer, "Alienist and Neurol.," Apr., 1882.
† "Jl. Ment. Sci.," Oct., 1875, p. 421. Present writer, *Ibid.*, Jan., 1876, p. 583.
‡ Billod, "Ann. Méd.-Psych.," May, 1877, p. 339.—Cullerre, *Ibid.*, 1878.

of this or that muscle of the larynx, and altered position and action of the vocal cords, or from local physical changes, as of congestion, swelling, degeneration of vocal cords, or changes in other parts of the vocal apparatus, may arise a series of phonopathic conditions. The affection of phonation is explained by bulbar lesions, or by secondary (or primary) peripheral lesions (nerve, muscle, &c.).

Finally, the patient may become almost speechless or mute from extreme cerebral changes; or from extreme degeneration in the nerves, nerve-nuclei, or muscles of the articulatory mechanism (medulla oblongata, larynx, tongue, lips, etc.).

Resurvey. Of the speech disorders, the articulatory have been held by some to be entirely somatic in origin and ataxic in nature, and due to lesions, direct or indirect, of nerve-centres in the medulla oblongata. Nevertheless, when slight, and at first, in the early stages, they are probably of cortical origin, the facial nerve having relationships with the lower part of both central gyri, and even with (among other parts) the back part of the second and third frontal; the hypoglossal nerve with the back part of the third frontal and adjoining portion of both central gyri; and several highly specialized and co-ordinate actions of the muscles animated by those nerves being impaired by disease of these cortical fields. Thus, at the very outset of g.p. there may be a syllabic stumbling and stuttering coexistent with the early mental affection. These symptoms, thus early, are often associated with a slight twitch of face and tongue, which is in appearance ataxic or quasi-convulsive rather than paralytic. It is not here so much a mere motor paresis that is produced, or an impairment of cortex as representing muscles, as it is an impairment, particularly, of the cerebral cortex viewed as a complex structure containing centres or nervous arrangements for regulating the motor adjustments necessary for the higher and more complex sort of acts, such as speech, writing, and various arts and trades, and especially for the more voluntary, intellectual, and more highly differentiated and specialized, parts or examples of these.

Thus, the paretic element in the affection of speech may undoubtedly derive from the lesion of cerebral cortex. But may not, as well, the form of inco-ordinate action observed in the speech early in g.p. as a rule, be also of cortical origin? For, as I have long ago said (1st Ed., pp. 151 & 155), at first there is fitful and irregular discharge of the nerve-cells, and an activity excessive in amount but irregular in rhythm, and inharmoniously adjusted to normal functional aims. Not only is transmission of the impulses to movement embarrassed and obstructed by the morbid changes, but by the same means is the action of the brain as an organ of volition made irregular, fitful, disharmonious.

And still another point in favour of the view advanced is the circumstance that, in some cases of g.p. where death occurred very early, and where speech only was affected, without any other indication of hypoglossal disease or participation, the bulbar centre of the hypoglossal nerve showed no marked changes; whilst, on the other hand, in some cases of long duration with advanced signs of tongue-palsy there were extreme atrophy and fatty pigmentary degeneration of the cells of that bulbar centre.

But more than the cortical disease is required to explain all the clinical facts in the later stages (at least in most cases), and hence the lesions of vessels and of some nerves and nerve-centres (7th and 9th) in the medulla oblongata must here be brought into requisition, and morbid changes in the pons and basal ganglia may be contrbiutory.

The cortical speech affections of g.p. cannot be attributed to lesion of one convolution; a single and simple language-centre or speech-convolution does not exist in the brain. Numerous centres, widely separated, complexly interconnected, and subserving sensory, motor, and intellectual functions—and probably not concerned in speech alone—do yet co-operate and focus their energies in the production of articulate language.

Why should speech be so strikingly affected in g.p.? The answer is that through language, written and spoken, are gained all the highest and latest mental acquisitions, the moral, æsthetic qualities; the love of the true, the good, the beautiful; in the race and in the individual. The speech sums up the mental life in the individual, and in the race, and is the great point of blending for the intellectual acts of the greatest diversity, complexity and delicacy, with the most complex and delicate of motor co-ordinations guided by sensory impressions; and from cortical disease and destruction, the highest and most delicate and latest acquired accomplishments fail first. The speech, therefore, must be one of the very first to fail in g.p., an affection characterized by loss of the highest intellectual, and of the most delicate motor, activities.

Gait. Paresis and Hemiplegia. The gait in g.p. varies with the relative proportions of ataxy and paresis present. Yet, as an incident of some spinal changes, the gait may present spastic features. An act like that of walking, with the numerous adjustments involved therein, in order to be normal, requires a healthy state of :—(1) Systems of afferent fibres; of (2) co-ordinating centre or centres; and of (3) systems of efferent tracts in connection with the muscular apparatus concerned. Disease or interference with the function of any one of these will disturb the power of maintaining the erect station and equilibrium of the body, and the power of locomotion. Equilibration and locomotion are usually supposed to have their co-ordinating centres particularly in the corpora quadrigemina, pons Varolii, cerebellum and middle cerebellar crura. These may be influenced and disordered from above. A defective power of equilibration and an inco-ordination of the locomotor acts may, of course, not be due to disease or disorder of the co-ordinating centres at all; for, as we have just seen, these centres and parts form only one of the three groups of factors concerned in the production of the co-ordinated motor adjustments. Thus, a notable cause of ataxic gait is the spinal disease observed in tabes dorsalis. Looking, then, at the phenomena of tabes dorsalis, and at its customary lesion, and observing a frequent posterior situation of disease of the spinal meninges and cord in g.p., one might at first be inclined to attribute to the spinal lesions of G.Ps. their usual troubles of gait, and defective power of equilibration. There are facts which prevent one from fully accepting this view. The affections of the limbs and trunk, in greater or less degree, are an almost essential part of g.p.; I have never met with a case of g.p. without marked cerebral disease; I have seen cases with no, or with comparatively slight, disease of the spinal cord.

Yet are there cases of g.p. in which spinal lesions appear simultaneously with, or even precede, the cephalic; and perhaps with a history of severe protracted pain in back and limbs. In these cases the gait is usually ataxiform, its impairment and the patient's helplessness are comparatively early; and the spinal disease, extending, may eventually cause, also, true paralysis, which further increases the helplessness. Inasmuch as in the cases just referred to I have usually found the evidences of chronic spinal meningitis, and in some cases grey degeneration, or white-hued sclerosis, or myelitis and lessened consistence, of the posterior spinal columns, I attribute the ataxic or tabic form of gait of g.p., when of well-marked tabic characters, to the spinal lesions; although there are in some other parts symptoms we are accustomed to call ataxic which are not of spinal origin.

True as this may be with respect to the well-marked tabic form of gait, it cannot, I think, be accepted as explanatory of all the disorders of gait in and throughout the course of g.p. And here, in some points, the views urged with reference to the part taken by the cerebral cortex in speech-disorders might also be urged with respect to locomotion in g.p., and to the connection of disorder of the finer locomotor actions and adjustments with cerebral disease. Thus supporting the thesis that both early inco-ordination and paresis in gait may depend on cerebral disease. But we need not travel over any part of that ground again. Nevertheless, in this relation, several facts, drawn from experimental physiology and from clinical observation, might, had we space, be cited here in support of this view; such as some experiments on the brain of lower animals. But the higher the animal in the animal scale the greater is the importance of the cerebral hemispheres in all that relates to station and locomotion; and in man some cortical cerebral lesions produce paresis or paralysis,—I, myself, have met with a number of examples of the kind—and thus far may interfere with the capacity for this erect station and for locomotion. What is stated elsewhere on the local pareses occurring in g.p. bears on this question whether the paretic gait in g.p. is ever, or in any degree, due to cortical brain-lesion. And the possibility of the connection between paresis in g.p. and cortical disease gains in probability from :—

1. The fluctuating character of the paresis, its rapid changes in extent and intensity.—2. The accompanying and following phenomena of motor irritation, spasms, cortical epilepsy, etc.—3. The existence of well pronounced changes in brain-cortex, in acute and sub-acute cases, while, then, the lower centres are far less affected.—4. The marked disease of, so-called, cortical motor zone in g.p.

The basis of early and temporary unilateral paresis is circulatory or exudative change; usually with some form of "seizure;" and temporary exhaustion of nerve-cells after their convulsive discharge was the explanation long ago given by Dr. R. B. Todd for epileptic paresis. In the last stages, permanent or slowly augmenting hemiparesis is associated; rarely, with local gross encephalic lesions; more frequently, with a greater degree of meningitis and adhesion of the opposite cerebral hemisphere than of the other; most frequently, with its greater atrophy. I have rarely found it connected in g.p. with local hæmorrhage, or softening, or local periencephalitis, to one of which it is ordinarily attributed. This atrophy—an occasional cause of hemiplegia in g.p.—is not infrequent in my experience, and on the opposite side are often convulsions and transitory pareses.

Nevertheless, although the cortical changes, including circulatory, suffice to explain the commencement and early part of the paresis, one cannot thus explain all the kinds and degrees of paresis and paralysis in g.p. For when the disease is advanced the paresis is increased, and modifications arise; such as spastic gait, "contraction," choreiform movements, tremor; then also the basal ganglia, medulla oblongata, and even the peripheral nerves are diseased, and myelitic and cord-sclerosing conditions arise in various sites and tracts, which, when affecting the pyramidal tracts, are likely of a secondary descending nature. Since the above was written as a passage in a paper read by me at annual meeting of Brit. Med. Assoc. in Aug. 1883, and since a passage on the subject in 1st Ed., 1880, p. 154; M. Pitres has published some investigations on the degenerations of the cord secondary to disease or local destruction of brain in definite sites. In man the degeneration is marked, and extends low down the cord; in dogs and cats the degenerative band is very slender in the cord; in rabbits and guinea pigs it only attains the lower end of medulla oblongata; in fowls it is entirely non-existent. These results tend to confirm my views expressed above. See, also, statements on secondary cord-changes in the last Chapter.

The spastic form of gait particularly attends sclerosis of the cord's pyramidal tracts. With the chiefly superficial diffuse disease of the brain in g.p., secondary degeneration of lateral columns is usually slight or incipient only, and somewhat diffuse. Cases reported by myself * and others show that a local lesion of the internal capsule, or of still lower parts, is not necessary, to produce this secondary descending sclerosis, as some have asserted it to be; and that severe diffuse disease of brain-cortex is sufficient. But in g.p. the affection of the lateral columns may also follow local lesions lower

* "Alienist and Neurol.," April, 1882. "Jl. Mental Sci.," April, 1885.

down; or, possibly, may be primary. Leyden found spastic paralysis in various spinal affections—acute, sub-acute, or chronic—as, partly diffused ascending meningitis and perimyelitis, partly circumscribed perimyelitis; acute myelitis in 2nd stage; chronic myelitis in one or several foci; chronic leucomyelitis; chronic myelitis without sensory affection or vesical palsy. But spastic paralysis may arise from any condition which raises the reflex influence of sensory nerves, whether in consequence of reflex irritability of motor fibres or of sensory roots, or owing to interruption of conduction of voluntary impulses from the brain; reflex conduction through the grey matter being preserved or even increased.

Therefore, the basis of the paretic and spasmodic affections of the limbs and trunk in the *early* stages of g.p. may often, or usually, be sought in the cerebrum; as examples of which I may refer to the early paretic form of gait and the failure of manual and digital aptitude and dexterity; and *later on*, the brain-base, pons, bulb and cord are variously concerned in different cases.

Apoplectiform attacks. Whether the attack results from active congestion of the brain, or whether it is conditionated by a discharge of nervous elements, I believe the state of brain actually coexistent with the acute apoplectiform state itself is one of congestion, at first active, even if afterwards passive, and probably some œdema; and certainly (although it was partly due to the mode of dying) in all cases in which I have made a necropsy when G.Ps. have died, struck down by active and fatal apoplectiform symptoms distinctly shortening life, I have found more or less extreme congestion, in some cases intense engorgement, of the vessels of the brain, its meninges, and the skull; or else meningeal hæmorrhage, with or without some remaining congestion. An associated thrombotic element is sometimes an important factor.

3. *More or less Local Motor symptoms.* We may now for a brief space follow the fortunes, in general paralysis, of single, special, so-called cortical centres.

In examining the brains of general paralytics with reference to the doctrine of localization, it is necessary in the first place to ascertain the portions of nervous tissue more particularly diseased in each case. In this inquiry, aid is usually derived from the individual distribution of the cerebromeningeal *adhesions*. Habitually forming verbal or diagrammatic charts of these, we soon learn to recognize anything unusual or peculiar in their extent or distribution, in a given instance. In several cases, microscopically examined by the present writer with reference to this point, the portions of cortex underlying the adhesions were more diseased than the cortex of neighbouring gyral summits, free from meningeal adhesions. In the investigation of this phase of the question of localization I have elsewhere * insisted upon the value of the precise distribution of these local adhesions of the pia-mater to the cerebral cortex, and of the changes in the latter associated therewith.

Of the cortical centres, the *psycho-motor* offer the most frequent and convenient opportunities of investigation with relation to this point. When I began to study this subject most of the motor symptoms in g.p. were almost universally attributed to the spinal and bulbar lesions. This was either explicitly avowed, or implicitly contained in the doctrines then current. But in the "Journal of Mental Science," January, 1876, I attempted to trace a relation between the special distribution and range of the convulsive symptoms, the pareses and the speech-troubles in g.p., on the one hand, and on the other, the distribution of the cortical changes, especially of that associated with adhesion to the meninges. In this, as well as in every case examined for years before, and since, I have kept minute records of all the cerebro-meningeal adhesions, in order to test various views as to the localization of cerebral functions, by comparing the differences in the distribution of the adhesions with the differences observed in the clinical symptoms, in different cases.†

In the "Journ. Mental Sci." for Apr., 1878, and for Oct., 1881, and in the 1st edition of this work (p. 179), I stated that as far as I knew these were "the first investigations into the exact localization and distribution of the adhesions in various cases of general paralysis, for the purpose of throwing light upon certain clinical features of the disease." Having thus thrice publicly claimed priority as the pioneer in this matter, and this position having also been ascribed to me by Dr. Hack Tuke in his Presidential address before the Medico-Psychological Association in London in 1881,‡ and no evidence of an adverse nature having been brought

---

* "Journal of Mental Science," Jan., 1876, p. 567. *Ibid.*, April, 1878. *Ibid.*, Oct., 1881; and 1st Ed. this work.
† "Jl. Ment. Sci.," April, 1878. See Chapter on Varieties of g.p., and cases at end of this work (both Eds.).
‡ "Journ. Mental Science," Oct., 1881, p. 329.

forward, I conclude, provisionally, that no such evidence is, or can be, forthcoming, and that I justly claimed to have been the pioneer in this matter, and first contributor to the solution of the questions raised, and that by a new and special method of investigation. And here I would point out that my paper (written May, 1875) was not, as were the principal papers, by others, which, in a year or two, followed it on this subject, a theoretical sketch of the possible result of the lesion of this or of that convolution in g.p.; or theory of this kind, if with delineations of brain-adhesions, yet without even any attempt to explain the symptoms by the lesions in the individual cases depicted; an explanation which, in those cases, would have been almost purely hypothetical. My paper, on the contrary, gave the symptoms, and the precise distribution of the lesions and adhesions, in one out of a number of cases in which I had carefully recorded both; and it embodied an attempt by this means to test definite physiological doctrines as to the localization of cerebral functions, published shortly before that time.

That which I wished to test, Dr. Foville subsequently assumed. Taking the cortical lesion in the fronto-parietal convolutions to be constant, he believed that the progressive disorders and defects of the motor functions in g.p.—such as speech-affection, spasm, convulsion, ataxy, teeth-grinding, contraction, paralyses—are explained; at first, by the excitation produced in the different motor centres situated in these convolutions, and due to the hyperæmia; then, by the successive congestive attacks; and, finally, by the progressive sclerosis in the period of decline. These views were purely hypothetical, but were those naturally flowing out of researches on the localization of cerebral function. Similar views, carried further and applied to the explanation of the mental and sensory symptoms, were, afterwards, published by Dr. (now Sir) Crichton Browne, who enriched his paper with plates exemplifying the distribution of the adhesions. Still later, Dr. Dufour traced out the adhesions and atrophies in several cases of g.p., and endeavoured therefrom to explain the functional localizations in various forms of the disease. See also Tamburini and Ricci, above.

Convulsions. Of the motor symptoms in g.p., the convulsions and spasms, when present, have the greatest prominence and most dramatic character. At first attributed to secondary inflammation with softening, and occasionally, meningeal adhesion of the cerebral cortex; or, rarely, to rapid serous effusion, they were at a later period attributed to seizures of cerebral congestion,—still later, to sudden anæmia from contraction of cerebral blood-vessels induced by various "irritations," and probably connected with an increased reflex-excitability of the medulla oblongata; and, later still, to any one of several causes such as meningeal congestion or ecchymosis, pachymeningitic turgescence or hæmorrhage; intense cerebral congestion, or even minute hæmorrhages, either capillary, or from rupture of ampullary dilatations of arterioles; or, by Oraggi, to stasis in the meningeal veins, easily brought about by disturbances of the circulation when the meninges are thickened. I have attempted to explain convulsion, in some cases, by the cerebro-meningeal adhesions; while others aver that the fixation of the vessels at these adhesions, together with the changes in the vessel-walls may, under very slight and ordinarily inoperative influences, lead to passive hyperæmia, and rapidly developed œdema, which may also account for the residual pareses; in connection with which are cited experimental production of epileptoid attacks by rapid changes of cerebral pressure; by irritation of dura, etc. Explained by vaso-motor spasm, the attack has been compared with normal rigor. Convulsion in g.p. has also been deemed to be a cortical epilepsy, brought about by simple nutritive disorder; and the increased paralysis and mental weakness after the seizures to be due to affection of the nerve-fibres.

To abnormal excitation of cortical centres may be referred the local twitches, spasms and convulsions, the general convulsions, the general muscular agitation. These can be very closely reproduced by electrical excitation of portions of the brain-cortex in the lower animals; most are transitory, and, perhaps, recurrent; others are more frequent, or more persistent. Practically, however, it is only in the case of the more localized convulsive twitch and spasm that one can tell the probable site of the lesion which brings about excitation of motor elements. When spasm spreads, when convulsion becomes more generalized, it matters not in what part it begins, or where alone it may for a time be manifest, or whether the parts first involved in convulsion are those which in health have been most in voluntary use; the lesion —producing that state of nerve-element which eventuates in abnormal excessive discharge, and is translated outwardly in convulsion—may in such a case occupy any one of a number of localities, or may be widely diffused. In fact it may not be in the cortex at all, but in the medullary tracts of the brain, or lower still. I have paid much attention to the above point, in g.p., and with the result of confirming the view I expressed some ten years ago.

Nevertheless, many symptoms of the groups now under consideration can with more confidence be referred to circumscribed local excitation of the cerebral cortex. Such are, many at least of, the local paralyses and rigidities; such, also, are the local twitches and spasms without loss of consciousness, now ceasing, now possibly transferred to another part, and anon returning to their primary site. Convulsion here is circumscribed, and is reduced to local convulsive twitch, or spasmodic distortion of a few muscles, or of one,—convulsive twitch which may continue for hours or days without ceasing; may alternate with equally localized twitch or spasm elsewhere, and which, when present, constitutes a striking feature in general paralysis. These partial convulsions, as Dr. Louis Landouzy points out, may be produced by many diseased conditions affecting the fronto-parietal region, such as tubercular meningitis, softening, local hæmorrhage or encephalitis, abscess, tumour, or injury. In this aspect the *nature* of the lesions is secondary, their *seat* everything. The more extensive of these convulsions for the most part accord with the description by Bravais, and later, by Dr. Jackson, of three kinds of unilateral epilepsy : the convulsion beginning in one at the face, in another at the hand, and in a third at the foot, consciousness being usually retained, the voluntary power over the part affected often being not wholly annulled during the "fit"; and paralysis, confined to the same limits as the convulsive movement, ordinarily following the latter as the shadow follows the body.

The cortical localized *paralyses* often seen in the affections mentioned above, like the local paralyses more commonly observed in g.p., have the following characters. They are, at least at first, I., partial; II., incomplete; III., transitory; and IV., variable. Any of these characters may be absent in a given case.

In the above groups of partial convulsions, spasms, and paralyses, it would appear as if the experiments of the physiologist were reproduced by disease, and one looks for some definite local, and probably cortical, lesion in explanation. Vain search, too often, in general paralysis! Many cortical centres, as a rule, are markedly involved in obvious change, and when those extremely diseased are comparatively few in number they often are not homologous with those by irritation of which similar results are produced in the lower animals.

By augmenting the duration and intensity of an experimental excitation of a cortical centre in the motor zone, the convulsive movement, at first limited, may become general; and the same thing may occur in g.p. And truly it must be confessed that the centres actually discharged, as represented outwardly in the convulsion or spasm of g.p., are not necessarily those most obviously diseased; for, without effecting important structural changes in them, the intra-cranial morbid process may utterly disorder their functions by the disturbance it necessarily causes in their circulation and nutrition. The convulsions, also, are not always of cortical origin: and it may be added that modifications of the blood itself may here suffice to rouse unstable centres to convulsive action.

I might relate at length the exact march, and distribution of many convulsive, spasmodic, and paralytic seizures in general paralysis, and compare them with the post-mortem records. But the recital would be tedious. The difficulty has been that in the vast majority of cases the cerebral lesions are very extensive. In some of the cases wherein the lesions apparently producing convulsions and spasms were more localized I have been unable to trace a harmony between these and the results of physiological experiment; in other cases they have seemed to harmonize fairly. I conclude* that convulsion and spasm, in general paralysis, when of cortical origin, are sometimes due to localized cortical lesions, to which the distribution of the cerebro-meningeal *adhesions* often affords the clue, but that sometimes there is diffused cortical "irritation," which expresses itself outwardly in convulsions which follow one or other of several favourite courses, and begin in parts which are highly differentiated and whose movements are of the more special and voluntary kind; and that very similar convulsions may sometimes be produced by irritation commencing in different parts.

Rigidity and contractions of the limbs in the early stages of g.p., and especially those associated with convulsions or palsies, may be assigned to lesions affecting the cerebral cortex or the cerebral medulla;—in the later stages they may be due to the same, or to similar spinal lesions. Never-

* As also in "Journal of Mental Science," Jan., 1876, p. 575, 576.

theless, in explanation of palsied, rigidly contracted, and spasmodic states of the limbs in the latest stages one may, in some cases, invoke the aid of the secondary descending spinal sclerosis, intimately related to lesions of the internal capsule or its irradiations, as described by Türck, Vulpian, Charcot, Flechsig, and Pierret, and which follows lesions pertaining to several groups, one of which consists of cortical lesions of some extent and depth situate in the central gyri and contiguous parts, or in the paracentral lobule. Or the spinal myelitis may follow spinal meningitis, or be primary.

Again, contractions of the limbs frequently supervene in ordinary meningo-encephalitic affections, of a limited local nature. A similar cortical irritation occurring in g.p., may account for some of the more transitory recurring "contractions" therein, such as those following "paralytic" attacks. Voisin assumes that "late" contractions, when occurring in g.p., are a result of cerebral or meningeal hæmorrhage. But I have so often seen all degrees of "contraction" and rigidity in g.p., from the most transitory to the permanent, from the slight to the most resistant, without any basis in cerebral or meningeal hæmorrhage, that I attribute them as a rule to the causes already mentioned; cerebral and spinal congestion, meningo-encephalitic conditions, if transitory, or even sometimes if more permanent, and in the early and middle stages;—to secondary spinal or neural changes when more persistent and in the later stages. Occasionally, the deformity depends on contraction of one set of muscles owing to palsy of the opposing set. Dr. A. Waller, in 1851, first described the wasting of *nerves* consecutive to various injuries and lesions; and similar changes play their part in g.p.

Bowed head and back. Possibly due to irritation of cortical centres, the persistent bending forwards of the head and neck, sometimes seen, is perhaps due, rather, to changes in the upper part of the spinal cord, and therefore, like the rigidity and contraction of the limbs in the latest stages, may be the result of secondary spinal and neural changes.

Conjugated deviation of head and eyes. Of the local motor symptoms of cortical origin, occasional in g.p., one of the most interesting is the rotation of head and conjugated deviation of eyes—a symptom in cerebral affections to which Prévost and Vulpian first directed attention. Probably, in some cases this is paralytic in nature—in others spasmodic; —being due, in the one case to the unopposed action of the healthy rotator centre in one hemisphere, that of the opposite hemisphere being temporarily put out of use by some lesion or inhibition; and, in the other case, to morbid unilateral excitation of the rotator centre.

Tremor. The ordinary tremor and twitch of the face, lips, tongue, especially during speech, and of the limbs, can be accounted for during the greater part of the course of g.p. by the weakened and interrupted motor impulses, deriving from the early-diseased psycho-motor centres, causing an analysis of the normally fused contractile waves into separate component elements, and thus substitution of one sub-continuous contraction by several interrupted ones; or that, in this condition of weakened nervo-muscular apparatus, muscular twitch and trembling is produced by a moderate or slight effort as it would be by a severe, prolonged, and straining effort in health. In those cases where there comes at last a far more constant tremor this may be due to degeneration of the nerve-nuclei of the nerves supplying the muscles affected, together with secondary degeneration of these nerves and muscles, bringing about a tremor, like as after experimental section of motor nerves.

Tremor cöactus; choreiform movements; *general constant tremor;* are dealt with in Chapter V.

Teeth grinding. Widely differing explanations have been given of this symptom in g.p. That it is simply a habit (Voisin), that it is a convulsive symptom (Baillarger), or from morbid excitation of the cortical centre, related to movements of the lower jaw (Foville, Browne), or from spasm of the internal pterygoid muscle innervated by the fifth nerve (Mendel). Spasm in the distribution of this nerve would, however, produce spasm, also, of the external pterygoid muscles, important agents in trituration of food; also of masseter and temporal muscles; and this is the explanation I prefer.

Fæcal incontinence. "*Dirty habits.*" At first fluctuating, latterly becoming permanent, this condition may arise from the inattention, indifference and apathy of dementia; from anæsthesia associated with other factors; from paresis of sphincter, at first with congestive conditions of brain and cord, later with degeneration of cord and spinal nerves, and associated with some degree of anæsthesia.

Urinary incontinence. "*Wet habits.*" Similar remarks apply here as to the last symptom. The inattention and shamelessness of dementia; the local impairment of sensibility, and relaxation, paresis or paralysis of the sphincter vesicæ, are blameworthy in various degrees in different cases. Congestive states of brain usually, epileptiform attacks occasionally, account for urinary incontinence in the earlier stages. The organic changes in the cord are often added to the above in the later stages. If the muscles of the bladder are paretic, but the sphincter retains power, there is retention of urine. If with anæsthesia the expulsive muscles of the bladder are paretic, there is retention of urine until the bladder is over-full; or if all the muscles of the bladder are equally paretic the urine dribbles away almost constantly from a partially over-full bladder. I do not agree with Dr. J. Christian that the wet and dirty habits of G.Ps. and their difficulty in swallowing are merely due to forgetfulness and mental failure.

Dysphagia, a late phenomenon, is probably due more to changes in the central bulbar nuclei of the nerves, and, finally, in the nerves and muscles themselves which constitute the apparatus engaged in the act of deglutition, although the cerebral condition has a distinct influence on it. The nervous mechanism engaged in this series of operations is a very complex one; and in g.p. the series may fail because of anæsthesia in the distribution of the trigeminus and glosso-pharyngeal nerves; or of paresis of muscles animated by branches of the vagus, hypoglossal, trigeminus, facial, and probably spinal accessory. Either the complex act is not roused by the usual afferent impressions; or the efferent impulses fail and cease; or these conditions are linked.

So-called reflexes; knee-jerk, etc. In Chapter V., I gave the spinal and cerebral lesions in cases of exaggerated and of absent knee-jerk. With *exaggerated knee-jerk in g.p.* I have found the spinal changes vary. In some, they are chiefly meningeal; the pia or arachnoid being thickened and granulated at the upper part of the cord; or the spinal dura adherent to surrounding parts. In one case I found inflammatory hyperæmia of lower part, right side, of medulla oblongata. In other cases there is diffuse myelitis. In some, myelitis or degeneration has softened the upper, and indurated the lower, part of the cord. In others, there is more or less sclerosis of the lateral columns, and, usually, of other parts; or of the crossed pyramidal tracts solely or chiefly.

In cases of *absent or slight knee-jerk in g.p.*, also, the

morbid changes vary:—chronic spinal meningitis, thickenings, adhesions, granulations, congestion; or grey degeneration of posterior and posterior median columns of cord; or slight degrees of myelitis, chiefly affecting the posterior columns; atrophy of spinal cord, or some of its grey cornua; or softening of lower part of cord.

The following do not apply to g.p., solely. Leyden connects with perimyelo-meningitis, spastic paralysis and exaggerated knee-jerk; but Dr. J. C. Shaw, usually, absent knee-jerk. Westphal held that there is absence of knee-jerk in degeneration of dorsal part of posterior columns of cord, even when the lateral are also affected. This view is not altogether supported by cases published by Strümpell * and J. C. Shaw, and others observed by myself, and some of them seem to be opposed to it. Claus † found absence of knee-jerk in pronounced posterior sclerosis; except in some cases where the sclerosis was not below the sixth dorsal-nerve level, or (in 2) even where slight sclerosis extended to the lumbar cord—but here only if the innermost portions of the posterior column, solely, were affected. This forms a complement to Westphal's assertion, that with degeneration of external part of posterior columns the knee-jerk always disappears. With combined sclerosis of posterior and lateral columns extending low down, Zacher found no tendon-reflex after "paralytic" attacks; Claus found knee-jerk, but here the posterior columns were not affected low down. Westphal stated that any G.P. with affected gait and constant absence of knee-jerk, or with the latter only, has degeneration of posterior columns, extending down to lumbar region; and that increased reflexes may result from primary disease of posterior and lateral columns, or from disease of the antero-lateral, and lower part of the posterior.

Pupils. That, as Prof. Tamburini believes, and as Dr. Foville suggests, the pupillary modifications in g.p. are due to excitation, to exhaustion, and to final destruction, of some cortical centre, is true in some cases, and Viel has made a clinical and experimental study on the pupillary changes caused by artificially produced, local, meningo-encephalitis; while functional excitation of cortex causes pupillary change. The iridal sluggishness to light—and, incidentally, the pupillary inequality—was attributed by Dr. O. von Linstow ‡ to

---
* "Brain," Jan., 1881, p. 563.
† Cited "American Jl. Neurol. and Psychiatry," May, 1882.
‡ "Allg. Zeitschr. für Psych.," xxiv Bd., p. 436.

defect or cessation of the reflex from optic to oculo-motor nerves mediated in the brain.

Inequality of pupils. Dilatation of pupil is readily effected by irritation of great sympathetic, or by palsy of motor oculi nerve; and, conversely, contraction of pupil, by irritation of motor oculi, or by paralysis of sympathetic. Prof. Förster * refers to Dr. C. Wernicke's † hypothesis. If the wider pupil is associated with lessened activity in accommodation, the pupil not reacting in convergence, or to light, or contracting less than that of the other eye—there is an obstacle affecting, or a defective innervation of, the oculo-motor nerve. But if, on the contrary, the pupillary difference is inconsiderable, is lessened by light-impression, and disappears in strong convergence, the pupillary diameter regaining the size normal to it under that condition ($\frac{1}{2}$-$\frac{3}{4}$'''), a paresis of the oculo-motor may safely be excluded, but the widened pupil probably is so from an irritation of the sympathetic nerve, or centrum ciliospinale of Budge, or central path of the sympathetic. In a third group, the relatively, but not abnormally, narrow pupil (1$\frac{1}{2}$-2''') shows loss of reaction to light, but prompt activity in strong convergence-position, the wider pupil reacting well under both conditions. An affection of the fibres connecting the optic and oculo-motorius of the eye possessing the narrower pupil gives only a partial explanation of the facts. Pupillary inequality in g.p. was ascribed by Doutrebente ‡ to congestion or sclerosis of great sympathetic.

Myosis, may be from irritation of third nerve; or from paresis of sympathetic, and then often with narrowness of the chink between the eyelids. It has been attributed to hyperæmia of the brain generally, or of the optic nerves in particular; or it may accompany hæmorrhagic pachymeningitis, and this especially with large and fatal effusion of blood. Seifert stated that if myosis occurs in acute mania, g.p. is tolerably certain to supervene sooner or later. Dr. C. B. Radcliffe insisted upon narrowness of pupils in some nervous affections as being associated with turgescence of cerebral circulation; and dilated pupils with the contrary state; and Prof. Rouget has made myosis an erectile condition, due to spasm of vessels, arrested return of blood, and gorged iridal veins. Sander attributed it to constant irritation of sphincter iridis.

---

* " Augenkrankheiten und Allgemeinerkrankung."
† " Das Verhalten der Pupillen bei Geisteskranken."
‡ "La France Méd.," May 5, 1880; cited " Jl. Nerv. and Ment. Dis."

Mydriasis. If there is loss of, or obstacle to, retinal function, it is usually associated with dilatation of the pupils. The pupillary widening follows when the optic nerve ceases to conduct impressions of light. Mydriasis is often due to paresis of motor oculi nerve; less often to irritation of sympathetic ganglia, or of their fibres to the iris. But affections of cilio-spinal centre, or hæmorrhage into cervical part of cord, may give rise to mydriasis: if transient, it may arise from hyperæmia of nucleus of third cranial nerve; if persistent, from lesions thereof.

4. *Summary of other experience as to motor symptoms and speech.* Only a few general conclusions need be added here. —In several cases where locomotor symptoms were very marked, and especially so relatively to the speech-affection, the spinal cord was, or it and parts of meninges and encephalon at the brain-base were, much affected relatively to the more usual incidence of the disease and distribution of its lesions. In some cases of markedly ataxic gait, the spinal cord and meninges were considerably diseased; or, again, with considerable disease of cord and its meninges were very awkward, swerving, unsafe gait; temporary unilateral rigidity of limbs; hemipareses; hemispasm; analgesia and anæsthesia of feet. In some there had been facial hemiparesis, apoplectiform seizure and hemiplegia, death; and extreme hyperæmia of opposite cerebral hemisphere. Or excessive tremor and very extensive and extreme adhesion and decortication; or tremor cöactus, greater on side of body opposite to the cerebral hemisphere chiefly diseased.

Intercurrent aphasia, undergoing exacerbations, remissions, intermissions in several cases, with atrophy chiefly affecting the left cerebral hemisphere. Speech *relatively* not much affected in early stages, nor limbs; and cortical disease not so extreme in frontal region and central gyri as in many othe rcases. Yet, again, tongue, speech and writing early and much affected, but no spasmodic or paralytic complication; disease of central gyri and third frontal, moderate, only.

5. *Vaso-motor and trophic* phenomena have been fully considered as to their pathological physiology in Chapter VIII., p. 133; Chapter IX., p. 138; and Chapter XVII., p. 335.

6. *Sensory.* *Anæsthesia of general senses; also tactile a.* Anæsthesia on one side of the body, brings to mind the results produced by section of the posterior part of the

opposite internal capsule (Veyssière, Raymond, Charcot, Rendu, Carville, Duret, *et al.*). But there are rarely any local lesions of this part in g.p.; and, indeed, the persistent anæsthesia, unconnected with special attacks, is usually bilateral. Yet hemianæsthesia may arise from some inhibiting influence upon, or temporarily embarrassing change in, the internal capsule. *Early* cutaneous anæsthesia has been attributed to encephalic congestion. Goltz found removal of any part of one cerebral hemisphere of the dog, followed by anæsthesia on the opposite side of the body, partly transient, but partly permanent; and both muscular movement and tactile sensibility of skin being more affected by parietal than by occipital lesions. The cortical fields of the tactile sensibility appear, according to Exner's researches and Meynert's views, to be nearly identical with the motor areas for the muscles of the same periphery, the tactile areas in the dog's brain corresponding to the motor, and thus identity in situation of the tactile with the motor areas is in harmony with the presence of a granular cortical layer, deemed sensory by Meynert, with the large ganglionic cortical nerve-cells (motor). The cortical field of tactile sensibility is also better endowed and more active in the right than in the left ($\frac{1}{2}$) cerebrum.

Attempts to locate the tactile sense have ended in the several views that its centre is the hippocampus major and gyrus uncinatus, or g. fornicatus; or in the optic thalamus, or in the crus cerebri, or external portion of pons, or in the occipital lobe. If the hippocampus major be proved to be the centre for tactile sensation, or, again, of the olfactory and not of the tactile sense (Munk), it would be of importance to note in relation to this, and to the failure of tactile and of olfactory sensibility, and the frequent convulsions, in the later stages of g.p., that Delaye long ago found this part markedly indurated and atrophied in many cases of g.p., and that its induration and degeneration in chronic epilepsy have been described by Bouchet, Meynert, Hemkes, and others. In relation to this, are further experiments of Ferrier (1884), confirming the location of the tactile centre in the hippocampus major and uncinate gyrus; and Messrs. Schäfer and Horsley's experiments on the gyrus fornicatus. Local anæsthesiæ in g.p. may be due to changes in the local sensory nerves or their nuclei.—At p. 129, I gave a necropsy-abstract of a case with extreme late anæsthesia.

Hyperæsthesia and hyperalgesia. Hyperæsthesia of skin, muscles and viscera at an early stage, was attributed by Voisin to congestion of spinal cord; and, later, to hyperæmia of spinal meninges, posteriorly, and to neoplastic products thereof, compressing and irritating the posterior roots, the electro-muscular sensibility being intact so long as there is not softening of the cord. Encephalic congestion in the early stages may give rise to the same result.

Headache. Sensation of compression of head. These have been vaguely attributed to cerebro-meningeal congestion; more precisely to defective vaso-motor innervation leading to dilatation of the *vasa nervorum;* and this to compression of the nerve-trunks where they leave the cranial cavity; and this to sensation of head-compression, or of headache. Similarly Dr. F. Runge * explained sensations of head-pressure and head-pains, in some diseases, by circulatory disorder propagated by the openings of the bony skull, and leading to compression of the sensory nerves at their exit therefrom, so that the causation of these sensations is thus immediately of peripheral origin. Severe headaches, in g.p., when not of purely neuralgic sort, are more often due to chronic meningitic, and especially pachymeningitic, states. Or the residual products of these inflammatory conditions may press on nerve-roots about the base of the skull, occasioning pain in the district of the nerves affected; or vaso-motor failure may lead to local hyperæmia of nerve-sheaths, and this to pain; cerebral congestion may, at intervals, cause a dull ache; while temporary brain-anæmia may occasion a weighty sensation at the vertex.

Neuralgia. Pain. Severe pains of the neck, trunk, or limbs, in g.p., are sometimes referable to central spinal-cord changes, as where the posterior root-zones of the spinal cord are affected by a sclerotic process; or when contractions of the limbs attend similar processes in the lateral columns, and passive movement, also, of the affected limbs is painful. But the chronic spinal meningitis, with its thickening and other changes, which so often attends g.p., is apt to exercise a pressure upon the sensory nerve-fibres entering the spinal cord, and thus give rise to pain. Acute spinal meningitis, in g.p., is credited by Voisin with the production of severe pain in part of the vertebral column, increased by pressure or cold applied locally. The severe lightning-pains sometimes found in g.p. have, no doubt, the same origin as

* " Ueber Kopfdruck." " Arch. für Psych.," vi Bd., p. 627.

those of tabes-dorsalis, even when the latter in g.p. does not become fully pronounced; that is to say, intra-spinal irritation of internal root-fibres of posterior nerve-roots. The influence of hyperæsthesia, neuralgia, and anæsthesia upon the delusions and actions of G.Ps. has been set forth in previous chapters.

Smell. Failure in the sense of smell is associated with alterations in the olfactory bulbs and tracts; and roots in the temporo-sphenoidal lobe; the last, a part which I have found apt to become more affected by the ordinary lesion of g.p., and by maceration, than is usually acknowledged. Illusion and hallucination of smell, and of the other special senses, will be included in a separate special section.

Sight. Various lesions have been held to account for blindness in g.p.; optic nerve atrophy; optic neuritis; or atrophy of optic chiasm; general atrophy of brain with atrophy of thalami; or lesion of corp. geniculata or of corp. quadrig.; or, breaking down of the last from serous distension of ventricles with a pull on optic nerves; or much implication of both angular gyri in the usual cortical change (Crichton Browne); or temporary blindness as due to encephalic congestion, permanent blindness to secondary neuro-retinitis or compression of the nervous elements. Lesions of occipital gyri should suffice according to Munk's conclusions. Defect of sight has been attributed to a less degree of the same changes, particularly of optic neuritis, wasting of optic disc, nerve-trunk, commissure, and tracts, or of optic thalami, or corp. quadrig.; or to pupillary dilatation. As Dr. Gowers remarks, the optic-nerve atrophy comes with chronic degeneration, and not with acute disease, of the spinal cord.

Failure or loss of sight, in g.p., I connect with the state of the optic nerve in the vast majority of cases; either an atrophy of the nerve, primary and perhaps commencing at the disc; or, usually, an optic neuritis, ending in atrophy. In some cases, where meningitis has been well-marked and has spent its force to an unusual degree on the anterior part of the brain-base; and where the resulting thickenings, adhesions and pseudomembranes have compressed or tied down, and impaired the nutrition of, the optic nerves or tracts; or where that inflammatory action has extended to the optic nerves, or has included their meninges in its first assault; one finds sometimes a sufficient source for the failure, or even loss, of sight, without going to lesions of the brain itself for an explanation (see also Chapter VI).—In another set of cases the blindness may be immediately dependent ôn optic nerve atrophy or degeneration, secondary to brain-lesions affecting the parts recipient of impressions from the retina.

Again, an unusual degree of destruction of these latter would abrogate the more intelligent psychical side of vision, though not absolutely precluding the perception of light-impressions as light. Which are the parts of the cortex whose lesions may thus affect the sight in g.p.? First observed in g.p. by Fürstner, Reinhard, Tamburini, and Luciani, these defects of vision were connected by the first-named with lesion of occipital lobes in some cases; Ferrier would indicate the angular gyri; Munk located vision in the occipital lobes; whereas Goltz by removal of a large portion of one side of cerebrum in any part, but chiefly in the occipital, caused temporary diminution or loss of sight in the opposite eye; and on removal of a similar portion on both sides, both eyes were blinded or nearly so, for a time only, the permanent result being visual impairment of the kind already described, in Chapter VI., as psychical blindness. Dalton* found extirpation of angular gyrus in dogs followed by loss of visual perception in opposite eye, but no effect on sensibility of retina or conjunctiva, on winking, or pupillary reaction to light; and Pierson * found destruction of both the occipital lobes and angular gyri necessary to produce persistent blindness. In one case (not g.p.) of complete blindness (from eye-disease), lasting for very many years, I † found extreme atrophy of optic nerves and tracts and of right geniculate bodies, slight atrophy of both optic thalami and of left corpus striatum; wasting of supra-marginal gyri, and slight of angular; whilst acute red softening of second (and slightly of first) occipital gyri on both sides, perhaps gave some evidence of change in them and tendency to obstruction of vessels. Exner finds the parts chiefly affected, when visual defects are attributable to the cortex, to be the first and second occipital gyri, the posterior part of cuneus, and to a less degree, the posterior half of parietal lobe, upper end of ant. central g., and a zone around the cuneus.

Hearing is comparatively little impaired in g.p. (deafness); but cerebral auditory defect may occur.

* Cited " Archives de Neurologie," July, 1881, pp. 122-128.
† " Medical Times and Gazette," Jan. 28, 1882, p. 89.

Taste. The impairment of smell weakens the power to perceive savours—odorous sapid qualities—impressions in which the sense of smell co-operates with that of taste; and finally nuclear or cerebral disease sometimes impairs the functions of the glosso-pharyngeal and fifth nerves, and still further enfeebles the sense of taste.

*Hallucinations and Illusions in g.p.** The portions of cortex at the cerebro-meningeal adhesions in g.p., I† have often found more altered than those free therefrom. Thus, the adhesions seem, in·part, to mark the points of greatest activity of the cerebro-meningeal change in g.p. Supposing the change to affect a true cortical sensory centre we would expect its earlier steps to be attended with violent functional perturbation and disorder of the centre, such as hyperæsthesia, hallucination, illusion. But in many cases of g.p. no hallucinations of special sense are observed; in other cases these hallucinations are vivid, and are perhaps revealed more or less frequently during a considerable space of time. Therefore, one is led to examine whether or not the cases in which hallucinations occur in g.p. are the same as those in which the adhesions and other marked cortical changes affect the supposed cortical sensory centres. And this without prejudice to the view that hallucinations in g.p. may also be the outcome of the general activity and fulness of the cerebral circulation, or of a tumultuous, disorderly reaction of disturbed ideational centres upon sensorial.

Let me not be understood here as taking any narrow and limited view of the production of hallucinations. So far from this, rather, let it be granted that hallucination may take its starting point in a morbid impression upon, or a morbid condition of, any part of the sensorial apparatus. The morbid condition forming the starting point of hallucination may have its seat in the recipient peripheral expansions of the sensory nervous mechanism, or in the conducting fibres passing thence, centripetally, towards the sensory basal ganglia, and sometimes renewed or reinforced on the way by intervening bulbar, or pontine, or more removed quadrigeminal centres; or, again, the morbid focus may be in these sensory ganglia themselves, situate at the brain-base, somewhat elementary and simple in endowment, and making early appearance in the order of evolution; or, again, the morbid condition may be in the nervous tissue forming the paths of

---

* Mickle, "Jl. Mental Sci.," Oct., 1881. Jan. and Apr., 1882.
† " Jl. Mental.Sci.," Jan., 1876, and Apr., 1878. 1st Ed., pp. 157, 178.

sensory vibrations between these ganglia and the higher perceptive cortical sensory centres of the cerebral hemispheres; or, finally, may be in these cortical centres themselves. And, indeed, these last-named and cortical sensory centres appear to be, of all, the most important in the genesis of hallucination. Without their participation in derangement it is difficult to apprehend how hallucination can exist, at least in its higher and more developed forms. In hallucination there is evidence of pathological activity in the supreme centres concerned with the co-ordination of sensory impressions into complex perceptions, and of the arousing into present consciousness of the results of the highest mental integrations—of the treasures of past experience—with the acquisition of which the highest mental faculties are obviously concerned. If the cerebral cortex minister to the psychical powers it must be deranged when true hallucination exists.

It may readily be conceded that in ordinary insanity * the starting point of hallucination is by no means limited to the cerebral cortex, that the peripheral sensory mechanism, the nerves, the lower centres, the basal ganglia, the cerebral medullary fibres, may each also be the focus of original perturbation; though in each instance a special modification of the supreme centres must exist to permit of the birth of hallucination. To be fruitful the seed must fall upon congenial soil.

The same view as to the genesis of hallucination may also readily be conceded as obtaining in general paralysis. But in it the symptoms directly or indirectly flow from definite structural lesions of the tissues making up the substance of the nervous organs, and in the most frequent case those lesions primarily, or even mainly, affect the periphery of the brain. Pathological anatomy, therefore, reinforces the view that in general paralysis, hallucinations, for the most part, depend upon the effects of the morbid process invading the cerebral cortex and laying it waste; for, while disordering the functions of the cortex, this process also destroys the more delicate and subtly-woven portions of its structure. In this relation, one may well bear in mind how often, in general paralysis, especially in the less early stages, hallucination is of a crude, coarse, confused nature; and the correlative fact as to the coarse, gross character of the morbid

* See also Prof. Tamburini, "Sulla Teoria delle Allucinazioni." "Estratto dai Rendiconti del R. Istituto Lombardo," V. xiii.

process here supposed efficient in initiating its production. And while admitting that in general paralysis hallucination may be immediately originated by the morbid state of a nerve of sense, as a starting point, yet it seems right to assign hallucinations far more frequently to a primary morbid condition of cortical sensory centres; and the more so inasmuch as the relative frequency with which the several nerves of special sense are structurally affected in general paralysis is not directly proportional to the relative frequency with which the corresponding several forms of hallucination are observed therein.

The optic thalami are the centres most likely to be put forward to contest with centres of the cerebral cortex for the leading place in the production of hallucination; but in my experience thalamic disease plays a far less important part in general paralysis than cortical.

In short, while fully recognizing that hallucinations in general paralysis sometimes take, as it were, their starting point in the nerves, lower centres, or basal ganglia, I think that, in the great majority of cases, even their points of departure are in cortical centres specially concerned with sensory perception, taking part in the higher mental operations, and intimately associated or blended with other centres, in equal rank, ministering to more purely intellectual function. The visible lesions, although not the most immediate excitants of hallucination, yet do condition a functional disturbance of the cortical centres of such kind that in reacting thereto conformably with the modes of energizing established in the course of their evolution and growth, and by past experience, these centres give birth to hallucination.

Olfactory hallucinations have been found with various diseased conditions of olfactory bulb or tract, and this may occur in g.p. Auditory hallucinations have been assigned to lesions or functional disorders of thalamus, insula, acoustic nerve, floor of fourth ventricle,* some gyri.

Of the brains of deceased G.Ps. examined by me † with reference to the morbid anatomy of auditory and visual hallucinations, some were opposed to, and some coincided with, the localization of the auditory and visual centres made by Ferrier; and in some the lesions were too diffuse to form any indication as to this matter. Among the general conclusions I formed were the following:—

* Luys " Gazette des Hôpitaux," 1880, Nos. 40-9.
† "Journ. Mental Sci.," Jan. and Apr. 1882.

"That in cases of visual hallucination in g.p., the angular gyrus is not affected in the marked manner one would anticipate on the theory that it is the sole cortical visual centre; nor, in cases of auditory hallucinations, is the first temporo-sphenoidal, viewing it as the sole cortical auditory centre. Thus the morbid anatomy of g.p. fails to support the exclusive view that these gyri are, or contain, respectively, the sole cortical centres of sight and hearing.

"Taking the cases together, we find that the supramarginal convolution is affected more than the angular in those with visual hallucinations, and the adhesions are often well-marked on the postero-parietal lobule. Also, that the second temporo-sphenoidal gyrus seems to suffer more than the first in the cases with auditory hallucinations, taken collectively."

7. *In conclusion.* How far, in general paralysis, the principal mental symptoms can be referred to the organic changes in some frontal (and parietal) convolutions; the motor to those of the so-called cortical motor zone; the sensory to those of portions of the occipital, temporo-sphenoidal and parietal; must remain a matter of question. And all the more so as, with some points of general accord, the conclusions of Fritsch and Hitzig, Ferrier, Carville and Duret, Charcot and Pitres, Schiff, Albertoni, Michieli, Tamburini, Fürstner, Lussana and Lemoigne, Munk, Goltz, and Luciani, show considerable, and in some cases irreconcilable, divergencies the one from the other.

That there is a localization of cerebral function is indubitable, but the rigid delimitation attempted by some investigators does not appear to be in harmony with the facts of nature. The action of one part of the cortex can be supplemented by that of another more than some of them are willing to allow; there is more alliance than they admit between different cortical loci or centres which can operate towards the same result,—more of a capacity for the loose, flexible, yet effective, association of units, as of an army of men—not a rabble,—an association for the accomplishment of a given purpose. This or that one may fall out of the ranks, but the march of the host is not arrested nor its purpose stayed.

The purely anatomical investigations of Prof. Golgi * of Pavia are of some interest here, as showing how widely branching must be the connections of most of the cortical

* Trans. Dr. Jos. Workman, "Alienist and Neurologist," July, 1883.

ganglion bodies. "The several nervous fibres, far from being found in isolate individual relations with a corresponding gangliar cell are, on the contrary, in the great majority of cases, found in connection with extensive groups of cells; but the opposite fact also is verified, that is to say, every (?) gangliar cell of the centres may be in relation with several nervous fibres, which have different destination and function."

Moreover, I think it cannot be without meaning that the mental symptoms usually differ so much between themselves when the morbid process is earlier, and more severe, extensive, and persistent in one or the other cerebral hemisphere, and it may be inferred that the functions of the right hemisphere differ considerably from those of the left, although they are similar to so very great an extent. This, at least, is the result of an analysis of my own cases, a result not anticipated, and which came somewhat in the nature of a surprise to me. As to the different pathological relationships of the two cerebral hemispheres; besides the group of facts relating to aphasia and the more usual situation of the lesions producing it, there are those upon which Brown-Séquard has based his view that of the two cerebral hemispheres the lesions of the right are (*ceteris paribus*) more frequent and fatal; give rise to more marked hemiplegia; more frequently are complicated with nutritive disorders (acute bedsore, sloughing, etc.) in the palsied parts; more often are evidenced by convulsion and tonic spasms of the limbs, and by conjugated deviation of the eyes, and more often, also, by *direct* paralysis or convulsion (*i.e.*, of same side as lesion). Hysterical paralysis is more frequent in the left limbs, and therefore generally with affections of the right brain. In cauterising definite parts of the cortex of some of the lower animals he found the narrowing of the palpebral fissure, caused thereby, to be constant only when the right was the hemisphere operated on, and the results were quite different according to the side cauterised.

In regard to the relative fatality of lesions of the right hemisphere, and their more frequent association with convulsions and spasms, the view perhaps originated with Mr. G. W. Callender. Excluding those of the ventricles, basal ganglia, and peripheral brain-surface, he found hæmorrhage and other sudden lesions of the right cerebral substance to be followed by convulsion in a much higher ratio than similar lesions of corresponding ubity in the left hemisphere.

In lesions of the parts specified he found sinistral paralysis almost always accompanied by convulsion or rigidity; dextral paralysis almost never so accompanied; and, concluded that the district the lesions of which more particularly gave rise to convulsion was that supplied by the middle cerebral arteries, and that some part of the *right* cerebral hemisphere was especially connected with the occurrence of convulsions. Among the cases collected by him, of 48 cases of palsy of the right side, only 7 also presented convulsion or rigidity; while of 61 cases of left-side palsy 39 presented either convulsion or rigidity. But excluding the cases in which the opto-striate bodies were the parts affected, the difference was still more striking, for then of 37 cases of lesion of the left hemisphere only 7 were associated with convulsion or rigidity; and of 47 cases of lesion of the right hemisphere as many as 39 were accompanied by either of these symptoms. Callender also found the *average* duration of life after sanguineous apoplectic effusion in the right hemisphere to be only about one-eighth of the *average* duration of life after similar effusions into the left hemisphere. In epileptic seizures, Dr. E. H. Sieveking observed that the left side is the one most frequently affected. Thereby almost certainly implying a point of departure in the right side of the brain, or medulla oblongata above the decussation.

Of the two cerebral hemispheres, the left has been regarded as the more active in mental processes, the right the more passive. Dr. Hughlings Jackson suggested that the right hemisphere was the one leading in Perception; the left the one leading in Expression. Moreover, that the left posterior parts of the cerebrum and the right anterior parts were the substrata of *subject* consciousness; and the right posterior and left anterior parts were the substrata of *object* consciousness. Further, that lesions of the right posterior lobe impair the intelligence more than similar lesions of the left. But this last is connected with his view that the posterior cerebral lobes are the seat of the most intellectual processes, a view which is in direct opposition with the results of post-mortem examination of the brains of the insane.

I have found that when one cerebral hemisphere principally is attacked in general paralysis, the difference in relative damage to the two hemispheres being great, then the clinical phenomena almost always vary extremely with the particular hemisphere more affected. When it is the *right* hemisphere, exalted delusion, gaiety, expansive delirium, and maniacal excitement, predominate; when it is the *left*, either emotional depression, melancholic ideas and, perhaps, hallucinations :—or else an extreme and early dementia, are unusually obvious, if not predominant. Thus the character of ideation, emotion, organic sensation, special sensation, of motor activity, and hence of general conduct, are considerably different in the two cases, as a rule. Exceptions exist, possibly because in them it may be that the morbid process is long drawn out in one hemisphere; and is active, and the cause of striking disorder, though less severe and of shorter duration in the other, thus masking the effects of the former and graver lesion, until, or partly until, the time arrives when the progressive lesion induces a profound dementia, and ataxic and paretic helplessness, in which all, or most, of the other mental and motor symptoms are extinguished; or, again, perhaps because the parts to lesions of which characteristic symptoms are specially due are less involved than usual. These are merely suggestions, on which no stress is laid, in explanation of the exceptional cases. Other exceptions are capable of explanation by the predominance of some one or more of those other conditions that rule the mental phenomena of the disease, and are discussed in a preceding section. Since my conclusions were published in the "Journal of Mental Science," April, 1878, and in the 1st Ed. of this work, Dr. Magnan has published cases of different unilateral auditory hallucinations on opposite sides; disagreeable on one side, expansive on the other. This goes to support the view as to the twofold actions, and some degree of functional independence, of the two hemispheres.

Though not markedly diseased, as a rule, in general paralysis, the *cerebellum* is often involved; its meninges become hyperæmic, thickened, and opaque; and near the median line it is sometimes the site of slight adhesion and of cortical degeneration. How far this diseased condition, when present, may give origin to the vague and purposeless movements, contribute to the general failure of power, and conduce to the imperfect maintenance of equilibrium, seen in general paralysis, or, possibly, even sometimes produce its convulsions, or its occasional headache and failure of vision, —it would not be easy to decide. Cerebellar changes, however, are not needed to explain the symptoms, and are probably of little importance here.

Nor is it useless to reiterate that some of the clinical

phenomena in the course of g.p. depend, or partly depend, upon disorder or impairment of the brain-medulla, basal ganglia, pons, medulla oblongata, spinal cord and nerves, and of the sympathetic ganglia, especially the cervical.

Yet is general paralysis primarily or principally a disease of the cerebral cortex.

CHAPTER XIX.

PROGNOSIS.

The prognosis is necessarily the very gravest. This was partially recognized even before " general paralysis " was separated in nosology, and when it was lost in the throng of "paralytic" affections. Thus, so far back as 1820 M. Georget wrote, "la folie compliquée de paralysie ne guérit jamais;" even in 1812 M. Ph. Pinel spoke of the form of "adynamia" corresponding partially to what we now call g.p. as one "qui se termine souvent d'une manière funeste;" and in 1814 M. Esquirol wrote "la complication de la démence avec les convulsions, l'épilepsie, et la paralysie résiste à tous les moyens curatifs, et ne laisse pas l'espoir d'une longue existence."

When he is fully satisfied of the presence of true g.p. the physician knows that the case is almost without hope, and *curative* art without reliable and permanent efficacy therein. Cures or recoveries of g.p. have been reported it is true; but, in view of the obscurities that may surround the diagnosis, the question has very justly and very often been raised whether the cases said to be cured or recovered were or were not genuine examples of the disease.* In some cases the so-called cures or recoveries have, in reality, been only temporary remissions of the relentless malady, which before long has attained the natural consummation of its course in death. To justify the gravest doubt and most cautious circumspection of reported cases of recovery from g.p., one need only bear in mind how many, among the cases published as g.p. during the past sixty years, have not been genuine examples of that affection; and one need only examine the mortality tables of some asylums in which, through faulty diagnosis, g.p. evidently stands accused of

* As a "recovery," probably not g.p., "Lond. Med. Gaz.," Aug. 25, 1848, p. 326.

ravages not its own. And yet it is not at all rare to find instances of highly pronounced remission, or, perhaps, of apparent complete disappearance of all mental symptoms; and sometimes, on a superficial examination, of the disorders and defects of motility as well. I say, on a superficial examination—for almost always, after active exertion, and upon a protracted and accurate investigation being made, the speech, or the tongue, or the lips, will betray that fatal ataxia, which to the practised ear and eye cries "no recovery."

Cases of Recovery from g.p. Bayle, Esquirol, Calmeil, Lélut, Ferrus, Falret, Trélat, Guislain, Bonnefous, Fabre, Foville, Laffitte, Combes, Lunier (6 cases), C. Pinel (2 cases), Bulard, Brierre de Boismont (5 cases), Rodriguez, and others, believed they saw cases of recovery or cure, and in one of C. Pinel's cases the recovery was still maintained three years afterwards. Baillarger mentioned nine recoveries, including some of those collected above, and also cases by Earle, Renaudin, Baume, and Morel; also six cases later, and four cited from Delasiauve. Billod described, in detail, the case of a man aged 33 whom he thought to be cured; and two cases since. Two G.Ps. mentioned by Dr. R. Boyd were discharged recovered "but after several months both relapsed." Several cases were tabulated by Dr. J. W. Burman as having been discharged on recovery. From the histories given there is no clear proof that these were recoveries; some clearly were remissions only, and the patients returned to die; some were lost sight of. Dr. S. W. D. Williams* mentioned a case in a male aged 25, and the fact of recovery was confirmed by examination nearly three years afterwards. Mercury had a place in the treatment. Dr. Hack Tuke quoted two cases of apparent recovery under the care of Dr. Wm. Macleod. Guttstadt† returned three recoveries as occurring in Germany in one year. Cures were reported by Dr. Moriz Gauster, 2 cases, one remained well five years afterwards;— by Dr. H. Schüle,‡ a male aged 44, classical form of g.p., well eight years afterwards;—and by Flemming, whose case died seven years afterwards of another affection. Dr. Böttger reported a patient discharged recovered, or almost so, and who followed his trade of master-mason successfully for two years, then died of cholera. Meynert had a case of recovery,

---

* "Medical Times and Gazette," May 30, 1868, p. 576.
† "Allgem. Zeitschr. für Psych.," xxxiv Bd., p. 243.
‡ "Allgem. Zeitschr. für Psych.," xxxii Bd., p. 627.

enduring at least three years; and Stölzner, cases enduring one year and five years, respectively. Oebeke found on record only seventeen cases in which recovery had proved itself to be permanent, taking, as such, only cases that had lived and had remained perfectly well for upwards of three years. Eight cases of recovery of g.p. under active treatment, were reported by Prof. L. Meyer; one of these relapsed in two years. Dr. Doutrebente mentioned original cases of what he calls complete remission; but of some of them the remission was short, and in some, slowness of speech apparently remained. He cited cases of supposed recovery by Bouillaud, Marcé, Delasiauve, Dubuisson, Thos. Willis, Védie, and S. Roy. Writing immediately after the patient's discharge, Dr. Wm. Macleod recorded a case as one of mental recovery enduring at least six months, but the still remaining affection of speech showed the case to be one of remission only. Two examples of recovery, and three of more or less prolonged arrest of the progress of the disease, were mentioned by Voisin; all being treated chiefly with cold baths; also, recovery, after the application of actual cautery over the spine, of a patient "threatened" with g.p., and affected with spinal meningitis. In another case, one of acute g.p., recovery also took place four or five months after the onset. Ideler reported a case in which improvement began six weeks after admission; and ten years subsequently the patient was following his occupation, and seemingly with no disease. Dr. Oebeke saw a recovery, perhaps not absolutely complete, in a merchant 37 years old on admission. Three-and-a-half years afterwards the condition continued to be the same. Galceran* reported cure in two cases of g.p. A case supposed to be due to lead poisoning and undergoing recovery was recorded by Doutrebente. Rendu reported two cases of recovery of g.p. of syphilitic origin. A case of recovery, in a male aged 36, remaining well nearly six years later, was observed by Tuczek.

Among my patients, one was discharged, apparently recovered, but was lost sight of. Another left the asylum, completely recovered mentally, and only the faintest occasional traces of some of the physical signs existed. He remained well for some time, but finally mental symptoms again came on. Another, discharged, mentally recovered, and bearing only the slightest and very doubtful trace of the motor order of symptoms, continued thus for ten months

* " Independencia Medica."

or more. This had been the second attack, the first having been followed by a highly pronounced remission or recovery of shorter duration than the one above-mentioned. Again he relapsed, and into his third attack of g.p. Of a fourth patient, discharged in a similar condition, nothing has since been heard. All traces have been lost of several others who were discharged with less complete indications of recovery than existed in the preceding patients.

Among the above recoveries in my own experience, I have not included several cases of psycho-somatic disease believed to be of syphilitic origin, which were, or which so closely resembled, g.p., that one's diagnosis was in doubt, but which underwent recovery under treatment, and were in due course discharged. One such I published in the "British and Foreign Medico-Chir. Review," July, 1876, p. 180, Case IV.; and another such in the "Journal of Mental Science," Oct., 1879, p. 392, Case II. Nor have I referred, here, to the not infrequent cases, in my practice, of temporary mental remission or recovery, the physical symptoms continuing.

It is noteworthy that several of the above-mentioned instances of recovery, or of very prolonged remission, have supervened on accidents, violent injuries, or diseases of such a kind as to produce lively, so-called, revulsive effects. In exemplification of this, one may refer to cases of apparent recovery following upon erysipelas (several cases); or profuse suppuration, or burns, or lumbar abscess, or liver-abscess;* or amputation of the thigh (2 cases), or fracture of the tibia and phlegmonous inflammation of the thigh, or crops of furuncles. Attacks of variola are said, as by Schlager, to have exercised a favourable influence; or even to have been followed by recovery, as in a case by Nasse; and (according to Mendel) Köstl reported the same as regards cow-pox. In one case by C. Pinel,† recovery followed, commencing some weeks after severe illness from dysentery, intense bronchitis, œdema of legs, erysipelas of left leg; but coincidently with the use of cautery to nape, baths with cool affusions, and tonics. A wild and troublesome G.P. under my care was much better in mind when, and for a time after, he was laid up with a compound fracture of the tibia, which he incurred by jumping from a wall in an effort to escape. Another was better for a time after an interscapular carbuncle; others after inflammation following the application of irritants to scalp or to nape.

* Morel, "Annales Méd.-Psych.," 1858, p. 388.
† Société Méd.-Psych., June, 1858; cited Voisin, *op. cit.*, p. 197.

Hence it has been thought beneficial to create large ulcers or suppurating surfaces, especially on the lower limbs of G.Ps., and to maintain a prolonged and free discharge therefrom.

CHAPTER XX.

THERAPEUTICS AND HYGIENE.

I. PROPHYLACTIC TREATMENT.

(a). *Prevention of hereditary predisposition.* The true prevention of g.p. would consist in a regulation of marriage by the pressure of enlightened public opinion, so that it should be esteemed a grave social offence to intermarry with those whose hereditary tendencies are undeniably neuropathic, and whose children would be predisposed to g.p.—Here, also, must be mentioned the psychical inheritances of the children of G.Ps. Many G.Ps. beget or conceive children in the early, or occasionally in the middle, stages, or during remissions. The experience of Simon, of Giraud, of König, shows with what untoward results:—early nervous disorders; g.p.; mental alienation or failure of various forms; spasm; epilepsy; also, club-foot, cleft-palate, etc., in many instances.

(b). *Prevention of individual predisposition.* For this a requisite is good education, both moral and intellectual, so as to render the various psychical powers strong, well-balanced, and harmonious in action; adapted to resist hostile influences; and to promote that self-restraint and self-contained reaction upon the social, religious, and intellectual environment, which conserve the mental energy, and fitly direct the dynamic powers of the whole being. It goes without saying that the above implies an avoidance of all excesses, of over-strained and extreme emotion, or of excessive intellectual labour. It is easy to prescribe all this; it is more than difficult to live up to it, and to induce others to do so. But intellectual labour of some severity may be carried on with safety to the mental health if the individual mental organization is sound to begin with, and the incidence of life's worries and struggles not too unfortunate or too severe.

(c). *Prevention of threatening attack.* Of more practical use is it to devise a prophylaxis and therapeusis adapted to arrest the onset or course of an already impendent g.p. A patient threatened with g.p., one who has shown symptoms of the prodromic period of mental alteration not yet amounting to decided mental alienation, should be at once freed, as far as possible, from the circumstances, or causes, under which symptoms of so sadly prophetic a nature have arisen. A perfectly regular life, early hours, moderate and regular bodily exercise, a total disuse of alcohol in any form, and of tobacco; discontinuance of cöitus; the use of bathing and friction of the skin, the application of cold to the head when it becomes unduly heated, while the feet are kept warm, and, if necessary, mustard, or hot, pediluvia are used; together with the maintenance of a perfectly free state of the bowels, —all these should be enjoined. Every source of mental worry, anxiety, annoyance, fear, chagrin, should be scrupulously avoided, at almost any cost. All intellectual labour should cease, and just such an amount of reading, of conversation, and of thought, should be undertaken as will afford the most gentle of intellectual and emotional exercise. The society should be that of the patient's family or intimates. In a word, the patient must, for a time, go out of his ordinary life, must retire from his duties, labours, contrarieties, turmoils, and ambitions, and in repose seek a renewal of nervous tone, and of power of resistance to hostile influences. If light and free from anxiety, the occupation may be resumed after a long interval; but not if it involves mental or physical overstrain. Sent travelling, pleasure-seeking, "to take the waters," or undergo a "hydropathic course," these patients become worse.

II. TREATMENT OF ESTABLISHED GENERAL PARALYSIS.

A. *General management, nursing, feeding, etc.* For the most part, the means suggested in the treatment of the prodromic period will still conduce to amelioration. In so malignantly fatal an affection, general management and nursing hold the chief place in the treatment. Most G.Ps. should be removed from home for a time, either to an asylum, or to some house of which part can be set aside for their use. To retain them at home, where, if depressed, they may refuse food and treatment or attempt suicide, where they may squander their means, or, as in many cases, are most dictatorial, fly into furious passion if not obeyed, and where everything rouses their desire to alter, sell, or destroy, is to promote angry and disturbing scenes, prejudicial to the patients, and painful to those about them. When there is quietude and dementia, the matter should be decided by the answer to the question—where, under the circumstances of the individual cases, good nursing, and skilled medical treatment, can best be secured and retained for the patients. Tact and gentleness are very necessary in their management: —by tact, gentleness, and *bonhomie* many of them may be kept in good-humour; or, at least, those angry outbursts which, too often, follow the slightest crossing or thwarting, may be avoided, as far as possible. In the early stages, peace and mental rest should be sought by every means, but too often are unattainable, and when furious paroxysms of destructiveness or violence occur a judicious isolation or seclusion may occasionally prove to be beneficial. In the early stages, also, and especially in the forms characterised clinically by much mental excitement, by expansive delirium, or by so-called congestive tendencies, the nourishment should be light, easily digestible, limited, and absolutely free from alcohol. This limitation of diet is far from being always easy to carry into effect, inasmuch as the patient's appetite is usually large, nay, often voracious, although when this is more apparent than real—the patient eating over-plentifully through inattention—it may be prevented by judicious management. In such cases, also, the excretory organs, and especially the bowels, should be kept in free action by the usual dietetic and other means. As far as possible the patient should live in the open air and take a moderate amount of exercise, while daily warm baths to the body, simultaneously with cold applied to the head, are often of service.

With the advancing progress of the disease the diet may be made more generous, and in the third and fourth periods may with advantage be extremely liberal. Alcohol, which, as a rule, does so much harm in the first and second periods, will sometimes aid in prolonging life when administered during the last weeks or months of existence. When the patient becomes unable to walk by himself with safety he must have the assistance of attendants in taking regular daily open-air exercise; at a later time he may recline in an easy-chair, sitting on air-cushions; and thence, either by a gradual decline, or after convulsive or paralytic attacks, he must sink a step lower in the scale, and pass his days on a water-bed. But usually for a short time only. Bedsores tend to form, pulmonary affections and diarrhœa afflict his feeble and now attenuated frame, the excrements all pass involuntarily beneath him, and, very often, the end is not far.

Hence, in the later stages, the paramount importance, next to good feeding, of perfect cleanliness and of the prevention of bedsores. These are only attainable by the most constant and scrupulous care in the instant removal of all excreta, the use of "railway" or other urinals, where possible, and of a water-bed, frequent changes of position, and the employment of a hardening, or protective, or other application on the parts exposed to pressure. After a comparative trial of a number of such applications I prefer a strong saturnine lotion.* Details of the treatment bedsores require are given below. Involuntary fæcal passages may be lessened by placing the patient regularly on a close-stool, after, or not, a simple enema. Throughout, the action of the bowels should be maintained. If they are inert, strychnia may be used as recommended by Verga, and I have found it useful, but in the later stages diarrhœa (for which Moreau advised atropine), is often troublesome and obstinate, and, like other incidents and complications, must be treated on general principles. By these means, and by those detailed below, a patient will sometimes be enabled to survive for more than a year after the commencement of a continuously bedridden state. I refer to a bedridden state entirely due to the disease itself, and not to any accident or complication. Prevention of inhalation of food, and the use of nutritive enemata, are discussed below.

B. Little need be said as to the *medicinal* treatment of g.p. The treatment of the disease itself, will first be referred to; and then that appropriate to special conditions, sometimes present.

Pharmaceutical, and other, Treatment of General Paralysis. The earlier writers on this subject enjoined the employment of an active antiphlogistic therapeusis. Those were especially the days of its treatment by low diet, bleeding, leeching, cupping, purgatives, moxas, vesicatories or setons to the nape or elsewhere; mercurial inunctions, antimony, diuretics, cold to head with warm bath to general frame. Possibly, in this country, treatment of this kind has suffered undue neglect. Bleeding, for example, might prove useful in some cases in which very obvious cerebral hyperæmia is present in the early stages, or where violent congestive seizures occur. Yet venesection could only be employed exceptionally here, and I have no experience of its use. Counter-irritation or

* Plumbi Acet. ʒii-ʒi; aquæ, oi : or Liq. Plumbi Subacet. (B.P.) sufficiently diluted.

revulsion by blisters, suppurants, cauteries or setons to the nape, the spine, or the scalp, have also, perhaps, fallen into too great desuetude in this country, and are sometimes of service in the middle and later stages of the disease. Some foreign physicians still make active use of these means, and of venesection, and of leeching the lower extremities. In 1871, and afterwards, I blistered the napes, or scalps, or spines, of some G.Ps., and kept the sores open by stimulating ointments. This plan then fell into disuse in my practice. Latterly, I resumed the old system, using, however, Ung. Antim. Tart., as recommended by Prof. L. Meyer, of Göttingen, to keep up constant suppuration and revulsive effects at the vertex, and over the larger fontanelle, and eventually employing some stimulating ointment (Meyer advised to use Basilicon u.). With this, and K.I. internally, he claims to have cured several cases. Sometimes the result is, or appears to be, beneficial; but in many cases the pain, discomfort, and disfigurement suffered by the patient are the only obvious results; and there is always the disadvantage in its use in restless, meddlesome patients that they may so irritate the part as to give rise to severe inflammation, and, possibly, untoward effects. Occasionally, may be revived with advantage the plan of leeching the nose, anus, or vulva, when habitual epistaxis, hæmorrhoids, or the menses, are checked at, or before, the incidence of g.p.

To speak more particularly:—

Venesection and leeching. Bleeding from the arm (17 to 20ozs.) has been held by Voisin, within recent years, to be useful at the onset, or in early congestive conditions or apoplectiform seizures in the robust and young. Or, in plethoric persons with prodromic signs as of g.p., smaller bleedings (3 to 7ozs.) repeated from time to time; say once a month, or else leeching at the anus regularly and systematically. Leeching the lower limbs has been recommended, or leeching the nostrils, or the mastoid processes. All of these means should be rejected in the later stages, except leeching in the more severe apoplectiform attacks.

Blistering the head and spine in *acute* cases, renewed as long as necessary; periodical blistering of the occiput and nape in chronic cases; and blisters along the spinal column in the "spinal form" of the disease, have all been employed; my experience in the use of these measures has already been stated.

Repeated cautery to the head and spine is recommended

by Voisin; or to the head only, if cerebral symptoms predominate; to the spine, if spinal predominate, or precede the cerebral; or vesicatories to the spine if acute spinal meningitis coexists; cauteries, if sub-acute spinal meningitis.

Seton, to the nape has been recommended, especially in cases with headache; a slow, continual derivative.

Cold to the head, either in the ordinary course; or on the occurrence of "congestive" or maniacal symptoms, may be applied by evaporating lotion, or by ice-bag, to the elevated head.

Cold baths, are very strongly urged by Voisin. At first used only in cases with excitement, much ataxy, and high temperature, cases of acute or sub-acute g.p., the baths continued to be beneficial to the same patients later on, and in no case did ill-result of any kind occur; thus he was led to use cold baths systematically in all cases of g.p. in the first stage, and then to recommend them in every stage. If considerable fever accompanies the g.p., he orders baths at 54° F. (12° C.) for ten minutes; if less fever, at 60° F. (18° C.) for five minutes; the patients in both cases being well wrapped up in warm coverings for three-quarters of an hour after the bath, until the pale face, blue lips, and *cutis anserina* give place to the signs of complete reaction. In winter the bath-room must be kept well-warmed. These baths are given daily; occasionally twice a day; later on, every two days, according to circumstances. The duration of this treatment is months or years. He claims that the cold baths improve the general state, lower temperature, lessen the ataxia, diminish the tendency to bedsores; in females sometimes restore the menses, if these are suppressed; as regards the mental symptoms, are especially beneficial in their effect on states of stupor, depression, dementia. Their action is at once antiphlogistic, tonic, and derivative. When a remission occurs under their use they should be continued for years (three, in one case). The contra-indications are, intercurrent diseases of some kinds; violent resistance by patients; menstrual periods; excessive marasmus and weakness. Only skilled and absolutely trustworthy persons must be allowed to bathe the patients.

Prolonged bath. Recovery of a G.P. under a bath of comfortable temperature for fifteen days was reported by M. Bonnefous. But here the therapeusis was complicated by a burn. Three-and-a-half years afterwards the patient returned suffering from g.p., of which he died seven months later. *Lukewarm baths* may be systematically employed.

Sinapisms to lower limbs and hot pediluvia, to avert inflammatory recrudescence, and apoplectiform attacks, have been recommended (Voisin) in any, but especially in the earlier, stages.

Purgatives, simple or aperient enemata, are often extremely useful in mentally depressed conditions of G.Ps., and where food is refused. A purge or an enema will often clear up these symptoms as if by magic. They often act well, also, when the face becomes flushed, and has a dull suffused look, and, under these circumstances, appear to avert impendent apoplectiform seizures. I have also already called attention* to the great abatement of the tendency to epileptiform convulsions, in some cases, by the regular or frequent use of simple or of aperient enemata; and the avoidance by this, or by other, means of the not infrequent tendency to constipation. Voisin advises, in the prodromic stage, to purge for revulsive effects; in the "intermedial" and first and second stages, every few days (8-15); in the third, only during congestion of encephalon.

Electricity. In the initial stage of g.p., Hitzig saw galvanic currents applied to the medulla oblongata and upper part of spinal cord with asserted temporary benefit, especially to speech; but Mendel had no success in g.p. from either constant or induced electricity. Dr. R. Arndt recommended electricity for the prevention of impending g.p., and mentioned a case immensely improved by a short galvanic treatment with very strong descending currents; so that the patient was discharged; but he died some months later, in an apoplectiform attack.

Potassium iodide. At one time I had not found K.I. very successful in somewhat advanced cases of pure g.p.; although several recoveries under its use had occurred, in my practice, in syphilitic cases closely resembling g.p. (references under "prognosis" *q.v.*). I have for some years given it a fresh trial, and in cases more recent, and with more maniacal excitement than many of those formerly treated therewith. In several, it has apparently exerted a control over some of the phenomena of the disease; and I have thought the mental, as a whole, more improved than the physical, state. It has more or less controlling power over pain in the head and limbs; and mental excitement

* "Journal Mental Science," April, 1884, p. 27.

and restlessness, epileptiform and apoplectiform attacks, are, I think, distinctly lessened by it in some cases. Beginning with doses of 5 grains to males thrice daily, I usually increase it gradually to 10 grains, or even higher doses (15, rarely 20 or 30), thrice a day. The earlier it is employed, the better the effects. Where the patient has been syphilitic, the iodide should be vigorously employed for a long time. If coryza, or an iodide-rash, appears the drug should be discontinued, or the dose be much reduced. Rendu reported two cases as g.p. cured by K.I.; and Meyer, cures in which it had a part.—I[*] have also recommended K.I. for the *prevention* of some cases of cerebral and mental disease, including g.p., produced by cranial injury.

Mercury, I have not employed in cases believed to be entirely free from syphilis. Should the patient be also syphilitic, or should the differential diagnosis between cerebral syphilis and g.p. be uncertain in a given instance, it is well to treat the patient for a short time with mercurials and for a longer with the iodides; carefully watching for, and energetically following up, any resulting improvement that, happily, may shed its ray of light athwart the gloomy prognosis or dark perplexity.

Bromides of Potassium and Ammonium, have been employed in g.p. to lessen encephalic congestive tendencies; to prevent or abate epileptiform attacks; to diminish mental excitement and motor restlessness; to procure sleep; to appease sexual ardour or irregular excitation. I have frequently combined small doses of ammonium or potassium bromide with potassium iodide when the tendency to epileptiform seizures was unusually well marked. Lunier was, perhaps, the first to use K.Br. extensively in g.p. with K.I., or both with martial preparations. A caution must be expressed here against the undue use of K.Br., particularly in the later stages. It concerns the property of the drug to cause anæsthesia of the fauces and throat, and thereby augment the already existing tendency, in g.p., of food to collect at the back of the mouth, or find its way into the bronchi.

Veratrum Viride, I have frequently employed when, together with moderate excitement and evident cerebral hyperæmia, the general frame and visceral health are robust. The tinct. (B.Ph.) in ℥. x doses, thrice daily, has usually been given.

Tonics. Iron. In the absence of great excitement and of

---

[*] " American Psychological Journal," April, 1883, p. 46.

sleeplessness, or of the maniacal paroxysms, I have usually given tonics if the patient is calm, quiet, or demented from the first, or becomes so during the course of the disease, and particularly if he is enfeebled, emaciated, exhausted, or phthisical; and then, indeed, whether he is agitated or not. But in most cases, even when their use is not judicious at first, the time for tonics and restoratives is not long delayed, and during the whole of asylum life the system of most G.Ps. responds favourably to their effects. Of those I have used the best results have been derived from liquor ferri perchloridi * (♏. x-xx thrice daily) taken immediately after each meal. If any tendency to constipation manifests itself twenty grains of magnesium-sulphate may be added to each dose; or relief may be obtained by enemata; or by the use of podophyllin occasionally, or of one-fifth of a grain of it every morning or more often, the podophyllin being employed either alone or with strychnia and belladonna; or, by the substitution of citrate of iron and ammonia for the perchloride. To the chalybeate, cod-liver-oil may often be added with advantage, especially in the later stages; so may arsenic, eulogized by Dr. Adriani. The later the stage, the more distinct the benefit from ferric perchloride and its congeners.

Ergot of Rye. Ergotine. Ergot of rye, or ergotine, in continuous administration in g.p., appears not to have been of any very decided value in the hands of others. In g.p., I have only used it, among other means, for the treatment of apoplectiform attacks; or of maladies merely intercurrent or incidental.

Physostigma. With some good results, I cannot, on the whole, say that physostigma has succeeded in my practice. It was recommended some years ago by Drs. (now Sir) C. Browne and G. Thompson. No results were obtained from it by Voisin and Mendel.

Digitalis. Digitaline. Digitalis tincture I † have employed for maniacal excitement in g.p., and in some cases with good result. Digitaline has been employed by Voisin for the congestive tendencies with rise of temperature; but, like him, I would deprecate the *continuous* use in g.p. of any of these preparations; which, too, are not without their dangers, as well as their drawbacks.

Quinine is recommended by Voisin for the treatment of

* Liq. ferri perchloridi (B. Ph.).—Or, solid perchloride, ℨii; water *fl.* ℨxvi.
† "Journal Mental Science," July, 1873, pp. 186 and 201.

congestive tendencies in g.p. *Papaverine*, recommended by some, has failed in the hands of others; and the disadvantages of *apomorphia* appear to outweigh its benefits. *Nitrate of Silver*. Bouchut (quoted by Voisin, but not found at place cited) claimed that in g.p. he saw recovery in some cases, and improvement in another case, of the motor symptoms under *arg. nitr.*

Rectal feeding, or *the stomach tube*, should be employed in the last stages, if, and when, the patient is paralysed about the mouth and throat, or is comatose.

III. TREATMENT OF THE MORE IMPORTANT CONDITIONS AND COMPLICATIONS IN GENERAL PARALYSIS.

Maniacal excitement in g.p. In the early stages, if the patient is acutely maniacal, or if paroxysmal outbursts occur, benefit may be derived from the use of quickly-acting purgatives, and of warm baths of from fifteen to thirty minutes' duration, with cold applied to the head simultaneously by ice-cap or cold-water cloths or cold-water douche. In these conditions, potassium or ammonium bromide, veratrum viride, and physostigma sometimes are of value; and æther-spray to the head may have a calming effect, as I have found. In other cases Tr. of digitalis (m. xv-xx every four to six hours) acts well; (but see caution above). I have found that opiates, though often extolled, do mischief in this phase of the disease. Austin advocated large doses of extract of hyoscyamus here. I cannot speak in favour of the tincture in large doses: with potassium-bromide it occasionally suffices. Blistering the shaven scalp has been recommended.

The remedies just mentioned are for the most part employed in the daytime. But during the high tide of excitement and sleeplessness in g.p. a moderate or a full dose of ammonium- or potassium-bromide, or of chloral-hydrate, or of a mixture of chloral-hydrate and potassium-bromide, at night, will often prove to be beneficially hypnotic. Even in the advanced stages of g.p. chloral may be given with benefit if the patient is very sleepless and restless. Thus, in a paper on another subject, I long ago remarked,* incidentally, that the patient " being restless and sleepless at night, chloral hydr. ʒß h.s.s. was ordered, and proved to be fully hypnotic. For many months before this drug was taken, the night-attendants had never found him asleep." 30 grains I never exceed, a second similar dose, not less than 4

* " Journ. Mental Science," Apr., 1872, p. 44-5.

hours afterwards, being given on extremely rare occasions. The patient must be kept very warm; the drug given for a few nights, only. A warm bath with cold to the head, beforehand, will sometimes render a dose hypnotic, that would otherwise fail. Rashes, sweating, flushing of face, early bedsore, gastric catarrh and icterus, heart-failure, and sudden death, are ill-effects attributed to chloral in g.p.— Heart-disease, extreme prostration and asthenia, are contra-indications to its use. To chloral in small amount some add small doses of Battley's sedative, or of morphia, for hypnotic effect. The mixture of chloral hydrate and K.Br. may occasionally be given more frequently. I * have observed that it does not necessarily lower the *average* temperature. Thus, in one case of extreme, almost constant paralytic excitement, kept somewhat in check by moderate doses for six weeks, the *average* a.m. temp. was 98·2°, and *p.m.*, 98·36°. When the excited stage had partially subsided these *average* temps. were ·6° and 1·° lower, respectively. As I stated in 1874 (*loc. cit.*), "the mixture of potassic-bromide and chloral-hydrate . . . often checks the nocturnal sleeplessness, noisy excitement, and destructiveness, of some G.Ps."—A full dose (ʒ1-ʒ1½) of Am. Br. or K.Br., alone, will sometimes procure sleep.

I do not like either morphia or hyoscyamine as hypnotics in g.p., and I have ceased to employ even the latter for that purpose (Merck's amorphous hyoscyamine gr. ¼ :—much smaller dose of crystals). I doubt the propriety of the large doses (¾ gr.) sometimes recommended; and a drug that in full doses, in my experience, widens the pupils, accelerates the pulse, causes reeling, and perversion of cutaneous sensibility, with associated illusions and hallucinations, and which lowers the body weight, cannot be desirable in g.p., except, perhaps, to meet an occasional emergency. On the whole, the hypnotic and sedative old vegetable neurotics, and many other nervines, are undesirable in g.p.

Temporary seclusion in a "single room" will sometimes calm the patient. *Exercise*, in a large space, under a special attendant, will often allow the mental storm to blow over.

Depression in g.p. has been treated by blistering the nape, by baths, and generous diet. Baths and generous diet are useful, so, occasionally, is blistering the nape; but the most efficacious means is to put, and keep, the bowels in free

---

* "Practitioner," June, 1874, p. 429.

gentle action, and them and the other viscera in a healthy state.

The Epileptiform Seizures of General Paralysis. In the severe epileptiform seizures it is well to give chloral-hydrate, either alone or with potassic bromide, and usually by enema, but sometimes by the mouth. If by enema, the parts should be well plugged to prevent its escape. If the patient can swallow it is better to give the chloral by mouth; this plan I have used with epileptics from time to time since 1871.*
If the *status epilepticus* is established, inhalations of chloroform may be resorted to, or the enemata just mentioned; and subsequently the latter with brandy. In any of these cases I generally use thirty, and never exceed forty, grains of chloral-hydrate in a single enema, but watch closely, and repeat the enema as may be required. Cold may also be applied to the head by ice-bag or evaporating lotion; the bowels may often with advantage be first cleared out by a clyster of sodium-chloride or of magnesium-sulphate and turpentine, and the patient be afterwards supported by nutritive enemata of peptonised milk or beef-essence, and, if necessary, of brandy.

Nitrite of amyl has failed in my hands here. Dr. R. Lawson recommended hyoscyamine; Foville, large enemata containing assafœtida, also vesicatory or cautery to nape. On the evils of allowing the patient to attempt to swallow food, see below.

The Apoplectiform Seizures of General Paralysis. In the apoplectiform attacks, elevation of the head; free purgation; cold applied by ice, or in various other ways, to the head with, or without, a prolonged warm bath to the rest of the body; and ammonium or potassium bromide and ergot in full doses, will often prove to be favourable ordinances.

In these conditions, also, and whenever cerebral congestion is marked, it is often suggested to resort to venesection, to apply leeches freely to the nostrils, mastoid processes, anus, vulva, or lower limbs; to cup the nape, give an emetic or calomel; and, when the congestive tendency is obvious, but apoplectiform symptoms have not yet arisen, to use very hot or mustard pediluvia. Here, also, digitalis, quinine, and cold baths have been extolled. Camphor injections may delay death for several hours (Mendel) when the pulse is small and frequent.

Feeding by rectum, or by stomach-tube. The following im-

* For this, in g.p., see Dr. J. A. M. Wallis, " West Rid. Rep.," v, p. 257.

portant point was, I believe, first brought before the profession in a practical form by the present writer. In the comatose, semi-comatose, paralytic and anæsthetic conditions attending the apoplectiform and also the epileptiform seizures in g.p., an extremely important part of the management consists in not allowing the patient to attempt to swallow food, but to administer it either by an œsophageal tube or by rectum, in order to avert the inhalation of food and the consequent pulmonary congestion and inflammation. If the stomach-tube can be employed without causing any untoward symptom, such as vomiting or dyspnœa, good and well; but, as I * mentioned in a paper on rectal feeding and medication, in some of these cases " the use of the stomach-tube causes vomiting, or gives rise to severe dyspnœa and threatening asphyxia; in others there is vomiting independently of the passage of any tube. Here the use of the stomach-tube introduces an element of danger; the patient, helpless, in stupor or comatose, paralysed, or convulsed, or locally anæsthetic, as he may be according to the circumstances in each case, and eructating or vomiting ineffectually the incoming food, is apt to inhale portions of it into the lungs; by strong inspirations the inhaled food is drawn into the far-distant ramifications of the bronchi and into the alveoli; and a destructive, traumatic, form of lobular pneumonia ensues." The practice of allowing G.Ps. to swallow when they are comatose, or are paralysed and insentient about the throat and mouth, should be absolutely prohibited. The advanced paralytic, and especially bulbar paralytic, the epileptiform and apoplectiform, conditions, supply the cases in which the above method of management is requisite. I always employ peptonised milk for enemata, prepared as described at p. 24 of my article quoted. The anus must often be plugged, to retain the enema.

Bedsores in g.p. Prevent them as far as possible by perfect cleanliness, by placing the patient on a water-bed, by frequent changes of his position, and stratagems devised to avert pressure upon threatened points, and by the application of strong lead-lotion. By keeping patients constantly in bed, enveloped in buffers of cotton wool and flannel bandages, and under the exclusive and constant care of special nurses, Dr. Wm. Macleod prevented bedsores. If, in spite of these precautions bedsores form, or if they are *acute*, and therefore not preventable, apply a solution of carbolic acid or of potas-

* "Journ. Mental Science," April, 1884, p. 21.

sium-permanganate and over it a linseed poultice, cut away sloughs as soon as they are loosened, syringe out regularly, frequently, and thoroughly with the carbolic, permanganate, thymol, or chloralum lotions, then dress with the same or with a solution of boracic acid, or with powdered zinc oxide; stimulate the granulations if necessary; and let the breach attempt to close under zinc ointment, or boracic acid lint.

Brown-Séquard* proposed a plan for the prevention of that sloughing over the sacrum and nates, found in some cases of spinal disease or injury, and closely allied in its pathology with many an acute bedsore in g.p. Two poultices were alternately applied, the one of pounded ice, the other a very warm bread or linseed poultice. The pounded ice, in a bladder, was applied for eight or ten minutes; the warm poultice for an hour or two, or even longer. In similar cases he also suggested the use of powerful galvanic currents. Dr. Warren of Dublin described the treatment by galvanism of the bedsores in protracted spinal cases, first employed by Dr. Crusel of St. Petersburg. "The method of applying it was simple; a clean silver plate of a size corresponding to the ulcer was placed over the sore, a zinc plate, under which was placed a piece of chamois-leather wet in vinegar, having been laid on some part of the neighbouring healthy skin. The current was completed by a copper-wire joining the zinc and silver plates." These several procedures deserve a fair trial for the prevention or cure of the bedsores occurring in g.p.

Urinary disorders. When retention of urine is due to loss of power in the detruding muscles of the bladder I have found strychnia useful. The bladder can often be relieved of its load by the simple expedient of firm pressure by the palm of the hand on the hypogastrium, the lumbar region being firmly supported simultaneously. *Catheterization* should be avoided if possible; yet the bladder should not be allowed to remain distended, as it may be in the "seizures." Before using the catheter a stream of strong carbolic lotion should be poured through it, and its exterior should be smeared with carbolized oil. When the urine is turbid, ammoniacal, and mucous, the bladder may be washed out with a solution of quinine; or with *very weak* nitric acid or carbolic acid or ferric perchloride injections; and at the same time boracic acid or sodium-biborate may be given internally.

* "Phys. and Path. of the Nerous Centres," Am. Ed. 1860, p. 260.

Refusal of food. All sorts of expedients have been tried here by various physicians; nutritive injections, amyl nitrite, etc. These have failed in the hands of others. I have found the point of greatest importance here is to secure free movement of the bowels by enemata or by other means, and to treat and cure any morbid state of the chylopoietic viscera.

Othæmatoma, I have usually treated by the constant application of very strong lead-lotion. By acetum cantharidis, applied to the inner surface of the pinna, Dr. G. J. Hearder* checked the hæmatoma, and lessened or prevented the usual subsequent deformity; only 1 of the 6 cases he reported was a G.P. I have occasionally blistered the cranial side of the pinna with good effect. If the ear becomes semi-globular and fluctuating, the fluid should be evacuated, and the auricle compressed by pad and strap.

CHAPTER XXI.

VARIETIES OF GENERAL PARALYSIS OF THE INSANE.

Of the subdivisions made of g.p., the majority are merely semeiological, and do not correspond to pathological varieties of the disease. Such are the divisions made by Billod, Brierre de Boismont, Falret, Marcé, Mendel; and (with some pathological admixture) those of Calmeil and Voisin were chiefly of the same nature. The basis of Lionet's subdivision was practically etiological.

This chapter is an abstract of a paper I published in the "Journal of Mental Science" for April, 1878. In any large group of cases of general paralysis, so great are the differences in the mental and physical symptoms, the mode of onset, the course, duration, mobility of symptoms, and pathological anatomy, of the several cases, that one must conclude that there are varieties of the disease. This has led to subdivisions of the affection, or grouping of its cases, scarcely any of which have had a pathological basis. True it is that Bayle, who was the first to separate and clearly describe g.p., was led to place his cases in five series; in the first of which were simply the lesions of chronic meningitis; in the second abundant serous effusion was added to these; while, in the third, consecutive inflammation of the grey cortex, in the fourth, arachnoid cysts, and in the fifth, various cerebral affections, complicated the chronic men-

* "Journ. Mental Science," April, 1876, p. 91.

ingitis. Like others, holding that g.p. occurred either with or without insanity, Baillarger further spoke, and in this was followed by Lunier, of the symptoms of general paralysis as being produced by (1) chronic meningo-encephalitis, and by (2) chronic hydrocephalus, or serous effusion following hæmorrhage, etc. Schüle also attempted an anatomico-pathological division. But nearly all the attempts hitherto made* to delineate shades or varieties of general paralysis, have but set forth and emphasized certain symptoms, and especially certain mental features. It is evident that this cannot correspond to any real division of the disease into true pathological varieties, and can, at the best, but play the *rôle* of a measure of clinical and descriptive convenience. Nosological divisions, based upon mental symptoms or psychological differences, as in the systems of Arnold, Pinel, and Esquirol, are confessedly of but temporary use, and not in correspondence with true pathological varieties. Applied to general paralysis, they are even more inexact. For in the latter affection, even more than in ordinary insanity, the mental symptoms often undergo most decisive and, perhaps, sudden changes. There is in it a great variability, so that he who suffers from expansive delirium to-day, may to-morrow exhibit maniacal symptoms, may soon afterwards be plunged into the depths of hypochondriacal woe, and ere long have left all hope behind, having irrevocably entered the portals of a progressive dementia. True it is that many retain almost throughout some predominating general character of mental symptoms; there are those who throughout specially exhibit either grandiose delirium, or symptoms allied to active mania, or to dementia, to hypochondria, to melancholia, or to circular insanity. True it is, also, that in cases such as these, there is a general tendency to differences in the morbid anatomy of the several groups. At least the facts of the morbid histology observed for several years past seem to warrant me in making this assertion. Yet, for the reasons above mentioned, it seems that one cannot establish any real varieties based upon the differences in the mental symptoms. Whether we shall ever arrive at them or not, I think the only true classification, subdivision, and terminology in mental diseases must be based upon the morbid alterations of tissue, or of function of tissue, which engender the symptoms—that they must be anatomico-pathological. For these purposes, there can be

* For examples see " Journ. Mental Science," Apr., 1878, p. 26.

no real finality, no pathological exactitude, in semeiology or in ætiology. With reference to the great diversities one finds in cases of general paralysis, it is desirable to inquire whether groups of cases cannot be formed corresponding to definite differences in, or modifications of, the pathological lesions proper to the disease, each group of which shall present its more or less characteristic *ensemble* of symptoms, or its individual method of association and relation of symptoms. General paralysis varies much in different cases as to the parts of the encephalon primarily or especially attacked; the disease has in this case spent its force with greater severity upon limited localities, in that case upon other definite foci, and in still another upon a third locality or group of localities. In endeavouring to localize the points of principal morbid implication, I began, early in 1873, to specially observe and record the exact localization of all the principal adhesions of the pia to the cerebral cortex. This adhesion I had always looked for, and recorded in outline, in earlier necropsies, and had come to view it as the most characteristic naked-eye appearance in g.p. In the "Journal of Mental Science," Jan., 1876, I gave details of one of the cases in which each adhesion was thus recorded, and an attempt was made in that communication to see how far some of the symptoms observed during life could be explained by the irregular distribution of the adhesion and decortication, and how far, in that case, the results of the experiments of disease tallied with those of the physiological laboratory. Perhaps, and as far as I know, the above were the first investigations into the exact localization and distribution of the adhesions in various cases of general paralysis, for the purpose of throwing light upon clinical features of the disease. Believing that valuable results are to be obtained from this line of inquiry, I have described the localization of these adhesions with minuteness. But the adhesions in question, and the associated changes wrought in the corresponding portions of the grey cortex, are not invariably present in g.p., and they constitute but one of the points to be examined in investigating the anatomical distribution of the more extreme degrees of morbid change in individual cases of that disease. The other conditions of the grey cortex must also be particularly observed, the portions of it first or chiefly attacked must be sought for, and its state, both general and local, as to vascularity, consistence, bulk, pathological products, and degeneration, examined. An investigation of the same kind must be directed to the ganglionic masses at the base of the brain, to its medullary substance, to the bulbar tissues, to the cerebellum, and to the spinal cord and its meninges, as well as the sympathetic ganglia. So, also, must be taken into account such secondary products and complications as excessive serosity, hæmatoma, pachymeningitis, local ramollissement. Finally, there is required a careful microscopical examination of the entire nervous systems of a large number of G.Ps., carefully compared with detailed records of the clinical features observed in each. My leisure has been insufficient for this examination. Yet I have observed such marked differences visible to the naked eye, or obvious upon making a less elaborate or extensive microscopical examination, that it is desirable to place some of the cases on record, grouping them according to differences in the pathological lesions of the encephalon, and to indicate the clinical features which in their totality constitute, as it were, the garb each group severally wears. I have long thought that under the name general paralysis were included several pathological varieties, in a manner analogous to, but less decided than, that by which formerly the name "pulmonary phthisis" shielded several varieties of pulmonary disease; and by which, again, "Bright's disease" included a number of more or less distinct morbid states. I do not offer the following groups as representing proved varieties of general paralysis, or even as by any means covering the whole of the ground in their clinical aspects. But if we can show that in g.p. cases can be placed in groups, the members of each of which have a considerable similarity to each other in their symptoms, course, duration and pathological anatomy, then has a step been taken in the direction of establishing pathological varieties of that disease. Some accident or intercurrent malady may cut off a general paralytic before his time, and the lesions may appear different from what they would have been had the affection run its usual course. This is, of course, true of all organic maladies. Avoiding such fallacies as would arise, for example, from considering that in the same case ending untimely or ending maturely we had two varieties, it is possible, I believe, to indicate several groups which possess an individual distinctness. Five of these will now be referred to. It is not meant that the essential pathological process is distinct in each. The separation between some of them is mainly based upon the differences in the encephalic localities affected in each (3 and 4). In another instance (group 2) the difference seems to be mainly one of chronicity and mildness; but also of locality. In another (5) the local cortical induration is a striking condition, differing somewhat from the more usual change of same part in g.p. The pathological and clinical features of the five groups are described in detail at pages 31 to 43 of the "Journal of Mental Science" for April, 1878. Of that description the following is a very brief and imperfect summary.

FIRST GROUP.

Principal changes. 1. Cerebral hyperæmia and softening are unusually generalized, but particularly affect the cortical substance of the superior, external, and, to a less extent, internal fronto-parietal regions. As a rule the cerebellum and basal ganglia are considerably affected; the mesocephale and spinal cord, less.

2. Adhesion and decortication are usually well-marked, are mostly confined in the cerebrum to the upper and external surfaces, are chiefly over the frontal lobes, are well seen on the parietal, less over the temporo-sphenoidal, and sometimes slightly or moderately over the internal and inferior surfaces. The cerebellum is often affected. The above changes are nearly, or quite, symmetrically disposed in the two cerebral hemispheres.

Principal clinical features. 1. The mental symptoms are mutable. Exalted delusions are the most marked feature, maniacal excitement and insomnia are frequently observed. Gaiety, self-satisfaction, benevolence or pride are evinced; or the patients are selfish, haughty, hostile, obstinate, abusive; or destructive, untidy. Transitory depression, or melancholia, sometimes comes on. Dementia is occasionally predominant from the first.

2. Motor ataxy and paresis are present, are sometimes well-marked; but in the earlier period are often masked by the maniacal state, or but imperfectly developed. Motor restlessness is frequent.

3. Occasionally, epileptiform or apoplectiform seizures, choreiform movements, or tremor cöactus are observed. Some have hallucinations of hearing or of sight. Later, are found defects of general or special sensation, or hypochondriacal sensations.

4. The *average* duration is short, being about 16 months.

SECOND GROUP.

Principal changes. 1. Atrophy of the brain, much intracranial serum, the ventricles dilated and much granulated.

The gyri of the brain are wasted, especially on the upper surface and at the frontal region, the corresponding grey cortex being either softened, or, occasionally, of about normal consistence, pale, watery, sodden, or at times of fair colour, or even mottled.

2. Adhesion and decortication, usually slight or moderate, are principally at (1) the Sylvian fissures, (2) upper frontal, and (3) parietal, surfaces, and (4) base (orbital or temporo-sphenoidal); and this is the order of their relative degree.

3. The white cerebral substance, softened in some cases, more or less indurated in others, usually tends to pallor. The basal ganglia are generally pale, soft, shrunken; the spinal cord may be more or less softened or indurated. The meningeal changes are very marked, extend to the base, and, like the other changes, are symmetrical.

Principal clinical features. 1. The mental symptoms in the earlier periods may consist partly or slightly in exaggerated notions, or paroxysmal excitement with strange demeanour; or, rarely, exclusive dementia predominates from the first. Subsequently, is protracted dementia, with which fitful outbursts of excitement, or hypochondria, may occur. The quiet self-satisfaction, or the unemotional state, of the early periods, is usually replaced by morose, peevish, distressed, or apprehensive states of feeling, and these by obliteration of the emotional life. The patients often are destructive, obstinate, abusive, degraded in habits and language.

2. Motor paresis is comparatively slight in the earlier stages, slowly becoming more marked, especially in the lower limbs. The patients are usually bedridden a long time, and often grinding their teeth.

3. There is peculiar absence of epileptiform and apoplectiform seizures, and of marked general tremulation. There is comparative absence of sensory symptoms, save for blunting of sensibility as the disease progresses.

4. The cases are of long duration, the *average* being four years.

THIRD GROUP.

Principal changes. 1. The *left* cerebral hemisphere is much more diseased than the right, and is atrophied. There is usually atrophy of the grey cortex, chiefly marked in the frontal lobes, but occasionally marked elsewhere. It is usually pale, or mottled by vascular redness, and is sometimes softened, at others indurated in a portion of its extent, either change being much more marked in the left hemisphere, and the frontal lobe being, usually, most affected. The white substance, pons, med. obl., and cerebellum, vary in consistence and vascularity; the basal ganglia are softened.

2. Adhesion and decortication are usually more marked on the left side, occur with equal frequency on the frontal and parietal lobes, while the temporo-sphenoidal suffer very considerably, and the changes in question may be well-marked on the inferior surfaces. The purely meningeal changes are usually well-marked, are either symmetrical or predominate over the left hemisphere, and are often well seen over the base.

Principal clinical features. 1. In the prodromic stage the patients often are very eccentric, odd, restless, fidgety, and occasionally excited.

2. Dementia, well-marked, early and predominant, is frequent.

3. Melancholic delusions of harm, annoyance, fear, suspicion, are equally frequent, and with them are feelings of alarm and apprehension, or the patients are querulous and irascible, or dejected and weeping.

4. Occasionally there is early maniacal excitement, with irritable outbursts; while exalted delusion, or some largeness of idea, may now and then occur either at an early or at a later period. The later course is mostly one of extreme dementia, sometimes with a melancholic, or even expansive tinge. Sometimes destructive, threatening or violent, the patients generally become tractable towards the last, but of degraded habits.

5. Muscular ataxy and paresis are well-marked, motor restlessness is frequent. *Hemiplegia* is more or less marked and frequent *in all*, and ordinarily is convulsive in origin, while temporary local pareses following local spasms, are frequent.

6. Epileptiform attacks, hemispasm, local spasm, are very frequent; and tremor cöactus is not unfrequent. Apoplectiform attacks and aphasia are sometimes observed. Occasionally, hallucinations, general obtuseness of sensibility, or local anæsthesia. The *average* duration is about 17 months.

Fourth Group.

Principal changes. The lesions are much more marked in the *right* than in the left cerebral hemisphere. The general description of the changes in the left hemisphere in the last group is here transferred to the right, and of the right in the third group to the left in this. The cerebral vascularity is, however, somewhat greater in this fourth group, and the basal meningeal changes somewhat less. The adhesion and decortication are usually more marked in the right hemisphere, occur mostly over the parietal lobe, often on the posterior part of the frontal and on the temporo-sphenoidal, occasionally upon points of the internal surface or base.

Principal clinical features. 1. Occasionally preceded by strangeness of conduct, there is usually early ambitious delirium, complacency or elation, with or without active maniacal agitation, and violent, destructive and dangerous tendencies. Now and then dementia, with fidgety, mischievous, restless, slovenly, or destructive tendencies.

2. Later, there are often exaggerated or exalted notions, alternating with conditions in which the patients are foul-mouthed, querulous, morose, irritable, depressed, or in dread; or the latter states come to predominate entirely. At first there may be the expression of an undiscerning generosity; later an abusive manner of address, often with degraded habits, and destructive or dangerous tendencies.

3. The muscular ataxy and paresis are of the ordinary type. Occasionally there is great tremulousness, or, again, tremor cöactus. Hemiplegia is frequent, sometimes occurring as a simple paralytic seizure, sometimes following epileptiform attacks, and sometimes due to embolism, or hæmorrhage.

4. Epileptiform seizures are very frequent. Sensation is blunted in the later stages; occasionally there are hallucinations of sight and hearing, blindness, or hypochondriacal sensations.

5. The duration is, on the *average*, 24 months.

Thus it will be noticed that some of the clinical features are very dissimilar in the third and fourth groups; mental symptoms like those of dementia and of melancholia predominating when the *left* is the hemisphere principally diseased; and exalted delusion and maniacal agitation when the *right* is the hemisphere in which the morbid process is earlier, and more extensive, severe, persistent, and disorganising.

FIFTH GROUP.

Principal changes. 1. There is local reddish, occasionally pale, induration of the cerebral cortex, sometimes of wide distribution in its lesser degrees, most marked in the frontal lobes or their anterior portions, and affecting either one hemisphere or both. The indurated part is usually atrophied. The non-indurated is of ordinary colour, or pale.

2. The adhesion and decortication, absent in one case, are

in others unequal in the two hemispheres, occur mostly on the parietal, often on the posterior part of the frontal, and on the temporo-sphenoidal lobes; now and then on the internal surfaces, or highly marked on the inferior surface. The purely meningeal changes are marked and extensive. The white substance, usually slightly indurated, may be fairly vascular, or paler than usual. Usually, are changes in the parts at brain-base, and in cord.

Principal clinical features. 1. The mental symptoms are various and varying. Some suffer mainly from symptoms of mental depression; others, of dementia; and others, of maniacal agitation and emotional exaltation. Complacency, irascible outbursts, gloom, or apprehension are observed in some cases. All are indifferent to their degraded habits, some are docile throughout, but others are at times destructive, or quarrelsome and abusive.

2. Muscular ataxy and paresis are fairly marked. Epileptiform fits, hemispasm, often followed by hemiplegia, are very frequent. Local spasms, followed by pareses, are not infrequent. In some, tremor cöactus, or, again, choreiform movements are observed, or apoplectiform seizures, with, or without, convulsive movement. Besides these, several have less grave, but frequent, apoplectiform attacks. A few show hallucinations, or marked anæsthesia, or early headache. The *average* duration is 23 months.

Microscopic appearances. In the *first* group, the bloodvessels are distorted, dilated here and there, and their walls the site of degeneration, of deposits, and of nuclear hyperplasia; while interstitially there is hyperplasia of the nuclei of the neuroglia, or proliferation of embryoplastic nuclei; free blood corpuscles, and many scattered collections of blood-pigment. The nerve-cells are more or less degenerate.

In the *second* group, the nerve-cells suffer perhaps more than the other elements of the cerebral grey cortex. In one case the atrophy and destruction were so marked in the frontal lobes that there seemed to be an absence of the large nerve-cells; those present were comparatively small, atrophied, with rounded wasted outlines, and possessed of but few processes. Some had undergone marked granular degeneration. Some had fallen asunder and the granules, strewn about, looked like a downfallen heap of pebbles. Even the hyperplastic nuclei of the neuroglia had undergone degeneration.

In the *third* group, in which the *left* is the cerebral

hemisphere chiefly diseased, the fact is obvious under the microscope as well as the naked eye. Thus, for example, in one case the grey cortex of the third left frontal gyrus showed some granular degeneration of the nerve-cells, hyperplasia of the nuclei of the walls of its blood-vessels, and thickening of the coats of some of the small vessels, some of which had also dark grey deposits or growths in the walls; increase of neuroglia or of nuclei, and some colloid bodies. Here, as compared with the corresponding part on the right side, the nerve-cells were more granular, there was more increase of the neuroglia and its nuclei, and the colloid bodies, present here, were absent in the right side. In the spinal cord were granule-cells, colloid bodies, and some granular degeneration of some of its nerve-cells.

In the *fifth* group the microscopical indications of interstitial sclerosis are well-marked.

Remarks. Concerning the above groups of cases, a few words may be added :—

The *first* consists of cases of a very common kind, illustrations of which are abundant in medical literature. Bearing in mind that it represents the shorter and more acute cases of a larger group, it may be passed without further comment.

But not so the *second* group. Possibly, some would call these cases atrophy of brain, or chronic hydrocephalus, or chronic meningitis, but an attentive study of the clinical features and necroscopic records will justify the position in which they are placed here. Bayle fully described cases of this kind, and several of them may be found in his second series. Drs. Baillarger and Lunier distinctly assert a place among general paralytics for cases somewhat similar. But those to be detailed hereafter, as composing my second group, will be found to approach much nearer to the typical cases of g.p. than to the hydrocephalic, and other, cases just referred to. Out of a large number of cases, a series could be selected, which, by gentle gradations, would lead from this group up to the most characteristic and typical case of g.p.

The *third* group, with its less usual symptoms, is relatively infrequent. The symptoms are not always the same, nor is this to be wondered at, as the principal lesions may be brought about in different ways. Turning to Calmeil's work, I find that his cases with lesions somewhat like those of the third group, presented, also, clinical features, on the whole, very similar to those of the latter. [T. I., pp. 385, 554. T. II., pp. 27, 53, 76, 89.]

To the *fourth* group the same general remarks apply. The several cases in Calmeil's work, presenting somewhat similar lesions, were also, on the whole, manifested by very similar symptoms, except that in them attacks of "apoplexy" and stupor and of spasm were more frequently observed.

In the *fifth* group the interstitial changes tend to sclerosis. The pathological process is different in its results from that which produces the more ordinary softening of g.p. This group, however, is not so well defined, clinically, as the others, and I do not lay special stress upon it. I find that Calmeil's cases, with somewhat similar lesions, exhibited also a clinical similarity to those of the fifth group, but, as compared with the latter, presented more mania, délire ambitieux, and exaltation; less sadness, less convulsion followed by hemiplegia, and somewhat less spasmodic twitch and tremor cöactus. Otherwise, as in frequency of apoplectiform attacks and somnolence, they are very similar. [T. I., pp. 311, 431, 437, 519, 571, 581, 591, 658. T. II., pp. 5, 60.]

It may be said that the *mental* differences to which I have referred in the above groups do not mark any essential differences in the cases; that, for example, it may be urged that the grandiose delirium is, in reality, only a manifestation of that dementia which, on the mental side, appears to be of the essence of the affection. I am content to record the facts as I find them, believing that differences in the *mental* symptoms of general paralysis, though not essential, are yet of valuable import. The views expressed here are based simply upon the clinical and necroscopical observation of a number of cases; upon the fidelity of this they must stand or fall; nor need one be concerned to trace a harmony or discord between the facts mentioned above and the conflicting results of experiments on the localization of cerebral function.

CHAPTER XXII.

CASES OF GENERAL PARALYSIS.

In order to curtail what follows I have omitted many of the cases upon which my paper was founded, and have abbreviated the clinical and necroscopical descriptions of those recorded here; omitting many, and summarily stating others, of the numerous and voluminous notes made in every case by me. Yet it is trusted that sufficient fulness of detail has been retained, and that the number of cases is adequate, to place those who take any interest in the subject in the same point of view as occupied by the writer. Particularly has the *first* group been abbreviated, representing as it does comparatively accentuated and rapid cases of the most common form of g.p.; while only the heads of the cases of the *fifth* group have been finally retained. In all, very full records were also made as to the thoracic and abdominal viscera, but only very brief summaries of these are given below. The cases are more fully reported in the 1st Ed. of this work.

FIRST GROUP (*see* p. 408).

CASE 52 (in abstract).—*Severe and protracted maniacal excitement; extreme motor agitation, partly masking the paresis; ambitious delirium; usually pleasant good-humour and exaltation of feeling; transitory weeping.*

Widely distributed encephalic hyperæmia and softening; extensive adhesion and decortication of grey matter, especially over the supero-lateral fronto-parietal regions and uncinate gyri, and well-marked over the internal surfaces.

J. T. R., a driver in the Royal Horse Artillery. Admitted Feb. 13th, 1874. Age 34. Married. Length of service, 16$\frac{1}{12}$ years.

History.—Marked symptoms dated from Dec., 1878. This was the first attack of mental disease, and before coming here he had been treated at Birmingham and in the Royal Victoria Hospital at Netley. He was stated to be neither epileptic nor suicidal; and the causes assigned were vague. He was reported to have been suffering from g.p. whilst at Netley, and to have been in a constant state of noisy, restless, active excitement, with sleepless nights, incoherent language, and extravagant delusions. He also said his wife came in through the ventilator, and he tried to stuff it with a pillow. Emaciation steadily progressed; the habits were wet.

State on Admission. Physical Condition.—Height 5ft. 5in. Weight 133lbs. Pulse 72, regular; arteries somewhat tense; no cardiac bruit, or change in area of præcordial dulness. Scar of venereal sore on penis. A few furuncles on body. Slightly florid complexion. Pupils equal and acting fairly well. Gait a little awkward and unsteady, especially in turning. Viscera healthy.

Mental Condition.—He was self-satisfied, smiling, exhilarated, loquacious, irrational and incoherent in conversation. Thus he said " Mrs. S—— is my husband, the Lord Bishop." " My rank is about (that of) a Lieutenant-Colonel." " I've plenty of money—all the nation is my money." " My pay is a Bishop's pay—some millions." " I've £700,000,000 in the bank."

Subsequent Progress of Case.—Several days later he was noisy, and of changeable temper, now excited, now angry, now placid for a moment, and, again, weeping like April shower. He said that he " had educated the Queen to manage armies and control the nations." He busied himself in collecting imaginary " droves of bishops." He had hallucinations of sight and of hearing : becoming excited and refusing to be examined, he turned as if gazing at some apparition and said, " See ! Major T——" Then, as if repeating the advice given to him by the phantom Major, he shouted, " Kick him ? Yes I will, I'll do as you tell me." He pointed to the Major, " dressed in gold," and revealed the delusion that he had seen and spoken to a Mr. G. ten minutes previously. He gave expression to extravagant, absurd, mobile, varying and self-contradictory delusions as to possession of wealth and power, and often stood talking to the wall, or shouted while capering about and confronting it. There was an occasional slight pause in the speech, which was loud and excited, and there was perhaps a faint fibrillar tremor of the tongue on protrusion, but not of the face during speech. Tr. Digitalis and Potassic Bromide. The excitement and restlessness were so incessant and extreme that he was emaciating rapidly, although on full diet, and therefore Ol. Morrh. and Ferri Perchlor., were also ordered ; yet he was found to have lost 6lbs. in weight at the end of eighteen days. To take 1$\frac{1}{2}$lbs. of minced meat daily; also beef-tea and extra bread ; and in twenty-five days he gained 9lbs. in weight. During this period left othæmatoma had developed, and the patient continued to be restless, loquacious, incoherent and destructive.

In April, noisy and loquacious, he was still in almost constant motion, clattering with his feet on the floor, destroying his clothing, wet by day, dirty at night, and smearing his bedding and the walls of his room with fæces. His language continued to be incoherent and expressive of the most extravagant and utterly disconnected notions, and of their relics,—words such as "millions" and "bishops" forming the nuclei of a farrago of nonsense that he sang by the hour together with the utmost exaltation of feeling and manner. Pulse 60, bowels free. Weight on April 14, 137¼lbs. After this, Succ. Conii was ordered, but the weight fell to 133lbs. by May 18th, and then Ext. Physostigmatis was substituted for the Conium, and the oil, iron, and extra diet were continued. Mentally, unchanged. The pulse, usually rather slow, became more frequent under conium. July 27th. Had diarrhœa for two days. Skin becoming dingy. The facial and speech signs of g.p. were still only slight. Was again very restless at night. Sept. 7. Weight 142lbs., his highest weight whilst here. Sept. 14th. Had a syncopal seizure this morning. Omitted the Physostigma and other remedies. On Oct. 9th profuse diarrhœa returned, and continued with intestinal hæmorrhage. Both persisted, mental confusion and stupor supervened, these passed into coma, and this into death on Oct. 13th, 1874. To the last he was noisy, restless, and full of exalted delusions. At first good-humoured, later on, he was now angry and excited, now pleased, now weeping.

Abstract of Necropsy.—25 hours after death. Body somewhat emaciated. Calvaria of slightly worm-eaten appearance in parts. A small flattened exostosis on the internal surface of the right side of the frontal bone. 2 fl. ozs. of serum escaped on removal of the brain. Arteries at base of brain healthy. Cerebral meninges congested. The gyri were somewhat shrunken, and the slightly wide and rounded sulci were filled with serosity, which bathed the pia-mater on the superior surface except over the occipital lobes. The pia and arachnoid were somewhat thickened, and the former was hyperæmic. There was some opacity of arachnoid over the frontal and parietal lobes.

The superficial layer of cortical grey matter of the prominences of the gyri was stripped off, along with the membranes, over the whole of the superior and lateral aspects of the cerebrum in the fronto-parietal regions. Especially was this marked in front, where the entire outer layers of the grey matter were stripped off, but every convolution of the area just specified was extensively involved, and further detail is unnecessary. The same change affected the prominences of the first and second temporo-sphenoidal gyri to a moderate extent and degree, but tapered off here and spared the third gyrus. In the occipital lobe it was only slight, affecting the anterior portions of the first and second gyri. The uncinate gyri also suffered, and a few scattered points of adhesion were found elsewhere on the inferior surface of the brain. The grey matter was hyperæmic, mottled by sections of the contents of visible dilated vessels, slightly softened, of a deep grey and somewhat slaty hue, of fair depth, and of imperfectly marked stratification. The white substance of the brain was somewhat softened, and was mottled, pinkish, the puncta sanguinea were numerous, and clots dragged therefrom in making sections. All the above appearances were symmetrical in the two hemispheres. Fornix softened, and serum in lateral ventricles turbid. Left corpus striatum and optic thalamus, slightly shrunken. Pons and medulla oblongata lessened in consistence, and, like their meninges, hyperæmic. Cerebellum hyperæmic and slightly softened. Weight of cerebrum 40ozs., of cerebellum 5½ozs., of pons and med. obl. 1oz.

Thorax.—The *heart* weighed 10½ozs., its left ventricle was moderately contracted, and was one inch at its point of greatest thickness. Its appearance was that formerly called "concentric hypertrophy." The heart-muscle had a healthy appearance. *Right lung*, weight 25ozs.; posterior and basal congestion. *Left lung*, the same; weight 23½ozs.

Abdomen.—Old peritoneal adhesions were found, particularly in right iliac fossa, and in splenic region. *Liver*, 80ozs., dark on section. *Spleen*, 10ozs., diffluent, of a dirty brick colour, hobnailed and of irregular shape. *Right kidney*, 10ozs.; *left*, 9ozs.; both of variegated, mottled, appearance.

Remarks.—1. This is specially interesting as an illustration of the earlier stages of those cases in which the ataxy and paresis are *masked* by the effects of the maniacal excitement. It will be noticed how slight, comparatively, were the ataxic or paretic indications afforded by the condition of speech, tongue, or gait; and yet how well-marked the ambitious delirium and the maniacal agitation. Had the patient lived longer the motor symptoms would have become more and more marked. The motor affection was relatively more marked in the lower limbs than in the upper.

2. This case also illustrates that form of extreme restlessness and motor agitation in g.p., that Bayle called "convulsive" and that is exemplified in several cases of his third series.

3. The rapid emaciation of this patient was arrested only by the use of a very large amount of nutritious food, with Ol. Morrh., and Ferri Perchlor. No means employed had any permanent control over the nearly incessant mental and motor agitation. The heart and arteries were comparatively unaffected by the moderate renal change. The condition of the left ventricle was mainly due to the mode of dying.

CASE 53.—*Extravagant delusions, with much and early mental weakness. Early exaltation of feeling; later on, alternately self-pleased, dull, and lachrymose. Dementia, childishness, becoming more marked. Progressive paresis; bedridden; bedsores: hæmorrhagic maculæ. Apoplectiform seizures. Late tremor cŏactus; finally, choreiform movements.*

Extensive encephalic hyperæmia and softening; widely-spread adhesion and decortication, especially over the supero-external surfaces of the cerebrum; and affecting the cerebellum, also. Marked changes in spinal cord.

A. B. Private, 100th Regiment. Age 35. Military service $15\frac{1}{12}$ years. Fairly educated. Admitted Aug. 2, 1873.

History.—This, the first attack of mental disease, had existed since about April, and the patient had been under treatment at Portsmouth, and at Netley. The cause assigned was vague. Neither epileptic nor suicidal. The mental symptoms had come on gradually, and the patient was stated to have had "delusions of a religious type" at first, but the symptoms at Netley were said to be those of "dementia" coexisting with "a marked paralytic tendency," and it was further certified that A. B. was quite confused, restless, incapable of taking care of himself, frequently out of bed at night, and walking about, and that he gave indications of impairment of memory, confusion of thought, and failing powers of comprehension and of utterance.

State on admission into Grove Hall Asylum. Height 6ft. 1¾in. Well nourished, and well built. Thoracic viscera apparently healthy; pulse 96, arteries slightly tense. Pupils equal. The speech, and the condition of the lips face and tongue, were those usually observed in the comparatively early or middle periods of g.p. The patient could not tell his own age correctly, mistook the month and year, was confused and incoherent in continued conversation, displayed general mental enfeeblement, and asserted that he was heir to his uncle's estate, worth £19,000. His manner was simple and childish, his expression was usually smiling, pleased and satisfied, and he was still clean in person and habits. R. Ferri Perchlor.

Oct. 30, 1873. Recently, had an apoplectiform seizure. Aperient enemata brought away large stools, and the symptoms vanished. Subsequently, he took aperients as required, the iron being continued. Slight "congestive" symptoms had appeared on several other occasions. Jan. 30, 1874. The tremulous twitches of the lips and tongue during movement were now well marked; speech was slow, broken, hesitating, indistinct. Pupils irregular, the right vertically oval, the left large, and both fixed. The grasping power of the hands was impaired, and the gait unsteady, the left lower extremity was the weaker, the knees and hips were kept partially flexed as he walked, and as he spoke the head shook. He said he was well and strong, and he had exalted delusions as to his possessions, physical prowess, and procreative powers. Yet he was not much elated; being ordinarily dull, confused, and often wearing an apprehensive, anxious expression. Often restless and sleepless. Had Ergot and K. Br.; also *haust. noct.* Hyoscyamus and K. Br.

March 1. In bed with signs of bronchitis and congestion of the lungs since

Feb. 18th, when treatment was altered; there were also præcordial œdema and increased dulness, and the heart-sounds were heard more distantly. Now, he was often lachrymose, sometimes very dull and confused, at others brighter. He was slow to reply, and spoke with very marked tremor and quivering of the lips. There was marked tremor cöactus of the forearms and legs, the limbs quivered constantly therewith, and the eyelids were generally closed and quivering. Reflex action was nearly abolished in the feet, but he retained some voluntary power over the lower extremities. 15th. Bedsores had formed over the hips and sacrum, and on the body there were scattered minute hæmorrhagic elevations. His mutterings now and then revealed the shattered relics of delusions of wealth, and occasionally he sang snatches of song. 24th. Tremor cöactus still marked. Hypostatic pneumonia. Pressure on lower limbs easily caused bedsore. 26th. Occasional retention of urine latterly, and drowsiness. Legs œdematous. Rapid flexion movements at the wrist, fingers kept semi-flexed, much tremulous twitch about the arms, legs, feet, head, platysma, and cervical muscles. But the extremities were also affected by conspicuous choreiform movements. Death on March 28th, 1874.

Sectio cadaveris; 44 hours after death. *Head.*—Calvaria thin; dura hyperæmic; arachnoid at base somewhat thickened; frontal interlobar adhesions. Arachnoid thickened and opaque at vertex, and the pia thickened, hyperæmic, and bathed in serosity. The anfractuosities, wide and rounded on the upper surface of the cerebrum, afforded a lodgment to serosity, and meningeal shreds remained behind in them after the cerebrum was stripped. Marked adhesion and decortication affected the prominences of all the gyri of the frontal and parietal lobes on their superior and external surfaces; also the orbital and temporo-sphenoidal surfaces to a less degree. The cerebral grey cortex was hypervascular, pink, especially in its deeper layers, somewhat softened, of fairly marked stratification, and moderate thickness. The white cerebral substance was hyperæmic and softened, the fornix softened, the velum interpositum tough, the grey commissure small, the lateral ventricles of the brain were large but not containing abundant fluid, the corpora striata somewhat softened, the optic thalami hyperæmic. Some cerebellar adhesion and decortication, about the median line. Cerebellum softened. Cerebrum 39½ozs.; cerebellum 5ozs; pons V., and med. obl. 1oz. Serum from cranial and spinal cavities, 3½fl. ozs.

Spine.—The spinal meninges were hyperæmic about the mid-dorsal region. The spinal cord was softened and somewhat anæmic in the cervical and upper dorsal regions. The softening was more marked in the cervical region, and there principally affected the grey matter, which was almost diffluent, and the antero-lateral columns. Unduly wide anterior median fissure in cervical region.

Thorax.—Heart 10ozs., healthy; but there was 1½ozs. of pericardial fluid Aorta atheromatous. In both *lungs,* hypostatic pneumonia. *Abdomen.—Liver* of slightly "nutmeg" appearance, 67½ozs. *Kidneys,* R., 4¾ozs.; L., 5¼ozs.; their cortices were thin and granular, and their capsules adherent. The middle tract of the *ileum* was hyperæmic.

Remark.—G.p. running a somewhat rapid and severe course. Early loco-motor helplessness. Various trophic lesions. Tremor cöactus, and, finally, choreiform movements were superimposed on the marked tremulousness and twitch common in g.p. In relation to them we note the unusual amount of disease in the spinal cord and at the base of the brain. There was also an un-usual amount of adhesion and decortication.

CASE 54.—*Early exaltation with delusions, auditory hallucinations, and restlessness. Later, almost complete remission of the mental symptoms, con-tinuance of the motor. Return of slight ambitious delirium; self-satisfaction; almost constant singing and joyous good-humour. Later, restlessness, excite-ment, destructiveness. Softening of encephalon and hyperæmia. Extensive adhesion and decortication, especially over the superior surface of the frontal convolutions, and the lower tiers of the parietal. Changes in and about the basal ganglia.*

T. H., Drummer, Grenadier Guards, aged 34. Admitted Nov. 11th, 1872, after 15$\frac{4}{12}$ years' service in the army. Single.

History.—This, the first attack of mental disease, had existed since June, 1872, and T. H. had been previously under treatment in the Guards' Hospital, London, and at Netley. The cause was stated to be "moral." Patient had been promoted to the rank of drum-major of the 99th Regiment, but his ideas appeared to be so large and his self-importance so inflated that the colonel refused to retain his services. He then became restless at night, and one evening he packed up his kit, declaring that he had orders from God to come to London. He was admitted at Netley with slightly exalted delusions, and some failure of the powers of utterance. Afterwards he became worse, the delusions more prominent and associated with some excitement and with auditory hallucinations. He also undressed at unseasonable hours.

On admission. Height, 5ft. 10in. Large cranium. A fine, soldierly, well-built, handsome man. Well nourished, fresh complexion, light-blue irides, pupils somewhat small, the left slightly the larger, both sluggish and of slightly irregular outline. Skin delicate, healthy. The tongue is protruded fairly, but the aid of the teeth is often invoked to grasp and steady it. There is paresis of right side of mouth, the speech is now and then imperfect, the words being clipped, and there is an occasional pause, with thickness and indistinctness of utterance.

Mental State.—Although he says that he is a clever musician and that he left the 99th because the quarters provided for him were unsuitable, yet he denies all his former hallucinations and delusions, and the remission in the mental symptoms of g.p. is decided. He denies having ever indulged in any alcoholic or sexual excess. Later in Nov.; speech thick, slow, hesitating, with tremor of upper lip; handwriting shaky.

The subsequent notes of this case may be briefly condensed as follows:—There gradually supervened a state of more or less mental confusion, now and then accompanied with delusions that his friends were about to come for him, while his general expression and bearing were those of great self-satisfaction and joyous good-humour. The ataxic and paretic symptoms still progressed, and the treatment was with digitalis and iron. In Feb. and March 1873 there was a gradually increasing excitement, manifested by the reiterated singing of a narrative of various trifling events in his life, and of an enumeration of his possessions, intentions, and good qualities. Together with this monotonous sing-song, carried on for hours together, there were restlessness, destructiveness, and a disordered state of his dress. The physical signs of phthisis were observed; then emaciation appeared and was progressive, and due, in part, to the continued restlessness and excitement. At the end of March he was bedrid on account of the pulmonary condition, and for the restlessness and sleeplessness hypnotics were required. His habits were wet and dirty, he became very feeble, and tremulous, and died in a syncopal attack on April 16th, 1873.

Necropsy (in abstract), 21 hours after death. Cranium thick. About 7½ozs. of serum escaped from the cranial and spinal cavities. Cerebral interlobar adhesions. The arachnoid over the superior surface of the brain was somewhat thickened, opalescent, and tough. The convolutions of the brain were loosely packed on this aspect, and subarachnoid fluid filled many of the anfractuosities. The superior meningeal veins were full, and the pia mater was still hyperæmic. There was wide-spread adhesion and decortication especially seen on the superior and external surfaces of the cerebrum. The frontal convolutions, and the lower tiers of the parietal (asc. p., s. marg., ang.,), were those most markedly affected; the other convolutions in this area were less involved. Adhesion and decortication also affected the cerebellum, whose arachnoid membrane was opaque. The grey matter of the cerebral cortex was diminished in consistence. White substance of brain universally softened, and the site of considerable vascularity. The lateral ventricles of the brain were large, and contained 1oz. of serosity. Fornix almost diffluent; corpora striata and optic thalami soddened and softened; pons V. softened and of a pinkish hue; the posterior part, especially, of the medulla oblongata softened and hyperæmic. Cerebrum, 31½ozs. Cerebellum, 5½ozs.; pons and med. obl., 1oz.

All the viscera were carefully examined, but it is only necessary to state here that both lungs were phthisical; and that the muscular substance of the heart was unduly friable, and of a somewhat pale and yellowish hue. Heart, 11ozs.; incipient atheroma of aorta. Spleen 12 ozs., soft, diffluent, of prune-juice colour. Kidneys 8, and 7½ozs.

Remark.—This case is an example of that clinical form of g.p. in which there is self-satisfaction, joyousness, abundant good-humour, the recounting of trifling events with much gusto, and childishness and feebleness of mental power, rather than decided ambitious delirium. Atrophy, adhesion, and the basal changes were all well marked. Intra-cranial hyperæmia was lessened by the mode of death.

CASE 55.—*Convulsive seizure, followed by failure of the mental powers, and defective speech, hearing, and locomotor power. The ordinary labial and lingual ataxy and paresis not so well-marked as usual. Extreme dementia and asthenia. Unusually short duration. Softening and hyperæmia of encephalon and spinal cord. Extensive cerebral adhesion and decortication. Wasting of gyri, especially in the frontal, and in the anterior part of the parietal, area.*

It is, perhaps, unnecessary to relate the clinical history or more of the necropsy than refers to the brain.

Necropsy: 87 hours after death. Fairly nourished; bleb on right malleolus. Calvaria very thick and dense. Dura slightly thickened, and its adherence to calvaria slightly increased. Arteries at base of brain, healthy. Interlobar adhesions. Meninges congested. Visceral arachnoid and pia thickened, and the latter infiltrated with serum in parts; especially over the upper frontal, and anterior part of the superior parietal surfaces, where, also, the anfractuosities were wide and rounded.

Adhesion and decortication over the frontal gyri, those of the anterior part of the parietal lobe, and to a moderate degree over the temporo-sphenoidal gyri. The superior and external surfaces of the first two lobes, the external surface of the last, were those principally affected. Adhesion was found strewn in patches over the internal and inferior surfaces of the cerebrum, but did not affect the occipital lobes. It was nearly symmetrical in its disposition on the two hemispheres. The cerebral grey cortex was hyperæmic, of a mottled reddish pink colour, softened, of fair depth; of imperfect stratification. The white cerebral substance, universally softened, was hyperæmic. 3iii fluid in lateral ventricles. Corpora striata and optic thalami softened and hyperæmic. Meninges at the base, especially in front, slightly thickened and opaque. Cerebellum, pons V., med. obl., and upper part of spinal cord, all diminished in consistence. Right hemisphere, 19½ozs.; left, 19½ozs. Cerebellum, 5½ozs.; pons, and med. obl., 1oz. Fluid from brain, more than 1oz. Granular, and other, renal disease.

GROUP II. (*see* p. 408-9).

CASE 56.—*Protracted duration. At first, excitement with delusions, succeeded by a somewhat steadily progressing fatuity, with a degraded state of feeling. Impairment of speech and of locomotion. Patient bedridden for a long time at the last. Sensory impairment. Speech less affected than usual; incessant grinding of teeth and noisy sniffing. Wet and dirty habits. Marked atrophy; softening, and some anæmia of brain. Marked meningeal changes. Moderate amount of adhesion of pia to cerebral cortex, especially about lower border of Sylvian fissure. Much chronic disease of spinal cord, as well as recent spinal tuberculosis.*

C. C., Private 56th Regiment. Service, 8 years. Age, 28. Married. Admitted January 24, 1874.

History.—This was stated to be the first attack of mental disease, and to have existed since 1871. He had been under treatment in 1871-2-3 in India, and latterly at Netley for one month. The effects of tropical climate were

supposed to be a factor. At the onset of the disease he was the subject of delusions with excitement. At Netley his habits were dirty, he hoarded rubbish, had passing fits of excitement; was unable to comprehend a simple question properly, and seemed to know but little except his own name.
On Admission.—Height, 5ft. 4in. Weight, 148lbs. Well nourished. Sallow aspect. Languid surface circulation; pulse 66. Pupils equal and sluggish, irides blue. Tongue a little tremulous, speech thick, hesitating, and dwelling upon the words. He had a habit of loudly sucking his lips, which were livid. The gait was unsteady; and he wetted the bed at night. His *mind* was a wreck; but little response could be elicited from him, and when elicited his replies were irrational and incoherent. He was dull, apathetic, and exhibited neither gaiety nor sadness, neither interest nor aversion. R. Ferri Perchlor.

Shortly after his admission the habits became more constantly wet and dirty, and othæmatomata appeared. Early in 1875 he remained in the same unintelligent condition; to questions replying either not at all, or by inarticulate sounds, or by such words or phrases as, "Yes." "I have forgotten." To the former loud "sucking" sound made by him was added a loud "sniffing" sound. Signs of phthisis pulmonalis. Temp. somewhat high. July 7th, 1875. Pulse, 82. Resp., 24. Temp., 100·3°. Phthisis advancing. Add Ol. Morrhuae to Mixt. Ferri et Quass. Right pleuritic effusion. Profuse sweat, relieved by Belladonna. Cough, dry. Grinding of the teeth was frequent, and the habit of "sniffing" most persistent. He was for a long time bedridden, and rarely spoke. Speech though slow, impaired, and somewhat tremulous, was not affected so much as is usual in the stage at which dementia becomes so extreme as here, and the habits so extremely and constantly "wet and dirty." The facial expression, which at first was that of hebetude, became latterly, rather, that of apprehensiveness and fear. Oct. 3. Pulmonary excavation. Cough, teeth grinding, and night-sweats, continued. Impaired sensibility of skin. Oct. 11. Pulse 90, Resp. 35, R. cheek flushed. Pleurisy of left side, and pneumonic patches. Vomiting was troublesome, but, ceased on Oct. 12th, returned on the 13th, subsided on the 14th. On the 12th, also, the temp. was raised, respiration laboured, moaning, and the expression indicative of distress. Oct. 22. Heavy, dull, drowsy, since the 21st. Increasing subsultus of hands and arms, tremors of lips and face, and marked tremulousness, as he restlessly pulled the clothes about. Expression more than usually apprehensive. Pulse 100, weak. Oct. 23. Since 7 A.M. had been comatose, and unable to swallow. At 9.30 A.M. the pulse was 90, full, and fairly compressible. The respiration varied from 44 to 54 per minute, it was irregular in rhythm, in depth, and in frequency, from one half minute to another. Thus, perhaps, at first there were a few audible respirations with guttural sounds, then came a few quiet, easy, noiseless, and less frequent respirations, during which the heart could be heard distinctly, and its action was then of moderate strength, but its sounds still were feeble and short; *i.e.*, " up and down," or true "respiration of ascending and descending rhythm." The surface of the body was clammy, cool, flabby, relaxed. Temp. in right axilla 95·2°; in left axilla below 95°. Conjunctivæ suffused, watery, insensitive to touch; pupils equal, dilated, immobile; the eyes sometimes turned a little to one side or the other. Limbs quite flaccid, limp, and unresisting to passive motion. No distortion of the face. Saliva ran from either side of the mouth that was the lower. A slight, suffused, flush relieved the, lately, usual pallor of the face. Complete coma. Later, there was much mucous bubbling in the throat. At 1 P.M., pulse 96, weaker and softer than before; resp. 36 and of irregular rhythm, as in the morning; temp. below 95° in both axillæ. Left pupil the larger, both were insensitive to light and moderately dilated; conjunctival reflex absent. The eyes and head turned somewhat to the left side now. Face pale and slightly livid. All the limbs remained equally flaccid, relaxed, and motionless; inability to swallow, and rattling of bronchial mucus were as before. Death at 6 P.M., Oct. 23, 1875.

Necropsy, 73 hours after death. Fairly nourished; rigor mortis. Calvaria, small, thick, dense; of slightly worm-eaten appearance, in parts, internally. Dura very slightly thickened. Serum in arachnoid cavity, and at base. Firm interlobar adhesions of meninges. Arachnoid at base of brain, thickened, tough, and opaque. *Right cerebral hemisphere.* Membranes over the superior, external, and internal, surfaces, thick, tough, pale, and slightly opaque. The pia was œdematous, chiefly in the fronto-parietal region, and fluid filled the sulci, except beneath the occipital tip. Adhesion and decortication were especially marked in the posterior part of the frontal, and the anterior part of the parietal, regions; were considerable over both the external and the inferior surfaces of the temporo-sphenoidal lobes, as well as over the gyrus fornicatus, existed slightly, also, over the præcuneus and cuneus, and over the posterior part of the orbital gyri, but not at frontal tips. Gyri more or less wasted, especially on the superior and external surfaces; the tip of the occipital lobe escaping. Grey cortex thin, rather soft, its strata not well marked, its colour ordinary. Anfractuosities very shallow. Enlargement of lateral ventricles. *Left.* The general condition of the meninges was alike on the two sides, and the distribution of the adhesions and cortical erosions was also much the same, except that the left temporo-sphenoidal lobe was less involved thereby than the right, and that the internal surface of the left hemisphere was nearly free. The brain was flabby, the white substance was alike on the two sides, both as to universal softening and pale lilac colour. A few puncta cruenta. Fornix in parts semi-diffluent. Corpora striata (and optic thalami) atrophied, softened, and paler than usual; pons and medulla oblongata, softened and pale; membranes adherent to the med. obl., and thickened over it. Cerebellum, soft and pale, its arachnoid opaque. Right hemisphere, 12¾ozs. Left 13ozs. Cerebellum, 5ozs.; pons and med obl., 1oz. Serum from cranial cavity 4 fl. ozs. The spinal cord, universally anæmic, exhibited a tendency to softening, especially in the lower dorsal region. Strewn all over the internal surface of the spinal dura-arachnoid were beautiful, transparent, minute, grey, tubercular granulations. Portions taken from the cord did not harden well in chromic acid solution; and stained badly with carmine; under the microscope were found molecular *débris*, granule-masses, fatty molecules, thick-walled vessels, granular nerve-cells, and altered misshapen nerve fibres. *Heart*, 6½ozs.; slightly soft and friable. *Lungs.*—Left, 19ozs.; old adhesions, puckered cicatrices, and a vomica at apex. Recent pleuritic effusion. Recently-formed caseous patches in lungs. Right lung, old, strong, and close pleuritic adhesions; tubercular granulations, etc. Right *kidney* 5½ozs.; ordinary cystic degeneration: Left 4¾ozs., congested; some blood in tubuli uriniferi. *Spleen*, firm, 9½ozs. *Liver*, 54½ozs.

Remarks.—1. The marked chronic disease of the spinal cord was of interest, as for example in relation to the excessively wet and dirty habits of the patient during so long a time. At first dependent upon intellectual, sensory, and moral, failure, they were latterly, I take it, dependent in part upon paralysis of sphincters from spinal disease.

2. In relation to the spinal tubercular meningitis, and the softening of the cord, especially in the lower dorsal region, we find that the closing scenes of life constituted the following drama. (a). Death after twelve hours of complete coma supervening somewhat suddenly. (b). During this period total relaxation and flaccidity of the muscular system, without the slightest convulsion, spasm, rigidity, distortion, or appearance of local palsy predominating anywhere, the condition being a generalized spinal paralysis. (c). A low temperature; and moist, relaxed, flabby, pale skin. (d). Respiration of "up and down," rhythm; controlling the coexisting variation of the pulse. Possibly these symptoms and signs were in part occasioned by an extension of the tubercular formations to the meninges of the med. obl. and pons, or even to the base of the brain, and not yet perceptible to the naked eye.

CASE 57. *Duration long, five years or more. At first the physical signs of g.p., with exalted delusions, the patient, quiet as a rule, subject to gusty fits of excitement. Afterwards, quiet dementia, dulness, then hypochondria, refusal of food, and emotional agitation; later, occasional paroxysms in which fear predominated; and, lastly, greater dementia, often with an expression of confusion, anxiety, or even distress; paresis of movements, especially of articulation; great failure of deglutition. Right side of body often slightly more paretic than left.—Marked meningeal changes. Adhesion and decortication slight, and mainly of the second and third temporo-sphenoidal gyri. Atrophy of brain, including cortex, which was fairly vascular, and slightly firm. Much intra-cranial serum. Enlarged cerebral ventricles.*

W. T. Private, 9th Lancers. Single. Service $13\frac{1}{12}$ years; admitted April 27, 1872, aged 32 years.

History. This attack of mental disease was the first, of unknown causation, uncertain duration, insidious onset; and previously under treatment at Aldershot, and for a month at Netley. At the latter he exhibited exalted delusions, and gave expression to great and chimerical schemes for doing universal good; was occasionally incoherent in conversation, and liable to gusty fits of excitement.

On admission. Height 5ft. $8\frac{1}{4}$in., fairly nourished, chloasma, scars of acne, viscera healthy, grey irides, equal pupils. Tongue pale, flabby, a little tremulous; was protruded slightly to the right, and he frequently licked his lips with it. At times, during speech or during protrusion of the tongue, were tremors or slight twitches of the muscles of the lips or face. Sometimes the speech was hesitating and indistinct. There was some awkwardness and inco-ordination when he turned round, or walked on a straight line. The power of grasping with the hands was good, and he was able to write his name, yet not without hesitation. The expression was dull, and indicative of mental failure. At times there was a feeling of placid self-satisfaction, but no decided exaltation. His real age being 32, he stated in examination that he was only 24, that he had served in the army for 14 years after enlisting at the age of 22, that he had been four months at Netley (really one), that three hundred persons came here with him (really ten), that he could do anything, indeed was quite an athlete, that he gave away sovereigns to anyone who cared for them, and that he had "£4,000 or £5,000 in the bank," adding the words "more than that—any amount in fact."

During 1872 he betook himself to ward-duties, and for a long time afterwards continued to be industrious, quiet, answering all simple questions in a rational and coherent manner; but never asking to be allowed to leave the asylum, never giving any trouble, and yet exhibiting much more intelligence than he did when first admitted. The expression still was one of impaired mental power; it never evinced depression, and scarcely ever was smiling. Slow and deliberate in all his movements, and apparently of sluggish perception and comprehension, he was easily put to mental confusion on being questioned. In this condition he remained until the end of 1874.

Jan. 2nd, 1875. He refused food, and remained in bed; when forced to take food he became very restless, jerky and tremulous, and voluntary movement was attended with convulsive jerks. He declared that "his throat was bad;" "he had no swallow;" "was dying;" and he made hideous noises. He drank some milk when forcibly urged; and, in drinking, swore, and thrust his jaws into the mug and fluid at each mouthful he took. The costive bowels were relieved by enemata. Pupils equal and acting.—Jan. 4th. Subsultus and twitches of upper limbs. Would scarcely take any food. Face somewhat injected; pulse 81, full and throbbing. Left pupil now the larger, and irregular. Tongue furred, breath foul. Speech indistinct, and during it were facial and labial tremors and twitches.—5th. Same state. Stomach-pump. Jan. 8th. He was bellowing, moaning, crying out, "Oh! oh! dear," and anon making strange inarticulate noises. The face was trembling violently, the eyes being kept closed; and the muscles of the trunk and limbs were jerked. When the mouth was opened the tongue was seen rolling about, jerked hither and thither almost choreically, but voluntary efforts restrained it. Pulse full, about 110, but varying in frequency; face injected. He kept muttering incoherently about "poison," "wrong poison," "wrong dead poison;" "Oh dear—oh dear—the brute—work broken," and so on. On another day he said his teeth were all gone (a delusion). 10th. Excessive tremulousness in speaking, or in making any movement. 11th. Pulse 84, resp. 20, temp. 101·2°, face less flushed; diarrhœa for several days, tongue furred. Still great tremulousness of the face, tongue, and hands, even when at rest *quoad* voluntary movement; the right hand was affected with slight paralysis agitans. 26th. Better again.

April 7, 1875. Pulse 84 and soft; for many weeks it had varied from 80 to 96. He had been very incapable and bedridden for some time, but had now been up again for a week. April 13th. Weight 126 lbs. June 6, 1875. Childish, confused, amnesic. The habits, which had become dirty, were now improved. The fingers were extremely tremulous in any act. Aug. 1st, 1875. Tongue protruded slightly to right, the tongue and face, chiefly right face, were tremulous when in movement; speech much affected. Patient quiet, of dull appearance, never elated; childish. Oct. 28. Weight 152 lbs.; increase, 26 lbs. in 6½ months, on extra diet and physostigma. Dec. 8th. Very slightly greater paresis on right side of body. Left pupil twice the size of right, which was contracted. During exercise the left pupil grew a little larger, and the right more than doubled its previous size. Facial tremors during speech, more on right side. Quiet, disposed to be hypochondriacal, saying that his medicine did him harm, that he had lost flesh, and was weak. No albuminuria. Slight œdema. April 6th, 1876. On the day before had been extremely dull and stupid, and had dragged one foot. Gait very unsteady, facial and labial tremor as before. Speech shaky, hesitating, indistinct. Left pupil, the larger; the right, contracted; both sluggish. Paresis of face and lower limbs more marked on the right side than on the left. Face palish, pulse full, soft, 100. Weight, 145 lbs.; loss during 5½ months, 7 lbs. Omit physostigma; take perchloride of iron and quassia. Aug. 3, 1876. The mumbling speech, the tremors of the face and other parts had increased greatly; the gait was slow. He was dull, amnesic, demented, and often had a confused anxious distressed look. At times he was most obstinate, and resistant. Aug. 8th. He became violently excited when brought to be examined, struggling to escape, and shouting in terror at some product of his imagination, or of sensorial disorder; but was easily pacified.

Nov. 3, 1876. Feeble, helpless, almost unable to speak or to swallow, and the subject of pneumonia. Pulse 132; resp. 33, rigors; much tremor. 6th. Pulse 96, soft; resp. 38; temp. 102·2°. Signs of patches of pneumonia and destructive changes scattered through both lungs. 9th. P. 102, R. 45, T. 102·5°. Dysphagia remained extreme, fluids tended very strongly to pass into the larynx, and thence into the lungs. If raised up or induced to attempt to swallow, he became much perturbed in mind and made violent respiratory efforts and guttural sounds. Passing of the œsophageal tube in order to feed him almost produced asphyxia, and only enemata could be resorted to. He died, exhausted, on Nov. 11, 1876; but before decided emaciation was present, or dementia excessive. And he died mainly in consequence of lung-lesions.

Necropsy.—54 hours after death. Much serum at base of brain. Dura slightly thickened. Olfactory bulbs of ordinary size, slightly adherent to dura. Over the whole of the superior and external cerebral surfaces the pia and arachnoid were thickened and tough, the pia infiltrated with serum, the arachnoid white, opaque. These changes terminated upon the occipital lobes, were about equally well-marked upon the frontal and parietal lobes, much less on the temporo-sphenoidal, and were symmetrically disposed over the two hemispheres. The membranes were congested, were stripped off readily, but left shreds in some of the sulci. The only points of adhesion of meninges to cortex were scattered over the temporo-sphenoidal lobes on both sides, especially over their second and third gyri, were slightly more marked on the right side, and there affected the inferior surface particularly. The grey cortex was of a pinkish lilac colour, especially its deeper layers; the distribution of this colour was not uniform, there being scattered, small, darker patches, and the naked-eye vascularity being greater anteriorly than posteriorly. Grey matter slightly firmer than usual, and of fair depth. Gyri small. Lateral ventricles of brain, dilated, contained much serum; ependyma thick and opaque. White substance of brain, hyperæmic, reddish, its consistence somewhat firm. Basal ganglia alike on the two sides; pons and med. obl. of a pinkish lilac hue. Lining membrane of fourth ventricle, thickened and opaque, and the subjacent tissue of a dull lilac colour. Each cerebral hemisphere, 17½ozs.; cerebellum, 5¼ozs.; pons and med. obl., ¾oz. Fluid from cranial cavity, 8 fl. ozs.

Heart, 10¾ozs.; its muscular substance slightly friable and darkish. *Right lung*, 42ozs.; cheesy masses and granulations, and destructive pneumonia in patches; part of middle lobe semi-gangrenous. Bronchial mucosa, greenish-grey. *Left lung*, 34½ozs.; similar changes, less advanced, cicatrices in front, small vomicæ at apex. *Spleen* firmish, of a chocolate hue, 4¼ozs. Left *kidney*, 4¾ozs.; right, 5ozs.; both healthy. Liver palish, 48ozs.

Remarks.—1. In this case we find on the one hand a long duration and an unusual course; *(a)* expansive symptoms, succeeded by *(b)* quiet dementia, and this by *(c)* hypochondriacal delirium, and, finally, *(d)* persistent dementia, with a look of confusion, anxiety, distress. And, on the other hand, a wasted and water-logged brain, marked changes in the cerebral meninges, but adhesion and cortical erosion only slight, and limited to the temporo-sphenoidal lobes.

2. Here, also, towards the close of life, articulation and deglutition were much impaired, and inhalation of some food and destructive pneumonia took place. In relation thereto were the atrophic and other changes of the pons V. and medulla oblongata.

CASE 58.—*Early delusions of wealth, and physical prowess; desultory conversation, self-complacency. Later, confused at times, excited, and in transient fits of anger; still later, more confused and demented, ceased to work, suffered much and long from insomnia, was restless and noisy at night; was much disturbed emotionally, became secretive, obstinate, and excessively coarse, foul and abusive in language. Still later, he resisted every manipulation, and howled and roared if touched. Finally, utterly repugnant to any interference, dull, obstinate, morose, peevish, and brutish. For years, of foul habits. At first, facial twitch and tremor, great motor restlessness, and fidgetiness; finally, a bedridden state, but the ataxy and paresis comparatively slight or moderate during most of the disease. No apoplectiform or epileptiform seizures. Changes of cerebral meninges very marked, extensive, and well seen at the base. Great atrophy of brain, large ventricles, and much intra-cranial serum. Cerebral cortex pale; also atrophied, especially in the frontal lobes. Adhesion and decortication of grey substance, almost limited to the parts bordering on the Sylvian fissures. Some softening and pallor of rest of encephalon; spinal cord atrophied and somewhat softened.*

J. M. Private, A. S. Corps. Admitted June 15, 1872, aged 32, married.

History.—First attack of mental disease; said to have existed since Jan., 1872; previously under treatment at Woolwich; and at Netley since May 7. It came on insidiously, and at Netley he had delusions of wealth, and exhibited self-complacency, desultory conversation; and transient fits of excitement.

On admission.—Height 5ft. 6in., well nourished, pupils equal, irides blue, clean in person and habits. Viscera, healthy. Gleet. Tremors of facial muscles during speech, which was impaired and ataxic. He manifested frequent but transient emotional disturbance, especially when spoken to. He believed himself to be, physically, very powerful, and said he could lift half a ton weight. Usually self-complacent, often smiling, pleased at trifles, he yet was easily roused to become excited, angry and abusive. At times his ideas were somewhat confused. He did a little work.

It will suffice to transcribe a portion of the notes. March, 1873. For several months he had been slowly retrograding, his ideas being now more confused and limited. He had ceased to occupy himself in any way, and had become destructive to his clothing. His articulation and locomotion were more affected. Taking Ferri Perchlor.—May 22, 1873. Omit the mixture of iron and quassia. June, 1873. He was becoming worse, and was still very destructive to clothing, but denied his destructiveness in the coolest and most unflinching way. Dec. 27, 1873. Resume perchloride of iron and quassia. Jan., 1874. Left othæmatoma.

In summary, it may be stated that during the same year (1874) he was restless, untidy, and from time to time excessively destructive to clothing. During the second half of that year he was much less disposed to insomnia, restlessness, and noisiness at night, than during the first half. In consequence of the insomnia, etc., he took hypnotics from Mar. to July.

Feb., 1875. Furuncle and abscess of left buttock. In March, chloral hydrate, for insomnia. In April, iron resumed. In May 1875, œdema of the feet and ankles was present, and increased towards night. Urine, no albumen, healthy. Obstinate ulcers on leg, aggravated by being picked at and by the œdema, but healing under rest in bed. Sep. 29, 1875. No œdema. Very demented: when interfered with he shouted and made uproarious complaints. Still in bed, fumbling, restless, mischievous and of dirty habits. Oct. 8th. To get up. Oct. 28th. Weight 149 lbs. He often stood still with drooping figure and head bent towards the ground, or occasionally fell. When he was washed, or dressed, etc., he shouted in a loud, distinct voice, often without hesitation or tremulousness. Yet much of his ordinary speech was impaired, and at times somewhat hesitating, tremulous, and stammering, but there were no marked facial or lingual tremors during speech. He was utterly careless, had a fatuous look, his hair was dry and bristly, his skin rough and coarse, and his general appearance degraded. He would not grasp at request, but his power of grasping was good. With difficulty he buttoned his coat; often he pulled off his clothes or tore them. At night he was restless and noisy, or pulling the bedding about. Right pupil the larger, both sluggish. Gait slow, irregular, swerving; steps short; equilibrium uncertain. He was irritable, suspicious, and obstinate if interfered with. Delusions about his wife being here. When questioned, or urged to do anything, he called out angrily, and roared when being washed or dressed. When left untouched and unspoken to, he was utterly apathetic, and would stand or sit for hours in one place, if not moved, or pull to, and shove from, himself, any object. Often he declared he had nothing to eat, and after so saying would swear, or shout. Nov. 30. Gait much impaired; he staggered, and the legs straddled far apart. If not supported he would have fallen, though now and then he could run a few steps. When he shouted the utterance was distinct, and tremors were absent or slight. But speech became mumbling at times, and in opening the mouth widely the upper lip trembled much. Amnesic, and in his usual obstinate, irritable, restless, mischievous, state. On Dec. 25, again ordered to bed, where he remained until his death, constantly of wet and dirty habits, and affected with ill-conditioned boils and bedsores. Feb. 24, 1876. Pneumonia. P. 120. R. 30. Brandy and ammon. carb. Feb. 26. Pulse now very soft and rapid, becoming filiform and imperceptible; respiration 38, laboured; vomiting, semi-coma. Dull lividity, with some flushing, of the face, and deglutition much impaired. The movements of the face were mainly confined to the left side; the right eye reflexes failed. Skin moist. Died 2 A.M., Feb. 27, 1876.

Necropsy.—41 hours after death. Foul odour from cadaver. Body fairly nourished. Scalp thick, calvaria thick, especially in front. Pale clot in left vertebral artery. Dura thick and dense. Chiefly on the superior and external surfaces of the cerebrum the meninges were thickened, opaque, pale, and highly œdematous, and these changes were about equally marked over the frontal, parietal, and temporo-sphenoidal lobes; and were well-marked over base and cerebellum. Interlobar adhesions at base. The rounded anfractuosities contained shreds of membranes. The flabby brain sank under its own weight. Limited adhesion and decortication of grey matter were found; almost entirely confined to the region of the Sylvian fissures, rather more marked on the left side, and mainly involving the third frontal and first and second temporo-sphenoidal gyri. A few isolated patches of this morbid change also affected all the other convolutions which border on the Sylvian fissures. A few slight points of the same were seen on the orbital and inferior temp. sphen. surfaces. The grey cortex was thin and very pale, especially the external layers. The pallor was greatest in the frontal and in the occipital regions; the cortical atrophy was more marked in the frontal region, and slightly more in the right than left frontal lobe. Consistence fair; stratification not very obvious. The white matter was of a pale faintly pinkish tinge, and was universally soft and flabby. Lateral ventricles very large, and contained about six drachms of fluid. Opto-striate bodies, of diminished consistence; and alike on the two sides. Right hemisphere, 15$\frac{3}{4}$ozs.; left, 16$\frac{1}{2}$ozs. Both the grey and the white of the *cerebellum* were pale and softish. Weight 5$\frac{3}{8}$ozs. Pons and med. obl. 1$\frac{4}{5}$oz., and slightly pale and soft. 7 fl. ozs. of fluid from cranial cavity. The spinal cord was somewhat atrophied, and its consistence, perhaps, lessened.

Microscopical examination, later. In the third left frontal gyrus the larger pyramidal nerve-cells seemed to be singularly absent: the smaller nerve-cells were some of them granular, rounded, with wasted outlines, possessing but few branches, some had quite broken down, and the granules were strewn about. In parts the neuroglia appeared to be increased. The nuclei of the neuroglia were very numerous, and even some of these had a granular appearance. Large oval granular cells also seen.

Heart. A double perforating ulcer was seen on one of the aortic flaps, and some dirty whitish clot in the sinus of Valsalva behind it. Aortic sinuses, and mitral flaps, atheromatous. Arch of aorta extremely irregular, nodular and ridged. Advanced atheroma in the left coronary artery; and in the walls of the right coronary artery a firm yellow nodule was imbedded. Heart, with three inches of aorta, 14$\frac{1}{2}$ozs.; its muscles, pale, flabby, and friable; under microscope, some granulo-fatty degeneration of muscular elements, and a few free fat-globules. *Lungs.*—More or less hypostatic congestion and pneumonia of both. Left, 25ozs. Right, 24$\frac{1}{2}$ozs.; old pleuritic adhesions.—*Right kidney* 6$\frac{1}{2}$ozs., pale, especially in the cortex, and soft and flabby. Numerous embolismic changes in the renal cortex. The same changes in *left kidney;* also on its surface a depressed cicatrix extending $\frac{1}{4}$ inch into the gland. *Liver,* flabby, pale, too friable, 63$\frac{1}{2}$ozs. *Spleen* 14ozs., on its external surface a depressed cicatrix, slight fibrous bands extending therefrom $\frac{3}{4}$ inch into the spleen, and ending in a hard yellow nodule, the central parts of which were wax-like, firm, dry, white. Spleen softened and almost diffluent; palish, except one portion of a deep purplish colour.

Remarks.—1. There was profound change in the blood at the close of life. The odour of decomposition was premature and strong, the viscera were soft and flabby, perforating ulcerative endocarditis affected the aortic valves, the clots in the heart were of a dirty whitish grey colour, soft, friable, and as if rotted; numerous emboli were strewn through the kidneys, and embolism of the spleen had occurred and perhaps in the lungs also. These may be examined in a correlative relation with the symptoms observed from Feb. 24th until death on Feb. 27th.

2. The aortic endoarteritis, the nodule in the right coronary artery, and that in the spleen, gave rise to the question of syphilitic disease, but the splenic nodule had not the character of a syphilitic growth; that in the right coronary artery was not decisive, and, in my experience, aortic endoarteritis is a very equivocal change. There was no history of syphilis, and no clinical indications of it whilst the patient was here.

CASE 59.—*Early stages not under writer's observation. Later, excitement with delusions; still later, exalted notions as to physical and mental powers, incoherence, and mental weakness, garrulousness, quiet pleased self-satisfaction. Ataxy and paresis, of much the ordinary characters; gait considerably impaired latterly. Patient more and more demented, and finally speechless, and utterly helpless. Some wasting of brain; a considerable amount of intra-cranial serum; marked meningeal changes, especially over the frontal lobes; atrophy, pallor, and softening of cerebral grey cortex: diminished consistence of the white cerebral substance universally, as well as of basal ganglia, cerebellum, and cord. Cerebro-meningeal adhesions considerable.*

C. H. W. Gunner, Royal Horse Artillery. Admitted April 27, 1872, aged 27, widower.

History.—This, the first attack of mental disease, had existed since January, 1870, and before coming here the patient had been under treatment at Woolwich; and, from August 11, 1871, at Netley. "Domestic trouble" had the chief place in the assigned causation. There was almost no history of this patient's case prior to his admission here, but his delusions, excitement, irrational language, and thickness of utterance were stated to be on the increase at that time; and he was described as beginning to fall when he attempted to walk.

On admission.—Height 5ft. 7in., fairly nourished, fresh complexion, equal pupils, hazel irides, skin clear, viscera healthy, habits clean. His utterance was imperfect. His tongue was protruded at one's request, but in a jerky manner, and whilst held out was affected with fibrillar tremors and twitches. He walked unsteadily. His expression was usually one of quiet, smiling, self-satisfaction. He was garrulous, and unable to fix his attention for any length of time. He enumerated long lists of persons and of duties, connected with his service in the army. He believed himself to be of very great physical strength, calling attention to his limbs in proof of his assertion, whilst he was weak in reality. Similar notions related to his mental powers, and intellectual acquirements. Ordinary simple questions he answered correctly, but became incoherent in continued conversation.

In July, 1872, there was a steadily progressive deterioration in the mental and physical conditions. By Oct., 1872, he was still more completely demented, and was becoming dirty in habits, and helpless, requiring to be washed, dressed, and in every way attended to. The expression and bearing, which formerly denoted a sense of well-being, were now usually indicative of dulness, confusion, bewilderment, fatuity. His replies were either utterly irrelevant or merely consisted of "yes," in response to every inquiry. He spoke in a slow drawling quasi-stammering and muffled manner. He could manage to stand, and to take short steps, with the feet kept widely apart, scarcely raised from the floor during progression, and their soles planted flatly on the ground. His manual grasp was feeble. Unable to button his clothing, he was frequently and restlessly fumbling with it. Labial tremors and twitches were observed, especially during speech, but he would not protrude his tongue. He was still somewhat stout, the features were flabby, expressionless, and his lips had lost all their natural flexibility and shapeliness of contour. Slight left external strabismus was observed, the pupils were about equal, somewhat sluggish, the right pupil becoming the larger in the shade, and assuming an oval irregular shape. By Dec. he was unable to articulate a single word intelligibly; but, without quoting from the various notes made, it may be briefly stated that after the date of the above note he gradually became more and more demented, helpless, wet and dirty in habits, emaciated and weak; and died on March 26, 1873.

Necropsy.—35 hours after death. Skull thick and dense; dura healthy. Serosity, 6 fl. ozs. Surface of convolutions, somewhat flattened. Arachnoid thickened, opaque, and of a milky hue, chiefly on the superior and external frontal surfaces, less over the parietal, and scarcely at all over the occipital. Meningeal veins full posteriorly. Over nearly the whole superior and external aspects of the brain was subarachnoid serous infiltration. The membranes were somewhat friable. Adhesion and decortication of summits of gyri affected the superior and external surfaces of the frontal and parietal lobes. The upper surface of the frontal lobe was more affected on the right side; on the other hand, the external surface of the left frontal and parietal lobes had more adhesions than the right. The cortical grey of the cerebrum was thin, atrophied, palish, universally softened. The white substance of the brain was softened, as also were the opto-striate bodies. Much serum in lateral ventricles. The cerebellum and spinal cord were of diminished consistence. Cerebrum, 40ozs.; cerebellum. 6ozs.; pons and med. obl. 1oz.

Right *lung* 24ozs.; left 20ozs.; their posterior portions were heavy, friable, darkly congested, œdematous. *Heart.*—10ozs. Pericardium over right ventricle, white and opaque in patches. Walls of left ventricle thick, its muscular tissue slightly yellowish. No atheroma. Tricuspid valves slightly thickened.

Remark. By its long duration, absence of seizures, slowly developed motor signs, and other clinical aspects, this case falls into the second group; but had less cerebral atrophy, and more adhesion and decortication, than other cases of the group.

GROUP III. (*See* p. 409-10).

CASE 60.—*Loss of memory, dementia, delusions of annoyance and persecution, with depression; quietness at first; later, more marked depression, weeping, and occasional irascibility; apoplectiform attacks. Motor signs of g.p., and somewhat tabic gait, slight right hemiplegia. Later, extreme motor failure; bedridden; dextral convulsions, followed by right hemiplegia; slight conjugated deviation of head and eyes, and left ptosis; respiration of "up and down" rhythm, somewhat excurvated belly, death. Arachnoidal changes symmetrical. Adhesion and decortication very moderate, mainly on the left hemisphere, and principally affecting tips of frontal lobes, left temporo-sphenoidal lobe, and posterior part of left inferior and left internal surfaces. Marked redness of nearly the whole grey cortex of left hemisphere, and of greater part of right frontal lobe, and part of right temporo-sphenoidal. Induration of left frontal grey, diminishing backwards. Spinal cord softened, its meninges thickened posteriorly, and the site of yellowish nodules.*

C. E. Private, 11th Regt., age 32, service 14½ years. Admitted June 4th, 1875.

History.—Said to be the first attack, insidious in its commencement, and existent before first recognized as mental disease in Feb., 1875. The cause assigned was "disappointment in love;" but heredity was a factor; a maternal aunt had been insane. He had been under treatment at Devonport, and, for a month, at Netley. Neither epileptic nor suicidal, of good conduct, latterly of temperate habits, and of indifferent education. When admitted at Netley he was suffering from the physical signs of g.p., and steadily advancing fatuity, the exalted delusions and *bien-être* being wanting. He decidedly lost ground whilst there, the paretic signs became much more marked, and to the report of his case it was added that " though the evidences of the present attack are said to have been first *noticed* in Feb., they must have existed for a considerable time before that." The medical certificates testified to the existence of dementia, loss of memory, extreme incoherence, delusions as to annoyances by other persons, and as to the desire of his comrades to murder him, with g.p., but as yet without dirty habits.

On admission.—Height 5 feet 10 inches: weight 157lbs. Cicatrix on glans penis; bubo-cicatrix in left groin. The patient said he had had " the venereal." Traces of ancient double iritis, the pupillary edges of the irides being frayed, and adherent to the lens. Leucomatous patch on left cornea. Some brownish semi-cicatricial spots on the body, of which one or two were in front of the left tibia. Lungs healthy, heart sounds full and loud, the second accentuated, yet the pulse quick. Area of liver-dulness slightly full; other viscera healthy. Tongue tremulous, and protruded unsteadily; speech very shaky; twitches of the lips and face. His hands trembled much when in use, and the gait was feeble and tottering. Extreme dementia; *great* loss of memory. There were no exalted delusions at that time. He said that "the men of his regiment were against him," and began to cry whenever his friends were mentioned to him. He was capable of very little conversation, his ideas being extremely limited.

After admission, the pulse varied from 108 to 72. The patient continued to be inapt for exercise, prone to sit about listlessly, and the grasping power of the hands was impaired. He soon became "wet and dirty," and took no interest in anything, but wore a dull, anxious, melancholy expression. His powers of comprehension became more impaired; and his utterance more embarrassed. On July 7th he was very feeble and tottering, dull, drowsy, and stupid, the pulse was 112, there was difficulty in swallowing, and the extremities were cold. He remained in bed for a time, and whilst there was often noisy and restless at night. By the end of the month he was able to leave his bed altogether, tonics having been substituted, with advantage, on July 7th, for the K.I. and physostigma, taken previously. He was stronger and took plenty of exercise for some weeks, but by the end of Sept. had again become weaker. In Oct. there was diarrhœa, followed by constipation. Again he was stronger. The right leg seemed to be the weaker of the two, the tongue was protruded very slightly to the right, but there was no decided hemiplegia. In Nov. he again became bedridden. The gait was extremely unsteady, trembling, jerky, of a somewhat tabic character, the feet were planted with extreme tremor and jerky incertitude. The right toes turned out very much; the left did not. The heel was the part of the left foot first brought down in walking, and the ball of the right great toe was, to some extent, dragged along the floor. Left to himself, he swayed towards the right side. The grasping power was feeble, but apparently equal in the two hands. On the slightest exposure he shivered much, but independently of that there was extreme tremulousness of the face and lips when in movement, and of the tongue when protruded, and after standing for a few minutes there was general tremor and twitch of the body and limbs. Circulation weak, hands and feet somewhat livid, the act of swallowing difficult. He was extremely demented, and his replies were unintelligible. Tapeworm. 19th. Right leg still the more helpless. 25th. Icteroid. 28th. Slight apoplectiform symptoms. During Dec. there was increased weakness of the right limbs. Jan. 22nd, 1876. This patient had been frequently falling into heavy, drowsy, inattentive conditions, and on this day he was dull in the afternoon, after having been very restless in the morning. Jan. 23rd. Dysphagia. Pupils irregular, small, and sluggish. He resisted passive motion, did nothing he was told to do, took little notice, but was not decidedly comatose. Tremulousness and subsultus of all the limbs. Pulse 102, regular, soft, and somewhat feeble; mucous râles over bases of lungs behind. Jan. 25th. A convulsive seizure affected the face, mouth, and eyes, especially on the right side (1) beginning equally in right eyelid and right side of mouth, simultaneously with turning of the head and eyes to the right, but affecting those parts also on the other side. The eyelids opened and closed rapidly, the face and mouth were spasmodically jerked, the mouth being drawn upward and outward to the right. The clonic spasms of the eyelids and face were equal in point of time on the two sides, but greater in degree on the right. The occipito-frontalis was affected about equally on the two sides. The left sterno-mastoid was firm and contracted, the right flaccid. The diaphragm and tongue were spasmodically twitched. Then (2) the head and eyes turned gradually to the left side, the spasm abated, then increased, and the right upper limb became stiff and straight, the hand, level with the hip, being convulsively jerked. Then (3) a stage of more general and more tonic spasm came on, a new convulsion, as it were, beginning, and the head and eyes again turned to the right. The right arm was held out straight and forwards, its fingers being rigid and extended, then severe clonic facial spasm, mostly of the right side, and clonic spasm of right arm and of diaphragm came on; and then a return to the condition in stage "(1)." Other changes in the convulsions occurred, especially in the right upper limb. Thus, the head and eyes turned to the right, the right hand was jerked, the thumb was straight and thrust between the fingers, then the right hand opened, the fingers were momentarily extended, somewhat to a position as in the "main en griffe." Then the elbow was flexed at an acute angle; and there was right facial spasm. This quasi *status epilepticus* lasted 2½ hours. Later in the day, other convulsions, and right hemiparesis, the patient remaining in a state of semi-stupor. 26th. Convulsions again occurred, mainly on right side, and followed by right hemiparesis, with slight deviation of head and eyes to left. Temp., *right* axilla 90·4°, *left* axilla 99°; pulse 90, breath fœtid, gums spongy. Patient drowsy, dull. 27th. Two convulsive attacks occurred on this day, one lasting ¾ of an hour, and the other 1¾ hour. Right hemiplegia continued; head and eyes deviated to the left. Dysphagia. 29th. Convulsive attacks, occasionally. Temp., right axilla 97·8°, left axilla 98·3°. 31st. Occasional spasmodic jerk about the face and mouth, especially right; and marked right hemiplegia with ptosis of left eyelid. Pulse 114, resp. 28. Loaded colon relieved by enema. Feb. 1st. Not wholly unconscious; pulse 135, feeble and thready; respiration 50, at times noisy, but variable, and of true "ascending and descending rhythm." The belly had gradually been becoming excurvated of late. Respiration mainly thoracic; basal crepitation. Checks flushed. During two days the right eye was occasionally opened very widely and staring. Increasing pulmonary mischief; recurring convulsions. Death on Feb. 3rd, 1876. Principal treatment.— June 9th, 1875, to July 7th, iodide of potassium and physostigma; July 7th until death, ferri perchlor. and quass.; Oct. 19th, pot. iod. and ammon. carb.; wine towards the close.

Necropsy, 51 hours after death. Body not plump. Calvaria thin, dura ordinary; arachnoid slightly opaque and thickened at the base, where, also, were marked interlobar adhesions. Olfactory bulbs wasted. Patchy thickening and opacity of meninges in front of optic chiasm. On both sides were small, symmetrically placed patches of similar yellowish-white thickening on the cerebral surface of the arachnoid, viz.: at the outer angle of the orbital surface; between the third temp.-sph. and uncinate gyri; and at the outer angle, and at the end, of the Sylvian fissures. Meningeal veins full. The pia and arachnoid at the superior and external surfaces of cerebrum were thick, tough, and faintly opaque, chiefly over the frontal and parietal lobes, fairly over the temporo-sphenoidal lobes, and internal surfaces of the hemispheres. Membranes at occipital tip, unaffected. The membranes, as a rule, separated fairly, leaving shreds in the sulci. Arachnoidal villi, well developed. Pia slightly œdematous in fronto-parietal region. Slight adhesion and decortication at frontal tips, more marked on left first frontal than on right. Adhesion was seen on the first, second and third left temporo-sphenoidal gyri, but only very slightly on the first, and that in front; of the right temporo-sphenoidal lobe only on second gyrus; also on upper part of posterior half of internal surface of left hemisphere; very slightly on that of the right; on upper end of right ascending parietal gyrus. A few scattered adhesions on inferior surface of right hemisphere; more decided adhesions disseminated over posterior third of inferior surface of left hemisphere. Grey cortical substance mostly of a dull pinkish hue, of fair thickness, convolutions of fair size, stratification obvious. In the *right* hemisphere the grey of the superior surface of frontal lobe was generally reddish, but that of the ascending frontal g. and of the posterior part of third frontal was of a dull yellowish fawn colour, and its strata were very obvious. Behind the frontal lobe the colour was fairly natural. Grey cortex at base of brain of natural hue, except the red anterior inch-and-a-half of temporo-sphenoidal lobe. In the *left* hemisphere the grey cortex of the superior and external surfaces was of a dull pinkish tinge, except the upper ends of the ascending gyri. Grey cortex firmer in left than in right frontal lobe. This undue firmness of the left grey cortex diminished from before backwards to a natural consistence at the occiput. The cortex was of a darker reddish hue here than in the right hemisphere. The grey cortex of the lower posterior half of the external surface, and of the posterior half of the inferior surface, was pale in the right hemisphere, and, for the most part, dark red in the left. On the inferior surface of the left hemisphere the grey cortex was firmer than natural in front, slightly so behind, but slightly softened in the temporo-sphenoidal region. White substance of brain moderately vascular, more so on left side, its consistence lessened. Fornix soft, ependyma ventriculorum firm, thalami mottled. On the second, and adjoining the first, frontal gyrus, 1½ inch from the tip, was a flattened firm yellowish-white nodule in the pia, and at that point the pia, arachnoid, and cortex cohered. Right hemisphere, 20¾ozs. ; left ditto, 20¼ozs. ; cerebellum, 5ozs. ; pons and med. obl., ⅔oz. Fluid from cranial cavity, 2½ fl. ozs. The spinal cord was softened, chiefly in its upper portions. The spinal arachnoid, over the posterior surface was greatly thickened, increased by adventitious layers, and adherent to the pia of the cord. About 4 inches and 6 inches, from the lower end of the medulla oblongata, the meninges were highly vascular, bulged behind, and contained two flattened, firm, yellowish-white nodules, which came off with the arachnoid, and were separable from the pia.

Heart, 10½ozs. Slight aortic atheroma. Valves, healthy. Muscle of heart slightly softened. Left *lung*, 25½ozs.; old adhesions; posterior part of lower lobe, inflamed ; apex slightly puckered, its summit tubercular. Right *lung*, 26¼ ozs. ; much like left, but without tubercle. *Spleen*, 6¼ozs. Cartilaginoid patch on capsule. *Kidneys* congested, left 6ozs., right 5¾ozs. *Liver*, 56ozs. A little turbid serum in right side of peritoneal cavity. Small intestine, gathered into pelvic cavity ; congested in parts.

(A.) In microscopical examination, the grey cortex of the posterior part of the *left* third frontal gyrus showed hyperplasia of neuroglia, some colloid bodies ; slight granular degeneration of some of the nerve-cells ; hyperplasia of the nuclei of the walls of the vessels, thickened parietes of some of the small vessels, and some of these had also dark grey deposits in their walls.

(B.) In the corresponding part of the *right* third frontal gyrus, where the grey cortex was much paler, and was of natural consistence, and not slightly indurated as on the left side, the microscopical appearances of the vessels were much the same. The microscopical differences were—1. Partly in the nerve cells. In the *right* gyrus they were scarcely granular, or were much less so than in the corresponding part on the left side. A large proportion of small pyramidal cells was seen, but only very few of the somewhat square-shaped nerve-cells. 2. Partly in the smaller number of neurogliar nuclei, and less observable connective tissue. 3. Partly in the fact that colloid bodies were absent here, but present on the left side.

(C.) In the spinal cord were granule-masses and colloid bodies, and some of the nerve-cells were more or less granular. (D.) The nodules from the spinal meninges were vascular, and contained much connective tissue ; a few of the vessels were thickened; there were pale, small, oval, nuclei or corpuscles.

Remarks.—1. In support of the view that syphilis was present, there were during life only, (I) the patient's assertion that he had had "venereal," (II) the penile and inguinal scars, and (III) the traces of old double iritis. Of these the first was vague and unreliable owing to the mental state of the patient, the second bespoke the non-infecting venereal sore rather than syphilis, and the third was equivocal. Then after death there were the firm yellowish-white flattened nodules in the spinal and cerebral meninges. These, perhaps syphilitic, were not unlike yellow tubercle. The patchy opacity of the cerebral arachnoid favoured the hypothesis of a syphilitic origin.

2. The chronic spinal meningitis and other changes in the spinal meninges, and softening, *etc.*, of the spinal cord, were probably in relation with the early and grave paresis. It was a question whether the diseased condition of the posterior spinal meninges, and the nodules in them, caused the tabic gait by pressure on the posterior columns of the cord, or whether that form of gait was due to disease in these columns.

3. Another problem was, whether any, and if so *what*, relation subsisted between the less usual character of the mental symptoms, namely, the complete absence of expansive symptoms, and the presence of much dementia, with delusions of fear and persecution, on the one hand ; and, on the other hand, either the predominance of all the morbid changes in the *left* hemisphere ; or the unusual involvement of the *posterior* part of this *left* hemisphere in the marked adhesive change ; a comparatively infrequent condition.

4. Still another question was the relation (if any) between the lesions in the left grey cortex, and the convulsions, which were mainly unilateral and

dextral. Or, the relation (if any) to the same convulsive movements, of the nodule in the meninges adherent to the left second frontal gyrus. The connection, I take it, was between the convulsions and the former change ; rather than between the convulsions and local irritation of cortex by the nodule specified. For the latter escaped what is called the motor region, and the beginning and march of the spasm, the *order* in which the several parts were affected thereby, and the eye and arm movements, did not correspond to what would have been anticipated, if one viewed the nodule as the focus of cortical irritation leading to convulsions, and expected to see the latter follow the course of those produced by stimulation of the corresponding region in monkeys.

5. The slight, varying right hemiplegia, which lasted for months, was dependent upon the somewhat diffused lesions in the left hemispheral cortex. The highly marked right hemiplegia of the last few days of life was evidently " post-convulsive."

CASE 61.—*Mental disease insidious in onset ; eccentricity of conduct, and then marked mental disorder, culminating in a resolute suicidal attempt. Subsequently, melancholic delusions, incoherence, loss of memory, depression, anxiety, alarm, and hallucinations of sight. Later on, garrulous, querulous, worried, dejected, and lachrymose (but now and then displayed some largeness of idea) ; restless at night, destructive, and even violent. Later on, confused and childish. Motor signs of g.p. of the first and second stages ; much motor restlessness ; apoplectiform cerebral congestion with hemiplegia, ending in death after four days. Intense meningeal engorgement. Meningeal changes not very well marked. Encephalon engorged by mode of death, especially on right side. Cerebral grey cortex thinner in left frontal region than in right, but of about the same depth on the two sides in the posterior part of parietal and in occipital lobes. Grey matter somewhat firm, especially in the upper layers on the right side, and chiefly in the frontal lobe on the left. Left hemisphere 1oz. less in weight.*

R. T., Sergeant 48th Regt. Admitted Aug. 13th, 1875, then aged 37 years, service, $17\frac{7}{12}$ years, married.

History.—This was said to be the first attack of mental disease, to have been marked from the beginning of July, 1875, but its origin to have been insidious, and some time further back. It was attributed to disease contracted in India. He was said to be suicidal. Invalided home from India, and transferred from the surgical to the lunatic wards at Netley on June 30th owing to peculiarities of conduct, he deteriorated both in mental and in physical condition. He seemed to be harmless, but on the night of Aug. 7th, made a determined attempt at self-destruction by hanging with sheets. He was, therefore, transferred to Grove Hall, certified as being the subject of hallucinations of sight at night, and of delusions as to everyone being hostile to him, and as to poison being placed in his food. Also, that he was incoherent in conversation, could not settle to any work, showed a propensity to wander about, was suspicious of harm from all those about him, had on the same night attempted suicide and violently assaulted an orderly ; and that he showed twitch of the muscles of the lips and tongue, and thickness of speech, indicating an early stage of g.p.

On Admission. Height, 6ft. 1in. ; weight, 156lbs. ; spare frame. Head narrow, dolichocephalic, and sagittal suture high and ridged ; these helped to give a carinate and pentagonal form to the skull. Pupils nearly normal, face slightly pale and thin. Speech thick, tremulous, and very characteristic of g.p. ; the patient was loquacious, and spoke rapidly, but the affection of speech, and tremors of face, lips, and tongue were greatly increased after active exercise. Pulse 120, quick, compressible ; heart's action irritable, its second sound clear, but not accentuated. Viscera apparently normal. Sudamina on chest. The right side of the mouth acted slightly better than the left, and the hearing was defective in the left ear, from which the patient said there had been a discharge. There was a history of a venereal sore on the penis about fifteen years before, of which no decided trace remained ; in the left groin were two large bubonic cicatrices. No indications of syphilis.

F F

No headache, or pain, or numbness of any part; no decided local palsy. Veins of legs varicose. He grasped very powerfully with the hands; the gait was firm and confident, but there was some unsteadiness in turning round, and he marched rather heavily. He declared that "everyone was against him" and put poison into his food. He did not feel safe, and begged for protection against his enemies. He was restless, excited, garrulous, frequently rambling in conversation, and gave long disconnected accounts about himself consisting of delusions as to his treatment, both in India and since, querulously bringing charges against all who had had the care of him in any way. His expression was anxious, agitated, and dejected.

Aug. 23, 1875. The patient had been restless since his admission, and full of complaints about his detention. Two days before this date he had been very restless, excited, and loquacious; smashed a window in his dormitory, and was sent to sleep in a padded room. He had taken chloral hydr. and K.Br., and had become calmer. Pulse 114, head somewhat warm, face pale, speech much affected. He was suffering from much mental malaise, anxiety, and "worry," and yet contemporaneously with these were sometimes statements which evinced a feeling of well-being, and largish ideas. During Sept. my notes mention temporary gastric disturbance, occasional restlessness, excitement, and undressing. At night the restlessness was greater, and he then had hallucinations of sight. In conversation he was childish, confused, and now and then broke into tears. He was taking digitalis, and the pulse varied from 76 to 90 on different days. Early in Oct. the pulse was tense, full and unduly frequent, the patient was less restless and excited, but worried himself a good deal, and was disquieted about various subjects.

Oct. 21st. The patient had been violent the day before, and heavy and stupid at night. He was now lying in bed, in a state of stupor, but could be roused slightly; respiration slightly stertorous. Pupils sluggish, of medium size, the right very slightly the larger, the left having been slightly the larger, as a rule. The left arm appeared to be more helpless, non-resistive, and flaccid than the right, and the left lower limb, also, was slightly palsied. The face and eyes were diffusely injected. Temp., left axilla 103·2°, right 103·1°; pulse 84, resp. 17. Heart sounds ringing, forcible beat. Ordered saline aperients and enemata, croton oil, bromide of potassium, veratrum viride, and ice to the head. *At midnight* pulse 84; pupils equal, smallish, and quite immobile; resp. 18, quiet, regular, slight puffing of cheeks in expiration. The left arm was the more paralysed, there was no palsy of the face, the left lower limb was slightly the less resistive to passive motion. Stupor had continued during the evening; ice had been kept to the head, the temples and nostrils had been leeched. For a time the left arm was rigidly bent at a right angle at the elbow joint, the hand being pronated, the thumb and forefinger stiff and straight, and the outer three fingers flexed. Left foot cold. Continue mixture and ice; sinapisms to the feet. Oct. 22nd, 9 A.M. Five pints of urine were drawn off. 1 P.M. Pulse 86, rather full, hard and long. Respiration 18, the cheeks puffed a little in expiration, and saliva was blown upon the lips. Bowels moved 7 or 8 times during the past 28 hours; face still red and suffused, the body warm, but the feet, especially the left, slightly cold, the left arm limp and palsied, the left leg less so; mouth somewhat drawn to the right; slight temporary conjugated deviation of head and eyes. There had been occasional convulsive jerk and tremors of the right arm. Now and then the patient would protrude the tongue momentarily at request, or in reply to a question would mutter "all right" or "no." Temp., left axilla, 103·2°; right, 102·9°. Treatment continued. Oct. 23rd. Much the same. Pulse 96, smaller. Resp. 26, slightly stertorous. He could still be partially roused as on the 22nd, but the tongue was not protruded. Temp., left axilla, 101·8°; right, 101·7°. Left hemiplegia; right arm resistive. Signs of pulmonary congestion and œdema. Oct. 24th. The urine was acid, and contained excess of urates and of phosphates. Much hiccough. Pulse 78, softer and smaller; resp. 28, and more noisy; temp., right axilla, 100·3°; left, 100·2°. Much flatus was passed downwards. The pulmonary signs were increased.

Lips and eyelids dusky. At 6 P.M. was more comatose, swallowed very badly, had great gaseous distension of abdomen. Temp., left axilla, 100·2. Left, then right, pupil the larger. Oct. 25th., 10 A.M. More comatose, sweating freely, face and ears livid and purplish; pulse 130, full and bounding; resp. 49, loud and laboured ; severe hiccough, mucous rattling in throat, pupils sluggish, smallish. Temp., right axilla, 101·3°, left, 102·5°; general venous congestion; abdomen as on 24th. Death at 5 P.M., Oct. 25th, 1875.

Necropsy, 21 hours after death. Calvaria congested. Sinuses of dura, and meningeal veins, gorged with dark fluid blood and clot. *Right hemisphere of cerebrum,* 23½ozs. Convolutions fairly plump. The membranes were separated with some difficulty, left shreds in the sulci, were slightly œdematous over part of the parietal and frontal lobes, were slightly thickened and tough, faintly opaque at the vertex, and were not adherent except at a few points at the upper part of the tips of the first and second frontal gyri, especially of the latter. The grey matter was congested, of a deep pinkish hue, particularly the frontal. In the upper parietal convolutions the colour was much paler, especially in the deeper layers. The colour was again deeper in the occipital region. The grey cortex was of ordinary depth, and was of increased consistence throughout this hemisphere, and particularly so in its superficial layers, and in the gyri at the vertex, and in the frontal lobe. Strata indistinct. Grey cortex of the inferior surface firmer than usual in the anterior region, but less so in the middle and posterior. White matter firm, of a mottled violet and lilac hue, and hyperæmic. Old adhesion between corpus striatum and wall of ventricle. Opto-striate body plump, firm, with numerous puncta cruenta and dilated vessels. Lateral ventricles nearly empty.

Left hemisphere, 22½oz. Membranes as on right side. Adhesion almost limited to upper part of tip of first frontal convolution. Convolutions depressed below the level of the adjoining ones over an area of about half an inch square, at the external parieto-occipital fissure. Grey cortex paler on this than on the right side, its layers not nearly so thick as those of the right, in the frontal lobe, but of about the same depth in the posterior part of the parietal and in the occipital lobe ; and gradually diminishing as one passed forward from the parietal lobe. It was usually firm in all the upper regions of the cerebrum, especially in the frontal lobe, where its colour was ordinary, with a faint slaty tint in the inner layers. Few vessels visible to the naked eye in it on the left side; a slight pinkish hue only in the posterior of the parietal and in the occipital convolutions; elsewhere the colour about ordinary. White matter much the same as on the right side; radicles of *venæ Galeni* equally engorged on the two sides. Opto-striate body firm, hyperæmic, but much paler than on the right side. Basal grey cortex thin, firm, and of ordinary hue in orbital region, but thicker, softer, and more vascular elsewhere ; and at posterior internal surface. Pons and med. obl. slightly congested. 1oz. Med. obl. very firm; tissues beneath the floor of the fourth ventricle, pinkish. Cerebellum 6ozs., its white slightly firmer than usual. *Heart* 11½ozs., its muscle healthy; 1oz. pericardial fluid ; left ventricle contracted ; mitral valves thickened, and the aortic slightly so. Above the aortic valves was an elevation, containing a calcareous mass ; and another of fibroid appearance. *Right lung,* 39½ozs. Hypostatic congestion, œdema, and pneumonia. *Left lung,* 37½ozs. The same ; part of the lower lobe was breaking down, and had a gangrenous tendency. Viscera, intestinal and abdominal veins, and gastric mucosa, congested. The *intestines,* distended with gas, thrust the diaphragm upwards. *Kidneys,* healthy. Right, 6ozs. ; left, 6½ozs. *Spleen,* 7¾ozs. A few old filamentous adhesions, and patches of slaty-grey thickening of capsule. *Liver,* 84½ozs. Old adhesions of its upper surface to diaphragm and abdominal walls ; at one of them, a large cicatrix, also passing to the depth of ¼ inch into the liver.

Tip of *right* second frontal convolution (hardened and stained sections): Nerve-cells abundant, slightly granular, a want of clearness in their contours. Nuclei of neuroglia, numerous. Many little vessels contained clot and were gorged with blood corpuscles, and one showed an appearance of rupture;

others, slight fusiform vascular dilatations. Tip of *left* second frontal: The want of clearness about the contours of the nerve-cells was more marked on this side. The other changes were much the same. *Right* ascending parietal: Appearances much the same. Some nerve-cells granular, and their nuclei obscured. Deposits on the walls of some of the vessels, and at one point a minute vessel seemed to have given way, and permitted of slight extravasation. Nothing special at tip of occipital lobe. The pia over the tip of the left frontal lobe, showed vessels with hypertrophied walls, some tortuous, and some bulging here and there with blood corpuscles.

Remarks.—1. Apoplectiform cerebral congestion, a common, and often a recurring, phenomenon in g.p., assumed a fatal intensity in this case; and was partly evidenced in left hemiplegia. On the one part may be seen an intense *symmetrical* meningeal congestion, and yet on the other marked hyperæmia and a normal depth of the cortical grey of the right hemisphere, and hyperæmia also of the right opto-striate body, contrasting with the paler corresponding parts in the left hemisphere, and with the thinner, atrophic, but equally firm, grey cortical substance of the anterior part of the left hemisphere. The morbid process, attacking the left (in advance of the right) hemisphere, had occasioned some atrophy, and lessened the vascular dilatability therein.

2. There were traces of old perihepatitis and perisplenitis. These had, perhaps, been connected with tropical diseases contracted in India, and all the more so as no traces of syphilis were observed, and no decided history of it was forthcoming.

3. Gaseous distention of the bowels hurried on the fatal termination by way of respiratory limitation and circulatory obstruction.

CASE 62.—*Mental impairment, defective memory; insomnia, fidgety excitement, listlessness and wandering habits. Transient, or doubtful, expansive notions. Later on, usually dull and demented in appearance, but sometimes nervous and alarmed, quiet by day, but at night shouting or hiding away in confusion, alarm and terror. Self-stripping, destructiveness, degraded habits. Occasional apoplectiform seizures. Finally, extreme dementia, helplessness, obtuse sensibility. Ordinary motor signs of g.p.; also more or less hemiparesis, growing worse paroxysmally. Latterly, great resistance to any manipulation. General tremulousness on exertion, nearly throughout. Greater adhesion, decortication, and atrophy of the left than of the right hemisphere, the left lateral ventricle being also the larger, and the left corpus striatum the smaller. Adhesion extreme and extensive, existing in the anfractuosities as well as on the summits of the gyri, and being unusually well marked over the base of the brain. Grey cortex palish, but reddish and vascular in parts, atrophied, of ordinary consistence. White cerebral substance pale. Meninges pale, thickened, in parts œdematous, and considerably changed over the base of the brain.*

J. J., Private Coldstream Guards. Admitted July 10, 1876, then aged 27, 9½ years, all "home" service. Single.

History.—This attack of mental disease was said to be the first, to have come on insidiously, to have existed for six months, and to have been treated at the Guards' hospital, London, and then at Netley since May 8. Cause "unknown." Not suicidal or epileptic. Of good character and temperate habits. No family history was procurable. There was no record in his medical history sheet of epilepsy, injury to the head, or sunstroke, but he was said to have suffered from constitutional syphilis, and there was a record of his admission to hospital during six days for "paralysis of the right side." Whilst the patient was at Netley there was impaired power of the right side, and a slight halt in the gait. He had, also, the signs and symptoms, physical and mental, of g.p., "with the exception of the delusions of grandeur." Whilst he was there, chloral hydrate and tincture of calabar bean had been employed without good effect. The certificates testified to: 1. Great confusion of thought, very defective memory, insomnia, a fidgety manner, occasional excitement, propensity to wander about in a listless objectless way, imperfect articulation, tremor of the muscles of the tongue and lips, incomplete paralysis of the right side, and advancing g.p. 2. Great hesitation in speech and nervousness, trembling of hands. Patient's statements that he had found large quantities of diamonds in the hospital-yard, and that a woman came at night and stole them; also that a woman came at night and made his bed.

State on admission.—Height 5ft. 11in., weight 168lbs. The tongue was protruded slightly to the left side, where the lips fell slightly together; but the right side of the mouth was the one less fully drawn up at request. Grasping power of the right, less than that of the left hand, but of both, fair; the right leg tired the more quickly of the two, and then the gait became limping in that limb. Pupils smallish, acting fairly. Slight tremor of tongue on protrusion; speech slow, hesitating, quasi-stammering, jumbled, shaky. Much labial, and with it some facial, tremor, especially just before and during speech, the writing was shaky, any muscular action was accompanied by marked tremor. He was restless and shaky. Features somewhat broad and of deficient expression. The heart's sounds were full and loud, the second was accentuated and heard widely, the pulse was of ordinary characters. There were no signs of syphilis, and both it and cranial pain were denied.

His mental operations were slow and impaired, his memory for leading events, and on simple matters, was still fair, but he forgot many things he should have recollected, as for example the name of his colonel. He exhibited no pleasant joy, no exaltation, and no grandiose delirium. With reference to what was mentioned in the certificates about diamonds, as noted above, he now said that a woman stole some valueless pebbles from him. His ideas were confused, he said he had been "on guard" the preceding day, and that he "was tormented by a woman at night." Sometimes he looked startled, nervous, confused; at others, dull, stupid, unemotional. Such was his usual condition during the day, but *at night* he would often get up in terror, declaring that "some persons were after him," or shouting "murder," and wildly endeavouring to get away, and on one such occasion he smashed the glass of his dormitory window. Sometimes he would hide his head under the pillow and huddle up underneath his bedding, in morbid fear and panic.

July 24th, 1876.—The patient was heavy, drowsy, had a dull and vacant look, and the face was flushed. This passed off. Aug. 3rd.—At times he was restless, confused, meddlesome; at others agitated, and in terror. Occasionally he was busy at night, pulling his bedding about, and once then was found bathed in profuse hot perspiration. His movements and speech were very tremulous, and he fumbled much, especially in handling anything. Aug. 8th.—The trace of former hemiplegia was now only recognizable in the face. He was so restless, wakeful and destructive at night that hypnotics were ordered. Aug. 16th.—Had again been restless and "bed-making" at night, and was destructive to clothing. He was confused, flushed, and heavy. 31st.—As last note; strips self day and night. Sept. 25th.—From time to time he partially lost power in the right limbs, and for some little time now had scarcely over entirely regained it. Nov. 1st.—He was very restless, seizing hold of and pulling at everything about him. Was still allowed to sit up, but was helpless, and in moving stumbled. Often he was restless, obstinate, with flushed face, and mental functions more than usually obscured. So demented was he, as well as restless, and resistive, that he was very difficult to manage. On Nov. 3rd, he became bedridden, and remained so until his death. Early in Nov. vesical catheterism was required, and slight bedsores formed with extreme readiness. Cutaneous sensibility had become obtuse. Jan. 1st., 1877.—He now was utterly fatuous and rarely spoke, but ground his teeth together with great constancy and vigour. There was slight right hemiplegia, which paroxysmally became worse, and there was always marked tremulousness of movement. By Jan. 14th there was hypostatic pneumonia, especially of the right lung, and a large abscess had formed also in the left thigh, connected with the ulcerated surface over the left hip. This was opened. R. Quinine and Sp. Vini Gallici. Emaciation became extreme, the pulse very weak and rapid, and exhaustion profound. Death on

Jan. 19th, 1877. At first potassic bromide and physostigma were given ; later, K.I. and Fe. I.; finally, Ferri perchlor.

Necropsy, 68 hours after death. Body somewhat emaciated. Calvaria thin. Dura, pale. Slight thickening and opacity of arachnoid, and marked interlobar adhesions, over base of brain. Arachnoidal villi of very full size. Arachnoid milky over the whole of the superior and external surfaces of the cerebrum, and the combined pia and arachnoid thick, tough, anæmic, and the former partially œdematous. The membranes were closely adherent to the surface of the cerebrum, and a considerable quantity of the cortical grey matter was stripped off with them when they were removed, not only from the summits of the gyri, but partially from the sides of the anfractuosities between the gyri as well. This adhesion and decortication occurred over every surface of the left hemisphere of the cerebrum, upon all its aspects,, especially over the parietal and part of the frontal lobe, and almost as much over the tip of the frontal, somewhat less over the occipital, and still less over the temporosphenoidal lobe. On the inferior surface of this last, there was more adhesion than on its external. The same were found in the right hemisphere, but less marked and less extensive than in the left. Grey cortex slightly reddish in parts, but as a rule rather pale; of ordinary consistence; somewhat thin; the sulci wide. White cerebral substance, palish, and perhaps unduly firm. The lateral ventricles were much dilated, particularly the left one, and contained more than 1 fl. oz. of fluid. Choroid plexuses pale. Left corpus striatum, slightly shrunken ; all the basal ganglia pale on section. Cerebellum, softened, its grey matter pale. Right hemisphere, 21½ ozs.; left 20¾ ozs.; cerebellum 5½ ozs. ; pons and med. obl., 1 oz. 6¾ fl. ozs. of serosity from cranial cavity.

Heart, 9¼ ozs. Blood nearly all fluid. Atheroma of both coronary arteries, especially of the left. Just above the valve, the inner surface of the aortic arch was irregular, and nodulated mainly by greyish tissue containing yellowish and opaque patches and spots, and, in parts, by superficial, easily detached, yellowish patches. Muscle of heart too friable. *Left lung*, 31½ ozs. In upper lobe, a few opaque, white, firm, and, in parts, cheesy nodules; the posterior portion, of a deep purple hue, firm, sank in water, and had patches of pneumonia. *Right lung*, 56 ozs., much congested. More advanced changes than left, and, at parts, excavation. *Spleen*, 15 ozs., of deep purplish chocolate hue. *Kidneys*, each 6¾ ozs., healthy. *Liver*, congested; 53 ozs. A starlike patch of thickening of capsule.

Remarks,—1. From time to time right hemiplegia, incomplete in degree, and, oftentimes, partial in extent, was observed. Now and then it almost disappeared, and anon it became paroxysmally worse. Its occurrence might be explained by several conditions, any one or all of which might be efficient, *a*. The great morbid surface change of the left cerebral hemisphere. *b*. The greater atrophy of the left cerebral hemisphere. *c*. The slight relative atrophy of the left corpus striatum. *d*. The reflex effect of irritation by the lesions observed.

2. The coexistence of aortic endoarteritis (nodulation) with the statement that the patient had had constitutional syphilis could not be lightly passed by. Yet I observed no clinical indications of syphilis, and no unequivocal indication of it after death. True it is, that the meningeal adhesions were unusually severe, extensive, and indicative of an active adhesive, and probably inflammatory process. But this change was of a kind common in g.p. True it is also, that the uniformity of the hepatic capsule was relieved by the star-like thickening in its upper surface ; but no trace of syphilitic gumma or inflammation was to be seen, there was no cicatrix extending into the liver; and milky circumscribed thickenings of this description may be due to some compression, and friction, during locomotion, owing to unaccustomed conditions of effort, as, for example, when soldiers are drilled invested with accoutrements.

CASE 63.—*Insidious origin, long duration. At first, eccentric conduct, gusty fits of passion and destructiveness, mistakes in, and unfitness for, performance of duty. Impaired perceptive, mnemonic, an d reasoning powers. Incoherence childishness, dementia, docility, obedience, the countenance generally expressive of placidity and pleasure. Speech much impaired.* Severe early headache, and, later, recurring headaches, followed by a dazed and confused state. Extensive incomplete anæsthesia. Tremor cōactus *of hands and arms. Unilateral convulsions and local spasms, long continued, changing from part to part, and chiefly dextral.* Right hemiplegia variable, recurring, somewhat persistent. Helplessness, rigid flexed contraction *of limbs, mainly of right.* Finally, vomiting, low temperature, slow feeble pulse. Brain pale. Left grey cortex paler than right, and especially so in frontal lobe, and of this, chiefly in the third gyrus. White substance, pale and very slightly indurated; mainly in left frontal lobe. Left hemisphere 1½ ozs. less than right. Adhesion and decortication, of superior and external surfaces of cerebrum, principally in the parietal, less in the temporo-sphenoidal and frontal lobes, slight in the anterior half of the frontal, invading the occipital, affecting the internal surface of left frontal lobe more than of right. Cerebral ventricles large.

F. H., Sergeant Royal Artillery, admitted March 17th, 1875, aged 31 years, service 13 12/12 years, 4 in India ; married.

History.—Stated to be the first attack, and to have become marked at the end of 1874, the mental disease had been existent and insidiously progressive for a long time previously. The causes assigned were "exposure to the sun and cerebral disease." No record of epilepsy, or of suicidal tendency. Habits temperate ; moderately well educated. No history of hereditary predisposition, syphilis, or apoplexy. He was said to have had a blow by a poker on the right side of the head many years ago. Four years before admission he was at Singapore, and a good deal exposed to the sun. Two years later, in Ireland, he was reduced from the post of Sergeant-Major; this was said to have preyed upon his mind, and his wife stated that since that time he had become sullen and taciturn ; and had suffered from headaches, which affected him so severely in June, 1874, that he was obliged to enter the hospital, but he performed his duties up to Oct., 1874, when he was sent to Gibraltar. Whilst there, his eccentricities became more marked, his memory quite failed, and his speech became thick and hesitating. He was then sent home. His father informed me that the patient had a sunstroke several years ago in India and was carried, insensible, into hospital; that he was never the same again ; was stupid, dazed, and occasionally violent, chasing his wife out of the house with a bayonet; that his fellow N. C. officers, with kindly intent, concealed his vagaries from the superior officers, but that, at times, he was " quite a maniac." He added spontaneously, that he thought the patient had really been insane for three or four years, and that when he lost his rank, because he improperly signed some document, he was in reality too confused to understand the latter. The medical certificates testified that F. H. had impairment of perception, reasoning faculties, and speech ; failure of memory, especially as to recent occurrences ; apathy, inability to take care of himself, forgetfulness of words, loss of the normal connection and association of ideas, and incoherence.

On admission.—Height 5ft. 10in., weight 162lbs. ; large head ; pupils equal, the right one irregular. The muscles of the face were jerked spasmodically, and twitched, even occasionally when not in voluntary movement, but this condition was increased when the tongue was protruded. There was also tremor cōactus of the pronators and flexors of both forearms and hands, one or both being affected, and to a degree which varied from moment to moment. The flabby tongue was jerked in the same rhythmical manner when it was protruded. The facial expression was less fatuous than might be expected. The speech was slow, interspersed with long pauses, quasi-stammering, the words at times being tremulous and broken, but for the most part elided in portions and sometimes uttered explosively. At other times speech was very much better. During speech tremor and twitch of the facial muscles occurred. The tactile sensibility of the skin was diminished, at least this was noted on the neck, chest, abdomen and extremities. Cardiac impulse very obvious near the epigastrium, less so near the left nipple; very strong and wide. First sound clear and full, a cardiac bruit, heard, appeared to be diastolic. Pulse full and slightly jerking. A psoriasis spot on left arm, a wart on buccal mucous membrane, no scar on penis; the patient said he had had "venereal" in 1862. He was incoherent, childish, quiet, obedient, demented, extremely amnesic. His countenance was always either smiling or void of emotional expression. R. Potassii iodid. grs. viii, et hydrarg. perchlor. gr. $\frac{1}{16}$, ter in die. May 24th.—Increase each dose of iodide to grs. xii. June 19th.—The fingers of the right hand were held straight and, as he stood, jerked the forearm rhythmically from the side of the thigh, and there was a similar but slighter rhythmical spasmodic tremor of the lower lip. The tongue was tremulous on protrusion. The speech was shaky, and during it there was slight trembling of lips and face; it was mumbling, pausing, hesitating, and this condition was succeeded by explosive utterance. The tremor cöactus occasionally affected the left upper limb also, particularly the left forefinger. The tremor and the dementia had increased of late. Diastolic bruit; apex beat of heart lowered; another impulse $1\frac{1}{2}$in. from zyphoid c. June 23rd. Confused and dazed. Headache. Pulse 114 and jerking, head warm, pupils equal and acting. Very paretic and helpless, he was ordered to bed and to take his mixture every three hours (96 grains of K. I. in 24 hours). June 28th. Better, pulse 102. The tremor cöactus now affected the right forearm only. (Omit Hg. P., continue frequent doses of K. I.). July 18th. Much improved. Aug. 3rd. Gait unsteady, tongue jerky. Speech indistinct, broken and shaky. Slight tremor cöactus of fingers of right hand, ceasing temporarily when attention was drawn to it; its movements varied from 60 to 80 per minute. Pulse 118, soft, pupils equal. Mental perception, attention and comprehension enfeebled. There was simply a disintegration or dying-out of mind. (30th, K. I. grs. xx. t. d.) Oct. 2nd. Again very heavy and confused (add Hg. P. to the K. I., and take every four hours). After this date he often complained of his head, and on these occasions was unusually demented, and occasionally wetted himself. The paretic state and the jerky tremulousness again increased, he could walk only slowly and unsteadily. Later on, there was some "iodide coryza." During Oct. and the beginning of Nov. he was gradually failing, but varying in state from time to time, paresis affecting the right limbs more than the left. Gait, awkward and somewhat stiff. Nov. 9th. Sudden dextral hemiplegia came on without convulsions, or loss of consciousness, or falling, or syncope, but the patient was found to be more dull and confused, and the face wore an expression of "shock." Dysphagia, occasional vomiting. Temp., right axilla, 98·9°; left, 98·7°; pulse varying from 60 to 70, full, soft, rather quick; respiration normal. 10th. Omit mercurial perchloride. (Mist. K.I., 4tis. horis sumend.) Nov. 11th. Temp., right axilla, 99·3°; left, 99·1°. Slight conjugated deviation of head and eyes to left. Mental state as on 9th and 10th. Nov. 13th. Temp., right axilla, 98·8°; left, 98·6°. Same state. Nov. 16th. Temp., right axilla, 98·3°; left, 97·4°; pulse 84, very soft, full, quick and bounding. Fluids were swallowed well, the tongue could not be protruded. The right hemiplegia and mental state were much the same. Nov. 19th. Some dysphagia. Right orbicularis palpebrarum less resistive than left. Later on, some spasm affected the mouth, and the tongue. Then, spasmodic jerking of the masseters and other muscles of jaws on both sides, and involuntary grasping, forward, movements of left hand. Nov. 21st. Temp., right axilla, 97·7°; left, 96·7°; pulse 74. Nov. 23rd. Had a number of spasms limited to the right side of the face, and to tongue, every day, and lasting from $\frac{1}{2}$ minute to 10 minutes. 25th. Temp., right axilla, 98°; left, 97°. Could not reply intelligibly. 26th. Temp., left axilla, 98°. Facial spasm continued. Nov. 27th. Synergic contraction of head and eyes to the left. Right face palsied and flattened, right arm palsied, but resistive to passive motion. The mouth gaped, then spasm of the lips was succeeded by tremor extending to the muscles of the cheeks, then the mouth was in spasm, the tongue protruded and bitten, next the mouth was agape and the tongue retracted; and, throughout, were wavy fibrillar tremors of the face, nostrils, and lower eyelids, especially on the right side,—also of the right platysma; which spread to the right corrugator supercilii before the left. The spasmodic tremor was intermittent. Temp., right axilla, 99·6°; left, 99·2°. Fed by stomach-pump. After this day the spasms subsided, the right hemiplegia gradually lessened, the power of movement returning first at the proximal parts, the patient grew brighter again in appearance, the pulse varied from 102 to 108, and by Jan., 1876, the patient was able to walk a little by himself, and to be regularly exercised by an attendant.

Jan. 29th, 1876. Sudden right hemiplegia occurred, chiefly in lower limbs; ordered to bed, which he did not again leave during life. Feb. 6th. Drowsy, swallowed badly. Feb. 12th. Pupils equal, wide, and sluggish; dysphagia; knees somewhat flexed and contracted. In the evening, face and limbs cold; pulse slow, vomiting (stimulants, hot bottles to feet). Feb. 13th. Vomiting, cold face and limbs, moaning; pulse 44, soft, of medium size. Heels drawn up to buttocks; arms slightly contracted, especially the right; right limbs resisted passive motion less than left. Conjunctivæ injected. And so, until death on Feb. 14th, 1876.

Necropsy, 38 hours after death. Body well-nourished. At the left temple a scar, adherent to a depression in the bone. Calvaria capacious, thick and dense. Dura somewhat too adherent to calvaria, especially at the left middle *fossa basis cranii;* slightly thickened and hyperæmic. Sinuses gorged with fluid blood. Marked basal interlobar adhesions, arachnoid at base, thickened, opaque. Cranial nerves adherent to meninges. Over the superior and external fronto-parietal surfaces, the upper half of the temporo-sphenoidal, the anterior part of the occipital, and the anterior two-thirds of the internal, the membranes were extremely thick, opaque and adherent; the sulci full of serum, and the membranes separated with difficulty.

Adhesion and decortication affected more or less the prominences of the gyri on the superior and external surfaces of both cerebral hemispheres, affecting the parietal lobes most, the temporo-sphenoidal and occipital slightly. In front, adhesions were almost absent, especially were the anterior halves of the second and third frontal gyri comparatively free. The posterior part of right third frontal was slightly more affected than that of left. The inner surface of left frontal lobe was more affected than of right. The upper parts of the ascending frontal gyri, and the postero-parietal lobules, were not much affected. The cerebral base was free from adhesions.

In the *right* cerebral hemisphere the grey cortex was pale. Its deeper layers were slightly firm: this firmness extended through the fronto-parietal region, was very slight in the occipital, and absent in the temporo-sphenoidal. Strata not very obvious. Grey, of fair depth, but, proportionally, of unusual depth in the occipital region. Its colour externally was a slaty grey; internally of a dirty whitish and yellowish hue. In the *left* hemisphere the grey cortex had much the same appearance as in the right, but was paler in front. It was still firmer than on the right side. The grey matter of the lower part of the left, and to a less degree of the right, frontal lobe was firm, very pale and homogeneous in appearance; strata best marked in asc. frontal gyrus.

White cerebral substance of increased consistence; its colour pale, and paler in left than right frontal region. The lateral ventricles, very large, contained much serum; their ependyma was thickened. Basal ganglia palish, of ordinary consistence. Cerebellar cortex pale. Pons and med. obl. slightly firm. Left hemisphere, 23¼ozs.; right, 24¾ozs.; cerebellum, 6½ozs.; pons and med. obl., 1¼ozs. Fluid, 6½ fl. ozs. No thrombosis, embolism, or strictly local softening.

Heart, 12¾ozs. Aortic semilunar valves, thickened, unduly opaque, rough, especially on their ventricular aspect; and one was slightly incompetent. Mitral valves large and thickened; other valves healthy. Some atheroma of aortic arch. Left ventricle contracted, hypertrophied; its apex tip 3½in. below L. nipple. Left *lung*, 29½ozs.; right, 43ozs., emphysema in front; some hypostatic congestion and pneumonia posteriorly. Right *kidney*, 6ozs.; left, 7¼ozs. *Spleen*, 4¾ozs., some old perisplenitic adhesions. *Liver*, 54½ozs.

Remarks.—1. The g.p. apparently arose from insolation. The moral cause suggested for the malady seems to have come into operation, after the prodromic stage of the disease. The blow on the head was too remote to merit consideration here.

2. With the mainly dextral convulsions and spasms, and right hemiplegia, too persistent to be purely "post-convulsive," the left cerebral hemisphere was the one more diseased, its atrophy being greater, and its frontal grey cortex paler and firmer than that of the right.

3. The brain, naturally a very large and heavy one, was sclerotic and anæmic (long duration, no exalted symptom).

4. Characteristic syphilitic changes were absent after death, though some of the changes (*e.g.*, the meningeal) may have been syphilitic.

CASE 64.—In the "Journal of Mental Science" (January, 1876, p. 567), I have already described this case in detail, and have appended an elaborate commentary. A very brief summary, therefore, will suffice here.

Extreme maniacal excitement, incoherence, noisy, destructive, violent: grandiose delusions. Intense irritability and furious outbursts of anger, the patient threatening, or even attacking others, proud and haughty in bearing. Auditory hallucinations: early epileptiform convulsions: insomnia. Later, tractable, quiet, dull, inert, with failing mental powers. Motor signs of g.p. at first comparatively slight: great motor restlessness. Later, right unilateral epileptiform convulsions, followed by right hemiplegia and temporary aphasia. Finally, bed-rid, recurrence of epileptiform convulsions and of right hemiplegia. Death in quasi-status epilepticus. Encephalon somewhat soft. Left hemisphere 2½ozs. less than right. Meningeal changes more marked over left side. Left hemisphere the softer; grey cortical matter, atrophied on both sides, more so on the left, and especially on the superior and external surfaces. Adhesion and decortication much more marked in left hemisphere, and observed especially over the posterior part of the parietal lobe, and posterior part of the frontal, but of wide distribution. Grey cortex of ordinary hue. Medullary substance hyperæmic. Marked granulations of fourth ventricle. Meninges thick and hyperæmic over pons and med. obl., and these parts hyperæmic.

GROUP IV. (See p. 410).

CASE 65.—*Early acute maniacal symptoms, and exalted delusions. Frequent paroxysms of dangerous excitement, violence, destructiveness. Later on, the same mania and delusions, the patient sometimes elated, and sometimes irritable and dictatorial. Disorder of muscular sense. Hallucinations of sight, hearing, touch, taste, smell. Shivering tremulousness, excessive motor restlessness. The impairment of speech masked by the state of agitation. Later on, subsidence of excitement, appearance of motor signs. Later still, hypochondriacal delusions, lachrymose depression, and hypochondriacal sensations. Death after violent epileptiform convulsions lasting 40 hours. Venous engorgement of meninges from mode of dying. Other meningeal changes ordinary. Brain hyperæmic, of diminished consistence. Some atrophy of grey cortex in anterior regions. Right cerebral hemisphere 2ozs. less in weight than left. Adhesions much more decided and extensive over the right than over the left hemisphere, and their more marked degrees confined to the external surfaces and to the anterior part of the superior surfaces.*

H. R., Private 4th Hussars. Admitted Feb. 26, 1876, aged 36 years, 18¾ years service, married. Neither epileptic nor suicidal.

History.—This attack of mental disease was said to be the first, and to have lasted from Jan. 12, 1876, and its alleged cause was his reduction to the ranks. He had previously been under care at Canterbury, and at Netley, into which he was admitted 18 days before his transfer to Grove Hall. There was no record of epilepsy, palsy, or hereditary tendency. He had been degraded from troop sergeant-major to private, and sentenced to six months' imprisonment, for deficiencies in his accounts. This had preyed upon his mind, and was believed to be the exciting cause of his mental disease, which, it was said, began suddenly on Jan. 12, with symptoms of "acute mania." His wife denied that he had been addicted to sexual or alcoholic excess, or that he had misappropriated the missing money to his own use. *Inter alia*, the medical certificates stated that: —" He is eminently irrational in his conversation and manner; has delusions of being possessed of immense wealth, and exalted ideas of his connections and importance. He is subject to frequent fits of excitement, during which he offers violence to the attendants and patients; is most destructive in his habits, and by his noise disturbs the inmates of the hospital. He is a dangerous lunatic."

State on admission.—Rather short and muscular, pupils equal, cicatrix in left groin, pulse 90, full; viscera normal. For several days he was either elated, or irritable and dictatorial, very excited, restless, noisy, and threatening, giving utterance to delusions, such as that he was possessed of extraordinary strength, that he could travel many miles in a minute, had been forty-three years in the army, and was the first man in it. The head remained heated, the face flushed very readily. He had a bilious sallow hue of the skin, a furred tongue, a frequent soft pulse. The excitement was constant, the insomnia complete, indeed no sleep was obtained until the fifth night. (Warm baths, cold to the head, aperients, night-draught of chloral hydrate.) The motor signs of g.p. were not present during this time. Subsequently, when the excitement had subsided in some degree, the tongue was slightly tremulous and twitching, and the whole muscular system was agitated by fine movements, as if the patient was shivering slightly from head to foot from the effects of cold, though he was manifestly warm and comfortable, and, on the other hand, was not the subject of febrile rigor. If the eyelids were closed the pupils *dilated* when the lids were again opened. When he was quiet, his words were at times slightly clipped. The cheirography was fairly natural. He abounded in extravagant delusions. " Has millions of money, and it doubles every minute; has hundreds of thousands of wives and innumerable children; is extremely strong; gives the schools £10,000 a day ; has been married and in the service forty-three years, but is only thirty-five years old." He averred that at the moment of speaking he was riding in a railway saloon carriage with his wife, and mentioned to me the stations as they arrived at them. " Now we are at Suez," he said ; and next moment, " we are at Paris; " and the moment after, " at Australia." There were hallucinations of sight and hearing. The motor restlessness and meddlesomeness were extreme. In his excessive restlessness and violence he fractured a metacarpal bone by striking his fist against the wall of his padded room. Each night a warm bath was administered at bed-time, and followed by a draught of chloral hydrate, grs. xl. By Mar. 4th he had become quiet, eating and sleeping well, but the shivering tremulousness and the mental condition were unchanged. Mar. 27th. Had continued to sleep well since, and was now quiet, and exercised regularly. Sometimes he was weeping, sometimes elated, and still had expansive delusions. At this date the mouth was drawn towards the left side, the pupils were small, irregular, somewhat sluggish, and the right dilated partially when the closed lids were raised. The tongue and lips trembled much when the tongue was protruded ; utterance was clear, but there was a faint under-tremor of voice like that of a person who is shivering, and the universal shivering-like tremor continued at times. Incipient left othæmatoma. No signs of increased arterial tension. To omit the night-draught of chloral, and to take physostigma and perchloride of iron. Apr. 6th. Was occasionally restless, or showing emotional distress. On the 4th, he had dashed his fist through a pane of glass. On May 30th he tripped, fell, and fractured the right fibula. The notes of June refer to his extravagant delusions. July 24th. Refusing food. He said, " I can't open my mouth," " I'm nearly dead," " I can't eat," " Nothing comes from me." Then he wept. The pulse was small, compressible, 74, but two minutes afterwards it was 104 per minute. Heart sounds ordinary. Hands cold. Tongue flabby, breath foul. By the time these notes were made the patient added : " I can't pass water," " I'm as mad as a March hare." Ordered an aperient draught, which he swallowed after much coaxing and reassurance, but, immediately, he bellowed out that " he was dead or nearly so," " had nothing in him to pass,'' and " was a duffer." The bowels were purged freely, and on the next day (25th) he was cheerful, was reading, and had exalted delusions, saying that "he had £1,100,000 a minute, and constant telegrams from his heavenly father." The pulse was now 102, soft, the head cool, tongue clean, breath sweet, and pupils as usual. July 26th. Was again depressed and refusing food, but was silent During Aug. and Sept. he frequently expressed hypochondriacal delusions like the above, and was at times depressed and weeping. Occasionally, he gave expression to expansive delusions, yet without any indication of pleasure or elation in his tone or countenance. The pulse, frequent, soft, smallish, latterly became of normal frequency, and, later, slightly too frequent; the arterial tension was diminished. On Aug. 16th the physostigma had been omitted, the iron being continued. Oct. 1st. He was now less hypochondriacal, and was stouter and stronger. The face was of a dull reddish sallow hue. Epistaxis occurred spontaneously. The motor indications of g.p. in the speech, tongue, lips, etc., persisted.

Oct. 13th. He was sitting down at 4 P.M., talking quietly to an attendant, when suddenly he uttered a cry, stiffened, and passed into violent convulsions. Immediately I visited him and found him in the *status epilepticus*. Each fresh convulsion that supervened on the continuous violent clonic spasms, began with stiffening and straightening of the right upper limb, and violent spasm drawing the head and eyes to the right, but quickly making way for general clonic convulsions, the face being turgid, and profuse perspiration coming on. When these general clonic convulsions subsided, there were sinistral unilateral spasms, the head being drawn to the left, but the eyes upwards and to the right, while the right upper limb was paralysed. The pulse and respiration were rapid. The convulsions lessened in two hours, but recurred from time to time, beginning with tonic spasm, mainly of right side, and then becoming general. Pupils sluggish, of medium size. The respiration rose to 52, and again fell below 30. In the evening, and all night, were spasms, mainly of the left limbs and of both sides of the face, especially the right. The bowels were cleared out by enema of magnes. sulph., then several enemata of chloral hydrate were given,—also milk, brandy, potassic bromide, and inhalations of amyl nitrite.

Oct. 14th. He still had twitches of the arms; also, slightly, of the legs, especially the right, and well-marked twitches of the face on both sides. The mouth was jerked somewhat to the left, the head to the left, but the eyeballs upwards and to the right, and there were frequent spasmodic jerks of the upper limbs, chiefly of the left, and of the thighs. Reflex movements were more impaired in the right lower limb; reaction to painful sensory impressions failed in both legs; paralysis affected all the limbs, particularly the left. Respiration 27, stertorous. More or less coma throughout. At 1 P.M., temp., 99·5°: resp. 30, laboured. Had persistent twitches of the face, arms, and legs, all especially of the right. Had had five protracted severe, chiefly dextral, convulsions this morning. Then, in a fit, which came on whilst this note was being made, the head and eyes turned to the right, and after it was over they turned to the left again—their previous position. Some abdominal tympanitis; lungs congested; bowels freely moved; catheterisation necessary. Later in the day, convulsions continued. Intermediately were twitches of face, arm, and trunk, the last-named being, throughout, much affected so. At midnight, pulse 125, respiration 45. The fits then taking place began with spasm of the right side of the face and of the right limbs, and became general. Both arms paralysed. During the night were fifty or sixty epileptiform seizures, which continued until death at 8 A.M. on Oct. 15th, 1876.

Necropsy, thirty hours after death.—Body well nourished. Calvaria congested, thin, soft; sinuses gorged with dark fluid blood and clot. Dura congested, and slightly thickened. Pia and arachnoid much congested, thickened, unduly tough, and slightly milky, over the whole of the superior and external surfaces of the brain, except the back parts of the occipital; the temporo-sphenoidal moninges suffering much less; those over the anterior ⅔ of the cerebral internal surfaces, slightly. Pia slightly infiltrated with serum. The membranes separated difficultly. Adhesion and decortication were more marked in the right cerebral hemisphere than in the left, and in the right were chiefly over the lower half of the parietal lobe, the lateral surface of the temporo-sphenoidal lobe, the lower part of the right ascending frontal gyrus, the right third frontal; and, to a less marked degree and extent, over the anterior and posterior ends of the middle frontal g., leaving an intermediate oasis unaffected, and over the whole of the superior surface of the first frontal; were observed slightly on the orbital surface, and on the internal surface, but the occipital lobe escaped, with the exception of the anterior part of the third occipital gyrus. Thus, a large band on the posterior two-thirds of the vertical portion of the right hemisphere escaped this change, the breadth of this oasis varied from two to three inches, it was limited anteriorly by the first frontal gyrus. The adhesive change was much less advanced in the left hemisphere, but of somewhat similar distribution, except that the superior surface of the first and second frontal gyri escaped, the orbital surface, however, being as much affected here as on the right side. The temporo-sphenoidal lobe was equally or more affected on this side, and the third frontal gyrus, considerably.

Numerous puncta-cruenta in the cerebral grey cortex, which had a faint lilac hue, and was atrophied in front. In the right first frontal gyrus, examined in the fresh state with the microscope, a few of the large pyramidal nerve-cells showed some granular degeneration, but others were healthy. Numerous nucleated cells, and leucocytes were seen, and one deposit of blood-pigment.

The whole brain was of diminished consistence. The white matter had numerous puncta-cruenta, and was pasty. Fornix and grey commissure, soft. Considerable fluid in the lateral ventricles, their lining membrane thickened and granulated. Lilac-coloured opto-striate bodies. Hyperæmic median half of left corpus striatum. Left hemisphere, 23½ozs.; right, 21½ozs. Hyperæmia, slight toughness and opacity of membranes of cerebellum, its cortex slightly hyperæmic, especially on the right of the median line. Beneath its inferior vermiform process, a slight pial hæmorrhage. Weight, 5½ozs. Pons and med. obl., hyperæmic. Weight ¾oz. Ependyma of fourth ventricle slightly granular and thickened. Fluid from cranial cavity, fl. oz., 1¼.

Pericardial fluid, 1oz.; *Heart*, 11½ozs.; its muscular substance, purplish and yellowish. Coronary arteries, extremely atheromatous; aortic arch slightly so. *Right lung*, 31½ozs. *Left lung*, 26½ozs.; posterior congestion of both, and slight hypostatic pneumonia of right. Anterior emphysema. *Stomach*, distended by flatus. *Spleen*, 2⅔ozs., firm, and rather pale. *Left Kidney*, 4¾ozs., its capsule slightly adherent. *Right Kidney*, 4¾ozs. *Liver*, 57⅓ozs.

Remarks.—1. Here were much more adhesion and decortication in the right than in the left hemisphere, and the right hemisphere was 2ozs. below the left in weight. A wide vertical band, comprising part of the superior aspect of the right hemisphere, was free from adhesive change, except over the first frontal g. This change, less in degree and extent in the left hemisphere, had a somewhat similar distribution to that in the right.

2. The unusual distribution of the adhesive change, and of the cortical changes associated therewith, might be thought to have had relation either to:—(*a*) The comparative absence of the motor signs of g.p. in the early stages. But I take it that the early absence of these signs was due to the "masking" effect of the extreme maniacal furor then present. Or (*b*) one might try to trace some relation between the points of the cortex specially irritated by the adhesive change, and the distribution of the convulsions which cut short life. This distribution was mostly on the right side during the epileptiform convulsions, but more on the left side during the intermediate spasmodic twitches. If we transfer the results of experiments on the brains of monkeys to the elucidation of these convulsions, and spasms, we are constrained to say that, *on this basis,* no decisive relationship can be proved, in this case, between the course and site of the convulsions and spasms, and the distribution of the adhesive change.

CASE 66.—*Insidious in onset, patient eccentric and peculiar for a long time before decided mental disease was recognised. Exalted delusions, maniacal ex-*

citement, destructiveness, filthy habits. Later on, usually joyous, expressing the most exalted, changeable and self-contradictory delusions, and the most benevolent and generous intentions. Later still, the exaltation alternated with an ever-increasing and supplanting condition, in which the patient was depressed, hypochondriacal, morose and querulous. The language became abusive, threatening, and fouler than ever. Finally, dementia had advanced apace, the mind was tinctured by hypochondriacal feeling, and the patient, usually silent, was of dolorous, woebegone appearance. Motor signs, ordinary; finally much dysphagia and speechlessness. Apoplectiform attack. Severe recurring epileptiform convulsions, followed by left hemiplegia, deviation of head and eyes to right, higher left temperature; latterly, low temperature. Finally, frequently recurring convulsions, an apoplectiform state, left hemiplegia. On right cerebral hemisphere, very recent meningeal hæmorrhage. No adhesion to cortex. Right hemisphere, 1¾ozs. the less in weight: and grey cortex softer in the frontal region in this than in the left hemisphere.

G. S., Private, 52nd Regiment. Admitted Dec. 9th, 1874, aged 35 years: service 16$\frac{11}{12}$ years, single.

History.—This was said to be the first attack of mental disease, of uncertain causation, and previously treated at Gibraltar and Netley. The patient was said to have had an attack of "epilepsy" in June, 1866, but none since. He was of intemperate habits. Mental symptoms became marked at Gibraltar in August, 1874, whilst he was undergoing imprisonment, but he had been thought eccentric and peculiar for some time before. Whilst at Netley for five weeks he expressed exalted delusions as to wealth and power, and was filthy and destructive in his habits. The medical certificates spoke of occasional excitement, general incoherence in speaking or writing, exalted notions as to wealth and power, dirty habits, tendency to destroy clothing, and "paralysis of the insane." They specified delusions such as that he could speak seven languages, was fitted to be a military railway engineer, and could get £500 by writing an order for it.

On admission.—Weight, 167 lbs., fairly built and nourished, depression at vertex. Wide area of cardiac percussion dulness, forcible impulse, and intermittent irregular action; pulse small, weak, intermittent, 72. Other viscera healthy. Very slight irregularity of shins, and cicatrices on the legs. The patient said he had had syphilis, but there was no history of it, and no proof of its existence then. The condition of speech, and the tremulous twitch of lips and tongue were those usual to g.p. in the early stage. His conversation was rambling, and trifling, and he gave the most varied and contradictory accounts of his life, or wealth, and often expressed the most benevolent, generous designs. At times he was much excited, at others, in good-humour, quite joyous, and laughing heartily at trifles. He thought he had unlimited wealth, saying "I am possessed of millions of millions." He gave orders for immense quantities of various articles, and directed that the following should be sent to him immediately, "25¼lbs. of tobacco, half a dozen of Eau de Cologne, four concertinas, a paper shirt and a paper cravat, 60½ dozen of pocket-handkerchiefs, a field marshal's uniform and bâton, 1,009 boxes of hams, 26,000lbs. of currants, a stage, and a carpenter." To pay for all this, he said "Draw on Cox & Co. to amount of £150,000, or more if necessary." He also stated that God sent him all the money, that he had large estates, and kept hunters, one of which had won "the Derby." He said, "Before I went into the army I was a physician to Guy's Hospital with a salary of £10,000 a year." Again, "The colonel, major, and doctor all conspired against me, and I will get them all hanged." "My father made all the clothing for the army, my mother was a lady in her own right and took in washing." Then he contradicted his previous statement, saying: "Before I 'listed' I was an undertaker, kept a donkey, washed clothes, pawned, and washed them again, for a living." "In 1800 I was sixteen years old, went to America as a farmer, and sailed round the world." R. Potassii iodid. and ammon. carb.

March 20th, 1875. Delusions much the same. Omit the above mixture, and take perchloride of iron. May 6: Night draughts of potassic bromide. By

June he was quiet, and his delusions were partially in abeyance. Now and then he declined food. Palpitation and dyspnœa on exertion, pulse 90, intermitting 1 in 20. The manual grasping power was lessened. Night draught omitted; mixture of potassic iodide resumed, with the addition of hydrarg. perchlor. Dec. 9th, 1875. He sometimes destroyed clothing, and blamed other patients for it. He had now for some time been querulous, morose, grumbling, sullen, profane. His favourite saying, that "he was happy," was rarely heard now. Yet he said he was "a general and had ten shillings a day as pay." His memory was bad, speech somewhat hesitating and tremulous, and tongue not kept protruded well. The slight cardiac bruit had disappeared. Omit the mist. K.I., &c., continue the mist. Ferri. Weight 176lbs.; gain since admission, 9lbs. Jan. 3rd, 1876. On the preceding day the mouth had suddenly been drawn up on the right side, and the patient became heavy and stupid. On this day there was moderate left hemiplegia, and semi-coma, but the head and eyes were turned to the left (*sic*); later, however, they were turned to the right, and the pupils were dilated. At midnight the right limbs were chilly, and the left were recovering motor power. Jan. 4th. Same state; head and eyes to right. Right foot cool; the left warm. Temp., right axilla 100·2°, left, 101·2°. Jan. 5th. Occasional convulsions. Marked left hemiplegia persisted. No albuminuria; heart intermittent. Jan. 8th. Convulsive seizures, both day and night, principally affecting left side of face. Afterwards, head, eyes, and face, were drawn to the right, and left hemiplegia was marked, and of the common form. Right hand cold, left warm. Jan. 9th. Left hemiplegia marked; 4 during, but no fits since, the night. Temp., right axilla, 95·8°. The thermometer had risen to 96·6° in the left axilla, when the first of a series of convulsive seizures came on. The description of one will suffice: 1st stage, (*a*) closure of left eyelid, by orbicular muscle, with fibrillar tremor; (*b*) to this were added clonic movements of the mouth and of the cheeks, especially the left, the eyeballs being turned upward and outward to the left:—2nd stage, slow turning of head from right to left, spasm of left face increased, and that of left eyelid relaxed, tongue actively convulsed, but not protruded, muscles of neck became involved;—3rd stage, extension, rigidity, and rapid fibrillar quiverings of the muscles affected. Left arm now rigid and raised, wrist and elbow somewhat flexed; occiput boring backwards, mouth widely open, chin to left, left leg rigid;—4th stage, momentary rigidity of right limbs, and general clonic convulsions affecting right more than left arm. Then a condition as in the second stage, and with it tremulous movements of left limbs. A few minutes after, temp. left axilla, 95·5°. Jan. 10th. No convulsions since. T. right axilla, 96·2°, left, 96·5°. Jan. 11th. Only slight convulsions since. Hemiplegia continued, the patient could reply. T. left axilla, 97°. He made a marvellous recovery from the condition above described, after having been supported by nutritive enemata for several days. In Feb. up and about again. March. Heart intermittent and irregular. June 25th. Confined to bed, feeble, much dysphagia, and often an inability to speak. His language was often abusive, foul, and disgusting. Drowsy and stupid, he waved his hands about. Urine occasionally retained, and evacuations passed involuntarily. July 11th. Feeble pulse, faint systolic murmur at right of cardiac apex; cool skin, foul breath, refusal of food. Patient was dull, unable to speak, looked dolorous, often stared vacantly, or pointed as if to imaginary objects. Recently had had left othæmatoma, and had been up and about again. He became more depressed, hypochondriacal, morose, and querulous; dementia was also marked; the motor power improved.

Sept. 7th. During the past night the attendant noticed him breathe stertorously, and have muscular twitches, distorted features, rolling eyes, clenched fists, and flexed limbs. A convulsion at 5 A.M.; and afterwards left hemiplegia, with conjugated deviation of head and eyes to the right. The mouth was drawn to the right, and gaped there. Pulse 92, soft; respiration heavy, laboured; temp., right axilla, 102°; left, 102·5°; skin hot and dry; lungs inflamed posteriorly. Patient dull and drowsy. Sept. 8th. Had a general tonic convulsion, after which the hemiplegia and coma were more marked, and the respiration stertorous. 9th. Still, marked hemiplegia, conjugated deviation to right, stertor. Temp., right axilla, 99·8°; left, 99·6°. Patient almost unconscious. 10th. Sensation very much blunted, deglutition difficult, sensori-motor activity lessened on the left, the paralyzed, side. Head and eyes turned to the left. Coma. Enemata chloral hydrate. During the day, were twenty-four convulsions, either general, or mainly confined to the left side of the face; besides spasms of the left face. Catheterisation. 12th. Pulse 126, full and feeble; resp. 58, noisy and rattling; nineteen convulsions on this day. Enemata of chloral hydrate; potassic bromide. Marked left hemiplegia persisted. Coma deepened; death at 6 P.M., Sept. 12th, 1876.

Necropsy.—Calvaria, easily sawn through, slightly worm-eaten appearance on inner surface by sagittal suture. Dura deeply blood-stained; attached to its inner surface was a thin, soft, almost diffluent blood clot, overlying the right temporo-sphenoidal and occipital lobes, extending along the middle fossa of the skull-base, and covering the anterior surface of petrous portion of right temporal bone. Fluid blood was about the base of the brain in the arachnoid cavity, and the soft meninges covering the right hemisphere had deeply imbibed the blood hue. The clot was quite recent, the source of hæmorrhage not obvious. The dura, not unduly adherent, was of ordinary thickness and consistence. Faint milky opacity of arachnoid over the superior and external surfaces of frontal and parietal lobes, where also the conjoined pia and arachnoid were thickened and tough, and the pia was infiltrated with serum. These meningeal changes were fairly symmetrical over the two cerebral hemispheres, and the convolutions in the region just mentioned were somewhat wasted and the sulci rounded. Interlobar adhesions at the base. The membranes were stripped off readily from the cerebrum, and there were no adhesions to the cortex. The cortex was very slightly stained beneath the blood-stained parts of the meninges. It was softer in the right than in the left frontal lobe. Both hemispheres were well supplied with blood, but the right less so than the left. Brain mostly of about normal consistence. Lateral ventricles contained 3iii of serum; their lining membrane thickened and granulated. Basal ganglia, considerable vascularity, no special wasting or lesion. Left hemisphere, 19¾ozs.; right, 18ozs. Cerebellum 4¾ozs.; pons and med. obl. 1oz., slightly hyperæmic: fl.ʒiii fluid from cranium.

Heart, 10½ozs., flabby; mitral valves, coronary arteries, and aorta, atheromatous. *Lungs;* slight hypostatic pneumonia of bases. Left, 34ozs; patch of pulmonary collapse, upper lobe, trace of recent pleurisy. Nothing else calling for notice. *Spleen,* 9½ozs. *Liver,* 57ozs. *Left Kidney,* 6¼ ozs; *Right,* 6 ozs.

Remarks.—1. In this case a suspicion of syphilitic disease was roused at first, but neither during life nor after death was its presence demonstrated.

2. This is one of those infrequent cases in which indubitable g.p. occurs without any adhesions between the cerebral cortex and meninges. Yet the other chronic changes in the meninges, usual to g.p., were present.

3. In Jan., were frequently recurring convulsions of the left side, left hemiplegia, and a higher left temperature. Later, were recurring convulsions, followed by increase of left hemiplegia, and low body temperatures. Then, in Sept., the fatal attack; (*a*) a slight convulsion, (*b*) left hemiplegia, (*c*) frequent convulsive and spasmodic seizures, (*d*) increase of paralysis and of coma. With these were; (a). Some atrophy of the right cerebral hemisphere. (b). Right meningeal hæmorrhage of recent formation. The lesion leading to the former was apparently connected with the earlier left unilateral convulsion and palsy, the latter with the final lethal seizures.

4. The final speechlessness was of mental origin; no local lesion afforded any explanation of it at the necropsy.

CASE 67.—*Insidious onset; absurd exalted delusions. Later on, similar delusions, patient irritable, often obstinate or abusive, and evinced mingled exaltation and moroseness, selfishness, and jealousy. Language often foul and profane. Early blindness. Later on, morose, depressed, sullen, and, on occasion, dull, confused, and stricken with left hemiplegia, impaired sensation in left upper*

Cases.—Group IV. 449

limb, and lower left temperature. *Embolism; loss of consciousness, sudden interference with respiration and deglutition, right hemiplegia and paralysis of left third cranial nerve; diaphragmatic respiration; cold hands and feet; earthy sallow pallor. Right hemisphere 1½ozs. less in weight. Cortex of right side thinner, paler, and more uniformly of slightly increased consistence than in the left hemisphere. Adhesions over the superior and external surfaces, especially the posterior part of the frontal, and lower and anterior part of the parietal lobes, moderately on the temporo-sphenoidal, slight on the anterior two-thirds of the internal and inferior surfaces. Embolism of left middle cerebral artery, and of kidneys. Suppurative nephritis, pyelitis.*

H. L., Private 1 Bn., 7th Regiment. Admitted July 21st, 1875, aged 37, service 17$\frac{4}{11}$ years; married.

History.—First attack, insidious in onset, and probably of nearly a year's standing at the time of admission, but only under active treatment since Apr., 1875. Cause "unknown." His wife informed me that the patient had for many years been addicted to drink, especially during the six months preceding his admission into hospital, Apr., 1875. She also stated that sexually he was frigid, and mentioned particulars in support of her assertion. He had always been selfish, unkind to, and careless about, his wife and children, and was harsh and hostile to her when he was in liquor. He inherited a cold and selfish nature from his mother. No other family history procurable. No record of epilepsy, suicidal tendency, syphilis, or injury to the head. From April to July 10th he was treated at Dover; afterwards, at Netley. The medical certificates on which he was admitted here testified to his incoherence, self-satisfied complacent manner, exalted and absurd ideas of his rank, station, wealth, and physical power; such as that he was commander-in-chief of the army, or a member of the royal family, and fought at Waterloo. Also, to great impairment of sight, hesitation of speech, and g.p.

On Admission.—Height, 5ft. 6in. Weight, 133lbs., cranium small, florid complexion, pupils wide, especially the left, both sluggish, and of irregular shape; the fundus of the eye looked hazy, sight was impaired, the right eye apparently the weaker, but he would not allow ophthalmoscopic examination at that time. The tongue was protruded tremulously, and there were then very marked tremors of the lips and face. Speech was rapid as a rule, but occasionally hesitating and stammering. Gait slouching and wanting in firmness; hands and fingers tremulous and uncertain in buttoning clothing. Heart-sounds clear, second sound slightly accentuated, pulse 102. Varicose veins in legs. He denied having had syphilis, and there was no proof of its existence, past or present. Memory was impaired. He was restless, obstinate, ill-tempered, abusive, and difficult to examine, but soon became complacent and even good-humoured, when conversing on the subject of his delusions. He said that he was the commander-in-chief, that the Queen was his mother, that he went with her in a yacht to Russia, to see his sister who was married to the Czar, that he, with forty comrades, killed 10,000 Russians at the Malakoff tower, and on the same day stripped the corpses, dug a hole, buried them, and sold their clothing for £20. He also boasted of his imaginary exploits in killing bears and tigers, and stated that he was General M——, had been forty years in the army, was about to present new colours to his regiment, was a district inspector and owner of the whole of Dover. R. Potassii iodid. and physostigma.

For some time after admission he showed great impatience, and often indulged in denunciatory language when not addressed by one of his assumed titles. Now and then he threatened to do violence, but gradually became more quiet in demeanour; the delusions of pride and ambition remaining of the same nature, but varying in their expression from day to day.

Sept. 5th. The sight had gradually become worse since his admission. Liq. strychniæ m. iv—vi. to be injected beneath the skin of the mastoid or temporal regions each day. Sept. 19th. Left pupil dilated, very sluggish. Right pupil contracted, almost immobile, and of irregular shape. Sept. 29th. By the ophthalmoscope was seen a somewhat bright, white, atrophic pallor of the optic

G G

discs, with a somewhat irregular outline, and the vessels much diminished in size. Left pupil the larger, both of moderate size, irregular shape, and almost immobile. He was now nearly blind.

Nov. 2nd. The patient was heavy, confused, but able to reply; the face was flushed, the ears red and hot, the skin sallow, the temperature raised. He was in bed, but when supported upon his legs the left limb failed beneath him, and he kept pushing with his whole power towards the left. R. K. Br. and K. I., elevation of, and cold to, the head. Nov. 3rd. Still somewhat flushed. Feet, especially the left, chilly. Left hemiplegia, especially affected the arm, the sensibility of which was also diminished: the face was scarcely affected. Pulse 84, and full. Temp. right axilla, 100°; left, 99·4°. Pupils as on Sept. 29th. Better after free action of enema of Mag. Sulph. and Turpentine. Nov. 5th. The left hemiplegia was much less. He was now in his more usual mental state, recounting to himself his various exalted notions with much satisfaction, but becoming foul and abusive in language, on occasion. Nov. 6th. Right leg now the weaker. Nov. 10th. Left hemiplegia again present, but more marked in leg than arm. Urine still removed by catheter. Nov. 12th. The hemiplegia was again more marked in left upper limb, was well marked in the lower, and only slightly in the face. The body was usually bent to the left in standing, occasionally to the right. Stronger. Nov. 20th. More helpless and stupid. Pulse 124; resp. 27; temp. 100·5. Slight signs of tubercle of left lung. 21st. Temp. right axilla, 101·5°; left, 101·3°; pulse 105. Dec. 2nd. Taking iron. After this he sat up every day throughout, his skin was of a dingy, coarse, muddy appearance. He muttered fragments of his old delusions; at other times he was querulous and morose, or abusive, or lachrymose. His selfishness and jealousy were childishly prominent, and now, as always, when angry he indulged in profanity and obscenity.

Jan. 11th, 1876. He expressed delusions as to matter lodging in his head, and running down over his whole body, especially over his back, and into his left leg. Grumbled, whimpered. Jan. 14th.—Feet much swollen, ordered to bed. Shivering. Jan. 18th.—Feebleness, diarrhœa, dysphagia, drowsiness and langour; would scarcely reply. At night, the hands and feet cold; the pulse frequent, soft, feeble. 19th.—This morning he was comatose, but deglutition was fair, and respiration as usual. Suddenly, at 8 A.M., the breathing became laboured and quick, then failed, and he seemed as if dying,—then he revived. A similar occurrence took place at 9 A.M., and the respiration soon afterwards was 50. At 10 A.M., respiration 36, irregular and loud, pulse 116 and feeble, frame warm. Patient moaning and unconscious. Paralysis of right limbs and left third cranial nerve: left pupil moderately dilated, left ptosis, and strabismus, the left eyeball being turned upwards and outwards. The head turned to the right, but not fixedly so. The coma, rapidity of respiration, and dysphagia increased. Respiration, mainly diaphragmatic. Tongue, dry, brown, and caked. At 1 P.M. resp. 57, loud, moaning; pulse filiform, at times imperceptible. The palsies remained. He looked sallow, yellow, and thin, the skin was dry, rough, muddy in hue. Left pupil wider than at 10 A.M.; both pupils immobile. Temp. right axilla, 101·5°; left, 101·3°. Later, the breathing progressively rapid and loud. Persistent coma; death at 6·40 P.M. Jan. 19th, 1876.

Necropsy. 44 hours after death. Calvaria ordinary; dura slightly thickened. The left middle cerebral artery, and branches from it over the insula, were distended by an embolus half an inch in length, of a dull pale reddish colour, elastic, friable, of granular fracture. Left posterior communicating artery very small, right posterior cerebral artery smaller than its fellow, bifurcation of basilar artery displaced towards left side. The walls of the vessels at the base were of normal thickness, but those of the right middle cerebral, and termination of the right carotid, were atheromatous. Olfactory bulbs slightly wasted. Optic nerves softened and markedly atrophied. The membranes were thick, dense, and faintly opaque over the superior and external surfaces of the hemispheres, especially over the frontal lobes, where, also, the pia was infiltrated with serum; veins full.

Adhesions of pia to cortex were present on the superior and lateral surfaces, principally over the posterior part of the frontal lobes, and to a less extent over the anterior and lower portions of the left parietal lobe. Scarcely any adhesions over occipital lobe, and upper and posterior portions of the parietal; they were slight over the anterior two-thirds of the internal surface; the temporo-sphenoidal lobes were moderately affected. The distribution of adhesions was much the same on the two sides. The middle and anterior parts of the cerebral base were slightly affected. The angular gyrus, postero-parietal lobule, and occipital lobes were almost entirely free from this change.

The convolutions were somewhat wasted, and sulci wide, in the superior and external frontal-parietal regions. The cerebral grey cortex was palish, especially in the deeper layers, which were of a dull, slightly yellowish, hue and were firm, whereas the superficial layers were slightly soft. About the vertex, where its deeper layers were of ordinary consistence, the left grey cortex was less wasted than the right, and in a section made at the level of the corpus callosum the grey matter of the left hemisphere seemed to be of greater depth in all regions, and was of an uniform pale grey colour. The right cortex was paler than the left, and its deeper layers were more uniformly hardened. The white cerebral substance was considerably more vascular in the left hemisphere than in the right. Its consistence on both sides was somewhat diminished: fornix soft. The large lateral ventricles contained much serum, their lining membrane was granulated, especially that of the right. More blood in veins of Galen on left than on right side. At the middle of the ventricular aspect of left corpus striatum, closely, but not immediately, beneath its surface, was a small, softened, reddish patch, and a similar patch was seen in one of the orbital convolutions adjoining the insula. The left optic thalamus was of mottled hue, the right opto-striate bodies paler. Basal ganglia of diminished consistence. Right hemisphere, 18ozs; left, 19½ozs. Cerebellum palish, slightly soft, 5¾ozs. Pons and med. obl. the same; ¾oz. Lining membrane of fourth ventricle, granulated. Fl. ʒiii fluid from cranium.

Heart, 10¾ozs. Milky spot on heart. Some reddish granular clot, like that in the brain, was attached to the interstices of the right ventricle, and was found in the pulmonary artery, especially in its right branch. Thickening of the edge of the anterior mitral flap. Muscle of heart of a dull pale hue; slight atheroma of left coronary artery. *Left Lung*, 17¾ozs., shrunken. Old general pleuritic adhesions, stellate thickening of pleura behind; puckered cicatrices, and several encysted chalky masses at the apex, surrounded by fibroid induration, dirty-whitish granulations, and lung of a dull greenish-grey colour. *Right Lung*, no adhesions, 18½ozs. *Spleen*, 6¾ozs. A thick, irregular patch on the capsule, of cartilaginoid density. *Kidneys*, surrounded by adventitious membranes and thickened adherent capsules. Left, 6ozs. Right, 5½ozs. Both in a state of suppurative nephritis with pyelitis, and embolisms. *Liver*, 59¾ozs., pale, flabby. A little muco-purulent fluid in the *bladder*, and a small soft creamy collection in its wall, at the summit. Stomach hyperæmic.

Microscopical examination.—The softened portion of the *left corpus striatum* exhibited blood corpuscles, red and white, and granule-masses. Of the *optic nerves* the right was apparently the smaller, both were atrophied and softened. The nerve-tubes broke down under the glass, and fat globules, granule-masses, and free molecules were seen. *Heart.* Slight incipient granulo-fatty degeneration. *Liver.* Hepatic cells somewhat small, many of their angles unduly rounded. Cells highly fatty; free extra-cellular fat, also.

Remarks.—1. Amaurosis occurred at an early date, and gradually became complete. Blindness in g.p. has been attributed by some to lesion of both angular gyri. But in this case the angular gyri were amongst the parts least diseased on the convexity of the cerebrum. In other cases of g.p., also, blindness is dependent upon a lesion which begins in the optic nerves, or discs, and not upon primary disease of any part of the brain proper.

2. Apoplectiform and paralytic seizures with left hemiplegia, recurring for

months before death, had evident relationship to the greater disease of the right cerebral hemisphere.

3. On the last day of life were marked right hemiplegia and paralysis of left third cranial nerve, coming on suddenly, and associated with coma. With these were embolism of the left middle cerebral artery, embolic red softening of part of the left corpus striatum, and greater injection of the left white substance and basal ganglia than of the right. These conditions readily account for the sudden palsy of the *right* limbs and face, but how came about the palsy of the *left* third cranial nerve? The irregular distribution of the arteries at the base of the brain has been described, and, in looking for points of difference between the third cranial nerves, one could only note that the left was contained within the hook formed by the termination of the basilar artery, deflected to the left, and the commencement of the posterior cerebral artery. The left cavernous sinus contained a soft dark clot; the right, none. One was almost constrained to think, that when the embolism produced sudden local stoppage and disorder of circulation the distended and unusually-placed arteries may have compressed the left third nerve. The emboli may have been of renal origin.

CASE 68. *Attack insidious in onset, protracted in duration. Exalted delusions, complacency, expression of generous and benevolent designs. Later on, bien-être, high opinion of self, but exaltation and exalted delusions not so marked as before, and mingled with some grumbling; and the speech less impaired. Later on, the patient sometimes confused, dazed, sometimes excited and irritable, usually morose, querulous, sullen, grumbling, his language at times most foul and obscene; occasionally, delusions of annoyance, or hypochondria. Still later, some complacency mingled with, and alternated with, this last condition, finally, emotional expression was erased. Apoplectiform attacks. The motor signs of g.p., ordinary, irregularly, and slowly progressive; 4, 6, 8, and 11 months before death temporary left hemiplegia. Ten weeks before death, apoplectiform and epileptiform seizures, followed by coma, variable right hemiplegia, and Cheyne-Stokes's respiration; return of left hemiplegia, limbs, especially the left, somewhat rigid, and usually extended. Acute herpes. Right hemisphere 2⅜ozs. less in weight than left, and its grey cortex more atrophied and paler and firmer; slightly too firm in anterior regions. Right white substance decidedly indurated, especially in front. Similar, but less marked and extensive, induration in left hemisphere (grey and white). Adhesions, more marked over right hemisphere, and principally over the frontal, orbital, supra-marginal and second temp.-sphen. gyri. Meningeal changes extreme and extensive; less over cerebellum and cord. Basal ganglia alike on two sides; spinal cord softened, pale, degenerated.*

H. B., Private 4th Bn., Rifle Brigade. Admitted May 29th, 1874, aged 25 years. Service, 1$\frac{1}{4}$ years, single.

History.—This attack of mental disease, the first, insidious in onset, uncertain in duration. Previously under treatment in 1873-4 in India, and at Netley. "Predisposition and tropical climate" were the causes assigned. When he arrived at Netley, Apr. 27th, 1874, the articulation was affected as in g.p.; he was incoherent in conversation, complacent, and expressed various exalted delusions, saying that he was a "general officer," and had "enormous wealth."

On admission.—Of a broad frame, stout, muscles large but flabby, fresh complexion, pupils of natural size, susceptible, of vertically oval shape. Tongue protruded slightly to the right. Facial expression heavy. The speech was thick, and there were tremors and twitches of the face, lips, and tongue, when in action, slightly more marked on the left side of the face. Pulse 98, resp. 18. Slightly morbid signs at apex of right lung. In the groin was the scar of a bubo. He denied having had syphilis. He had extravagant and exalted delusions, saying that he was a general, a field-marshal, and had £12,000 a day. "Is as strong as a lion, and as sound as a church bell." "Can speak two Indian languages, and has a dozen pair of socks." Thus he propounded a sort of anti-climax—a descent from the magnificent to the trivial. He was self-satisfied, complacent, and expressed himself in a generous and benevolent manner.

July 22nd. Weight 175lbs., pupils as on admission, speech markedly impaired, the face and lips tremulous or twitching, especially during speech. The patient declared that he was well and strong, and had high notions about himself. He was respectful to the officials, and patronizing to the other patients. Taking perchloride of iron. Feb., 1875. He was now highly self-satisfied, and declared that he never had been insane. The memory was fair for simple matters, but the power of attention was lessened, he could not proceed easily from one subject of discussion to another, and in enforcing his views became loud and somewhat declamatory, and then one could observe the muscles of the face and lips tremble much and twitch, and there was a hesitating pause in the speech, and at times a decided stutter, with clipping of the words. But when he was perfectly calm the speech was better now than formerly. May 19th. He had been confused, and dull for two or three days. May 22nd. About the same; ears and face red and heated, pulse usually about 100 and sharp. R. aperient enemata, podophyllin, warm baths, and cold to the head. Under this he improved, but on May 24th he was again dull and heavy, wore a sour, sullen expression, would scarcely reply, and said " they are killing me." June 11th. Better since, but to-day thinner, looked dull, unhappy, and said that " he felt badly." He shook tremulously when standing, and was unsteady in gait. Ocular mucous hypersecretion. Oleum morrhuæ and liq. arsenicalis were added to the iron which he was now still taking, although for a time he had been on K.Br. and digitalis. Oct. 7th. The facial expression was heavy, dull, and sulky. He would reply to questions, but was averse from conversation. At times he now had delusions such as that he was about to be buried in the w.c.; would purposely dirty any clothing of which he disapproved, was often obstinate, refusing to parade, and would not occupy or amuse himself in any way, but grumbled and growled morosely, first about one thing, then about another. Feeble apex-beat of heart.

March 9th, 1876. Had been very peevish, morose, irritable, bad-tempered, swearing, grumbling, and using foul language, for many months; but notwithstanding this he occasionally said he "was all right," "there was nothing the matter with him," "he was as hearty as any man." Apices of lungs affected. March 28th. Acute herpes zoster over front, inner, and, partly, outer side of left thigh, with some pemphigus blebs. The herpes extended upwards from the groin, trending outwards above the crest of the ilium to the sacral and lower lumbar region. An isolated patch of it over inner side of head of tibia. No complaint of pain. This eruption left cicatrices. May 23rd. An attack of left hemiplegia occurred without convulsion or loss of consciousness, and disappeared in about two weeks. In July, his general health was improved, and he was taking a large amount of exercise; was often restless, noisy and irritable, abusive and foul in language.

Sept. 23rd. A seizure, during the night, of left hemiplegia, which became less decided in a few hours. Right pupil somewhat dilated, left slightly small, both sluggish; conjunctival reflex much less in left than right eye; and the sight of the left eye was bad. He wept, chattered to himself, and expressed delusions of injury. 24th. Attack over. Nov. 10th. At 4 A.M., his night attendant found that he had left hemiplegia again, but this diminished during the day, the paralysed thoracic limb being flexed, the pelvic limb extended, and both rigidly resistive to passive motion. Temp., right axilla, 97·8°; left, 99°. No loss of consciousness as far as known. Tongue protruded slightly to the right; no very decided facial paralysis. Ears injected, face of usual hue. Nov. 11th. He was regaining power in the left limbs, but paresis was detectable in the face. Temp., right axilla, 97·2°; left, 97·9°. He was childish and very amnesic.

Jan. 5th, 1877. Had been exercising regularly for 1½ months, was delighted with his power to do so, and repeated for hours together his favourite phrase, "I'm fit to walk about." He talked good-humouredly as a rule, but occasionally had been noisy and obstinate. This morning he fell down suddenly, and was found to be drawn to the left side, and to be stricken with left hemiplegia.

Soon this palsy diminished. Temp., right axilla, 97·3°; left, 97·8°. Jan. 11th. Able to walk, but still had a trace of the left hemiplegia.

March 16th. Up and about since last note until this day, when, after being for several days flushed in the face, he suddenly sank down from his seat and became unconscious for a short time, after which he was quiet, looked dazed, could not reply, was unable to swallow, but vigorously resisted any passive motion. Then, at 11 P.M., he had three epileptiform seizures, beginning in the right hand which, also, was much shaken by movements executed at the shoulder and elbow joints, and the fingers, though shaken by the spasm, were held rigidly extended. The convulsions were followed by *right* hemiplegia, and insensitive conjunctivæ, especially the right. After an enema of chloral hydrate, he fell into a heavy sleep, and the convulsions did not return on this day. March 17th. Right hemiplegia now only slight. Pulse 114, full, soft, bounding. Respiration 25, of the "modified Cheyne-Stokes" character, the cycle being short, the period of apnœa occupying about one-third of each cycle. Later on, after coughing, the rhythm became modified, and there were simply ascending and descending periods of greater and less frequency, depth, and loudness of respiration. Temp. 101·2°. Face, flushed; right conjunctiva again insensitive; the left sensitive. Pneumonia patches. Nutritive enemata. March 19th. Head boring over to the left side, eyes turning thitherward. Upper limbs flexed. Had had a general convulsion, beginning at the mouth, after which he was flushed. Pupils equal, of medium size, sluggish, right conjunctiva now again almost insensitive. Right upper extremity flaccid, the left rigid; conjugated deviation of head and eyes to left. March 20th. Swallowing a little food for the first time since the 15th (save a little on 18th), having been supported by nutritive enemata *interim*. A convulsion occurred again last night. Pulse 60. Temp., right axilla, 97·4°. Conjunctivæ now equally sensitive. Upper limbs, especially left, rigidly flexed. (Was having ol. morrh., ferri perchlor., vin. rubr., K. Br.)

April 8th. Bedridden of late, quiet, replied to simple questions. Face usually flushed. Slight left hemiplegia, left upper limb rigidly flexed, and pronated. No convulsions recently. May 5th. Bedridden since, dysphagic, often grinding the teeth, and having marked flushings of face and head. Sinistral hemiplegia, and left arm rigidly extended. The patient, of "foul" habits, indulged in obscene denunciation of imagined personal persecutions. May 8th. In much the same condition. Gangrenous fœtor of the breath; signs of hypostatic pneumonia; and pneumonia patches at apex of right lung in front. ℞. Terebinth. May 12. Bleb on right thenar eminence. May 15th. The palsy, more marked in the left limbs, was now well marked in the right also. An acute bedsore had appeared over the right first metatarso-phalangeal articulation: ordinary bedsores were forming also. Temp., right axilla, 100·6°; pulse, 98. May 17th. Left limbs thoroughly palsied, slightly rigid; the right less palsied. Slight left ptosis. Summary: after this the condition remained much the same, the pulse and resp. becoming very rapid on the last day of life. The mental life was much effaced. He changed colour often and suddenly on the day of death, May 25th, 1877.

Necropsy, 24 hours after death. Moderate emaciation. Firm, pale, clot occupying the first half inch of the left middle cerebral artery, not occluding it, but apparently formed before death. Olfactory bulbs wasted; considerable interlobar adhesions, slight thickening and opacity of membranes over base and internal surfaces of hemispheres. The pia and arachnoid over the superior and external surfaces of cerebrum, pale, thick, and somewhat opaque, especially in the frontal and parietal regions, where the membranes were infiltrated with serosity.

Adhesion and decortication extensive, and somewhat more marked over right than left hemisphere. On the *right* hemisphere, extremely well-marked over the greater part of the orbital surface, especially the rectus; over the middle and front of the first frontal g., but absent at its tip; over isolated spots at the same level on the second frontal g.; on the anterior part of the third frontal, avoiding the tip; over the supra-marginal and second temporo-sphenoidal gyri. Pretty well, but very much less, marked over the temporo-sphenoidal lobe generally, the rest of the parietal lobe not already mentioned, and the lower and anterior part of the occipital. Slight on the upper and internal surfaces adjoining the longitudinal fissure, and more so in this than in the left hemisphere.

In the *left* hemisphere the adhesive change was best marked on the central portions of the orbital surface, over the anterior part of the first and second frontal gyri, but avoiding the tip, and over the posterior part of the third frontal, the rest of which was intact. Much less, over the posterior part of left second frontal, and middle of ascending frontal; and affecting the upper end of ascending parietal, the postero-parietal lobule, the angular, the supra-marginal and 2nd temporo-sphenoidal very slightly; while the rest of the temporo-sphenoidal lobe, and the whole of the occipital lobe, were intact in the left,—in this respect differing from the right. The gyri were slightly wasted and pale, though in parts exhibiting a few dilated vessels. The grey cortex was somewhat thin; its consistence was slightly increased, especially in the fronto-parietal region; it was slightly paler, and thinner, in the right than in the left hemisphere. The white substance, especially adjoining the frontal cortex, was somewhat indurated, and was pale. In the left hemisphere the induration of the grey and white diminished backwards, consistence becoming normal in the posterior third of the brain. In the right hemisphere the induration, more marked, was at its acmé in the white substance of the frontal lobe, had a slightly cribriform appearance, and a pure brilliant white colour. Thence the induration diminished backwards, into the occipital lobe. The veins of Galen were fuller on the right than left side. Ependyma v. thickened and opaque, ventricles enlarged, and containing considerable serosity. The two corpora striata were almost alike. Basal ganglia not plump. Right cerebral hemisphere, 15½ozs.; left, 17¼ozs.—difference, 2⅜ozs. Pons Var. and med. obl. palish, and of slightly increased consistence; 1¼oz. Lining membrane of 4th ventricle, thickened and opaque. Cerebellum. Meninges slightly opaque; grey matter palish; consistence fair; weight, 5¾ozs. Serosity, 5½ fl. ozs. The spinal cord was slightly pale and soft. On microscopical examination, hæmatoidin pigment masses were seen on the surface of the cord, by the meninges; and colloid bodies —among other changes—were widely scattered throughout its substance.

Heart, 9½ozs., its muscle pale, friable, valves healthy, a few patches of atheroma in coronary arteries, and aortic arch. *Left lung*, 37½ozs. *Right*, 39¼ozs.; lobular and hypostatic lobar pneumonia and semi-gangrenous patches in both lungs. Some fibrosis of right apex. *Right Kidney*, 7¼ozs.; *left*, 6⅓ozs. *Spleen*, 4¼ozs. *Liver*, 50ozs., of pale pinkish hue, capsule thickened and adherent on upper surface.

Remarks.—1. Connected with the much more advanced disease of the right hemisphere was the left hemiplegia which recurred on several occasions,—finally returning in an extreme degree before death, and accompanied by rigidity of the left limbs. But that will not explain the apoplectiform seizures, dextral epileptiform convulsions, coma, *right* hemiplegia, with some anæsthesia, and Cheyne-Stokes's respiration, ten weeks before death.

2. Acute herpes zoster affected the left thigh, buttock, and lumbo-sacral region fourteen months before death; latterly were acute blebs on the right hand and foot, contemporaneous with rigidity of the extended left limbs. These conditions are to be connected with lesions, especially affecting the left side of the spinal cord, and spinal nerves.

GROUP V. (*see* p. 411).

So far have the intended limits of space been overpassed that the heads only of the remaining cases will be given, and some of the appended remarks.

CASE 69.—*Excitement, discursive conversation, exalted delusions, emotional exaltation and joy, alternating with anger and hostility; hallucinations of hearing; patient quarrelsome, reviling, and foul-mouthed. Later on, exalted delusions, with much maniacal restless violence, insomnia, and a boastful cynical disposition. Language more and more profane, obscene, and abusive, and habits* wet and dirty. Frequent attacks of stupor or semi-stupor, in which he resisted passive motion, but swallowed well, sometimes with these were spasmodic tremors of head, trunk, and limbs, and, at times, rigidity of the arms. Intermediately, very noisy and restless at night. Frequently recurring, epileptiform convulsions, followed, in the later stages, by unconsciousness, and afterwards by a character of speech and of mentation often observed in epileptic mania. These convulsions, occurring singly or in sets, latterly were very severe, frequent, and associated with marked stupor and local spasms of trunk and left limbs. Finally, rigid flexed contractions of all the limbs, and continuance of epileptiform seizures. Throughout; impairment of speech well-marked, an unusual amount of stuttering, very great motor restlessness, marked tremor. In right frontal lobe, the grey cortex of the superior and external surfaces markedly indurated; atrophied, red, hypervascular. This condition was mainly limited to this lobe, the rest of the right grey cortex being only slightly firm, and far more normal in appearance, as was, also, the whole of the grey cortex of the left hemisphere. White substance, injected in right frontal lobe, pale in rest of right hemisphere, of ordinary vascularity in left; and universally of equal and slightly increased consistence. Adhesion and decortication considerable, especially over parietal lobes, also well marked over frontal. Lateral ventricles, especially right, large; their ependyma firm. Right basal ganglia pale, right corpus striatum somewhat shrunken. Pons and med. obl. somewhat firm, their meninges thickened. Upper part of spinal cord somewhat softened, left posterior cornu atrophied in dorsal region.

Details as to the meningo-cerebral adhesions. In the *right* hemisphere, besides extensive adhesions over the other parietal gyri, were a few adhesions on the prominences of the ascending convolutions; also on the posterior part of the inner surface of the first frontal. The third frontal was affected. The external surface of the temporo-sphenoidal suffered considerably; the base of the temporo-sphenoidal and of the frontal lobes, slightly. In the *left* hemisphere, the anterior two-thirds of the parietal gyri suffered much more, the superior surface of the frontal gyri suffered a little more, and the external surface of the temporo-sphenoidal considerably less, than the corresponding parts on the right side. The left third frontal g. was quite free. The following were the parts most affected by this change, in the order of severity,—left parietal; right parietal; right temporo-sphenoidal; left temporo-sphenoidal; left frontal; and right frontal; lobes.

Microscopical examination.—*Fresh tissues.* Grey cortex of right first frontal convolution. Much neuroglia; hyperplasia of the neurogliar nuclei and of the vascular, with hypertrophy, also, of the latter. Some vessel-walls irregular and rather thick. Many small round and oval cells. Pyramidal nerve-cells small. In the left first frontal convolution the nerve-cells and vessels were more normal. In the fresh spinal cord the multiplication of nuclei and condition of cells were much the same. *Prepared sections.*—In the right first frontal gyrus these did not take the carmine stain well, and there were unstained patches of a ground-glass appearance. The walls of some vessels thick. The nerve-cells were atrophied, some were granular, some opaque, some disintegrating. Round and oval nucleated cells; scattered dark pigment grains. The left first frontal gyrus stained better than the right; its changes were similar, but less. The right third frontal more nearly approached the condition of the right first frontal. A number of unstained patches were seen here, also. The left third frontal gyrus resembled the right, except that the nerve-cells were more granular. The spinal cord did not stain very well; the walls of some of its vessels were thickened and their nuclei hyperplastic; there was increase of the connective tissue; some granule-cells; also numerous small round and oval cells.

Remarks.—1. It is probable that the convulsions were in relation with the change involving atrophy and induration of the right frontal cortex. Whatever the relation of the morbid process causing cerebro-meningeal adhesion was to the condition of brain conducing to convulsion, it had no obvious bearing upon the localization of the convulsions and spasms observed.

2. The left third frontal gyrus was free from adhesion; the right was affected thereby; its cortex sclerosed. The speech was much affected; the stuttering severe.

3. The mentation at times assumed, in part, the characteristics of that often associated with epilepsy. The special type of noisiness, following the attacks of stupor with spasms, was of this nature; in several points there was a coincidence with epileptic delirium.

4. The question of syphilis arose in this case, and the conditions of the basilar artery and of the spleen might readily be taken as traces of ancient syphilis; both, however, were slight and equivocal. The attacks of stupor were like those sometimes occurring with syphilitic disease of the cerebral arteries. Microscopically, the small arteries were noted as being thick-walled.

5. Coincidently, were atrophy of left posterior grey cornu of dorsal region of spinal cord, and long disuse of left leg following upon a rare injury:—*transverse fracture of the patella from direct violence.*

CASE 70.—*Failure, and loss, of mental powers. No connected conversation. Smiling, contented, obedient, docile, quiet. Later, some restlessness and resistance to manipulation, habits wet and dirty. At an early period, insensitive to discomfort; later, insensitive even to a painful operation; quite analgesic, apparently. Occasionally, heavy, drowsy states independently of the occurrence of convulsions. Impaired speech and co-ordinating power. Right hemiparesis. Right— and then left—and again right hemispasm. Increase of paresis, especially in the right limbs. Twitches of the right upper limb, and choreiform movements supervening. Later, much right, and some occasional partial left, hemispasm. Right upper limb occasionally rigid, and then less resistive to passive motion than left. Occasional attacks of right hemiplegia. Drowsiness, coma, slow and feeble pulse, low temperature, death. Slight induration of cerebral grey cortex in superior and external fronto-parietal region, gradually lessening below and behind these parts; also atrophy of it, especially in the frontal lobes, diminishing thence to tips of occipital lobes. Somewhat lessened vascularity of white substance, and meninges. Adhesion and decortication comparatively slight, slightly more in right than in left hemisphere, and mainly seen at the summit of the fissures of Rolando. Meningeal changes highly marked and extensive, especially over parietal lobes. White substance slightly too firm; lateral ventricles large, ependyma thickened, as also in fourth ventricle. Basal ganglia somewhat pale and shrunken; pons and medulla oblongata slightly too firm.*

On the *left* side, the cerebro-meningeal adhesions occurred at the summits of the gyri where the first frontal, and the ascending gyri converge, by the longitudinal fissure. On the *right* side, besides the ones corresponding to these, were a few adhesions affecting the posterior half of upper surface of the first frontal gyrus, the posterior half-inch of the second frontal, the upper fourth of the ascending parietal, and part of the postero-parietal lobule

Remarks.—1. As to the history of syphilis, we note here the absence of lesions distinctly syphilitic, after death, and the absence of syphilitic symptoms during the time the patient was under my care.

2. In a case presenting so many motor symptoms it was of interest to find the adhesion and decortication within the so-called "cortical motor zone." Attributing the convulsive and spasmodic symptoms to the adhesive lesion, which was mainly about the summits of the fissures of Rolando, the facts would support the localization views of Carville and Duret—to a less extent those of Charcot and Pitres. But that the spasm and convulsion predominated on the right side of the body, was not explained by this hypothesis.

CASE 71.—*Hypochondriacal and melancholic delusions, at first with gloomy depression, later with garrulous incoherence. Patient moaning and lachrymose; later, repeating, parrot-like, the words used by others, exhibiting fear, apprehension, anxiety, and attempting to commit suicide; but subsequently becoming far less distressed. Early hallucinations of hearing, and, later, of sight. Finally, very incoherent and demented. The ordinary motor signs developed* very late, long after decided mental disease. Finally helpless, bed-rid, of very wet and dirty habits; the limbs flexed and rigid. Stationary paresis of right 7th and 9th cranial nerves. Left othæmatoma. Left acute bedsore. Some induration of cerebral grey cortex in superior and external fronto-parietal regions, gradually shading off backwards and downwards, to normal consistence at tip of occipital lobe, and at base. Grey cortex pale, but streaked by visible vessels. White substance of slightly increased consistence, and pale, especially in front. Meningeal changes ordinary, with considerable œdema. No adhesions of meninges to brain. Lateral ventricles large, fornix firm; spinal cord slightly softened.

Microscopical Examination.—First and second right frontal convolutions near the tip. Many small round or oval cells containing several dark molecules. The nuclei of the vessel-walls somewhat increased in number, and an increased number of "Deiter's cells." Slight atrophy and degeneration of some of the nerve-cells, several of which were partly surrounded by vacuoles. Upper part of ascending gyri. Some nerve-cells of somewhat rounded outlines. Increased number and size of "Deiter's cells." Spinal cord, less changed than was anticipated. Some multipolar cells degenerating.

Remarks.—1. The physical signs of g.p. came on in a patient already more than a year and a half insane. Notwithstanding the statements of the earlier writers on the subject, I deem this to be of rare occurrence, and dependent upon the special primary localization of the morbid process in the cerebrum.

2. In this case we note, also, the *absence* of any adhesion of the meninges to the cerebral surface; of any marked changes in the dura; and of any encephalic softening. And on the other hand, the *presence* of slight diffused hardening of the grey cortex in the supero-external fronto-parietal regions.

3. There was an absence of unequivocal syphilitic lesions after death, unless, indeed, the incipient aortic endoarteritis be taken as such.

4. Acute bedsore appeared on the left buttock forty-five days before death, the position and course of which corresponded to those of bedsores sometimes following lesions of the right cerebral hemisphere; yet there was no circumscribed lesion of that hemisphere, and the morbid appearances were symmetrical in the two hemispheres.

INDEX.

A.

Abbreviations used, 1.
Abdomen, changes in, in g.p., 296, 142.
Abducens nerve in g.p., 106.
Acute ascending paralysis, 242.
Acute general paralysis, diagnosis of, 245; lesions in, 279.
Addison, A., Mr., on urine, 197.
Adhesion and decortication in g.p., 282 et seq.; table of, 284; 309, 340, 346, 363; present writer on, 364, 406.
Adler, Dr., on lesions in g.p., 302.
Æsthetic feelings in g.p., 12, 33.
Æther-spray to head, 399.
Affections in g.p., 12, 27, 28, 33.
Age and g.p., 248 et seq.
Alcoholic excess, 265. Alcoholic mental disease, 218. Alcohol in diet of G.Ps., 392.
Aldridge, Dr. C., on eyes in g.p., 118.
Allbutt, Dr. C., „ „ „ 117.
Amadei, 278
Amaurosis and amblyopia in g.p., 112.
Ambitious delirium (see Expansive).
Ammonium bromide, in g.p., 397, 399, 401.
Ammon's horn in g.p., 316.
Amnesia in g.p., 29, 31, 33 (see Dementia).
Amyloid bodies, 305, 310, 315, 320.
Amyloid degeneration, 303.
Amyl nitrite, 401, 404.
Anæsthesia in g.p., 8, 22, 28, 128, 374-5.
Analgesia in g.p., 128.
Anatomy, pathological of g.p., 278 et seq., 298 et seq.
Anosmia in g.p., 21, 120.
Antim. tart., ung., in g.p., 394. Aorta, 294-5.
Apoplectiform seizures in g.p., 10, 26, 167 et seq., 179, 237, 363, 401.

Appetites and desires in g.p., 12, 34, 194.
Apyrexial rigor, 21, 26, 96.
Aphasia in g.p., 357 (see Diction).
Arachnoid in g.p., 280.
Arachnoid cysts, 173, 279, 337.
Arms in g.p., 89 (see also Ataxy, etc.).
Arndt, Dr., 301, 396. Arrest of g.p., 210.
Arteries, basal, in g.p., 278.
Arterioles; brain, meninges, 298, 302.
Arthropathy in g.p., 155-6.
Articulation, 19, 24, 48, 72, 77, 357-8.
Ashe, Dr. I., 153, 260, 270.
Ataxy in g.p., 10, 20, 25, 28, 72, 82, 86, 328, 354, 358.
Ataxy after acute affections, 240.
Atkins, Dr. R., on cord, 323 et seq.
Athetosis in g.p., 26, 96-7.
Atrophy of brain and cord in g.p., 281, 288, to 291, 340, 362; of cerebral nerve-cells, 312; of medulla oblongata, 286, 291, 318; of cord, 287, 291, 320.
Atropine poisoning, 244. Aubanel, 337.
Auditory hallucinations, etc., 66. 70, 21, 123.
Austin, pupils, 107; causes, 270.
Automatic movements in g.p., 93.
Axis-cylinder changes in g.p., 315.

B.

Bacon, Dr. M., on handwriting, 80.
Baillarger, 52, 57, 207, 276, 292, 329, 331, 332.
Baths in g.p., 395-6, 399, 401.
Ball, Prof. B., on heredity in g.p., 251.
Bayle, A. L. J., 5, 201, 276, 280, 282, 293, 334, 343, 345, 404, 405.
Basal ganglia, 285. Baume, 292.
Bechterew, 186.
Bedsores, 142 et seq., 393, 402-3.

Belhomme on lesions, 293.
Belittlement, delusions of, 59.
Berstens, 275. Besser, on adhesions, 309.
Biante, 154. Bickerton, on eyes, 119.
Billod, Dr., on palsy of motor oculi, 330.
Bladder in g.p., 93, 142, 145-6, 298.
Blandford, Dr., on handwriting, 80; on diagnosis, 234.
Blending of delusions, 45, 55.
Blindness, 27, 112, 377.
Blisters (treatment), 394, 399, 400-1.
Blood in g.p., 157 et seq.
Blood-vessels in g.p., 278, 295, 298 et seq., 322.
Böens, 270. Boll, on granule-cells, 321 et seq.
Bones (fractures, etc.) in g.p., 150 et seq.
Bonnefous, remission in g.p., 209.
Bonnet, on lesions, 299 et seq.; theory, 335.
Bottex, 293. Böttger, 209, 243.
Borrysiekiewicz, 120.
Boucherean, 329. Bowed head and back, 369.
Boyd, Dr. R., 288, 290, 292. Bramwell, Dr. B., 244.
Bravais on unilateral epilepsy, 367.
Brierre de Boismont, 12, 200, 337.
Broussais's generalization, 200.
Brown-Séquard, 135-6, 232, 383, 403.
Browne-Crichton, Sir, adhesions, 365; blindness, 377; physostigma, 398.
Bruise-marks in g.p., 147.
Brunet, Dr. D., on blood, 158.
Bucknill, Dr. J. C., on diagnosis, 240.
Burdon-Sanderson, Dr., 339.
Burlureaux, Dr. C., 258, 308, 330-1.
Burman, Dr. J. W., 253.

C.
Cachexia in g.p., 27, 138.
Calmeil, Dr. L. V., 49, 201, 258, 283, 293, 413-4.
Calcification of nerve-cells, 314.
Calvaria in g.p., 278. Campani, 204.
Cases of g.p., 414 et seq., and scattered.
Catheter, 403. Catalepsy in g.p, 63.
Causes of g.p., 245 et seq.
Cavalier, 266. Cautery, 394-5, 401.
Capillaries of brain and meninges in g.p., 298 et seq., 302.
Callender, on hemispheres, 383-4.
Central canal of cord, 323.
Cerebellar diseases, 230.
Cerebellum in g.p., 286-9, 316-7, 385.

Cerebral hemispheres, different functions of, 383-5, 411. Cerebritis in g.p., Chaps. 16-17-18; 342.
Cerebrum in g.p., 281 et seq., 286 to 291; present writer on, 289; 298 et seq., 325 et seq., 332.
Chambard, Dr., on tremor, 90-1.
Character and g.p., 251. Charpentier, 225.
Chapman, Dr. T. A., sex, 246; marriage, 254.
Christie, Dr. T. B., remission, 209.
Choreiform movements, 26, 97-9.
Christian, Dr. J., 154, 371.
Cheyne-Stokes resp., 191.2.
Chloral hydrate in g.p., 399, 401; and K. Br., 399, 400, 401. Chloroform 401.
Circular form of g.p., 24, 63, 239.
Circulation in g.p., 186 et seq., 337.
Clarke, Dr. L., on lesions, 301 et seq.
Classes of society and g.p., 255.
Clapham, Mr. C., brain-weights, 286-9, 290.
Claus, Dr., 69, 258, 330, 372.
Climate and g.p., 259.
Climacteric period and g.p., 276.
Cleanliness, 393, 402.
Clouston, Dr. T. S., 152, 174.
Clymer, Dr. M., 244. Cohnheim, 322.
Cold to head in g.p., 395, 401.
Cold baths in g.p., 395, 399, 401.
Coldness, local or unilateral, 133-4, 167. Colin, M., 257, 331.
Coenæsthesis in g.p., 40, 70.
Colloid bodies, 305, 310, 315 (degen. 302).
Colour-blindness, 115.
Commissioners in Lunacy, Reports, 245-6, 249, 252-3-5-6, 262-3, 271-2-6.
Comparison of causes, g.p., "all cases," 264.
"Complicated" g.p., 3.
Congestion of brain and meninges in g.p., 281-2, 298, et seq., 332, 346 et seq., 351, 363; diagnosis 237.
Congestion of head in g.p., 17, 18, 133, 363.
Congestive nature of g.p., 332-3.
Conjugated deviation, etc., 112, 369.
Conjunctival reflex, 100. Conolly, Dr., 200, 256.
Contraction of limbs, 26, 28, 92.
Convulsion (see Epil.).
Contusion and concussion of brain, 271.
Copland, Dr. J., 204.
Corpus striatum, 285, 316. C. callosum, 316.

Index. 461

Cortex cerebri in g.p., 281 *et seq.*, 298 *et seq.*, 332, 344 *et seq.*, 349 *et seq.*, 358-9, 367, 380.
Course of g.p., 3, 199 *et seq.*
Counter-irritation, 389, 393-4.
Cowan, F. M., on lesions, 308.
Cranial injuries, 259, 271.
Crimes by G.Ps., 12, 34, 36, 38-9.
Crozant, 8.
Criminal charges, 13, 34. Crusel, 403.
Crura cerebri, 316. Cranial nerves, 318.
Cures, 216, 386 *et seq.* Cupping, 393, 401.
Cutaneous hæmorrhage, 134. Cystoid degeneration, 304.

D.

Dagonet and Dally, 329. Dalton on vision, 378. Daveau, 247.
Davey, Dr. J. G., on bones in g.p., 150.
Debove, 178. Deafness in g.p., 123-4.
Death in g.p., 213, 164, 170.
Definition of g.p., 1.
Degenerative nature of g.p., 324, 333-4.
Degeneration of blood-vessels, 298.
Déjerine, 317, 323, 324.
Delasiauve, 219, 243, 275. Delaye, J. B., 204, 283, 375.
Deliria in g.p., 41 *et seq.*
Dementia in g.p., 11, 13, 22, 27, 29, 349.
Dementia (diagnosis), 239.
Descending myelitis, 288. Devouges, 274.
De Witt, 198. Diabetes, 199.
Diagnosis and diff. d., 217 *et seq.*
Diarrhœa in g.p., 27, 142.
Diction, disorders of, 75, 356-7.
Diet, 270, 391 *et seq.*, 400.
Digestion, in g.p., 194.
Digitalis, digitaline, 398-9, 401.
Dilatation of cerebral vessels, 301.
Discovery of g.p., 2. Disposition, 251.
Divisions of g.p.; and into stages ; 3, 5.
Double hemiplegia and g.p., 240.
Doutrebente, 207-9, 247, 275, 373, 388.
Dreer, Dr. F., 225. Duchek, Dr., 272, 335.
Dufour, Dr., 365.
Duration, 211. Dura mater, 278.
Durhæmatoma (see Arach.).
Duterque, 120.
Dysphagia in g.p., 25, 28, 85, 371.

E.

Eating by G.Ps., 27, 85.
Electricity, in g.p., 396. Electrical reaction, 94. Electrical sense, 131.

Emaciation, 294. Emetics, 401.
Embryonic tissue, 340 *et seq.*
Emotions in g.p., 12, 14, 22-4, 31, 40-2-4.
Enemata, 396, 401, 404.
Ependyma in g.p., 285, 293, 310, 317.
Epilepsy (diagnosis), 237; (cause), 277.
Epileptiform seizures in g.p., 10, 26, 160 *et seq.*, 180, 237, 365 *et seq.*, 401.
Ergot, 398. Erlenmeyer, 157, 293.
Esquirol, E., 160, 228, 330, 386.
Erysipelas (as cause), 276, 331.
"État criblé" in g.p., 304. Eulenberg's experiments, 136.
Exalted delusions (see Expansive).
Exciting causes, 262. Excretion, 391.
Exercise, 391 *et seq.*, 400.
Expansive symptoms in g.p., 15, 22, 27, 42, 228, 351.
Exner, 375, 378. Exophthalmos, 112.
Eyebrows and Eyelashes, 106, 112.
Eye-symptoms in g.p., 106 *et seq.*

F.

Fabre on "circular" g.p., 63.
Face, 10, 20, 25-6-8, 82-3, 133 ; nerve, 320.
Fœcal incontinence, 12, 28, 93, 370.
Falret, Dr. J., 56, 329. Fauces, 85.
Fatty degeneration, 146, 295, 310, 320, 323.
Feeding in g.p., 391 *et seq.*
Ferrier on tactile centre, 375.
Females, g.p. in, 205, 212, 247, 256.
Fevers and g.p., 276. Fluid from cranium, 291.
Fifth cranial nerve, 84, 319.
Folie congestive, 258-9.
Forms of g.p., 29, 42, 47, 49, 52, 61, 63.
Fornix, 316. Förster on pupils, 373.
Fourth cranial nerve, 106, 319.
Foville, A. *(père)*, 194, 282. Foville, A. *(fils)*, 90-1, 224,241, 329, 365.
Fox, Dr. B., 219. Frauds (see Theft).
Frerichs, 293.
Fulgurant pains, 132, 376.
Functional exaggeration, 16.
Furnace heat (as cause), 275.
Fürstner, 113, 149, 378. Furuncles, 27, 141.

G.

Galcerau, 388. Gallopain, 318.
Gait in g.p., 10, 20, 25, 28, 86 *et seq.*, 102 *et seq.*, 360.
Gambus, 265. Gangrene in g.p., 53, 57, 296.
Ganster on remissions, 210.
General feeling in g.p., 40, 70.

Generalized palsy (diagnosis), 240.
G.p. in chronic insanity, 200. G.p. without insanity, 204. G.p. by extension by contiguity, 330.
Georget, 386. Goltz, 114, 378, 130, 375, 350.
Giraud, 390. Golgi on nerve-cells and fibres, 383.
Gowers, Dr. W. R., 119, 377.
Granular degen., nerve-cells, 311.
Granule-cells and masses, 315, 320.
Gray, Dr. J. P., on lesions, 301 *et seq.*
Grey plaques in g.p., 306.
Grey degeneration of cord, 287, 320.
Griesinger, 272, 293. Groups of G.Ps., 408 to end.
Gubler, A., 240, 353. Gudden, 153.
Guislain, 12, 13, 265. Gulliver, 320 *et seq.*
Gums in g.p., 141. Guntz, 181.

H.

Hæmorrhage, and g.p., 331; cerebral (diagnosis), 237; do., in g.p., 27, 134, 172.
Hæmatomyelia, 288. Hair, 140.
Hallucinations in g.p., 21, 64; present writer on, 65 *et seq.*, 379, 381.
Hammond, Dr. W. A., optic-nerve, 119, odour 141, temp. 179.
Hands in g.p., 89 (see Ataxy, etc.).
Handwriting, 10, 20, 79. Hayem, 305.
Headache, 8, 30, 127, 376. Hearder, Dr. G. J., 404.
Hearing, 9, 66, 70, 378. Heart in g.p., 186, 261, 294-5.
Hemiparetic attacks, 25, 83, 361.
Hemiplegia in g.p., 83, 90, 360.
Heredity, 207, 213, 251.
Hesitating speech (see Ideational).
Herpes Zoster in g.p., 27, 139.
Henle, J., 350. Heterophasia (see Diction).
Heubner, 223.
Hitchman, Dr. J., 235, 265, 341.
Hitzig, 293, 299, 396.
Hoffmann, 303 *et seq.*, 329.
Homology of g.p. and Bright's d., 341.
Horn, 329. Housebreaking by G.Ps., 39.
Höstermann, Dr. C. E., 259.
Horsley, Mr. V., on tactile sense, 375.
Hughes, Dr. C. H., on K.Br. and K.I. poisoning, 244.
Huguenin on lesions, 313 *et seq.*
Huppert, 198. Hygiene, 390 *et seq.*
Hyoscyamus and hyoscyamine, 399, 400-1.
Hyperplasia, etc., of neuroglia, 305.

Hyperæmia, brain, or meninges, 281-2, 298, 332, 346 *et seq.*, 351.
Hyperæsthesia, 21-2, 125, 376.
Hyperalgesia, 125, 376.
Hyperidrosis unilateralis, 195.
Hypochondria, 14, 23, 52, 70-1, 239, 352, 400.
Hypoglossal nerve, 84, 320, 359.
Hysteriform attacks, 172.

I AND J.

Ideational disorders of speech, 48, 73, 356.
Ideler, 209. Illusions, 21, 64, 132, 379 *et seq.*
Incendiarism, 38.
Incongruity of words and acts, 45.
Indecency, 12, 34-5. Imprisonment, 277.
Insomnia, 15, 23, 30. Insular sclerosis, 231, 331.
Insolation, 272. Inflammations, acute, 276.
Inflammatory nature of g.p., 324, 332-3, 338.
Inequality of hemispheres, 291.
„ of pupils, 373.
Intensity of symptoms, 203.
Intestines, 296, 27, 142, 392, 400, 404.
Invasion or onset of g.p., 18.
Iron, in treatment, 397 *et seq.*
Irritable temper, prodromic, 13, 16.
Jackson, Dr. H., 367, 384.
Jastrowitz, 321 *et seq.*
Jehn, 118, 122, 319. Jespersen, 273.
Jessen, on lesions, 318 *et seq.*
Joffe, on spinal cord, 320 *et seq.*
Joire, 293. Jolly, on tobacco, 275.
Jones, Bence, on urine, 197.
Jung, Dr., on g.p. in females, 247.

K.

Katz, 244. Kesteven, Dr. W. B., miliary d., 306.
Kidneys, changes of, in g.p., 297-8.
Kiernan, J. G., 15, 275. Kirn, L., 236.
Kjelberg, syph. and g.p., 273.
Knee-jerk in g.p., 101-2 *et seq.*, 371; do., exaggerated; symptoms with, 104.
Köhler, 258. Kölliker, 305.
König, 199, 390. Kownlewsky, 244.
Krafft-Ebing, 174, 234, 265, 271, 275, 283.
Krœmer, on'temp., 176, 181.

L.

Laehr, rarity of g.p. in gentlewomen, 256.

Index.

Lancereaux, syph. and g.p., 273.
Landois, vaso-motor, 136. Landouzy, convulsions, 367.
Larynx in g.p., 85. Lasègue, 113.
Laudahn on bones, 152.
Laufenauer on lesions, 318.
Lawford, J.B., optic nerve, 319, 116.
Lead-poisoning, 274 *et seq.*
Leah, Dr. Wm., on pathology of g.p., 342-3.
Legrand du Saulle, 35, 56, 210.
Leidesdorf, syph. and g.p., 223.
Lélut, 288, 330, 337.
Leeching, 384, 401.
Lefebvre, tobacco and g.p., 275.
Lewin on syphilis and g.p., 273.
Lewis, Mr. Bevan, 109, 187, 324.
Ley, Mr. Rooke, on fracture, 154.
Leyden, 363. Linas, A.J., 51, 123.
Limbs in g.p., 20, 25-6, 86 *et seq.*; 91 *et seq.*
Linstow, 199, 372-3. Lionet, 207, 213.
Liouville. H., on lesions, 318 *et seq.*
Lip-smacking, 83. Liver-changes, 296-7.
Locality, 259. Local brain-lesions, 232-3.
Localization, Chap. xviii: p. 343.
Loewe on neuroglia, 305.
Lolliot, 265. Löwenhardt, 186.
Löwenthal, 327.
Lubimoff on lesions, 299 *et seq.*
Luciani, 378. Lungs, 295-6, 194, 190.
Lunier, 49, 64, 208, 251, 397.
Lussana and Lemoigne, 354.
Luys, J., 249, 300 *et seq.*, 333, 354.
Lymph-spaces, 300 *et seq.*
Lying, 12, 33.

M.

Macleod, Dr. Wm., 63, 176, 272, 402.
Macphail, Dr. R., 158.
Management of g.p., 391 *et seq.*
Mastication (see Eating).
Matræmatoma (see Arachnoid c).
Mania in g.p., 23, 47, 399; 227 (diagnosis).
Marriage of G.Ps., 35. Marriage, 253.
Macdonald, A. E., g.p. in U.S., 260.
Marcé, L. V., 234-5, 265, 302, 331.
Major, H., on lesions, 300 *et seq.*
Magnan, V., 265, 300, 329, 330, 334.
Maudsley, 36, 266, 343.
Marchi on lesions, 301 *et seq.* Mairet, 353. McDowall, Dr. T. W., 290.
Medullary brain-substance, 285, 306, 315-6, 327, 386.
Medulla oblongata, 286, 289, 291, 317, 325, 354-5, 359, 386.

Melancholic symptoms, 14, 24, 49, 238, 352, 400.
Mendel, Dr. E., 5, 111, 178, 190, 198, 212, 244, 267, 274, 299.
Meningeal hæmorrhage, 172, 281.
Mental alteration (prodromic), 14, 15.
Mental symptoms, 11-18, 22, 27, 29, 71, 349 *et seq.*
Mental defect as a termination of g.p., 214.
Mental activity, 251.
Mental causes, 257, 268.
Meninges in g.p., 278 *et seq.*, 299, 287, 323-4-6.
Mercer, fracture, 153. Merson, urine, 197.
Mettenheimer, 303, 309.
Meynert on lesions, 307 *et seq.*, 375.
Meschede on lesions, 299 *et seq.*
Mercury, 397, 401.
Meyer, L., Prof., 174, 271-2, 275, 279, 300 *et seq.*, 334, 388, 394.
Meyer, Dr., 275.
Mills, C. K., 330. Michéa, 157, 244.
Micromania in g.p., 59.
Mierzejewski on lesions, 300 *et seq.*
Microscopical anatomy, 298 *et seq.*
Mobèche on eyes, 107, 330.
Mode of commencement, 2.
Monakow on lead and g.p., 275.
Monomania (diagnosis), 228.
Montyel, Mirandon de, 213.
Moral failure in g.p., 12, 28, 30, 32-3-4.
Moral causes of g.p., 268 *et seq.*
Moreau, 41, 106. Morel on heredity, 251.
Motet, tabes, 329. Motey, Renault, 229.
Motor signs, 9, 19, 24, 28, 71 *et seq.*, 354, 363, 374
Motor oc. (see Third). Moxon on epilepsy, 163.
Müller on syphilis and g.p., 223.
Muscular sensibility, 71, 131.
Muscles, electrical reaction, 94.
Muscular atrophy, prog. in g.p., 146.
Muscles, fatty degen., 146; hæmatoma, 147.
Mydriasis, 110, 374. Myosis, 110, 373.
Myelitis, 287 *et seq.*, 320 *et seq.*, 323.

N.

Nails in g.p., 140. Narrow lid-chink, 112.
Nasse, W., alcohol, 266; recovery, 389.
Nature, clinical, of g.p., 324.
Necromimesis, 59.
Nerve-cells and nerve-fibres of cerebral cortex, 311 *et seq.*, 314, 341-2, 346-8, 350-5.

Nerve-cells and fibres of cord, 322.
Nerve-cells of bulbar "nuclei," 318.
Nerves, changes in, 318, 323, 369, 386.
Nettleship on eyes, 119.
Neumann on sexual excess, 266.
Neuralgia, 8, 22, 125, 276, 376.
Neuroglia in g.p., 304 *et seq.*, 341.
Neurosis, g.p. as a, 336. New vessels, 302.
Newcombe, C. F., on convulsions, 165.
Nicol, P., 329. Nitrate of silver, 389.
Nuclei of vessels, 299 ; of neuroglia, 306; of nerve-cells, 312.
Nursing in g.p., 391 *et seq.*

O.

Obermeier on lesions, 322.
Obersteiner, 5 ; on lesions, 299 *et seq.*
Obliteration of vessels, 303.
Occupation and g.p., 254.
Ocular palsies (external), 111.
Odour, 140-1. Oebeke on recovery, 388.
Œsophagus, 85. Olfactory nerves, 281, 318.
Opiates in g.p., 399-400.
Optic nerve and disc, 106, 116 *et seq.*, 281, 319.
Optic thalami, 285, 381.
Oraggi, 366. "Orders of persons" and g.p., 256.
Ormerod, E. L., on bones, 151.
Othæmatoma, 147, 404.
Overstrain (or strain) as cause, 268-9, 271 *et seq.*, 346 *et seq.*

P. AND Q.

Pain, 8, 125, 132, 376. Pachymeningitis, 278-9.
Paralytic attacks, 159; simple, 170, 182, 367-8.
Paralysis agitans, 26, 96 ; diagnosis, 236.
Paresis and paralysis in g.p., 9, 20, 24-5, 28, 72, 82, 87-9 (gait), 328 *et seq.*, 354, 358, 360, 367.
Parchappe, 200, 211, 283, 330, 345.
Particular semeiography, 29 to 199.
Patellar tendon reflex, 101.
Pathological physiology, 343 *et seq.*
Pathological anatomy (see Anat.).
Pathology, etc., 324 *et seq.*
Pecuniary position and g.p., 254.
Pediluvia, 396, 401. Pemphigus, 27, 139.
Peptonized food and enemata, 401-2.
Pericellular changes, 313-4.
Perforating ulcer of foot, 157.
Periods of g.p., 4, 6, 18, 19, 24, 28.
Perivascular lymph-space, 300.

Persecution, delusions of, 52.
Pharynx, 85. Physostigma, 398-9.
Phonation, 79, 357.
Pia-mater, 281, 299. Pierson, 378.
Pinel, Ph., 386. Pinel, C., 52, 389.
Pitres, 327, 362.
Plantar reflex, 100. Plaxton, 329.
Plumbism, 243. Poincaré, 299 *et seq.*, 335.
Pons Varolii, 286, 289-91, 317.
Pontoppidan, 273.
Position of limbs, 92, 402.
Potassium iodide, 396-7 ; do. bromide, 397, 399, 401.
Precedence of symptoms, 199.
Predisposing causes, 245.
Prevention of g.p., 390, 397.
Priority of brain or cord lesions, 325, 360.
Processes of nerve-cells, 313.
Prognosis of g.p., 386 *et seq.*
Prophylaxis of g.p., 390 *et seq.*
Projection-fibres, 355 etc.
Prolonged bath, 395. Propagation, 325, 328.
Psycho-motor centres, 364 *et seq.*
Pulmonary diseases, 190, 194, 295.
Ptomaines in urine, 199.
Pulse, 186 *et seq.* Pupils, 9, 20, 25, 106 *et seq.*, 372.
Pyrexia, 182. Purgatives, 396, 399, 401, 404.
Quinine in g.p., 398, 401.

R.

Rabenau, 198, 316 *et seq.*, 337. Rabow, S., 197.
Race and g.p., 259. Radcliffe, Dr. C. B., 373.
Rayner, Dr., cranial injury, 271.
Recovery in g.p., 216, 386 *et seq.*
Rectal feeding, 399, 401-2. Rectum, 93.
Reflex action and reflexes in g.p., 27, 99, 371 (and Eye).
Refusal of food, 404.
Régis, E., 16, 243, 251. Reinhard, 181, 378.
Religious feeling, 33.
Remissions, 205 *et seq.* Renaudin, 64.
Rendu, 388, 397. Rey Ph., 283, 329.
Respiration in g.p., 190-1.
Revulsive effects, 389-90, 394.
Ricci, 344. Richter, 198.
Rigidity of limbs, 26-8, 93, 294, 368.
Rigor mortis, 294.
Rindenblindheit, 114, 378.
Rindfleisch, 316.
Ripping, L. H., on lesions, 304 *et seq.*
Riva, 176, 302 *et seq.* Robin, 305, 310.

Index.

Rodriguez, 200, 202. Rogers, T. L., on bones, 152-3.
Rokitansky, 293, 302 *et seq*. Rollet, 274.
Rouget, 373. Runge, 376. Rutherford, Dr., 306, 313.

S.

Saliva in g.p., 199.
Samt, P., 292. Salomon, E., 5, 293, 341.
Sander, W., 213, 248, 322 *et seq*., 373.
Sankey, Dr. W. H. O., 154-5, 210, 256, 266, 286, 302 *et seq*.
Sauze, A., 206, 266.
Savage, Dr. G. H., 104, 174, 189, 287, 304.
Schiff, 327. Schläger, 271, 389.
Schopfhagen, F., 294. Schuberg, W., 337.
Schultze, Fr., 330.
Schüle, H., 199, 299 *et seq*., 405.
Schäfer, 375.
Sclerosis in g.p.,—of brain, 281-5, 305 *et seq*.; of nerve-cells, 312; of cord, 320 *et seq*.; of pons and med. obl., 317; sclerosis in patches, 288, 306, 315.
Seat of g.p., 325. Séguin, 244.
"Seclusion," 400. Secondary g.p., 331.
Seelenblindheit, 114, 378.
Seizures in g.p., 159 *et seq*. Selmi, 199.
Self-consciousness, 39. Semeiography, 29.
Senile dementia, 234-5.
Sensory symptoms, 7, 9, 21, 26-7-8, 120, 355 *et seq*., 374.
Seppilli, 302 *et seq*. Septum lucidum, 316.
Setons, 394-5. Sequence of symptoms, 202.
Seventh cranial nerve, 320.
Sex and g.p., 245. Sexual power and appetite, 94; failure, 12, 34; excess, 12, 34, 266; ideas, 45.
Shaky speech, 78, 357.
Shaw, J. C., 104, 154-6, 323, 372.
Sheppard, E., 35, 266.
Shivering subsultus, 21, 26, 96.
Silent excitement, 25, 47, 91. Silence in g.p., 74-6, 356-8.
"Simple" g.p., 3. Sinapisms, 396.
Simon, Th., 303 *et seq*.; 321 *et seq*.; 334, 336, 390.
Sixth cranial nerve, 319. Skae, D., 203.
Skin in g.p., 27, 138 *et seq*. Skin reflexes, 100.
Smell, sense of, 21, 66-70, 120; writer on, 121-2, 375, 377.
Social position and g.p., 254.

Softening in g.p.,—of brain, 280-1-3-5; of pons and med. obl., 317; of cord, 287, 320 *et seq*.
Spastic symptoms, 25, 82, 88 (gait), 328, 360-2.
Speech in g.p., 10, 19, 24, 28, 48, 71 *et seq*., 242, 356 *et seq*.
Specific gravity of brain, 292.
Sphygmograms, 189 *et seq*.
Spinal cord, 287, 291, 320, 325, 327, 336, 386; nerves, 323.
Spinal symptoms, 99 *et seq*., 165; dissociation of, 105.
Spider-cells, 307. Spleen, 297.
Spitzka, E. C., 20, 261, 336, 354.
Stages of g.p., 3, 5, 6, 19, 24, 28.
Stammering, 77.
Starting-point of g.p., 325.
Steinthal, 236. Stenger, 114.
Stomach, 296. Stomach-tube, 399, 401.
Strain in causation (see Overstrain).
Strümpell, 372. Strychnia, 393, 403.
Stupor in g.p., 24, 61-3. Stuttering, 78.
Subcortical white of brain, 315, 350. 1-5.
Subadventitial lymph-space, 300.
Suicide by G.Ps., 38, 13-14.
Suppression of excretions, *etc.*, 275.
Sutherland, Dr. A. J., 197; Dr. H., 158, 337.
Sweat, 195-7. Swollen nerve-cells, 313.
Symmetry (bilateral) of adhesions, 283.
Sympathetic nerve, 2, 106; ganglia, 320, 335-6, 386.
Symptoms of g.p., in stages, 5 *et seq*.; physical, 7, 19, 24, 28, 71 *et seq*., 82 *et seq*.; mental, 11, 22, 27, 29 *et seq*., 71; after epileptiform s., 166; after apop. s., 169.
Synonyms, 1. Syphilis, 225-6, 272, 220.

T.

Tabes dorsalis, 235; writer on, 236; 328 *et seq*.
Table of causes, 263; of adhesions, 284.
Tamburini, 288, 344, 372, 378.
Taste, sense of, 27, 379, 66-70.
Tebaldi, 118. Teeth-grinding, 26, 29, 84, 370.
Teissier, 329.
Temperature (body), 174 *et seq*.; writer on, 175 *et seq*.
Temperament, 251. Terminations of g.p., 213.
Testamentary acts by G.Ps., 39.
Tetaniform seizures, 171.
Theft by G.Ps., 12, 36-7.

I I

Therapeutics, 390 et seq.
Thermal sensibility, 131. Thickened vessels, 301.
Thomeuf, L., 265. Thurnam, J., 290.
Theories of g.p., Chap. xvii.
Thompson, Dr. G., 189, 398.
Third cranial nerve, 106, 319. Thore, 288.
Tigges on electr., 195; lesions, 303 et seq.
Tiling, 186. Tizzoni, 194.
Tobacco (as cause), 275. Todd, R. B., 361.
Tongue in g.p., 10, 20, 24, 84, 357, 370.
Tonics, 397. Tortuosity of vessels, 302.
Touch, sense of, 66-70, 125, 374-5.
Toxæmia in g.p., 270-5,341, 352-3.
Transformation, insanity into g.p., 258.
Treatment, 390 et seq. Trance, 63.
Tremor, 9, 10, 21, 24, 91, 370. Tremor cöactus, 26, 96, 370. Tremor of age, 242.
Trophic changes, 138 et seq., 294.
Trunk, 25, 86, 89.
Tuczek, F., on lesions, 314, 327.
Tuke, Dr. Batty, on colour-blindness, 21; diagnosis, 219; miliary sclerosis, 306; nerve-cells, 313.
Tuke, Dr. Hack, address, 364; cures, 387.
Tumours, 229, 330-1.
Türck on cord, 320 et seq.

U AND V.

Ulrich, 186. Unilateral sweat, 195.
Urban life, 261.
Up and down respiration, 191, 193.
Urine, 197. Urinary incontinence, 28, 93, 370. Urinary disorders, 403.
Vacuoles, brain, 303-4, 310; nerve-cells, 314.
Varieties of g.p., writer on, 404-14.
Variola, 241, 389.
Vaso-motor symptoms, 9, 21, 133 et seq., 335, 340, 347, 352.
Veins of brain and meninges, 278, 281, 298.

Veratrum viride, 397, 399.
Verga, 251, 262, 393. Verneuil, 199.
Venesection, 393-4, 401.
Ventricles of brain, 285, 293, 310, 317. Ventricle, left of heart, 261, 294-5.
Vertigo, 9, 17, 18, 133. Victor, R., 275.
Viel, 372. Virchow, 272, 305.
Visceral feeling, 70, 131. Vision, 106, 113, 377-8.
Visual hallucinations, 66-70; imperception, 114, 378.
Vitreous degeneration, 306.
Voisin, Dr., 5, 21, 121, 158, 160, 189, 201, 224, 244, 258, 276, 287, 299 et seq., 349, 376, 394 et seq.

W.

Waller, A., on nerve-wasting, 369.
Warm baths, 399, 400-1. Warren, Dr., 403.
Wedl, 303, 309.
Weight, in g.p.; of body, 138, 294; of brain, 289 et seq.; unequal, of hemispheres, 291.
Weiss, 288. Wernicke, C., on pupils, 373.
Westphal, 184, 241, 303 et seq., 329, 372.
White substance (see Medullary).
Wiedemeister, 314. Wiesinger, 304.
Wiglesworth, eyes, 119; neuroglia, 305.
Wilks, Dr. S., on lesions, 391 et seq.
Williams, Dr. Rhys, 287. Williams, Dr. S, W. D., 152, 185.
Wirsch, 186. Witkowski, 270.
Wolff, O. J. B., on pulse, etc., 190.
Wood, Dr. Wm., 51, 208.
Word-deafness and blindness, 76, 124.
Work, exhausting, 268-9, 271.
Workman, Dr. J., 153, 195, 260.
Writing, 10, 20, 79.

Z.

Zacher, on vision, 115; reflex, 372.
Zambaco on syphilis, 224.
Zenker, on voice, 79; resp., 193.

www.ingramcontent.com/pod-product-compliance
Lightning Source LLC
Chambersburg PA
CBHW022102300426
44117CB00007B/559